£7.50

Julien 1 E[illegible]
5th October

Basic electrical engineering science

I. McKenzie Smith and K.T. Hosie

B.Sc., Dip.A.Ed., C.Eng., M.I.E.E., M.I.E.R.E., F.I.T.E.
Head of Department of Electrical Engineering
Stow College
Glasgow

B.Sc., C.Eng., M.I.E.E.,
Senior Teacher of
Electronics Engineering
Department of
Electrical Engineering
Bell College of Technology,
Hamilton

Basic electrical engineering science

Longman

LONGMAN GROUP LIMITED
Longman House,
Burnt Mill, Harlow, Essex CM20 2JE, England

Published in the United States of America by
Longman Inc., New York

First published 1972
Second impression 1973
Third impression 1978
Fourth impression (with corrections) 1980
Fifth impression 1982

ISBN 0 582 42429 1

Printed in Hong Kong by
Hing Yip Printing Co.

To

our wives,

Elspeth and Marion

Contents

Preface

That's a very true remark–
On the merits of their pleading,
We're at present in the dark!
W. S. GILBERT

The objective of this book is to provide an introduction to electrical engineering. During the past decade, this branch of science has undergone many modifications; the authors feel therefore that this text will aid the understanding of the present approach to the subject. The text covers the work of the student technician engineer and is based on lectures given to Ordinary National Certificate students over a number of years. In particular, the text follows the syllabus of the O.N.C. as drawn up by the Scottish Association for National Certificates and Diplomas but other sections have been added to cover all other O.N.C. courses as far as possible. The consequent broadening of the content is such that the text will prove useful to students following any technician course in electrical and electronics engineering.

Four important innovations have been introduced in the text. The first is that the International System of Units (SI) has been used throughout. The student may therefore concern himself with that system only, being the system generally recognised in engineering.

During the last decade, introductory texts written for a comparable level of work have dealt with the traditional approach to electrical and electronics engineering. It is now common practice in Scotland as well as many parts of England to use a systems approach in the teaching of electronics and also machines. A chapter has therefore been included which deals with electronic network systems as opposed to the traditional treatment of the components.

Also during the last decade, semiconductor electronics have increased in importance to a level that is at least comparable with that of machines. Four chapters, including that on electronics systems, have been given to the study of electronics and electronics circuitry. The authors feel that this innovation is required to redress the balance between electronics and power engineering.

The final innovation concerns the study of machines. In recent years, it has become apparent that it is not sufficient to study only particular types of machines as though each were a separate topic. A unified approach is required to show that all machines, particularly electromagnetic machines, are constructional variations about a single principle. This unified approach is used by the more enlightened examining bodies and the authors hope and believe that others will, in time, abandon the outmoded fragmental system as well as the diluted generalised machine theory which is beyond the capabilities of most technician engineers. The unified approach presented in this text is based on that prepared by a SANCAD committee under the chairmanship of Prof. M. G. Say. This approach has been used by one of the authors for several years and has been found to be of great advantage.

The symbols and abbreviations contained within the text conform to those recommended by the Institution of Electrical Engineers for use in electrical and electronics engineering courses. This recommendation was made in 1968. Symbols and abbreviations not contained therein conform to BSS 1991, BSS 3939 and PD 5686 published by the British Standards Institute. To avoid confusion, symbols and abbreviations are not mixed with one another.

A considerable number of worked examples are included in the text to help the student to understand the principles contained therein. At the end of each chapter, there are a number of examples against which the student may pit his wits. Many of these examples come from current examination papers presented by the Scottish Association for National Certificates and Diplomas; these examples are labelled SANCAD. The authors wish to acknowledge their gratitude to the Association for permission to use these examples.

Finally the authors would like to thank Mr. W. E. I. MacLeod of the Glasgow College of Technology for his advice and assistance in the preparation of this text. Mr. I. McKenzie Smith would like to thank Mrs. B. Rowland, of Coatbridge Technical College, Professor M. G. Say and Mr. M. Hawes, of Napier College of Science and Technology, for their invigorating discussions on Unified Machine Theory. Thanks are also due to other friends and associates for their assistance.

Milngavie,
September 1971

List of symbols and abbreviations

	Symbol	Unit abbreviation
Acceleration, linear	a	m/s^2
Admittance	Y	S
Area	A	m^2
Capacitance	C	F
Charge	Q	C
Conductance	G	S
Conductivity	γ (gamma)	S/m
Current	I	A
Current surface density	A	A/rad *or* A/m
Efficiency	η (eta)	–
Electric field strength	E	V/m
Electric flux	Q	C
Electric flux density	D	C/m^2
Electric potential	V	V
Electromotive force	E	V
Energy	W	J
Force	F, f	N
Form factor	k_f	–
Frequency	f	Hz
Frequency, angular	ω (omega)	rad/s
Frequency, resonant	f_r	Hz
Gain	G	–
Inductance	L	H
Inductance, mutual	M	H
Impedance	Z	Ω (omega)
Leakage factor	σ (sigma)	–
Length	l	m
Mass	m	kg
Magnetic field strength	H	At/m
Magnetic flux	Φ(phi)	Wb
Magnetic flux density	B	T
Magnetic flux linkage	Ψ(psi)	Wb t
Magnetic potential difference	F	At
Magnetomotive force	F	At
Period	T	s

Permeability	μ (mu)	H/m
Permeance	Λ (lambda)	H
Permittivity	ϵ (epsilon)	F/m
Phase angle	ϕ (phi)	rad
Power, active	P	W
Power, apparent	S	VA
Power, reactive	Q	var
Reactance, capacitive	X_C	Ω (omega)
Reactive, inductive	X_L	Ω (omega)
Reluctance	S	/H *or* At/Wb
Resistance	R	Ω (omega)
Resistivity	ρ (rho)	Ω m
Stacking factor	β (beta)	–
Susceptance	B	S
Temperature coefficient	α (alpha)	/K
Temperature difference	θ (theta)	°C
Time	t	s
Torque	T, M	N m
Torque angle	λ (lambda)	rad
Work	W	J
Velocity, angular	ω (omega)	rad/s
	n	rev/s
Velocity, linear	u	m/s
Volume	V	m^3

1

Introduction to electrical quantities

Electrical engineering is the application of electricity to the service of man. Electricity is a physical agent whose manifestations conform to the law of conservation of energy. All knowledge of its nature is based on experimental evidence.

1.1 Electric charge

Electricity appears in one of two forms which, by convention, are called negative and positive electricity. Electric charge is the excess of negative or positive electricity on a body or in space. If the excess is negative, the body is said to have a negative charge and vice versa.

It is generally known that all matter consists of large numbers of particles termed molecules, being the smallest pieces of matter that can exist separately. Molecules can be subdivided into atoms and comprise one or more atoms bonded together. An atom is the smallest particle of matter that can take part in a chemical change.

Not all atoms are the same, there being 98 different types that are naturally occurring, e.g. gold, iron, oxygen, etc. Each type is called an element. All atoms consist of three basic constituent parts:

1. electrons,
2. protons,
3. neutrons.

An electron is an elementary particle charged with a small and constant quantity of negative electricity. A proton is similarly defined but charged with positive electricity whilst the neutron is uncharged and is therefore neutral. In an atom, the number of electrons normally equals the number of protons; it is the number of protons that determines to which element type the atom belongs. An atom can have one or more electrons added to it or taken away. This does not change its elemental classification but it disturbs its electrical balance. If the atom has excess electrons, it is said to be negatively charged. A charged atom is called an ion.

A body containing a number of ionised atoms is also said to be electrically charged.

It can be shown that positively- and negatively-charged bodies are mutually attracted to one another while similarly charged bodies repel one another.

1.2 Movement of electrons

All electrons have a certain potential energy. Given a suitable medium in which to exist, they move freely from one energy level to another and this movement, when undertaken in a concerted manner, is termed an electric current flow. Conventionally it is said that the current flows from a point of high energy level to a point of low energy level. These points are said to have high potential and low potential respectively. For convenience the point of high potential is termed the positive and the point of low potential is termed the negative, hence conventionally a current is said to flow from positive to negative.

This convention was in general use long before the nature of electric charge was discovered. Unfortunately it was found that electrons move in the other direction since the negatively-charged electron is attracted to the positive potential. Thus conventional current flows in the opposite direction to that of electron current. Normally only conventional current is described by the term current and this will apply throughout the text.

The transfer of electrons takes place more readily in a medium in which atoms can readily release electrons, e.g. copper, aluminium, silver, etc. Such a material is termed a conductor. A material that does not readily permit electron flow is termed an insulator, e.g. porcelain, nylon, rubber, etc. There is also a family of materials termed semiconductors which have certain characteristics that belong to neither of the other groups. The process of conduction is further described in chapter 13.

1.3 Current flow in a circuit

For most practical applications, it is necessary that the current flow continues for as long as it is required; this will not happen unless the following conditions are fulfilled:

1. There must be a complete circuit around which the electrons may move. If the electrons cannot return to the point of starting, then eventually they will all congregate together and the flow will cease.

2. There must be a driving influence to cause the continuous flow. This influence is provided by the source which causes the current to leave at a high potential and to move round the circuit until it returns to the source at a low potential. This circuit arrangement is indicated in Fig. 1.1.

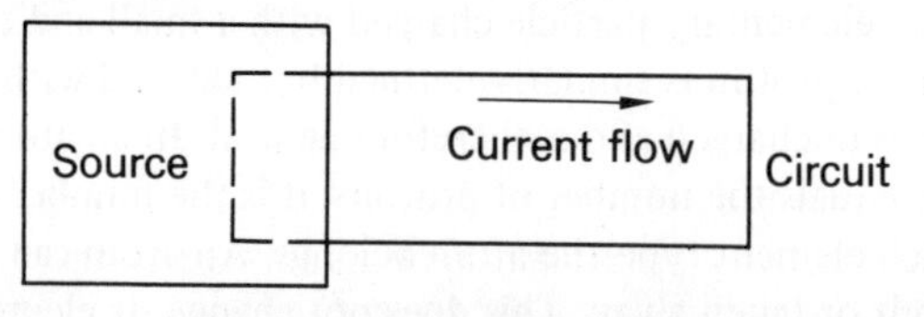

Fig. 1.1. Elementary circuit

The driving influence is termed the electromotive force hereafter called the e.m.f. Each time the charge passes through the source, more energy is provided by the source to permit it to continue round once more. This is a continuous process since the current flow is continuous. It should be noted that the current is the rate of flow of charge through a section of the circuit.

The basic electrical circuit has four constituent parts as shown in Fig. 1.2.

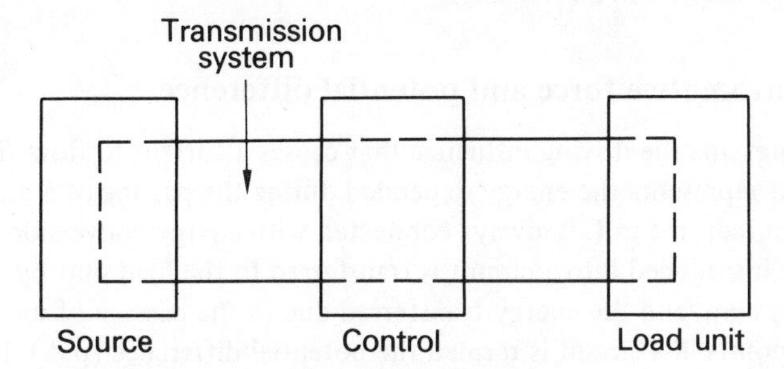

Fig. 1.2. Parts of an electric circuit

The source Its function has already been described. At this introductory level, the source may be considered to be a battery and other sources such as generators will be introduced later.

The load unit The function of this unit is to absorb and convert the electrical energy supplied by the source. Most electrical apparatus loads the system, i.e. accepts energy from the system, and common examples are lamps, heaters and motors.

The transmission system This is required to conduct the current between the source and the load unit.

The control apparatus As the name suggests, its function is to control. The most simple control is a switch which either permits current to flow or else interrupts it.

A simple circuit is shown in Fig. 1.3. In it, a battery is supplying a lamp bulb, whilst a switch is included to put the lamp on and off.

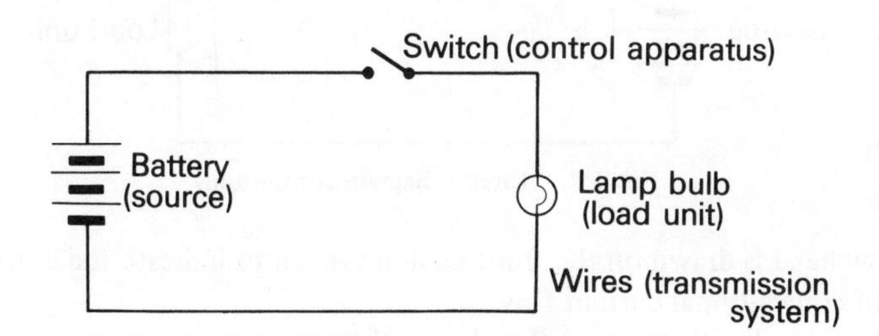

Fig. 1.3. Simple lamp circuit

This simple example serves to illustrate the fundamental function of any electric circuit–that function is to transmit energy from the input source to an energy-converting device. Electricity permits the source of energy to be remote from the point of conversion.

Before analysing this method of transmitting energy, it is necessary to discuss the relation between the energy and the charge flow in a circuit. The current may either flow continuously in one direction, in which case it is called direct current, or else it may continuously reverse its direction, in which case it is called alternating current. These names may be abbreviated to d.c. and a.c. when used to describe a subject, for example a d.c. motor or an a.c. signal.

1.4 Electromotive force and potential difference

The e.m.f. represents the driving influence that causes a current to flow. The e.m.f. is not a force but represents the energy expended during the passing of a unit charge through the source; an e.m.f. is always connected with energy conversion.

The energy introduced into a circuit is transferred to the load unit by the transmission system, and the energy transferred due to the passage of unit charge between two points in a circuit is termed the potential difference (p.d.). If all the energy is transferred to the load unit, the p.d. across the load unit is equal to the source e.m.f.

It will be observed that both e.m.f. and p.d. are similar quantities. However an e.m.f. is always active in that it tends to produce an electric current in a circuit whilst a p.d. may be either passive or active. A p.d. is passive whenever it has no tendency to create a current in a circuit.

Unless it is otherwise stated, it is usual to consider the transmission system of a circuit to be ideal, i.e. it transmits all the energy from the source to the load unit without loss. Appropriate imperfections will be considered later.

Certain conventions of representing the e.m.f. and p.d. in a circuit diagram should be noted. Each is indicated by an arrow as shown in Fig. 1.4. In each case, the arrow head points toward the point of high (or assumed higher) potential. It is misleading to show an arrow head at each end of the line as if it were a dimension line.

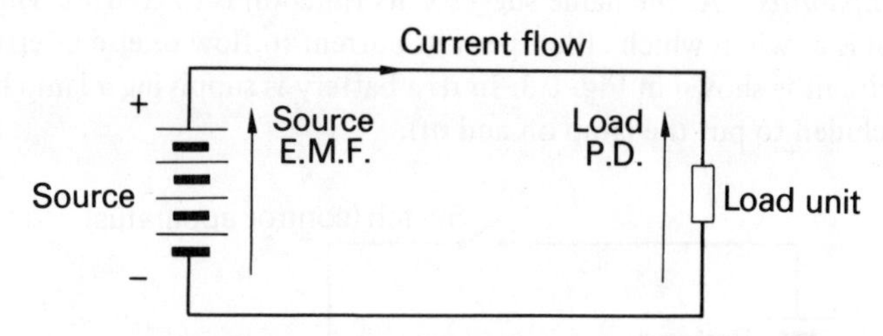

Fig. 1.4. Circuit diagram conventions

An arrow head is drawn on the transmission system to indicate the corresponding direction of conventional current flow.

It will be seen that the current flow leaves the source at the positive terminal and therefore moves in the same direction as indicated by the source e.m.f. arrow. The current flow enters the load at the positive terminal and therefore in the opposite direction to that indicated by the load p.d. arrow. Energy is converted within the load unit and, depending on the nature of this conversion, the p.d. may be constituted in a variety of ways. It is sufficient at first to consider the p.d. as the change in energy level across the terminals of the load unit. This is termed a volt drop since the p.d. (and e.m.f.) are measured in volts.

1.5 Symbols and abbreviations

So long as simple circuits with one source, one load and an ideal transmission system are being considered, there is no possibility of confusion since each quantity has a term defining it specifically, e.g. e.m.f. However, if an attempt is made to mathematically analyse the behaviour of a circuit, it is easier to manipulate symbols representing the defining terms, e.g. E instead of e.m.f., V instead of p.d., etc.

Each electrical quantity has a letter assigned to it. The choice of these symbols is not random but is that almost universally recognised, being recommended by the International Electrotechnical Commission and the British Standards Institution. A list of all symbols and abbreviations used in this text is given at the beginning of the book. Also, as soon as an electrical quantity is introduced and can be adequately defined, the corresponding symbol will be emphasised as in the following example:

Electromotive force | Symbol: E

It will be taken that thereafter the symbol need not be explained again. For distinction, subscripts can be added to the letter symbols, e.g. E_1 and E_2. In cases where it is desired to indicate that a maximum value is involved, the subscript m is used, e.g. E_m. Finally, in circuit theory, it is usual to indicate constant values by capital letters whilst lower-case letters are reserved for time-varying quantities, e.g. E is constant whilst e is time varying.

All physical quantities have to be measured in an appropriate unit system. The system appropriate to electrical engineering–and to most other sciences–is the Système International d'Unités (SI). The SI is a metric system and is described in para. 1.7 below.

At this time, however, it is appropriate to note that the names of the units can also be abbreviated and there are unit-symbols corresponding to the unit of each electrical quantity. When each electrical quantity is defined, the appropriate unit symbol will be emphasised as follows:

Electromotive force | Symbol: E | Unit: Volt (V)

Here the unit is given with the appropriate unit symbol in brackets after it. The abbreviated units are used only after numerical values, e.g. 25 V. Otherwise the full unit name must be given, e.g. a few volts. The unit symbol remains the same in both singular and plural, e.g. 1 V and 5 V. It will also be seen that there is no period after the unit symbol and that a space is set between the number and the unit symbol.

Finally certain names are abbreviated, e.g. e.m.f. for electromotive force. These will be introduced as necessary throughout the text.

1.6 Principle of Ohm's law

In the foregoing discussion concerning the simple circuit, the electrical action was discussed in general terms. One of the most important steps in the analysis of the circuit was undertaken by Georg Ohm, who found that the p.d. across the ends of many conductors is proportional to the current flowing between them. This, he found, was a direct proportionality provided that temperature remained constant. Since the symbol for current is I, this relationship may be expressed as

$$V \propto I \tag{1.1}$$

Relation (1.1) is the mathematical expression of what is termed Ohm's Law.

Subsequent experimental evidence has shown that many other factors affect this relationship and that in fact few conduction processes give a direct proportionality between p.d. and current. However, this relationship is almost constant for many electrical circuits and it is convenient at this introductory stage to consider only circuits in which the relationship is constant. The corresponding characteristic is shown in Fig. 1.5.

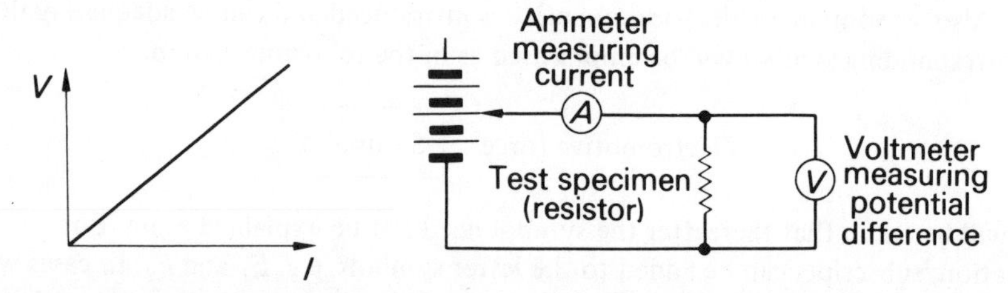

Fig. 1.5. Constant potential difference/current characteristic and the circuit from which it is obtained

Since the relationship is assumed constant, then

$$\frac{V}{I} = R$$

where R is a constant termed the resistance of the conductor. The zigzag line used in Fig. 1.5 is generally used to represent a load with resistance properties. The expression involving R is usually expressed as

$$V = IR \tag{1.2}$$

It should be noted that this relationship is derived from Ohm's Law and is not a symbolic expression for it. Ohm's Law only notes the constancy of p.d. to current provided that other physical factors remain unchanged, i.e. for a given p.d., the current will vary in consequence of variation of external physical factors.

1.7 SI units

The SI was formally introduced in 1960 and has been accepted by many countries as their only legal system of measurement. It has world-wide recognition and is generally accepted by almost all electrical engineers.

One of its most important advantages over its predecessors is that it is a coherent system wherever possible. A system is coherent if the product or quotient of any two quantities, as measured in the system, is the unit of the resultant quantity. For instance, unit area results when unit length is multiplied by unit length. Similarly, unit force is obtained when unit mass is multiplied by unit acceleration.

In any system of measurement, the magnitudes of certain physical quantities must be arbitrarily selected and declared to have unit value. These magnitudes result in a set of standards and are termed the basic units.

The SI selects seven physical quantities as its basis:

Mass
Length
Time
Amount of material
Electric current
Absolute temperature
Luminous intensity

All other units are derived units and are related to the basic units by definition.

It should not be assumed that all relationships are completely coherent and certain physical constants have to be introduced. These are reduced to an absolute limit. In electrical engineering, only the values of the permeability and permittivity of free space are of immediate importance and these will be introduced in chapters 4 and 6 respectively.

The SI also retains the advantages of previous metric systems in that the units may be grouped in multiples and sub-multiples. Appropriate to each, there is a prefix that may be added to the unit. Each prefix also has an abbreviation.

Multiplication factor		*Prefix*	*Symbol*
1 000 000 000 000	10^{12}	tera	T
1 000 000 000	10^{9}	giga	G
1 000 000	10^{6}	mega	M
1 000	10^{3}	kilo	k
0·001	10^{-3}	milli	m
0·000 001	10^{-6}	micro	μ
0·000 000 001	10^{-9}	nano	n
0·000 000 000 001	10^{-12}	pico	p

Consider a p.d. of 25 000 volts. In index notation, this may be expressed as 25×10^3 volts and hence 25 kilovolts. If the unit is abbreviated, then the p.d. is expressed as 25 kV.

Finally there are some points to note about the above table:

1. The prefix k for kilo may be replaced by K so that all positive powers of 10 are represented by capital letter prefixes.

2. The Greek letter μ is only one of many used due to the lack of letters in the ordinary alphabet.

3. There is also a multiplication factor of 10^{-2} which is reserved for use with the metre, i.e. the centimetre.

1.8 SI basic units

Of the six basic quantities, mass, length and time can be defined by example. It therefore remains only to give their symbols and units of measurement as follows.

Mass	Symbol: m	Unit: kilogramme (kg)

It should be noted that the megagramme is also known as the tonne.

Length	Symbol: l	Unit: metre (m)

The centimetre is additional to the normal multiple units.

Time	Symbol: t	Unit: second (s)

Although the standard sub-multiples of the second are used, the multiple units are often replaced by minutes (min), hours (h), days (d) and years (a).

Of the remaining basic quantities, only electric current is of immediate significance. However, before proceeding to define it, it is necessary to consider some of the derived units.

1.9 SI derived units

Although the physical quantities of area, volume, velocity, acceleration and angular velocity are generally understood, it is worth noting their symbols and units.

Area	Symbol: A	Unit: square metre (m^2)
Volume	Symbol: V	Unit: cubic metre (m^3)
Velocity	Symbol: u	Unit: metre per second (m/s)
Acceleration	Symbol: a	Unit: metre per second squared (m/s^2)
Angular velocity	Symbol: ω	Unit: radian per second (rad/s)

The unit of force, called the newton, is that force which, when applied to a body having a mass of one kilogramme, gives it an acceleration of one metre per second squared.

Force	Symbol: F	Unit: newton (N)

$$F = ma \tag{1.3}$$

The force of attraction between a body and the earth is termed its weight. The weight is obtained from the mass and the value of gravitational acceleration g which is 9·81 m/s^2. Thus the weight experienced by a mass of 1 kgf is equal to 9·81 N.

The unit of energy, called the joule, is the work done when the point of application of a force of one newton is displaced through a distance of one metre in the direction of the force.

Work or energy	Symbol: W	Unit: joule (J)

$$W = Fl \tag{1.4}$$

Note that energy is the capacity for doing work. Both energy and work are therefore measured in similar terms.

The unit of torque, called the newton metre, is given by the product of a force of one newton and a perpendicular distance of one metre from the line of action of the force to the axis of rotation.

Torque	Symbol: T (or M)	Unit: newton metre (N m)

If the perpendicular distance from the line of action to the axis of rotation is r, then

$$T = Fr \tag{1.5}$$

The symbol M is reserved for the torque of a rotating electrical machine. It is used extensively in chapter 12. Note that the unit of torque is the product of two other units. Whenever this occurs, the symbol abbreviations are put together as shown.

Power is the rate of doing work. The unit of power, called the watt, is equal to one joule per second.

Power	Symbol: P	Unit: watt (W)

$$P = \frac{W}{t}$$

$$= F \cdot \frac{l}{t} \tag{1.6}$$

$$P = Fu \tag{1.7}$$

In the case of a rotating electrical machine:

$$P = M\omega \tag{1.8}$$

There is another unit of energy which is used commercially–the kilowatthour (kWh). It represents the work done by working at the rate of one kilowatt for a period of one hour. Once known as the Board of Trade Unit, it is still referred to, by the supply authorities, as the unit.

$$1\ \text{kWh} = 1\,000\ \text{Wh}$$
$$= 1\,000 \times 60\ \text{W min}$$
$$= 1\,000 \times 60 \times 60\ \text{W s}$$

But working at the rate of one watt for a period of one second provides one joule of work, hence

$$= 1\,000 \times 60 \times 60\ \text{J}$$
$$= 3{\cdot}6 \times 10^6\ \text{J}$$

It should be noted that when a device converts or transforms energy, some of the input energy is consumed to make the device operate. The efficiency of this operation is defined as

$$\text{Efficiency} = \frac{\text{energy output in a given time}}{\text{energy input in the same time}} = \frac{W_o}{W_i}$$

$$= \frac{\text{power output}}{\text{power input}} = \frac{P_o}{P_i}$$

Efficiency	Symbol: η	Unit: none

Example 1.1 A generating station has a daily output of 280 MWh and uses 500 tonnes of coal in the process. The coal releases 7 MJ/kg when burnt. Calculate the overall efficiency of the station.

Input energy per day $W_i = 7 \times 10^6 \times 500 \times 1\,000$
$= 35{\cdot}0 \times 10^{11}$ J

Output energy per day $W_o = 280$ MWh
$= 280 \times 10^6 \times 3{\cdot}6 \times 10^3$
$= 10{\cdot}1 \times 10^{11}$ J

$$\eta = \frac{W_o}{W_i}$$
$$= \frac{10{\cdot}1 \times 10^{11}}{35{\cdot}0 \times 10^{11}}$$
$$= \underline{0{\cdot}288}$$

Example 1.2 A lift of 250 kg mass is raised with a velocity of 5 m/s. If the driving motor has an efficiency of 85%, calculate the input power to the motor.

Weight of lift $= F = mg = 250 \times 9{\cdot}81$
$= 2452$ N

Output power of motor $P_o = Fu = 2452 \times 5$
$= 12\,260$ W

$$\text{Input power to motor } P_{in} = \frac{P_0}{\eta}$$

$$= \frac{12260}{0{\cdot}85}$$

$$= 14450 \text{ W}$$

$$= 14{\cdot}45 \text{ kW}$$

Turning to the electrical quantities, it is necessary to define the basic one chosen – electric current. From this starting point, all the other electrical quantities can be defined. The unit of electric current, called the ampere, is that current which, when flowing through each of two parallel conductors spaced one metre apart in vacuo and of infinite length, give rise to a force between the conductors of 2×10^{-7} newtons per metre run.

Current	Symbol: I	Unit: ampere (A)

This definition is outstanding for its complexity. However, by using such a definition, most of the electrical units take on suitable magnitudes. The figure of 2×10^{-7} is therefore one of convenience and the definition will be explained in para. 4.8.

The unit of electric charge, called the coulomb, is the quantity of electricity transferred in one second by a current of one ampere.

Charge	Symbol: Q	Unit: coulomb (C)

$$Q = It \tag{1.9}$$

The unit of electric potential, called the volt, is the difference of potential between two points of a conducting wire carrying a constant current of one ampere, when the power dissipated between these two points is one watt.

Electric potential	Symbol: V	Unit: volt (V)

Potential difference has the same symbol and unit. Electromotive force has the symbol E but has the same unit. Because p.d.'s are measured in volts, they are also referred to as volt drops or voltages. By experiment, it can be shown that the relation corresponding to the definition is

$$V = \frac{P}{I} \tag{1.10}$$

This is better expressed as

$$P = VI \tag{1.10.1}$$

It also follows that

$$V = \frac{P}{I}$$

$$= \frac{W}{t} \cdot \frac{t}{Q}$$

$$= \frac{W}{Q}$$

That is, the p.d. is equal to the energy per unit charge. In this manner, the definition is related to the concepts of e.m.f. and p.d. put forward earlier.

Electric resistance is measured in ohms. Unit resistance is the resistance between two points in a conductor when a constant potential difference of one volt, applied between these two points, produces in this conductor a current of one ampere, this conductor not being the source of any electromotive force.

Electric resistance	Symbol: R	Unit: ohm (Ω)

Experimental evidence has shown that

$$R = \frac{V}{I} \tag{1.2}$$

Note that

$$P = VI$$

$$= (IR)I$$

$$= I^2 R \tag{1.11}$$

Example 1.3 A motor gives an output power of 20 kW and operates with an efficiency of 80%. If the constant input voltage to the motor is 200 V, what is the constant supply current?

$$P_o = 20000 \text{ W}$$

$$P_i = \frac{P_o}{\eta} = \frac{20000}{0{\cdot}8}$$

$$= 25000 \text{ W}$$

$$= VI$$

$$I = \frac{25000}{200} = \underline{125 \text{ A}}$$

Example 1.4 A 200-tonne train experiences wind resistance equivalent to 60 N/tonne. The operating efficiency of the driving motors is 0·87 and the cost of electrical energy is 0·5 p/kWh. What is the cost of the energy required to make the train travel 1 km?

If the train is supplied at a constant voltage of 1·5 kV and travels with a velocity of 80 km/h, what is the supply current?

In moving 1 km, $W_0 = Fl$

$$= 200 \times 60 \times 1\,000 = 12 \times 10^6 \text{ J}$$

$$W_i = \frac{W_0}{\eta} = \frac{12 \times 10^6}{0{\cdot}87} = 14{\cdot}4 \times 10^6 \text{ J}$$

But 1 kWh $= 3{\cdot}6 \times 10^6$ J, hence

$$W_i = \frac{14{\cdot}4 \times 10^6}{3{\cdot}6 \times 10^6} = 4{\cdot}0 \text{ kWh}$$

Cost of energy = 0·5 x 4·0 = 2·0 p

Work done in 1 h when moving with a velocity of 80 km/h

$$= 14{\cdot}4 \times 10^6 \times 80 \text{ J}$$

Work done per second, which is equivalent to the input power

$$= \frac{14{\cdot}4 \times 10^6 \times 80}{3600}$$

$$= 32 \times 10^4 \text{ W}$$

$$= P_i$$

But

$$P_{in} = VI$$

$$I = \frac{P_{in}}{V} = \frac{32 \times 10^4}{1{\cdot}5 \times 10^3}$$

$$= 214 \text{ A}$$

Some mention is required about temperature measurement, which is in the Celsius scale. Absolute temperature is measured in degrees Kelvin but for most electrical purposes at an introductory stage, it is sufficient to measure temperature in degrees Celsius. Reference will be made to the absolute scale of temperature in chapter 13.

It should be remembered that both degrees of temperature represent the same change in temperature–the difference lies in the reference zero.

Temperature	Symbol: θ	Unit: degree Celsius (°C)

A useful constant to note is that it takes 4 185 J to raise the temperature of 1 litre of water through 1°C.

Example 1.5 An electric heater contains 4·0 litre of water initially at a mean temperature of 15°C. 0·25 kWh is supplied to the water by the heater. Assuming no heat losses, what is the final mean temperature of the water?

$$W_i = 0{\cdot}25 \times 3{\cdot}6 \times 10^6 = 0{\cdot}9 \times 10^6 \text{ J}$$

Energy to raise temperature of 4·0 litre of water through 1°C

$$= 4{\cdot}0 \times 4185 \text{ J}$$

Therefore change in temperature is

$$\Delta\theta = \frac{0{\cdot}9 \times 10^6}{4{\cdot}0 \times 4185}$$
$$= 53{\cdot}8°\text{C}$$
$$\theta_2 = \theta_1 + \Delta\theta = 15 + 53{\cdot}8$$
$$= \underline{68{\cdot}8°\text{C}}$$

1.10 Circuit investigation

Whilst investigating the circuit in Fig. 1.5, reference was made to devices termed voltmeters and ammeters. The operating principles of these measuring instruments are described in chapter 17. However, it should be noted that:

1. The voltmeter is a device which measures e.m.f. or p.d. It is connected as indicated in Fig. 1.6 between the required points in the circuit.

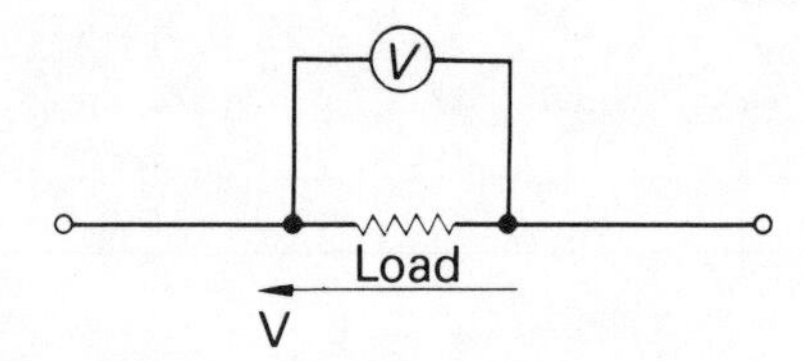

Fig. 1.6. Voltmeter connection

2. The ammeter is a device which measures current. It is connected into the conductor so that all the current passes through the meter. The method of connection is shown in Fig. 1.7.

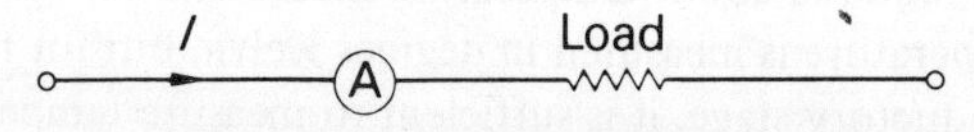

Fig. 1.7. Ammeter connection

It may be assumed for the present that both voltmeters and ammeters are perfect in that they may be added to the circuit without changing the circuit and yet indicate the desired quantities.

PROBLEMS ON ELECTRICAL QUANTITIES

1 Complete the following table:

V	I	R
10 V	5 A	
	20 A	0·1 Ω
8 V		2 Ω
12 V		3·3 kΩ
2 V	50 μA	
	15 mA	6 kΩ

2 Ω; 2 V; 4 A; 3·64 mA; 40 kΩ; 90 V

2 With reference to the basic electrical circuit, write a short essay about each of the following:

(*a*) The source
(*b*) The load unit
(*c*) The transmission system
(*d*) The control apparatus.

Describe an experiment used to illustrate Ohm's Law and relate the apparatus used to the above four subjects.

3 The following data was obtained from an investigation into the voltage/current characteristic of a lamp bulb.

Voltage (V)	100	150	200	250
Current (A)	0·12	0·19	0·28	0·40

Calculate the resistance of the lamp in each case and comment on the inconsistency of the resistance.

4 A 500-V, d.c. motor of efficiency 0·95 drives a pump of efficiency 0·80. The pump raises 1 200 m^3 of water per hour against a head of 25 m. Calculate the input current to the motor given that 1 m^3 of water has a mass of 1 000 kg.

215 A

5 A hydro-electric generating plant is supplied from a reservoir of capacity 20 000 000 m^3 with a head of 200 m. The hydraulic efficiency of the plant is 0·8 and the electric efficiency is 0·9. What is the total available energy?

7·85 GWh

6 The area of the reservoir in problem 5 is 3·0 km^2. The plant supplies a load of 12 MW for 3 h. Calculate the fall in the level of the reservoir during this period.

30·5 mm

7 The reservoir in problem 5 is supplied by a river at the rate of 2 m^3/s. Assuming constant head and efficiencies, what does this flow represent in terms of megawatts, megawatthours per day and gigawatthours per annum?

2·83 MW; 68 MWh/d; 24·8 GWh/a

8 A 1 500-V, d.c. locomotive draws a load of 100 tonne mass at 50 km/h. The tractive resistance of the load is 50 N/tonne, the motor efficiency is 0·9 and the train travels along a level track. Calculate the supply current to the locomotive.

51·4 A

9 A 460-V, d.c. motor drives a hoist which raises a load of 100 kg with a velocity of 15 m/s. Calculate:

(*a*) The power output of the motor assuming the hoist gearing to have an efficiency of 0·80

(*b*) The motor current assuming the motor efficiency to be 0·75.

18·4 kW; 53.3 A

10 A power station has a capacity of 400 MW and an overall efficiency of 0·31. The coal-handling plant has a maximum handling capacity of 200 tonne/h, and the station is to operate continuously at full load. What is the lowest possible energy value for the coal in joules per kilogramme.

23·2 MJ/kg

2

D. C. circuits

The concept of resistance should now be familiar to the reader. It is rare, however, that the resistance of a circuit is concentrated in one section; instead complicated arrangements exist and in order to analyse these arrangements, which are termed networks, it is now necessary that consideration be given to some of the physical factors that affect resistance and to network analysis.

2.1 Conductance in circuits

It has been noted that resistance is the ratio of potential difference to current, which derives from the relation

$$V = IR$$

This expression places emphasis on the potential difference, which at first sight might appear reasonable because most electrical supplies are defined in terms of the supply voltage. However, in many instances, network analysis, particularly in electronic networks, stems more readily from an expression placing emphasis on the current, thus

$$I = \frac{V}{R}$$

Reciprocal values such as $1/R$ are difficult to manipulate and therefore the reciprocal value of resistance is termed the conductance.

Conductance | Symbol: G | Unit: siemens (S)

From the definition of conductance as the reciprocal of resistance, it follows that

$$G = \frac{1}{R} \tag{2.1}$$

hence

$$I = VG \tag{2.2}$$

It may be observed that a circuit of high resistance is one of low conductance and vice versa. Unit conductance is the conductance measured between two points in a conductor when a constant potential difference of one volt, applied between these two points, produces in this conductor a current of one ampere; the conductor for this purpose must not be a source of e.m.f. This is an identical definition to that of unit resistance and it should be noted that unit resistance is also unit conductance. Both resistance and conductance will be used to relate voltage and current when considering circuit analysis.

2.2 Kirchhoff's Laws

Gustav Kirchhoff, a German physicist, observed two conditions fundamental to the analysis of any electric network. These may be stated as follows:

First (current) Law At any instant, the algebraic sum of the currents at a junction in a network is zero. Different signs are allocated to currents held to flow toward the junction and to those away from it.

Second (voltage) Law At any instant in a closed loop, the algebraic sum of the e.m.f.'s acting round the loop is equal to the algebraic sum of the potential drops round the loop.

Stated in such words, the concepts are difficult to grasp and they are more readily appreciated from Fig. 2.1. In Fig. 2.1(*a*), the currents flowing towards the junction

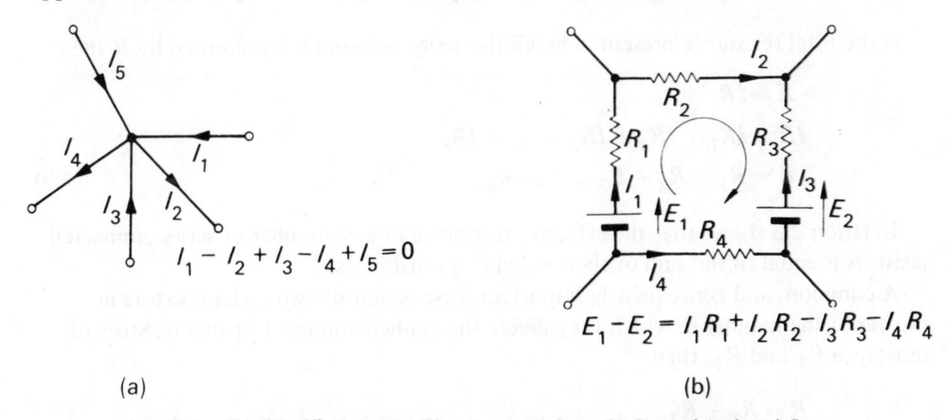

Fig. 2.1. Kirchhoff's Laws (*a*) First (current) Law (b) Second (voltage) Law

have been considered positive whilst those flowing from the junction negative. Had the opposite convention been taken, the algebraic expression would have remained the same since the application of −1 to all terms does not change its validity.

In Fig. 2.1(*b*), the sum of the e.m.f.'s is given by $E_1 - E_2$. E_1 attempts to pass current around the loop in the chosen direction and is therefore taken to be positive but E_2 opposes the flow and is taken to be negative. The volt drops depend on random currents that have been inserted. Those volt drops that stem from currents flowing in the chosen direction are positive, whilst the converse again applies. Thus I_1R_1 and I_2R_2 are positive whilst I_3R_3 and I_4R_4 are negative.

Both laws are logically self evident—for instance, if the current flow into a junction were greater than the current flow from the junction, there would rapidly be a great collection of charge at the junction. In the second case, an electron passing round the circuit would vary its level of energy but must return to its original level of energy when it returns to the starting point.

2.3 Resistors in series and in parallel

A resistor is a device that is designed to have a specific value of resistance. If any number of resistors are connected in such a way that they each pass the same flow of

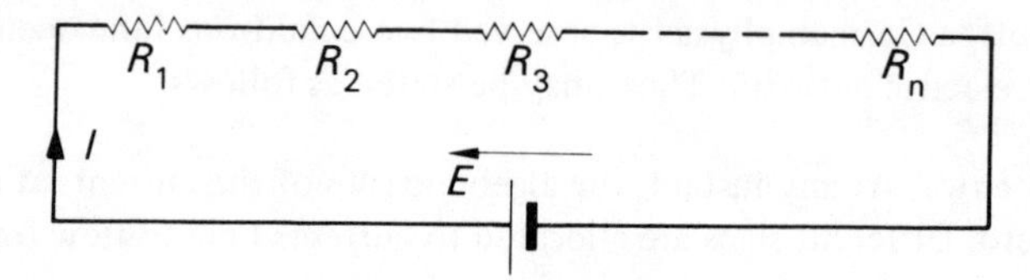

Fig. 2.2. Resistors in series

current then they are said to be connected in series. Figure 2.2 shows such an arrangement of resistors connected in series across a source of e.m.f.

Applying Kirchhoff's Second Law to the circuit in Fig. 2.2,

$$E = IR_1 + IR_2 + IR_3 + \ldots + IR_n$$

If the total resistance presented by all the series resistors is represented by R then

$$E = IR$$

$$IR = IR_1 + IR_2 + IR_3 + \ldots + IR_n$$

$$R = R_1 + R_2 + R_3 + \ldots + R_n \qquad (2.3)$$

Relation 2.3 shows that the effective resistance of any number of series-connected resistors is equal to the sum of their individual resistances.

A common, and consequently important, case is that of two series resistors in respect of the manner in which they divide the applied voltage. For two resistors of resistance R_1 and R_2, then

$$R = R_1 + R_2$$

$$I = \frac{V}{R}$$

$$= \frac{V}{R_1 + R_2}$$

The volt drop across R_1 is given by

$$V_1 = IR_1$$

$$= \frac{R_1}{R_1 + R_2} \cdot V \qquad (2.4)$$

The division of voltage is therefore a function of the resistances and may be readily obtained without reverting to calculation of the current.

Example 2.1 For the circuit shown in Fig. 2.3, calculate the potential difference across the 5-Ω resistor and the power supplied to the circuit.

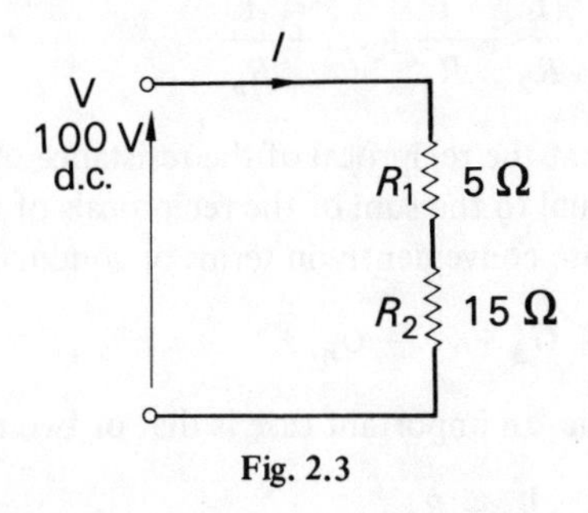

Fig. 2.3

$$V_1 = \frac{R_1}{R_1 + R_2} \cdot V = \frac{5}{5+15} \cdot 100 = \underline{25\ V}$$

$$R = R_1 + R_2 = 5 + 15 = 20\ \Omega$$

$$I = \frac{V}{R} = \frac{100}{20} = 5\ \text{A}$$

$$P = VI = 100 \times 5 = \underline{500\ \text{W}}$$

If any number of resistors are connected in such a way that they have the same potential difference across them they are said to be connected in parallel. Figure 2.4 shows such an arrangement.

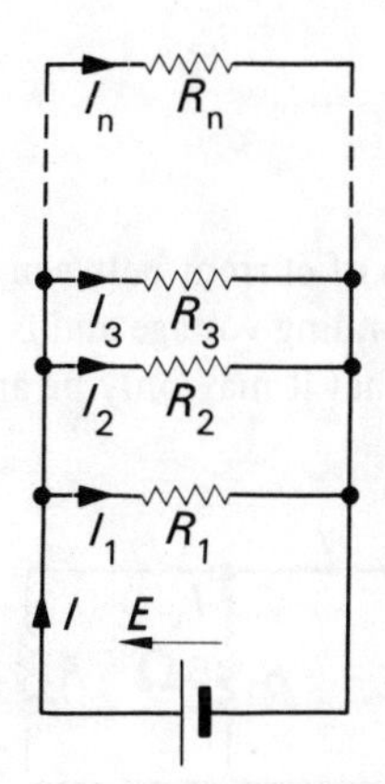

Fig. 2.4. Resistors in Parallel

Applying Kirchhoff's First Law to the circuit in Fig. 2.4:

$$I = I_1 + I_2 + I_3 + \ldots + I_n$$

If the resistance presented by all the resistors connected in parallel is represented by R then:

$$\frac{E}{R}=\frac{E}{R_1}+\frac{E}{R_2}+\frac{E}{R_3}+\ldots+\frac{E}{R_n}$$

$$\therefore\quad \frac{1}{R}=\frac{1}{R_1}+\frac{1}{R_2}+\frac{1}{R_3}+\ldots+\frac{1}{R_n} \tag{2.5}$$

Relation (2.4) shows that the reciprocal of the resistance of any number of resistors connected in parallel is equal to the sum of the reciprocals of the individual resistances. This may be expressed more conveniently in terms of conductance as follows:

$$G=G_1+G_2+G_3+\ldots+G_n \tag{2.6}$$

Because it is a common one, an important case is that of two resistors in parallel:

$$\frac{1}{R}=\frac{1}{R_1}+\frac{1}{R_2}=\frac{R_1+R_2}{R_1\,R_2}$$

$$\therefore\quad R=\frac{R_1\,R_2}{R_1+R_2} \tag{2.7}$$

If the two resistors R_1 and R_2 are connected in parallel and the currents flowing in them are I_1 and I_2 respectively, the total current being I, then:

$$E=I_1\,R_1=I_2\,R_2=I\times\frac{R_1\,R_2}{R_1+R_2}$$

$$\therefore\quad I_1=\frac{R_2}{R_1+R_2}I \tag{2.8}$$

and

$$I_2=\frac{R_1}{R_1+R_2}I \tag{2.8.1}$$

This expression for the division of current between two parallel paths avoids the necessity to calculate the corresponding voltage and is a useful shortcut in network analysis. It must be remembered that it may only be applied to two parallel resistors.

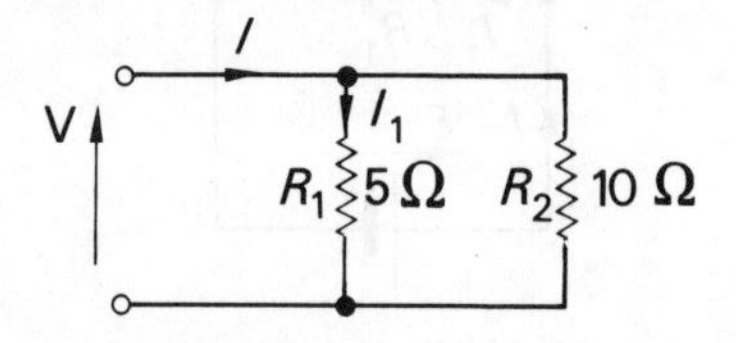

Fig. 2.5. Network for example 2.2

Example 2.2 For the network shown in Fig. 2.5, calculate the supply current I if R_1 dissipates energy at the rate of 20 W.

$$P_1 = I_1^2 R_1$$

$$\therefore \quad I_1 = \left(\frac{P_1}{R_1}\right)^{\frac{1}{2}} = \left(\frac{20}{5}\right)^{\frac{1}{2}} = 2\text{ A}$$

$$= \frac{R_2}{R_1 + R_2} I$$

$$\therefore \quad I = \frac{R_1 + R_2}{R_2} I_1 = \frac{5 + 10}{10} \times 2$$

$$= \underline{3\text{ A}}$$

2.4 Double-subscript notation

Figure 2.6 shows a resistor connected between two points A and B in a circuit. The potential difference between A and B can be specified by V_{AB}, which means the potential of A with respect to B. With a current flowing in the direction shown the

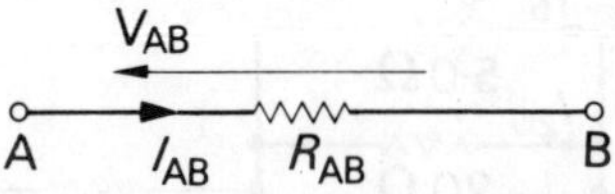

Fig. 2.6. **Double subscript notation**

potential of A must be positive with respect to that of B, therefore V_{AB} must be positive. Similarly V_{BA} means the potential of B with respect to A, and this must be equal in magnitude to V_{AB} but opposite in sign, i.e. $V_{BA} = -V_{AB}$. This system is known as double subscript notation and will be used throughout where appropriate.

If a current specified by V_{AB}/R_{AB} is positive, where R_{AB} is the resistance between A and B, it indicates that the direction of the current is from A to B. Similarly if negative its direction is from B to A.

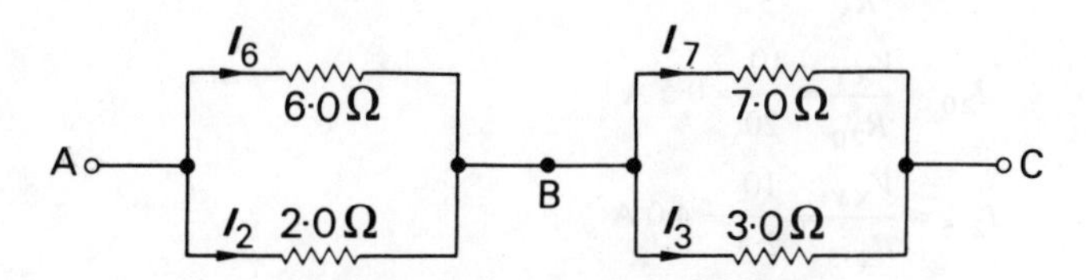

Fig. 2.7. Network for example 2.3

Example 2.3 For the network shown in Fig. 2.7, the potential difference between A and C is 10 V. Calculate the current in each resistor.

$$R_{AB} = \frac{6{\cdot}0 \times 2{\cdot}0}{6{\cdot}0 + 2{\cdot}0} = 1{\cdot}5\ \Omega$$

$$R_{BC} = \frac{7{\cdot}0 \times 3{\cdot}0}{7{\cdot}0 + 3{\cdot}0} = 2{\cdot}1\ \Omega$$

$$R_{AC} = R_{AB} + R_{BC} = 1{\cdot}5 + 2{\cdot}1 = 3{\cdot}6\ \Omega$$

$$\therefore \quad I_{AC} = \frac{V_{AC}}{R_{AC}} = \frac{10}{3{\cdot}6} = 2{\cdot}8 \text{ A}$$

$$\therefore \quad I_6 = \frac{2{\cdot}0}{6{\cdot}0 + 2{\cdot}0} \times 2{\cdot}8 = \underline{0{\cdot}7 \text{ A}}$$

$$I_2 = I_{AC} - I_6 = 2{\cdot}8 - 0{\cdot}7 = 2{\cdot}1 \text{ A}$$

$$I_7 = \frac{3{\cdot}0}{7{\cdot}0 + 3{\cdot}0} \times 2{\cdot}8 = \underline{0{\cdot}8 \text{ A}}$$

$$I_3 = I_{AC} - I_7 = 2{\cdot}8 - 0{\cdot}8 = \underline{2{\cdot}0 \text{ A}}$$

Example 2.4 For the network shown in Fig. 2.8, the power dissipated in the 5·0-Ω resistor is 20 W. Determine:

(*a*) the current in the 10-Ω resistor;
(*b*) the potential difference across XZ;
(*c*) the total power dissipated

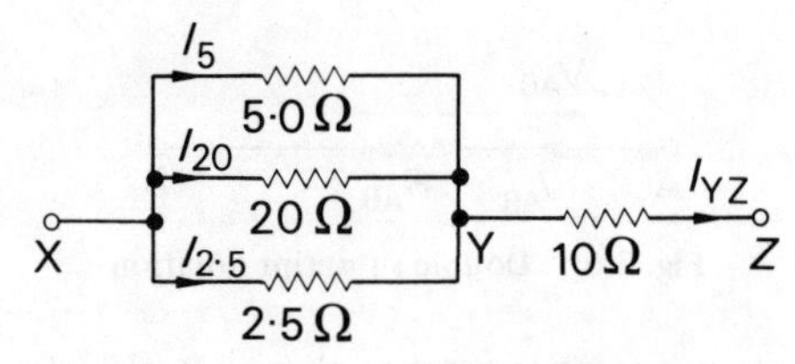

Fig. 2.8. Network for example 2.4

$$P = \frac{V_{XY}^2}{R_5} = \frac{V_{XY}^2}{5} = 20$$

$$\therefore \quad V_{XY} = 10 \text{ V}$$

$$I_5 = \frac{V_{XY}}{R_5} = \frac{10}{5} = 2{\cdot}0 \text{ A}$$

$$I_{20} = \frac{V_{XY}}{R_{20}} = \frac{10}{20} = 0{\cdot}5 \text{ A}$$

$$I_{2{\cdot}5} = \frac{V_{XY}}{R_{2{\cdot}5}} = \frac{10}{2{\cdot}5} = 4{\cdot}0 \text{ A}$$

$$\therefore \quad I_{XY} = I_5 + I_{20} + I_{2{\cdot}5} = 2{\cdot}0 + 0{\cdot}5 + 4{\cdot}0$$

$$= 6{\cdot}5 \text{ A}$$

$$= I_{YZ}$$

$$V_{YZ} = I_{YZ} R_{YZ} = 6{\cdot}5 \times 10$$

$$= 65 \text{ V}$$

$$V_{XZ} = V_{XY} + V_{YZ} = 10 + 65$$

$$= \underline{75 \text{ V}}$$

$$P = V_{XZ} I_{XZ} = 75 \times 6{\cdot}5$$

$$= \underline{488 \text{ W}}$$

2.5 Resistivity and conductivity

The resistance that any piece of material possesses may be shown experimentally to depend on:

(*a*) its length;
(*b*) its cross-sectional area;
(*c*) the material;
(*d*) ambient physical factors such as temperature.

Setting the ambient physical factors aside, the resistance may be shown to be directly proportional to the length of the conductor material and inversely proportional to its cross-sectional area.

$$R \propto \frac{l}{A}$$

$$R = \rho \frac{l}{A} \qquad (2.9)$$

where ρ is an appropriate constant of proportionality for the material at the given ambient conditions. It is known as the resistivity of the material.

Resistivity	Symbol: ρ	Unit: ohm metre (Ω m)

From relation (2.9) if l is 1 m and A is 1 m^2 then $\rho = R$, i.e. the resistivity of the material is numerically equal to the resistance measured across the opposite faces of a 1-m cube of the material.

The conductivity γ of a material is the reciprocal of the resistivity.

Conductivity	Symbol: γ	Unit: siemens per metre (S/m)

$$\gamma = \frac{1}{\rho} \qquad (2.10)$$

thus $$R = \frac{l}{\gamma A} \qquad (2.11)$$

and $$G = \frac{\gamma A}{l} \qquad (2.11.1)$$

Example 2.5 A copper wire is 100 m long and has a diameter of 1·0 mm. If the resistivity of copper is 0·017 2 $\mu\Omega$ m, calculate the resistance of the wire.

$$A = \pi r^2 = \pi \times 0{\cdot}5^2 \times 10^{-6}\,\mathrm{m}^2$$

$$R = \rho \frac{l}{A} = \frac{1{\cdot}72 \times 10^{-8} \times 10^2}{\pi \times 0{\cdot}5^2 \times 10^{-6}}$$

$$= \underline{2{\cdot}19\ \Omega}$$

Example 2.6 A copper conductor is connected in parallel with twice its length of aluminium conductor. The resistivity of copper is 0·017 2 μΩ m and that of aluminium is 0·025 4 μΩ m. If the diameter of the aluminium is 10 mm, determine the necessary diameter of the copper conductor in order that each conductor will carry the same current.

Let l be the length of the aluminium conductor,
R_a be the resistance of the aluminium conductor
and R_c be the resistance of the copper conductor.

$$R_a = \frac{\rho l}{A_a} = \frac{2{\cdot}54 \times 10^{-8} \times l}{\pi \times 5^2 \times 10^{-6}}$$

$$R_c = \frac{1{\cdot}72 \times 10^{-8} \times 0{\cdot}5\, l}{\pi r^2}$$

For equal currents,

$$R_a = R_c$$

$$r^2 = \frac{5^2 \times 10^{-6} \times 1{\cdot}72 \times 0{\cdot}5}{2{\cdot}54}$$

$$r = 2{\cdot}9 \times 10^{-3}\ \text{m} = 2{\cdot}9\ \text{mm}$$

$$\therefore \quad d = 5{\cdot}8\ \text{mm}$$

2.6 Temperature coefficient of resistance

It will be shown in para. 13.5 that the resistance of materials varies with temperature. For some materials (e.g. metals) the resistance increases with increasing temperature, while for other materials, including semi-conductors and some insulators, it decreases with increasing temperature. In all cases the resistance is a complex function of

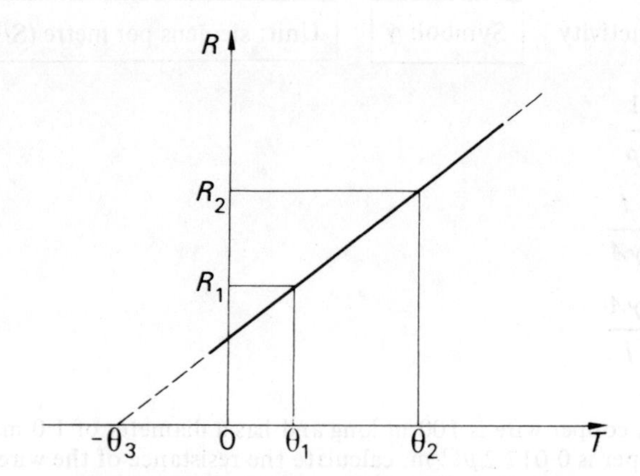

Fig. 2.9. Variation of resistance with temperature

temperature. However, over the limited temperature ranges usually met with in practice, little error is introduced in assuming that the function is a linear one, as shown in Fig. 2.9.

In order to estimate the change of resistance or the new values of resistance due to changes of temperature, a change coefficient must be introduced–this is the temperature coefficient of resistance which is defined as the change in resistance per kelvin (Celsius degree), expressed as a fraction of the resistance at the temperature being considered.

Temperature coefficient of resistance	Symbol: α	Unit: per kelvin (/K)

From Fig. 2.9 the temperature coefficient of resistance (α_1) at temperature θ_1 will be given by:

$$\alpha_1 = \frac{\dfrac{R_2 - R_1}{\theta_2 - \theta_1}}{R_1} \tag{2.12}$$

Similarly at temperature θ_2:

$$\alpha_2 = \frac{\dfrac{R_2 - R_1}{\theta_2 - \theta_1}}{R_2} \tag{2.12.1}$$

Re-arranging equation 2.12 gives:

$$\alpha_1 R_1(\theta_2 - \theta_1) = R_2 - R_1.$$

$$\therefore\ R_2 = R_1[1 + \alpha_1(\theta_2 - \theta_1)] \tag{2.13}$$

It is not always possible to ascertain α, readily without further calculation and therefore it is usual to give the temperature coefficient at a particular reference temperature–usually 0°C. If R_0 is the corresponding resistance, then

$$R_1 = R_0(1 + \alpha_0\,\theta_1)$$

and

$$R_2 = R_0(1 + \alpha_0\,\theta_2)$$

$$\therefore\ \frac{R_1}{R_2} = \frac{1 + \alpha_0\,\theta_1}{1 + \alpha_0\,\theta_2} \tag{2.13.1}$$

The temperature coefficient of resistance (α_n) at any other temperature θ_n can be determined as follows:

If the linear section of the resistance/temperature characteristic is produced so that it cuts the temperature axis at $-\theta_3$ as shown, then from similar triangles:

$$\frac{R_2 - R_1}{\theta_2 - \theta_1} = \frac{R_1}{\theta_1 + \theta_3}$$

therefore from equation (2.12)

$$\alpha_1 = \frac{\dfrac{R_1}{\theta_1 + \theta_3}}{R_1} = \frac{1}{\theta_1 + \theta_3}$$

$$\therefore\ \theta_3 = \frac{1 - \alpha_1\,\theta_1}{\alpha_1}$$

In general $\alpha_n = \dfrac{1}{\theta_n + \theta_3}$

Substituting for θ_3

$$\alpha_n = \frac{1}{\theta_n + \dfrac{1 - \alpha_1 \theta_1}{\alpha_1}}$$

$$\therefore \quad \alpha_n = \frac{1}{\theta_n - \theta_1 + \dfrac{1}{\alpha_1}} \tag{2.14}$$

Again for the common reference temperature of 0°C,

$$\alpha_0 = \frac{1}{\dfrac{1}{\alpha_1} - \theta_1} \tag{2.14.1}$$

Example 2.7 When the potential difference applied across a coil of copper wire at a mean temperature of 20°C is 10·0 V a current of 1·0 A flows in it. After a period of time the current falls to 0·95 A with the potential difference unaltered. Determine the mean temperature of the coil. The temperature coefficient of resistance of copper at 0°C as $4{\cdot}28 \times 10^{-3}$/K.

At 20°C, $\quad R_1 = \dfrac{V}{I_1} = \dfrac{10{\cdot}0}{1{\cdot}0} = 10{\cdot}0\ \Omega$

at θ_2, $\quad R_2 = \dfrac{V}{I_2} = \dfrac{10{\cdot}0}{0{\cdot}95} = 10{\cdot}53\ \Omega$

$$\frac{R_1}{R_2} = \frac{1 + \alpha_0 \theta_1}{1 + \alpha_0 \theta_2}$$

$$\therefore \quad \frac{10{\cdot}0}{10{\cdot}53} = \frac{1 + (4{\cdot}28 \times 10^{-3} \times 20)}{1 + (4{\cdot}28 \times 10^{-3} \times \theta_2)}$$

hence $\quad \underline{\theta_2 = 33{\cdot}4°\text{C}}$

Alternatively $\quad \alpha_0 = \dfrac{1}{\dfrac{1}{\alpha_1} - \theta_1}$

$$\therefore \quad 4{\cdot}28 \times 10^{-3} = \frac{1}{\dfrac{1}{\alpha_1} - 20}$$

$$\therefore \quad \frac{4{\cdot}28 \times 10^{-3}}{\alpha_1} - 8{\cdot}56 \times 10^{-2} = 1$$

$$\therefore \quad \alpha_1 = \frac{4{\cdot}28 \times 10^{-3}}{1{\cdot}08} = 3{\cdot}96 \times 10^{-3}$$

$$R_2 = R_1[1 + \alpha_1(\theta_2 - \theta_1)]$$

$$\therefore \quad 10{\cdot}53 = 10{\cdot}0[1 + 3{\cdot}96 \times 10^{-3}(\theta_2 - 20)]$$

$$\therefore \quad \theta_2 - 20 = \frac{0{\cdot}53}{3{\cdot}96 \times 10^{-2}} = 13{\cdot}4 \quad \therefore \quad \underline{\theta_2 = 33{\cdot}4°\text{C}.}$$

Example 2.8 A 240-V, 100-W incandescent filament lamp operates at a filament temperature of 2 000°C. If the temperature coefficient of resistance of the filament material is 5 x 10^{-3}/K at 15°C, determine the current taken by the lamp at the instant of switching on.

$$\text{Resistance of lamp at 2 000°C} = \frac{240^2}{100} = 576\ \Omega$$

$$R_2 = R_1[1 + \alpha_1(\theta_2 - \theta_1)]$$

$$\therefore \quad 576 = R_1[1 + 5 \times 10^{-3}(2000 - 15)]$$

$$= R_1(1 + 9{\cdot}92)$$

$$\therefore \quad R_1 = \frac{576}{10{\cdot}92} = 52{\cdot}7\ \Omega$$

$$\therefore \text{Current at instant of switching on} = \frac{V}{R_1} = \frac{240}{52{\cdot}7} = \underline{4{\cdot}55\text{ A.}}$$

2.7 Complex network analysis by Kirchhoff's Laws

While it is possible to solve problems associated with networks with the sole use of equations 2.3 and 2.5, more complex circuits call for other methods. The direct application of Kirchhoff's Laws, as defined in para. 2.2, is one of the more common methods. The examples below illustrate their use. In each case the first law is applied when allocating currents to the branches of the network.

Example 2.9 Calculate the current in each branch of the network shown in Fig. 2.10.

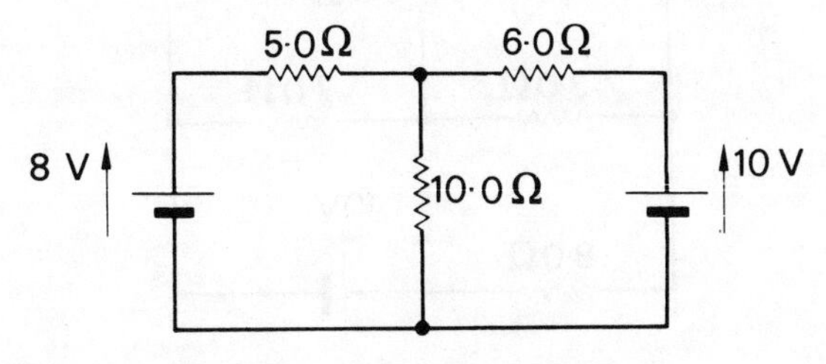

Fig. 2.10. Network for example 2.9

Let the currents flowing in the network be as shown in Fig. 2.11.

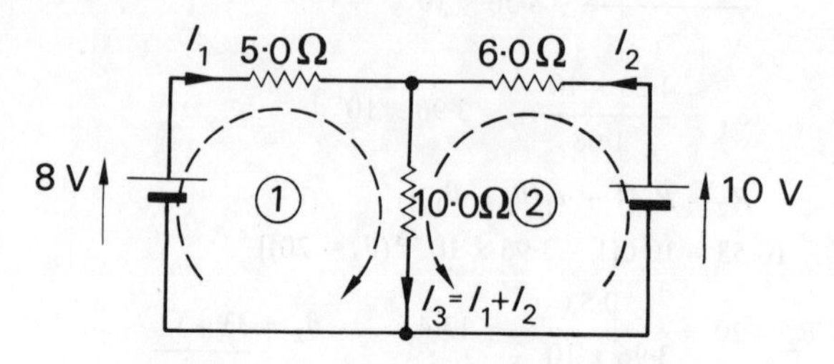

Fig. 2.11. Part of example 2.9

In Loop ① in direction shown:

$$8 = 5I_1 + 10(I_1 + I_2)$$

$$\therefore \quad 8 = 15I_1 + 10I_2 \qquad [1]$$

In Loop ② in direction shown:

$$10 = 6I_2 + 10(I_1 + I_2)$$

$$\therefore \quad 10 = 10I_1 + 16I_2 \qquad [2]$$

$$[1] \times 2, \quad [2] \times 3$$

$$16 = 30I_1 + 20I_2 \qquad [3]$$

$$30 = 30I_1 + 48I_2 \qquad [4]$$

$$[4] - [3]$$

$$\therefore \quad 14 = 28I_2 \quad \therefore \quad I_2 = \frac{14}{28} = \underline{0{\cdot}50\ \text{A}}$$

Substitute in [1],

$$8 = 15I_1 + 5{\cdot}0 \quad \therefore \quad I_1 = \frac{3{\cdot}0}{15{\cdot}0} = \underline{0{\cdot}20\ \text{A}}$$

$$I_3 = I_1 + I_2 = 0{\cdot}50 + 0{\cdot}20 = \underline{0{\cdot}70\ \text{A}}$$

Example 2.10 **Calculate the current in the 5·0-Ω resistor in the network shown in Fig. 2.12.**

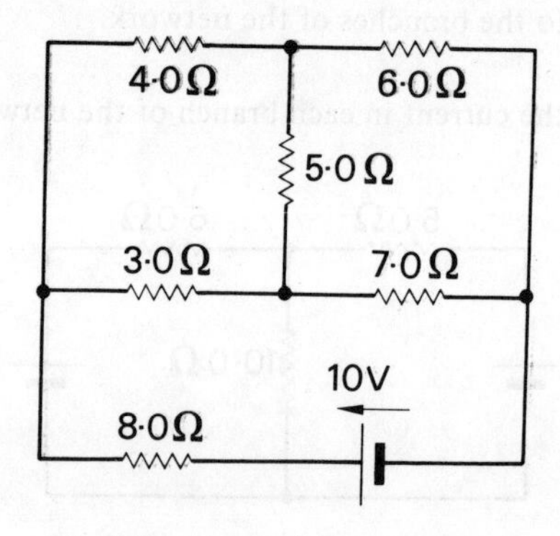

Fig. 2.12 Network for example 2.10

Let the currents flowing in the network be as shown in Fig. 2.13.

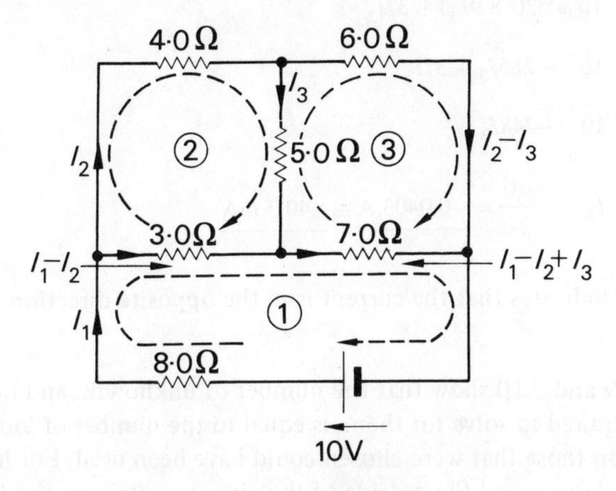

Fig. 2.13. Part of example 2.10

Note that the current in the required branch has been specified by one symbol only (i.e. I_3). This means that the equations need be solved only for this one unknown.

In Loop ① in direction shown:

$$10 = 8I_1 + 3(I_1 - I_2) + 7(I_1 - I_2 + I_3)$$

$$\therefore \quad 10 = 18I_1 - 10I_2 + 7I_3 \qquad [1]$$

In Loop ② in direction shown:

$$0 = 4I_2 + 5I_3 - 3(I_1 - I_2)$$

$$\therefore \quad 0 = -3I_1 + 7I_2 + 5I_3 \qquad [2]$$

In Loop ③ in direction shown:

$$0 = 6(I_2 - I_3) - 7(I_1 - I_2 + I_3) - 5I_3$$

$$\therefore \quad 0 = -7I_1 + 13I_2 - 18I_3 \qquad [3]$$

[2] × 6

$$0 = -18I_1 + 42I_2 + 30I_3 \qquad [4]$$

[1]+[4]

$$10 = 32I_2 + 37I_3 \qquad [5]$$

[2] × 7 [3] × 3

$$0 = -21I_1 + 49I_2 + 35I_3 \qquad [6]$$

$$0 = -21I_1 + 39I_2 - 54I_3 \qquad [7]$$

[6] − [7]

$$0 = 10I_2 + 89I_3$$

$$\therefore \quad I_2 = -8{\cdot}9I_3 \qquad [8]$$

Substitute [8] in [5]

$$\therefore \quad 10 = 32(-8{\cdot}9I_3) + 37I_3$$

$$\therefore \quad 10 = -285I_3 + 37I_3$$

$$\therefore \quad 10 = -248I_3$$

$$\therefore \quad I_3 = -\frac{10}{248} = \underline{-0{\cdot}0403\ \text{A}} \equiv \underline{-40{\cdot}3\ \text{mA}}$$

The minus sign indicates that the current is in the opposite direction to that shown.

Examples 2.9 and 2.10 show that the number of unknowns, and hence the number of equations required to solve for them, is equal to the number of loops in the circuit. Loops other than those that were chosen could have been used. For instance in Example 2.10 a loop round the outside of the circuit containing the 10-V source of e.m.f., the 8·0-Ω, 4·0-Ω and 6·0-Ω resistors could have been used equally well. The only criterion is that each time a new loop is used at least one branch which has not been included in other loops must be a constituent part of this new loop.

2.8 Network analysis by Maxwell's circulating currents

This method of circuit analysis is similar to that using Kirchhoff's Laws. Circulating currents are, however, allocated to the closed loops in the circuit rather than to the branches. The branch currents are then found by taking the algebraic sum of the loop currents which are common to the branch. The equations required are obtained by equating the algebraic sum of the e.m.f.s round the loop to the algebraic sum of the potential differences in the direction of the circulating loop current as per Kirchhoff's Second Law.

Example 2.11 Calculate the current in each branch of the network shown in Fig. 2.14.

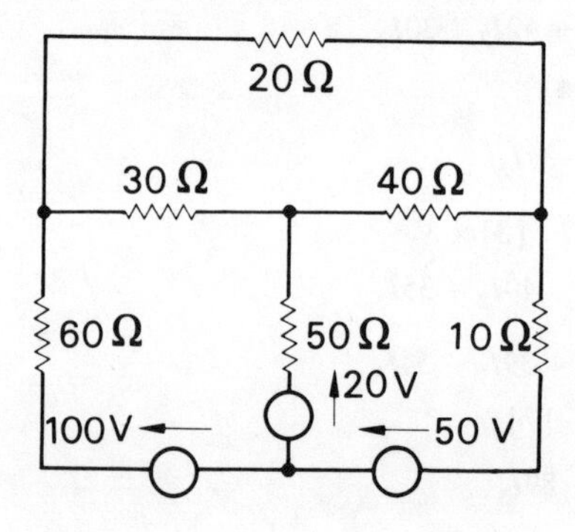

Fig. 2.14. Network for example 2.11

Let the circulating loop currents be as shown in Fig. 2.15.

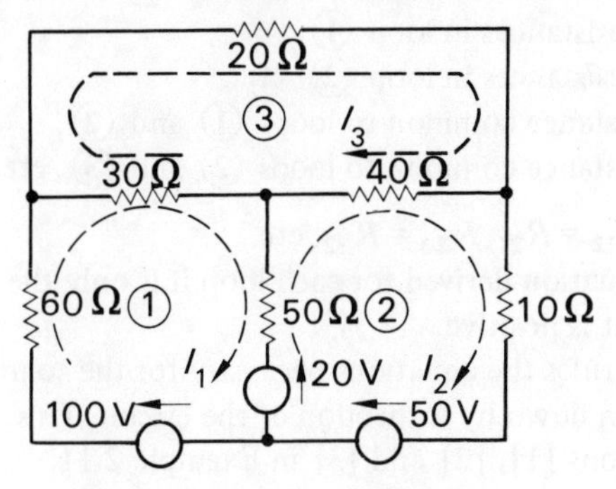

Fig. 2.15. Part of example 2.11

In Loop ①:

$$100 - 20 = I_1(60 + 30 + 50) - I_2\,50 - I_3\,30$$

$$\therefore \quad 80 = 140I_1 - 50I_2 - 30I_3 \qquad [1]$$

In Loop ②:

$$50 + 20 = I_2(50 + 40 + 10) - I_1\,50 - I_3\,40$$

$$\therefore \quad 70 = -50I_1 + 100I_2 - 40I_3 \qquad [2]$$

In Loop ③

$$0 = I_3(30 + 20 + 40) - I_1\,30 - I_2\,40$$

$$\therefore \quad 0 = -30I_1 - 40I_2 + 90I_3 \qquad [3]$$

Solving for these equations gives:

$$I_1 = 1{\cdot}65\text{ A}, \quad I_2 = 2{\cdot}16\text{ A}, \quad I_3 = 1{\cdot}50\text{ A}.$$

Current in 60 Ω = I_1 = 1·65 A in direction of I_1
Current in 30 Ω = $I_1 - I_3$ = 0·15 A in direction of I_1
Current in 50 Ω = $I_2 - I_1$ = 0·51 A in direction of I_2
Current in 40 Ω = $I_2 - I_3$ = 0·66 A in direction of I_2
Current in 10 Ω = I_2 = 2·16 A in direction of I_2
Current in 20 Ω = I_3 = 1·50 A in direction of I_3.

In Example 2.11 all the circulating loop currents have been taken in the same direction (i.e. clockwise). This is not essential when using this method, but if the same direction is adopted for the loop currents then the equations will always be of the form:

$$E_1 = R_{11}\,I_1 - R_{12}\,I_2 - R_{13}\,I_3 \ldots - R_{1n}\,I_n$$

$$E_2 = -R_{21}\,I_1 + R_{22}\,I_2 - R_{23}\,I_3 \ldots - R_{2n}\,I_n$$

$$E_3 = -R_{31}\,I_1 - R_{32}\,I_2 + R_{33}\,I_3 \ldots -R_{3n}\,I_n$$

$$E_n = -R_{n1}\,I_1 - R_{n2}\,I_2 - R_{n3}\,I_3 \ldots + R_{nn}\,I_n$$

where E_1 = the algebraic sum of the e.m.fs in loop ① in the direction of I_1,
E_2 = the algebraic sum of the e.m.fs in loop ② in the direction of I_2, etc.,
R_{11} = sum of resistances in loop ①,
R_{22} = sum of resistances in loop ②, etc.,
R_{12} = total resistance common to loops ① and ②,
R_{23} = total resistance common to loops ② and ③, etc.,

By their definitions $R_{12} = R_{21}, R_{23} = R_{32}$, etc.

Note that in the equation derived for each loop it is only the term in the loop's own circulating current that is positive.

By observing these rules the equations necessary for the solution of the circuit problem can be written down by inspection of the circuit. This can be confirmed by examination of equations [1], [2] and [3] in Example 2.11.

2.9 Cells with internal resistance

The representation of a source of electrical energy solely by a source of e.m.f. does not give a complete representation of it. When the source delivers current the potential difference across its output terminals drops. This is due to the internal resistance which all sources possess. For example, the electrolyte of a cell will present resistance to the flow of current in it. This means that not only will the potential difference across its output terminals drop when it supplies current, but that some of the chemical energy stored in the cell will be dissipated as heat within the cell itself. Thus a more accurate

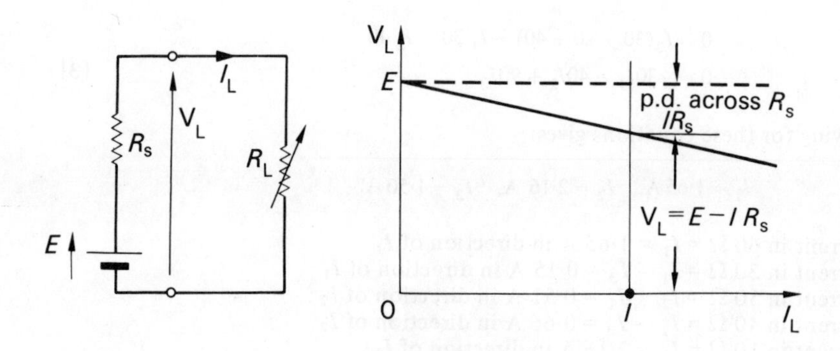

Fig. 2.16. Cell with internal resistance

representation would be obtained by showing resistance connected in series with the source of e.m.f. This is shown in Fig. 2.16 with a characteristic showing the variation of output terminal voltage with output current.

Example 2.12 A battery is formed from five identical cells in series. When an external resistance of 4 Ω is connected across the battery, the current is 1·5 A. If the external resistance is changed to 9 Ω, the current falls to 0·75 A. Find the e.m.f. of each cell and its internal resistance.

Let E = e.m.f. of each cell
and R = internal resistance of each cell.

$$\therefore \qquad 1{\cdot}5 = \frac{5E}{5R+4} \cdots \qquad [1]$$

$$0{\cdot}75 = \frac{5E}{5R+9} \cdots \qquad [2]$$

$$[1] \div [2]$$

$$\therefore \qquad 2{\cdot}0 = \frac{5R+9}{5R+4}$$

$$\therefore \quad 10R + 8 = 5R + 9 \quad \therefore \quad 5R = 1 \quad \therefore \quad \underline{R = 0{\cdot}2\ \Omega}$$

$$\therefore \qquad 1{\cdot}5 = \frac{5E}{5 \times 0{\cdot}2 + 4} = \frac{5E}{5}$$

$$\therefore \qquad 5E = 1{\cdot}5 \times 5 \quad \therefore \quad \underline{E = 1{\cdot}5\ \text{V}}$$

2.10 Thevenin's theorem

Thevenin's Theorem states that:

> The current which flows in any branch of a circuit is the same as that which would flow in the branch if it were connected across a source of electrical energy, the e.m.f. of which is equal to the potential difference which would appear across the branch if it were open circuited, and the internal resistance of which is equal to the resistance which appears across the open-circuited branch terminals.

In calculating the internal resistance, sources of e.m.f. are treated as short circuits.
Thevenin's theorem is a most useful tool in the solution of problems associated with networks, particularly electronics networks. It states that however complex the network, after the branch has been removed, it can be replaced by one source of e.m.f.

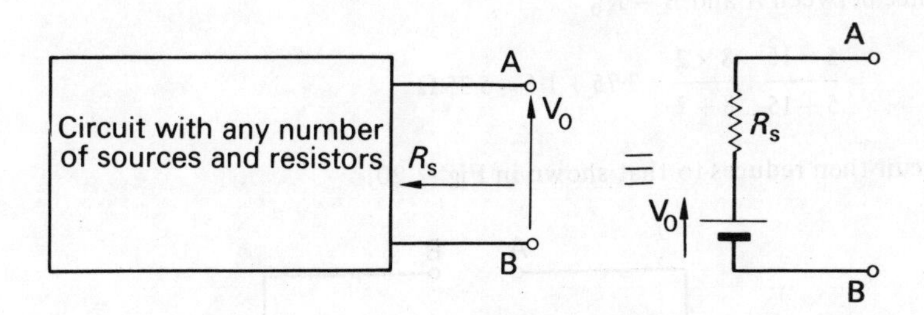

Fig. 2.17. Thevenin's theorem

in series with an internal resistance. This is illustrated in Fig. 2.17. Thevenin's theorem will be used to best advantage in networks where the removal of the branch concerned simplifies greatly the complexity of the circuit.

Example 2.13 Calculate the current in the 10-Ω resistor in the network shown in Fig. 2.18.

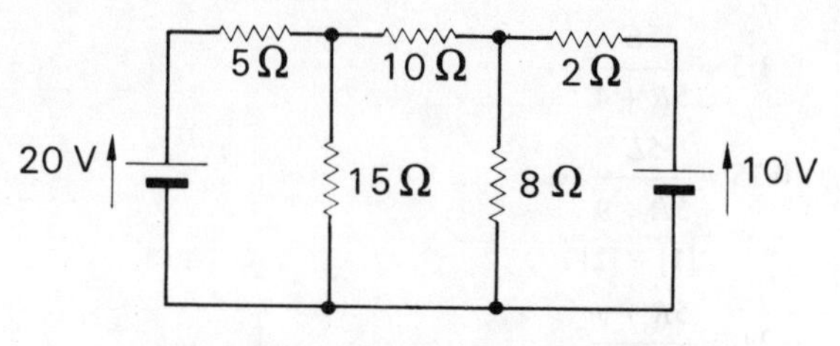

Fig. 2.18. Network for example 2.13

Open circuiting the branch containing the 10-Ω resistor gives the network shown in Fig. 2.19.

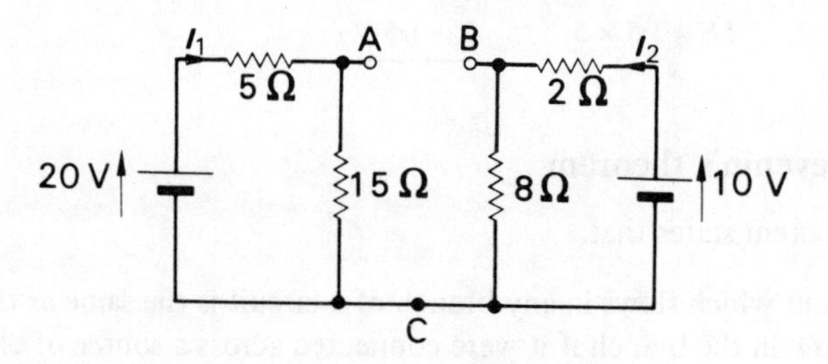

Fig. 2.19. Part of example 2.13

$$I_1 = \frac{20}{5+15} = 1{\cdot}0\text{ A} \quad \therefore \quad V_{AC} = 1{\cdot}0 \times 15 = 15\text{ V}$$

$$I_2 = \frac{10}{2+8} = 1{\cdot}0\text{ A} \quad \therefore \quad V_{BC} = 1{\cdot}0 \times 8 = 8\text{ V}$$

$$\therefore \quad V_{AB} = V_{AC} - V_{BC} = 15 - 8 = 7\text{ V}$$

Resistance between A and B = R_0

$$= \frac{5 \times 15}{5+15} + \frac{8 \times 2}{8+2} = 3{\cdot}75 + 1{\cdot}6 = 5{\cdot}35\ \Omega$$

The circuit then reduces to that shown in Fig. 2.20.

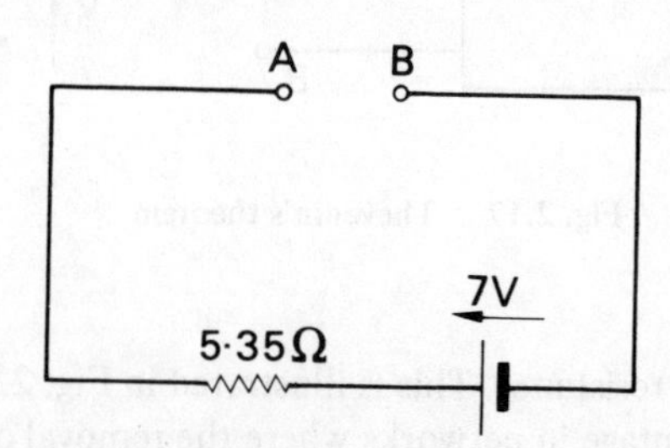

Fig. 2.20. Part of example 2.13

∴ Current in 10-Ω resistor across AB

$$= \frac{7}{5{\cdot}35 + 10} = 0{\cdot}46 \text{ A from A to B}$$

Example 2.14 Calculate the power which would be dissipated in a 50-Ω resistor connected across XY in the network shown in Fig. 2.21.

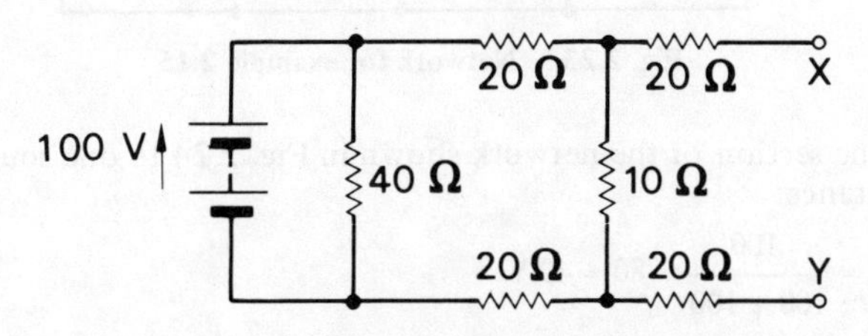

Fig. 2.21. Network for example 2.14

With XY on open circuit the current in the 10-Ω resistor is given by:

$$I = \frac{100}{20 + 10 + 20} = 2 \text{ A}$$

∴ open circuit voltage across XY $= V_0 = 2 \times 10 = 20$ V

$$\text{Resistance looking into XY} = R_0 = 20 + \frac{10(20 + 20)}{10 + (20 + 20)} + 20$$

$$= 20 + 8 + 20 = 48\ \Omega$$

The circuit then reduces to that shown in Fig. 2.22.

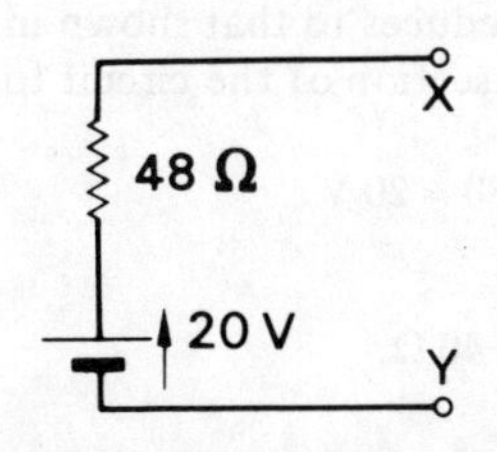

Fig. 2.22. Part of example 2.14

∴ current in 50-Ω resistor connected across XY

$$= I = \frac{20}{48 + 50}$$

$$P = I^2 R = \left(\frac{20}{98}\right)^2 \times 50 = 20{\cdot}8 \text{ W}$$

Note that the response across XY is unaffected by the resistance of the resistor connected across the 100-V supply.

Example 2.15 Calculate the current in the 50-Ω resistor in the network shown in Fig. 2.23.

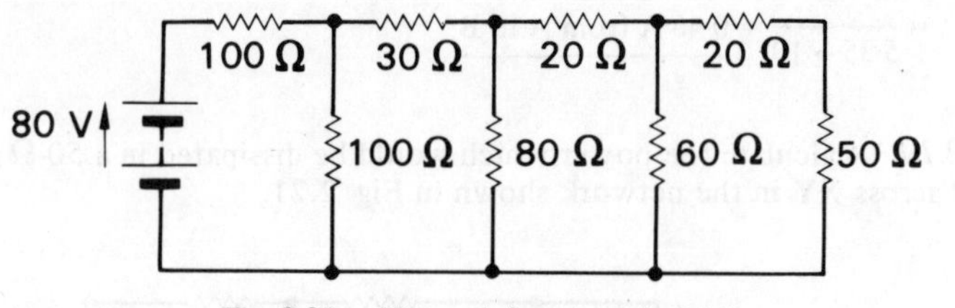

Fig. 2.23. Network for example 2.15

Converting the section of the network shown in Fig. 2.24 to one source of e.m.f. and an internal resistance:

$$V_{01} = \frac{100}{100+100} \times 80 = 40 \text{ V}$$

$$R_{01} = \frac{100 \times 100}{100+100} = 50 \ \Omega$$

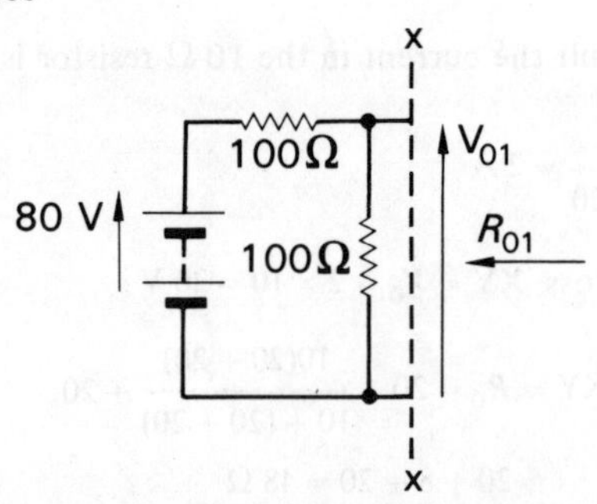

Fig. 2.24. Part of example 2.15

Therefore the original network reduces to that shown in Fig. 2.25.
Using the same procedure to the section of the circuit to the left of *yy*:

$$V_{02} = \frac{40}{50+30+80} \times 80 = 20 \text{ V}$$

$$R_{02} = \frac{80(50+30)}{80+(50+30)} = 40 \ \Omega$$

Fig. 2.25. Part of example 2.15

Therefore the original network reduces to that shown in Fig. 2.26.

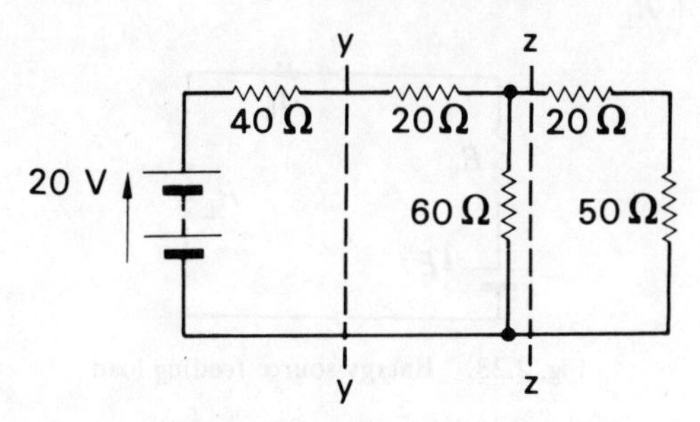

Fig. 2.26. Part of example 2.15

Using the same procedure to the section of the network to the left of zz:

$$V_{03} = \frac{20}{40+20+60} \times 60 = 10\text{ V}$$

$$R_{03} = \frac{60(20+40)}{60+(20+40)} = 30\ \Omega$$

Therefore the original network reduces to that shown in Fig. 2.27.

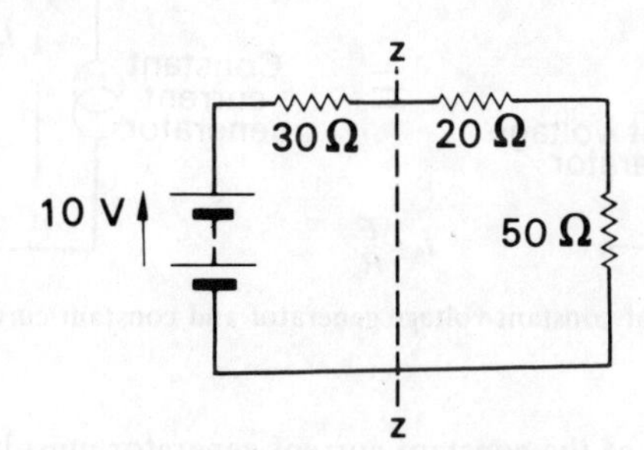

Fig. 2.27. Part of example 2.15

$$\text{Current in } 50\text{-}\Omega = I = \frac{10}{30+20+50} = \underline{0{\cdot}10\text{ A}}$$

2.11 The constant-current generator

It was shown in para. 2.10 that a source of electrical energy could be represented by a source of e.m.f. in series with a resistance. This is not, however, the only form of representation. Consider such a source feeding a load resistor R_L as shown in Fig. 2.28.

From this circuit:

$$I_L = \frac{E}{R_s + R_L} = \frac{\dfrac{E}{R_s}}{\dfrac{R_s + R_L}{R_s}}$$

therefore $$I_L = \frac{R_s}{R_s + R_L} \times I_s \tag{2.15}$$

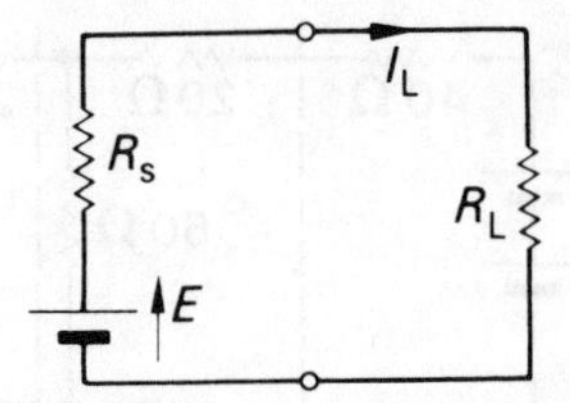

Fig. 2.28. Energy source feeding load

where $I_s = E/R_s$ = the current which would flow in a short circuit across the output terminals of the source.

Comparing relation 2.8 with relation 2.15 it can be seen that, when viewed from the load, the source appears as a source of current (I_s) which is dividing between the internal resistance (R_s) and the load resistor (R_L) connected in parallel. For the solution of problems either form of representation can be used. In many practical cases an easier solution is obtained using the current form. Figure 2.29 illustrates the equivalence of the two forms.

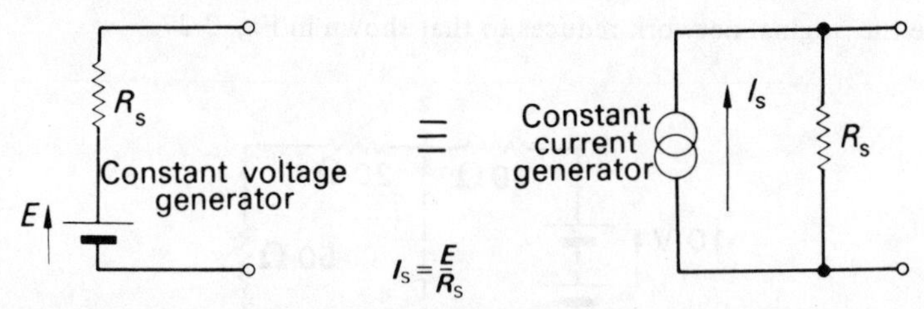

Fig. 2.29. Equivalence of constant-voltage generator and constant-current generator forms of representation

The internal resistance of the constant-current generator must be taken as infinite, since the resistance of the complete source must be R_S as is obtained with the constant-voltage form.

The ideal constant-voltage generator would be one with zero internal resistance so that it would supply the same voltage to all loads. Conversely the ideal constant-current generator would be one with infinite internal resistance so that it supplied the same current to all loads. These ideal conditions can be approached quite closely in practice.

Example 2.16 Represent the network shown in Fig. 2.30 by one source of e.m.f. in series with a resistance.

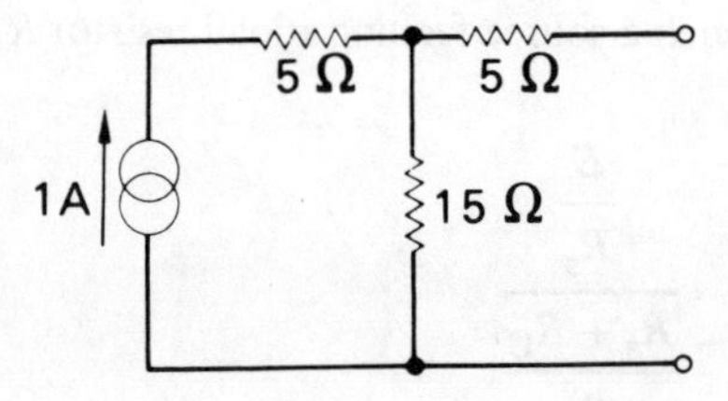

Fig. 2.30. Network for example 2.14

Potential difference across output terminals

$= V_0 = 1 \times 15 = 15$ V

Resistance looking into output terminals

$= 5 + 15 = 20\ \Omega$

therefore the circuit can be represented as shown in Fig. 2.31.

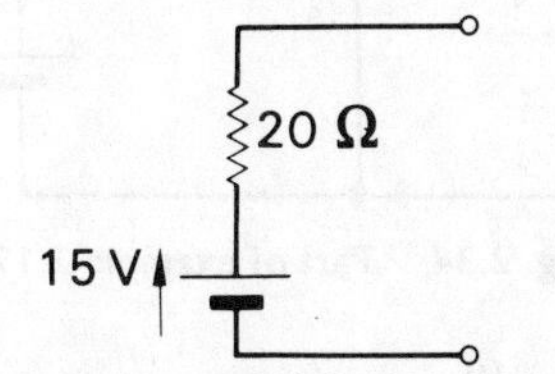

Fig. 2.31. Part of example 2.14

2.12 Norton's theorem

When a branch in a network is open-circuited the remainder of the network can be represented by one source of e.m.f. in series with a resistor; it follows from what has been said in para. 2.11 that it could equally well be represented by a source of current in parallel with the same resistor. Norton's theorem is therefore a restatement of Thevenin's theorem using an equivalent current-generator source instead of the equivalent voltage-generator source. It can therefore be stated that:

> The current which flows in any branch of a network is the same as that which would flow in the branch if it were connected across a source of electrical energy, the short-circuit current of which is equal to the current that would flow in a short-circuit across the branch, and the internal resistance of which is equal to the resistance which appears across the open-circuited branch terminals.

Norton's theorem is illustrated in Fig. 2.32.

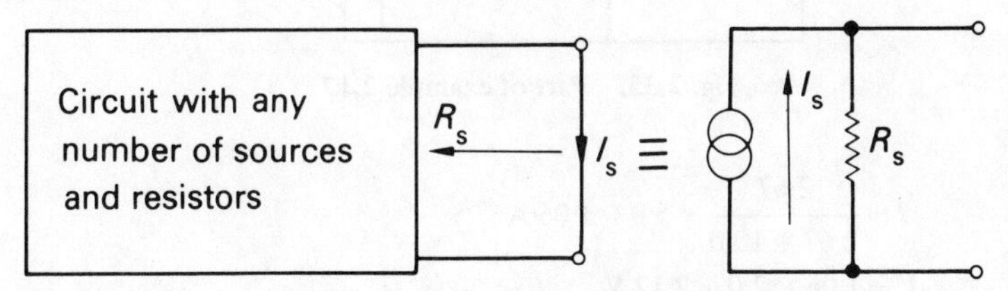

Fig. 2.32. Norton's theorem

Example 2.17 Calculate the potential difference across the 2·0-Ω resistor in the network shown in Fig. 2.33.

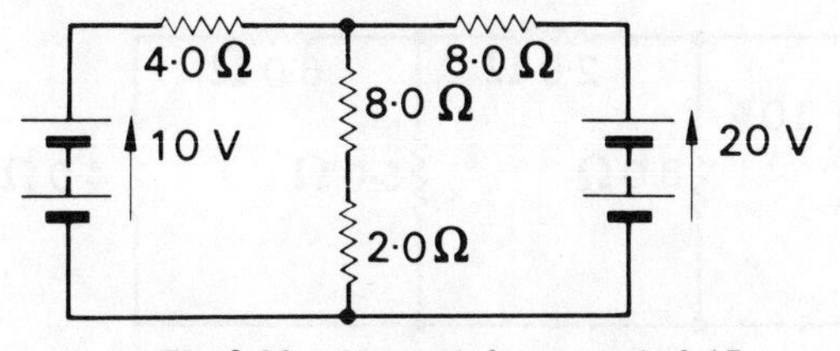

Fig. 2.33. Network for example 2.17

Short circuiting the branch containing the 2·0-Ω resistor gives the network shown in Fig. 2.34.

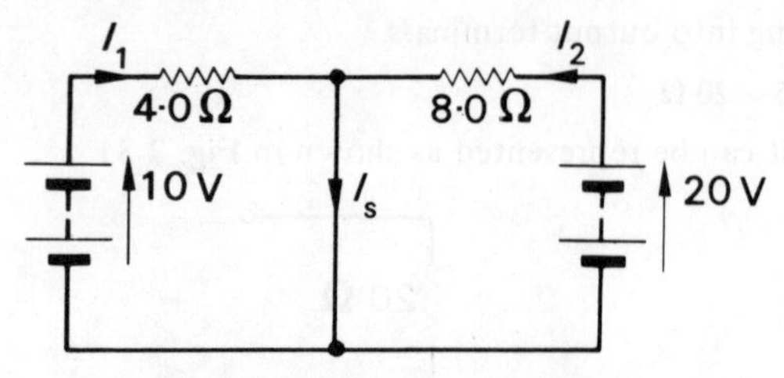

Fig. 2.34. Part of example 2.17

$$10 = 4{\cdot}0\, I_1 \quad \therefore \quad I_1 = \frac{10}{4{\cdot}0} = 2{\cdot}5 \text{ A}$$

$$20 = 8{\cdot}0\, I_2 \quad \therefore \quad I_2 = \frac{20}{8{\cdot}0} = 2{\cdot}5 \text{ A}$$

$$\therefore \quad I_s = I_1 + I_2 = 5{\cdot}0 \text{ A}$$

Resistance across open-circuited branch

$$= \frac{4{\cdot}0 \times 8{\cdot}0}{4{\cdot}0 + 8{\cdot}0} = 2{\cdot}67\ \Omega$$

therefore the circuit reduces to that shown in Fig. 2.35.

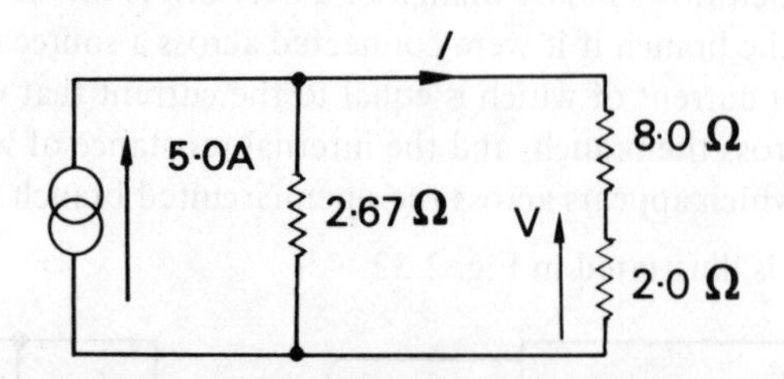

Fig. 2.35. Part of example 2.17

$$I = \frac{2{\cdot}67}{2{\cdot}67 + 10{\cdot}0} \times 5{\cdot}0 = 1{\cdot}06 \text{ A}$$

$$\therefore \quad V = 1{\cdot}06 \times 2{\cdot}0 = 2{\cdot}12 \text{ V}$$

Example 2.18 Calculate the current in the 5·0-Ω resistor in the network shown in Fig. 2.36.

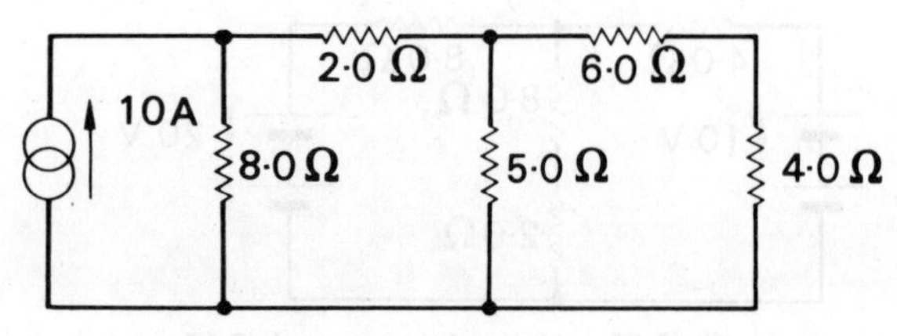

Fig. 2.36. Network for example 2.18

Short circuiting the branch containing the 5·0-Ω resistor gives the circuit shown in Fig. 2.37.

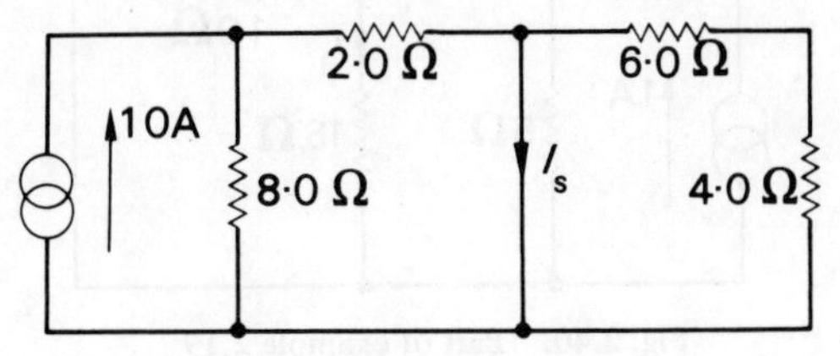

Fig. 2.37. Part of example 2.17

$$I_s = \frac{8{\cdot}0}{8{\cdot}0 + 2{\cdot}0} \times 10 = 8{\cdot}0 \text{ A}$$

Resistance looking into open circuited branch terminals

$$= \frac{(2{\cdot}0 + 8{\cdot}0)(6{\cdot}0 + 4{\cdot}0)}{(2{\cdot}0 + 8{\cdot}0) + (6{\cdot}0 + 4{\cdot}0)} = \frac{10 \times 10}{20} = 5{\cdot}0\ \Omega$$

therefore the circuit reduces to that shown in Fig. 2.38.

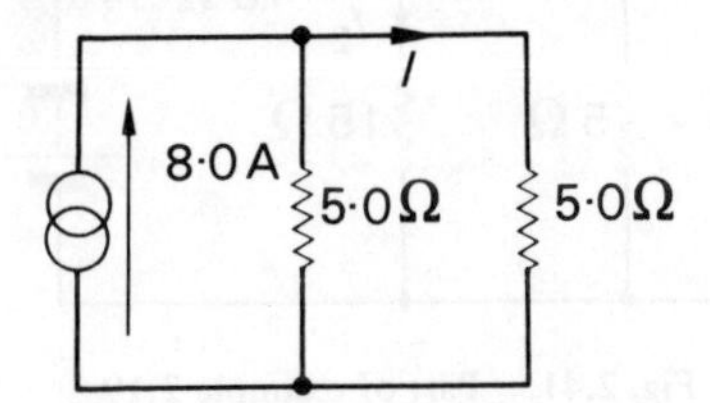

Fig. 2.38. Part of example 2.18

$$I = \frac{5{\cdot}0}{5{\cdot}0 + 5{\cdot}0} \times 8{\cdot}0 = \underline{4{\cdot}0 \text{ A}}$$

2.13 The superposition theorem

The superposition theorem states that:

> In any network containing more than one source (voltage and/or current), the current in or the potential difference across any branch can be found by considering each source separately and adding their effects. Omitted sources of e.m.f. are replaced by short circuits and omitted sources of current by open circuits.

Example 2.19 Calculate the current in the 15-Ω resistor in the network show in Fig. 2.39.

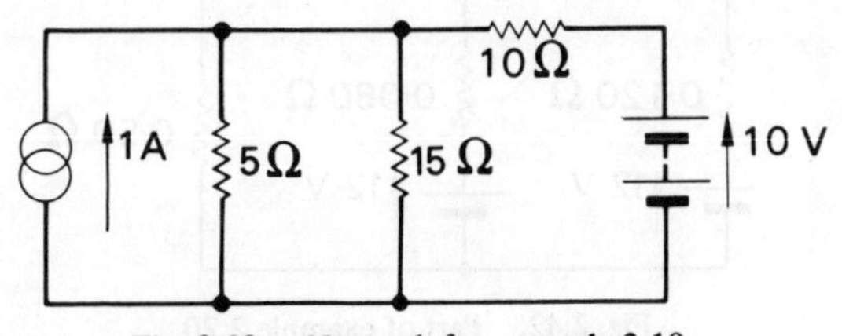

Fig. 2.39. Network for example 2.19

Consider the current source alone as shown in Fig. 2.40.

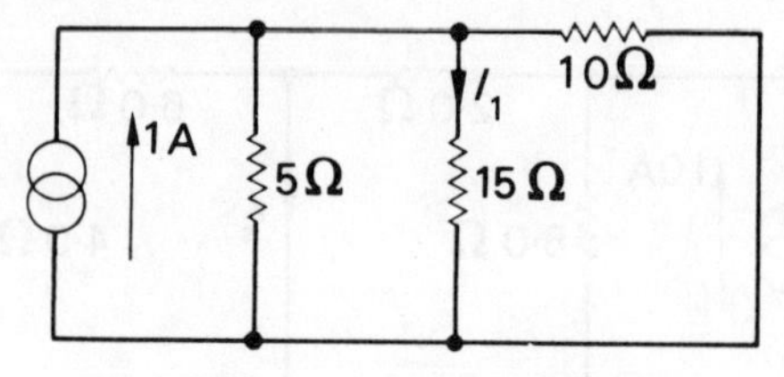

Fig. 2.40. Part of example 2.19

Resistance of 5 Ω in parallel with 10 Ω

$$= \frac{5 \times 10}{5 + 10} = 3{\cdot}33\ \Omega$$

$$\therefore \quad I_1 = \frac{3{\cdot}33}{3{\cdot}33 + 15} \times 1 = 0{\cdot}18\ \text{A}$$

Consider the voltage source alone as shown in Fig. 2.41.

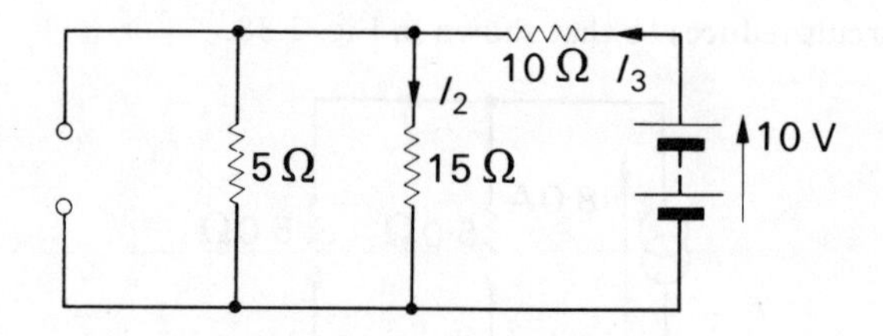

Fig. 2.41. Part of example 2.19

$$I_3 = \frac{10}{10 + \dfrac{15 \times 5}{15 + 5}} = \frac{10}{13{\cdot}75} = 0{\cdot}73\ \text{A}$$

$$\therefore \quad I_2 = \frac{5}{5 + 15} \times 0{\cdot}73 = 0{\cdot}18\ \text{A}$$

$\therefore$ Total current in 15 Ω = $I_1 + I_2$ = 0·18 + 0·18 = 0·36 A.

A combination of Norton's theorem and the superposition theorem can be very useful with certain circuits. This is illustrated in example 2.20.

Example 2.20 Two batteries, each of e.m.f. 12 V, are connected in parallel to supply a resistive load of 0·50 Ω. The internal resistances of the batteries are 0·120 Ω and 0·080 Ω. Calculate the current in the load and the current supplied by each battery.

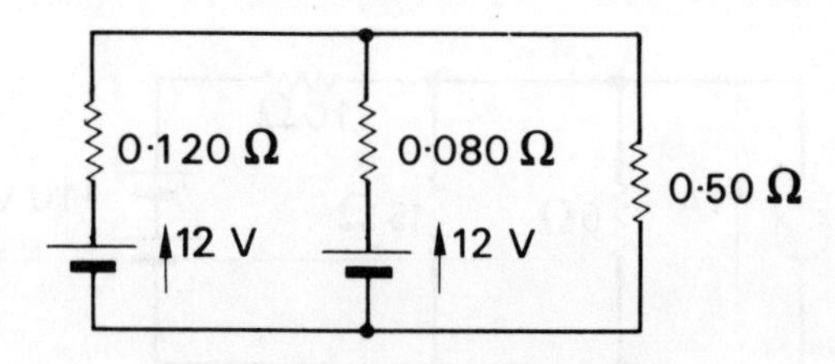

Fig. 2.42. Part of example 2.20

The circuit is shown in Fig. 2.42. If a short circuit is placed across the load the total short circuit current is given by:

$$I_s = \frac{12}{0{\cdot}120} + \frac{12}{0{\cdot}08} = 100 + 150 = 250 \text{ A}$$

The resistance looking into the open circuited load terminals is given by:

$$R = \frac{0{\cdot}120 \times 0{\cdot}080}{0{\cdot}120 + 0{\cdot}080} = 0{\cdot}048 \ \Omega$$

therefore the circuit reduces to that shown in Fig. 2.43.

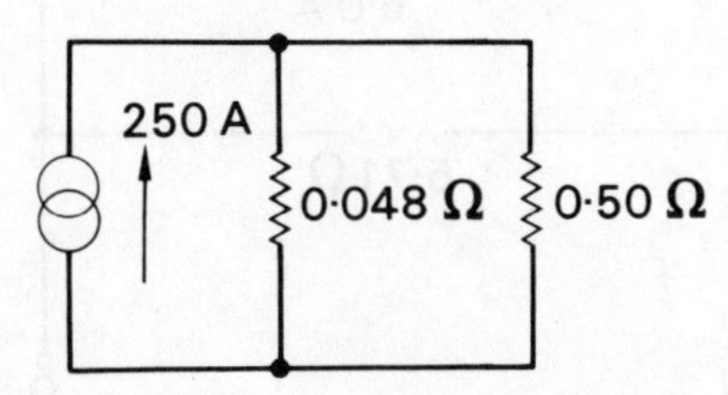

Fig. 2.43. Part of example 2.20

$$I = \frac{0{\cdot}048}{0{\cdot}048 + 0{\cdot}50} \times 250 = \underline{21{\cdot}9 \text{ A}}$$

Battery terminal voltage $= 21{\cdot}9 \times 0{\cdot}50 = 10{\cdot}95$ V

$\therefore$ Current in first battery $= \dfrac{12{\cdot}0 - 10{\cdot}95}{0{\cdot}120} = \underline{8{\cdot}8 \text{ A}}$

Current in second battery $= \dfrac{12{\cdot}0 - 10{\cdot}95}{0{\cdot}08} = \underline{13{\cdot}1 \text{ A}}$

Example 2.21 For the network shown in Fig. 2.44, calculate the potential difference between the points O and N and what current would flow in a 50-Ω resistor connected between these points.

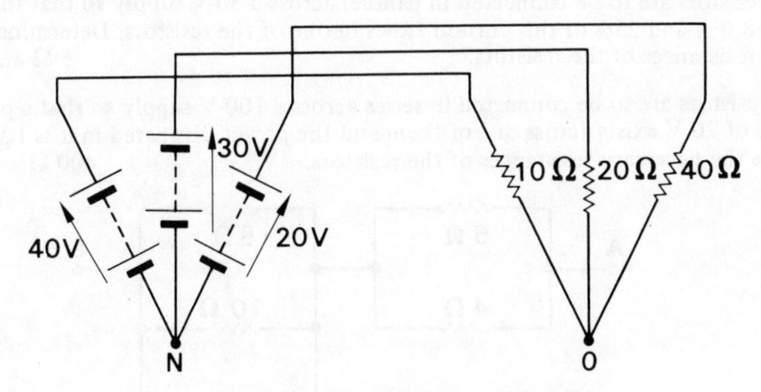

Fig. 2.44. Network for example 2.21

Total current in short circuit across ON

$$= I_s = \frac{40}{10} + \frac{30}{20} + \frac{20}{40} = 4{\cdot}0 + 1{\cdot}5 + 0{\cdot}5 = 6{\cdot}0 \text{ A}$$

If R = resistance between ON then

$$\frac{1}{R}=\frac{1}{10}+\frac{1}{20}+\frac{1}{40}=\frac{7}{40}$$

$$\therefore \quad R=\frac{40}{7}=5{\cdot}71\ \Omega$$

therefore the circuit reduces to that shown in Fig. 2.45.

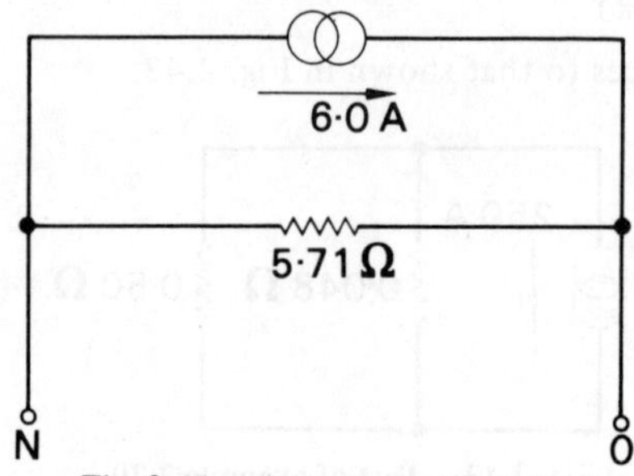

Fig. 2.45. Part of example 2.21

∴ open-circuit voltage across ON

$= 6{\cdot}0 \times 5{\cdot}71 = \underline{34{\cdot}26\ \text{V}}$

Current in 50 Ω connected between ON

$$=\frac{5{\cdot}71}{5{\cdot}71+50}\times 6{\cdot}0 = \underline{0{\cdot}62\ \text{A}}$$

PROBLEMS ON D.C. CIRCUITS

1 What is the combined resistance of three resistors of 10 Ω, 25 Ω and 50 Ω connected in (*a*) series and (*b*) parallel? (*a*) 85 Ω (*b*) 6·25 Ω

2 Two resistors are to be connected in parallel across a 30-V supply so that the total current is 8·0 A and 25% of this current flows in one of the resistors. Determine the necessary resistances of the resistors. 5 Ω and 15 Ω

3 Two resistors are to be connected in series across a 100-V supply so that a potential difference of 20 V exists across one of them and the power dissipated in it is 1 W. Determine the necessary resistances of the resistors. 400 Ω and 1600 Ω

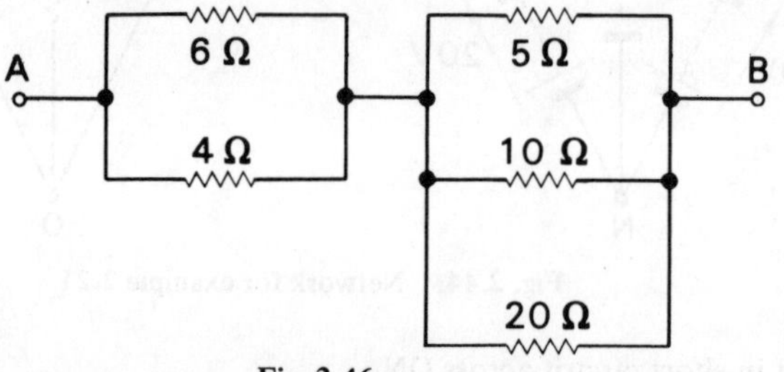

Fig. 2.46

4 What is the resistance between the points A and B in the network shown in Fig. 2.46. 5·3 Ω

5 If the power dissipated in the 4-Ω resistor shown in Fig. 2.47 is 1 W, determine the potential difference between points X and Y. 11 V

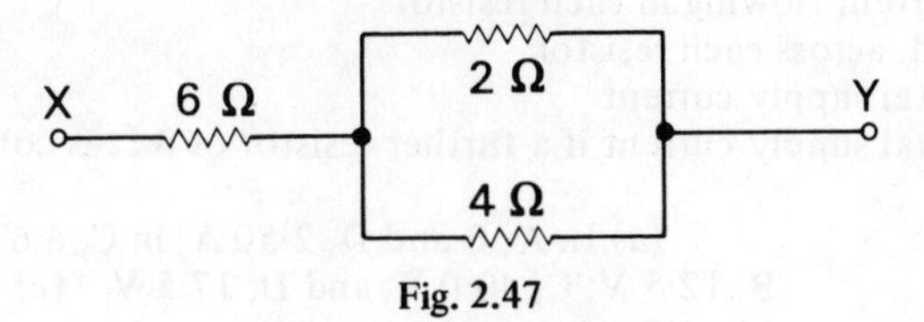

Fig. 2.47

6 Given that the total power dissipated in the network shown in Fig. 2.48 is 16 W, determine the value of R and the currents I_1, I_2, I_3 and I_4.
6 Ω; 1·2 A; 0·8 A; 1·6 A; 0·4 A

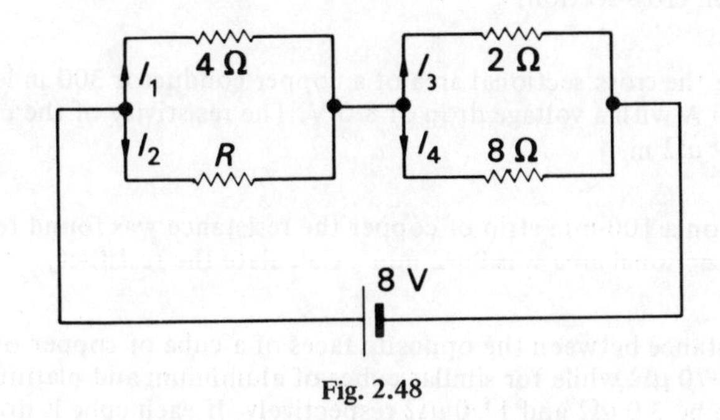

Fig. 2.48

7 For the network shown in Fig. 2.49, evaluate I_1, I_2, I_3 and E.
6·75 A; 4·80 A; 1·95 A; 40·5 V

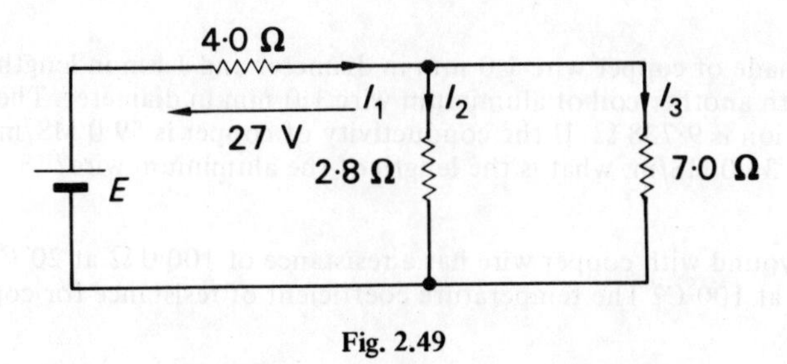

Fig. 2.49

8 For the network shown in Fig. 2.50, calculate the current in each resistor and the potential differences V_{ED} and V_{CD}. 1·69 A; 0·99 A; 0·70 A; −33·8 V; 35·0 V

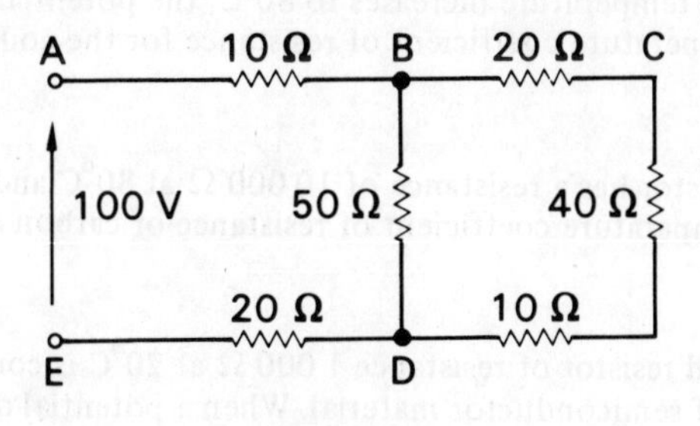

Fig. 2.50

9 Four resistors A, B, C and D, of 4, 5, 6 and 7 Ω respectively are joined to form a closed circuit in the form of a square. A d.c. supply of 40 V is connected across the ends of resistor C. Calculate:

(*a*) The current flowing in each resistor
(*b*) The p.d. across each resistor
(*c*) The total supply current
(*d*) The total supply current if a further resistor of 8 Ω is connected in parallel with resistor A.

(*a*) In A, B and D, 2·50 A; in C, 6·67 A (*b*) A, 10·0 V; B, 12·5 V; C, 40·0 V; and D, 17·5 V (*c*) 9·17 A (*d*) 9·40 A

10 A copper wire 1 m long and 1·0 mm^2 in cross section has a resistance of 0·019 Ω. What is the resistance of a wire made from the same copper, the wire being 250 m long and 60 mm^2 in cross section? 0·079 Ω

11 Calculate the cross sectional area of a copper conductor 300 m long such that it may carry 500 A with a voltage drop of 8·0 V. The resistivity of the copper may be taken as 0·019 μΩ m. 356 mm^2

12 In a test on a 100-mm strip of copper the resistance was found to be 171 μΩ. The average cross sectional area was 9·92 mm^2. Calculate the resistivity. 0·017 μΩ m

13 The resistance between the opposite faces of a cube of copper of 10 mm side was found to be 1·70 μΩ while for similar cubes of aluminium and platinum the resistances were found to be 3·0 μΩ and 11·0 μΩ respectively. If each cube is drawn out into a wire 1·00 mm^2 in section, determine the current taken from a cell having an e.m.f. of 2·0 V and internal resistance of 0·343 Ω, when the three wires are connected in series across the cell. 4·0 A

14 A coil made of copper wire 1·0 mm in diameter and 1 km in length is connected in parallel with another coil of aluminium wire 1·0 mm in diameter. The resistance of the combination is 9·738 Ω. If the conductivity of copper is 59·0 MS/m whilst that of aluminium is 36·0 MS/m, what is the length of the aluminium wire? 500 m

15 A coil wound with copper wire has a resistance of 100·0 Ω at 20°C. What will be its resistance at 100°C? The temperature coefficient of resistance for copper at 20°C is 0·004 3/K. 134·4 Ω

16 When a potential difference of 10·0 V is applied across a coil a current of 1·00 A flows in it, the mean temperature of the coil being 20°C. What current will flow in the coil when its mean temperature increases to 80°C, the potential difference remaining constant? The temperature coefficient of resistance for the coil conductor is $4{\cdot}28 \times 10^{-3}$/K at 0°C. 0·81 A

17 A carbon resistor has a resistance of 10 000 Ω at 80°C and 10 310 Ω at 20°C. Determine the temperature coefficient of resistance of carbon at 0°C.

$-5{\cdot}00 \times 10^{-4}$/K

18 A wire-wound resistor of resistance 1 000 Ω at 20°C is connected in series with another resistor of semiconductor material. When a potential difference of 1·20 V exists across the series combination a current of 1·00 mA flows irrespective of the

ambient temperature. If the temperature coefficient of resistance of the wire at 20°C is $2{\cdot}00 \times 10^{-4}$/K, what is the corresponding value for the semiconductor material? The heating effect of the current can be considered negligible. $-1{\cdot}00 \times 10^{-3}$/K

19 Calculate the current in each branch of the network shown in Fig. 2.51.

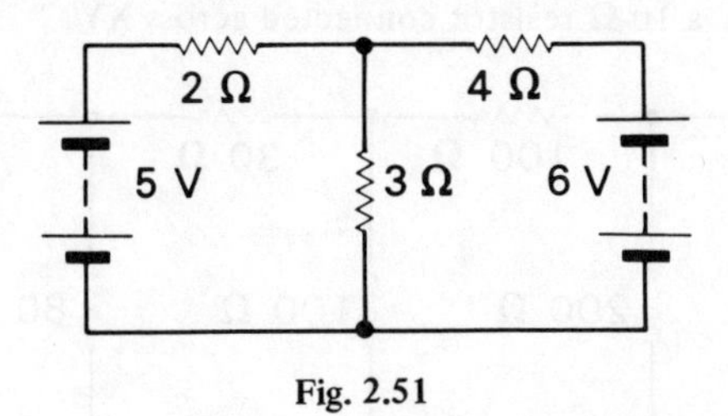

Fig. 2.51

0·654 A; 0·577 A; 1·231 A

20 Calculate the magnitude and direction of the current in the galvanometer (G) in the bridge network shown in Fig. 2.52. The resistance of the galvanometer is 80 Ω and the internal resistance of the cell is negligible.

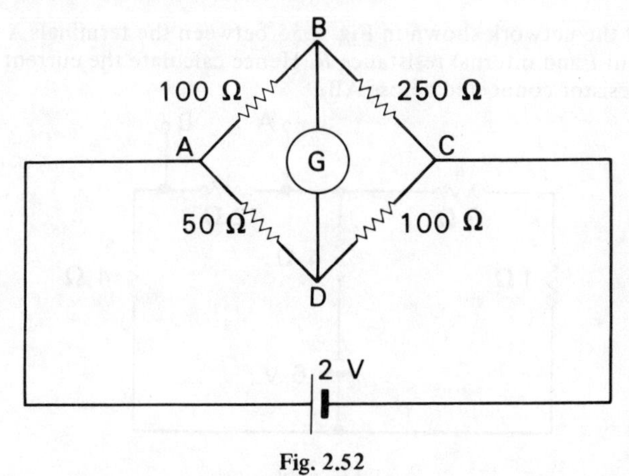

Fig. 2.52

0·52 mA from B to D

21 Calculate the current in the 10-Ω resistor in the network shown in Fig. 2.53.

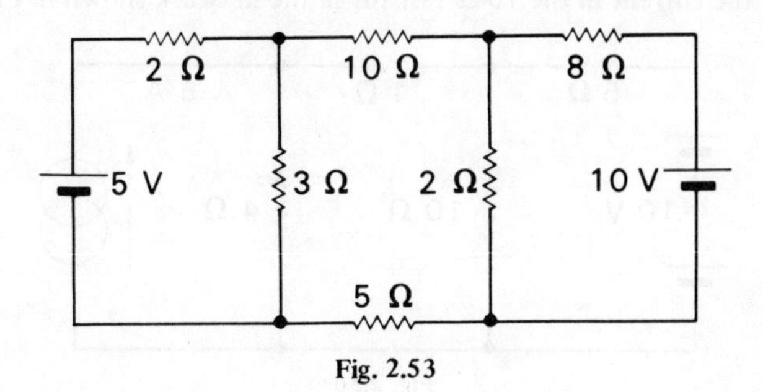

Fig. 2.53

56·2 mA

22 When a certain battery is loaded by a 60-Ω resistor its terminal voltage is 98·4 V. When it is loaded by a 90-Ω resistor its terminal voltage is 98·9 V. What resistance would give a terminal voltage of 98·0 V? 49·0 Ω

23 Derive an equivalent circuit consisting of one source of e.m.f. E in series with a resistance R for the network shown in Fig. 2.54. Hence determine the power that would be dissipated by a 10-Ω resistor connected across XY.

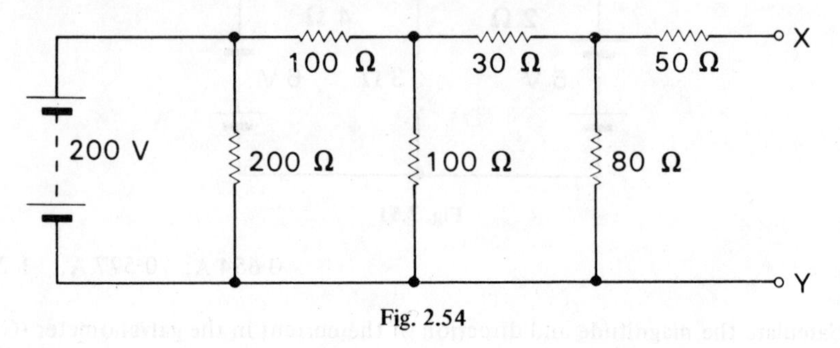

Fig. 2.54

E = 50 V; R = 90 Ω; 2·5 W

24 Represent the network shown in Fig. 2.55 between the terminals A and B by one source of current I and internal resistance R. Hence calculate the current that would flow in a 6-Ω resistor connected across AB.

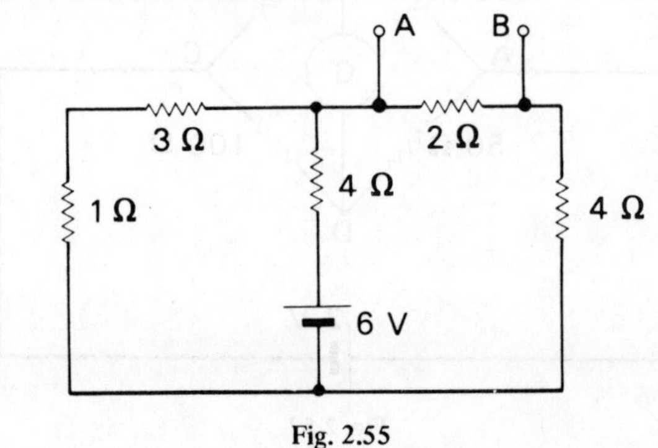

Fig. 2.55

I = 0·5 A; R = 1·5 Ω; 0·10 A

25 Calculate the current in the 10-Ω resistor in the network shown in Fig. 2.56.

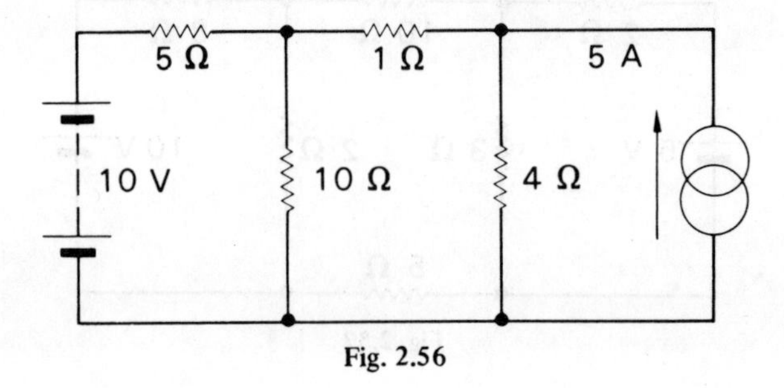

Fig. 2.56

1·2 A

26 For the network shown in Fig. 2.57, calculate the current in the 2-Ω resistor and the currents drawn from each battery.

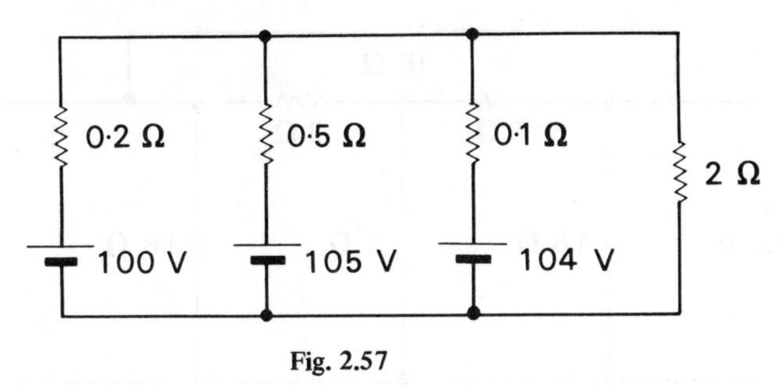

Fig. 2.57

50·0 A; 0 A; 10·0 A; 40·0 A

27 For the network shown in Fig. 2.58, calculate the potential difference V_{NO}. Calculate the resistance of a resistor connected across NO that would draw a current of 1·0 A?

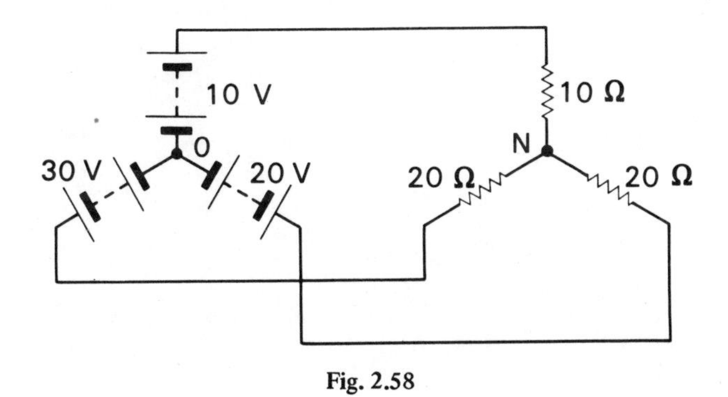

Fig. 2.58

17·5 V; 12·5 Ω

28 For the network shown in Fig. 2.59, calculate the potential difference V_{AB}. Hence determine the magnitude and direction of the current that would flow in a 3·10-Ω resistor connected across AB.

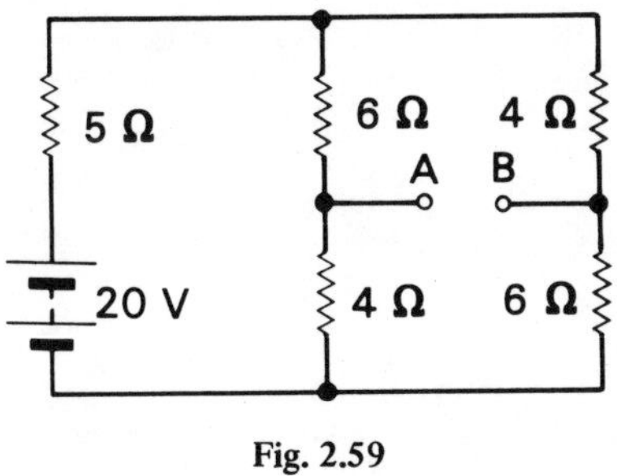

Fig. 2.59

$V_{AB} = -2{\cdot}0$ V; 0·25 A from B to A

29 Calculate the battery current and the current in the 4·0-Ω resistor for the network shown in Fig. 2.60.

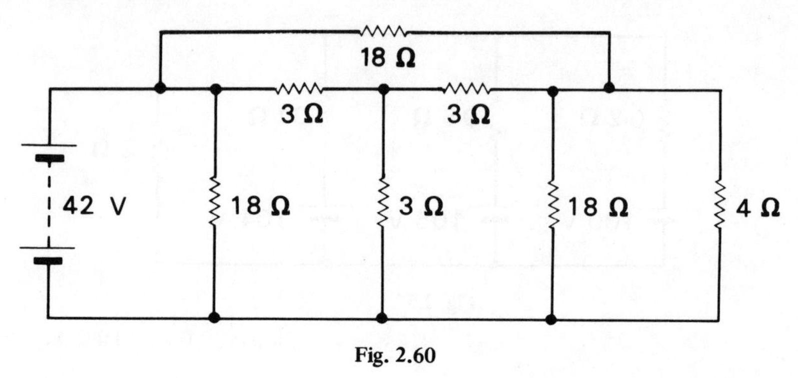

Fig. 2.60

12·0 A; 3·0 A

3

Alternating quantities

An alternating quantity is one that regularly acts first in one sense and then in the opposite sense. A typical alternating quantity is alternating current, hence an alternating current is one that flows first in one direction and then in the other. This reversal of flow occurs at regular intervals of time.

Alternating current circuits are more common than direct current circuits because the former can be transmitted, controlled and utilised more easily in general applications. Alternating current is not the only electrical example of an alternating quantity, thus alternating voltage is another typical example.

Before considering the response of a network to alternating currents and voltages, it is necessary to discuss the properties of an alternating quantity and to note the ways in which these may be defined. This chapter may therefore appear to be a mathematics lesson with electrical terms thrown in for good measure but it is important that the terms now introduced are clearly understood since most of the ensuing theory is based on them.

3.1 Frequency and period

If the instantaneous values of an alternating quantity are plotted to a base of time, the resulting graph is termed a waveform graph. A possible wave diagram is shown in Fig. 3.1.

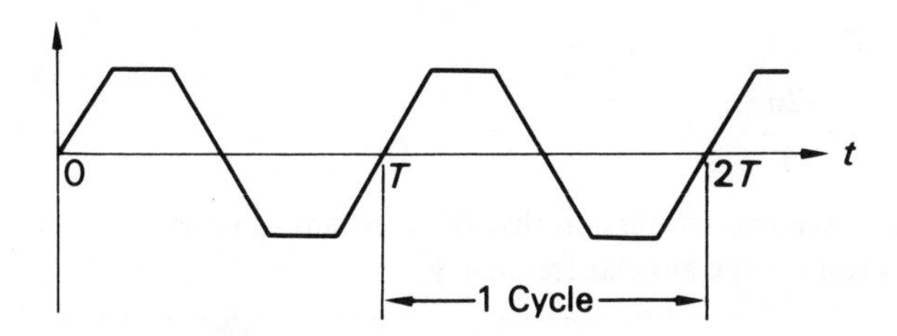

Fig. 3.1. Frequency and period of an alternating quantity

The wave diagram shows that after a period of time the wave starts to repeat. The sequence of events during this period is termed a cycle; one cycle is indicated in Fig. 3.1. The time taken for one cycle is called a period or the period of the waveform.

Period	Symbol: T	Unit: second (s)

Of greater interest is the number of cycles occurring per second; this rate of repetition is termed the frequency. It will be seen that the frequency is fundamental to all a.c. circuit analysis.

Frequency	Symbol: f	Unit: hertz (Hz)

The unit of frequency used to be the cycle per second (c/s) and this is still to be found. It follows from the definitions of period and frequency that

$$T = \frac{1}{f} \tag{3.1}$$

Of all possible alternating quantities in electrical engineering, that, which is most important, is the sinusoidal wave. It is important because most alternating voltages vary either sinusoidally or almost sinusoidally. Even in cases that are non-sinusoidal, it should still be noted that any alternating quantity approximates to a sinusoidal quantity. The practical implications of non-sinusoidal waveforms are discussed in paragraph 7.17.

3.2 Sinusoidal waves

A sinusoidal alternating quantity is one that can be expressed instantaneously by a sinusoidal function of time, e.g. a sinusoidally-varying voltage may be expressed as

$$v = V_m \sin \omega t$$

where V_m and ω are appropriate constants that may be interpreted as follows. The maximum value of $\sin \omega t$ is 1 hence V_m must be the maximum or peak value of the alternating voltage. That it is a maximum value is indicated by the subscript m. The use of a small letter for instantaneous voltage should be noted.

The other constant ω determines the rate of alternation of the voltage; the time taken for one cycle is ωt to produce a change from 0 to 2π radians. However, the time taken for one cycle is T seconds hence

$$\omega T = 2\pi$$

$$\therefore \quad \omega = \frac{2\pi}{T}$$

It can be seen from this expression that the constant ω must be measured in radians per second. ω is termed the angular frequency.

Angular frequency	Symbol: ω	Unit: radian per second (rad/s)

However, from relation (3.1)

$$f = \frac{1}{T}$$

Substituting this in the previous expression,

$$\omega = \frac{2\pi}{T}$$

$$\omega = 2\pi f \qquad (3.2)$$

Thus the sinusoidal voltage can be expressed as

$$v = V_{\mathrm{m}} \sin 2\pi f t$$

The appropriate waveform diagram is shown in Fig. 3.2.

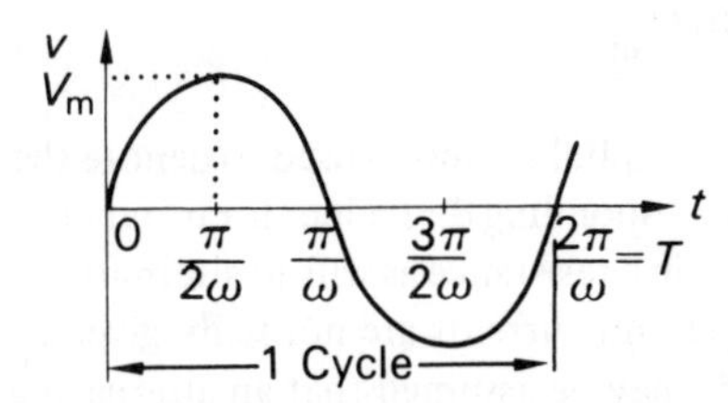

Fig. 3.2. Waveform of a sinusoidally alternating voltage

In a similar manner, an alternating current can be expressed as

$$i = I_{\mathrm{m}} \sin \omega t$$

3.3 Peak, r.m.s. and mean values of an alternating current

An alternating current varies from instant to instant. It is therefore necessary to find some manner of expressing the effective value of the current. Because the reason for creating an electric current is the transmission of energy and because current is a rate of flow of change, it follows that the effective value of the current should be related to the rate of energy transfer. Thus an alternating current may be defined in terms of the average power which it may cause.

Consider a current $i = I_{\mathrm{m}} \sin \omega t$ flowing through a resistor R. At any instant, the power p is given by

$$p = i^2 R$$

The average power P can be calculated over a period of one cycle.

$$P = R \times (\text{mean value of } i^2)$$

If the effective value of the current is I then

$$P = I^2 R$$

$$I^2 = (\text{mean value of } i^2)$$

$$= (\text{mean value of } I_m{}^2 \sin^2 \omega t)$$

$$= I_{\mathrm{m}}{}^2 \times (\text{mean value of } \sin^2 \omega t)$$

since I_m is a constant.

$$I^2 = I_m{}^2 \times (\text{mean value of } (\tfrac{1}{2} - \tfrac{1}{2}\cos 2\omega t))$$

The mean value of a cosine wave over one cycle is zero since the positive and negative loops are equal in area.

$$I^2 = \frac{I_m{}^2}{2}$$

$$I = \frac{I_m}{\sqrt{2}} = 0{\cdot}707 I_m \qquad (3.3)$$

This is the effective value of the current and is termed the root mean square (r.m.s.) current. Similarly, for a sinusoidal voltage $v = V_m \sin \omega t$,

$$V = \frac{V_m}{\sqrt{2}} = 0{\cdot}707 V_m$$

It will also be noted that a capital symbol is used to denote the r.m.s. value. The r.m.s. value is equivalent to the corresponding d.c. value: for instance, a direct current I will produce the same heating effect in a resistance as will an alternating current of r.m.s. value I.

In an a.c. circuit, voltages and currents are normally given in r.m.s. values unless otherwise stated. Further it may be assumed that an alternating quantity varies sinusoidally unless otherwise stated. Thus a.c. measuring devices are generally calibrated to read the r.m.s. values of the appropriate sinusoidal a.c. quantities.

The derivation of the r.m.s. value has been abbreviated since the reader may not have a sufficient knowledge of mathematics at this early stage to consider the complete analysis. For those readers who require a complete analysis, the derivation is as follows.

Again $\quad p = i^2 R$

Over one complete cycle, the average power is given by

$$P = \frac{\omega}{2\pi}\int_0^{\frac{2\pi}{\omega}} i^2 R \,.\, dt$$

$$= \frac{\omega}{2\pi}\int_0^{\frac{2\pi}{\omega}} I_m{}^2 R \,.\, \sin^2 \omega t \,.\, dt$$

$$= \frac{I_m{}^2 R\omega}{2\pi}\int_0^{\frac{2\pi}{\omega}} (\tfrac{1}{2} - \tfrac{1}{2}\cos 2\omega t)\, dt$$

$$= \frac{I_m{}^2 R\omega}{2\pi}\left[\frac{t}{2} - \frac{\sin 2\omega t}{4\omega}\right]_0^{\frac{2\pi}{\omega}}$$

$$= \frac{I_m{}^2 R\omega}{2\pi} \,.\, \left(\frac{\pi}{\omega} - 0 + 0 - 0\right)$$

$$= \frac{I_m{}^2 R}{2}$$

But if the effective circuit current is I,

$$P = I^2 R$$

$$I^2 = \frac{I_m^{\ 2}}{2}$$

$$I = \frac{I_m}{\sqrt{2}} \text{ as before.}$$

The power waveform diagram is shown in Fig. 3.3.

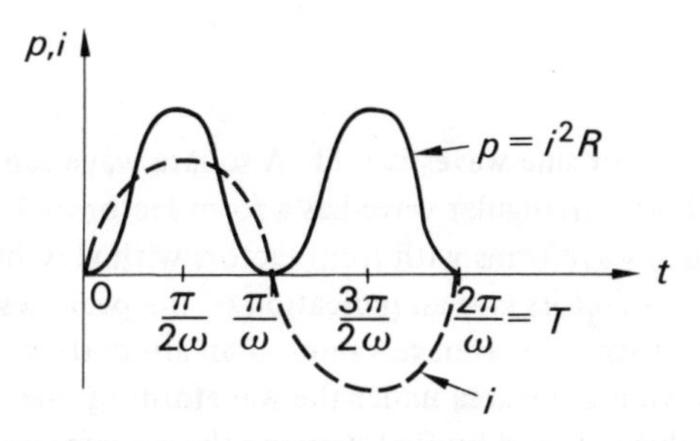

Fig. 3.3. Power waveform for current in a resistor

The average value of a sinusoidal alternating quantity over a complete cycle is zero, by the symmetry of the waveform. But over half a cycle, the average value is $2/\pi$ times the peak value. For a sinusoidal current, the average current I_{av} is given by

$$I_{av} = \frac{2}{\pi} \cdot I_m \tag{3.4}$$

The proof of this statement again involves further mathematics and the reader may, at this stage, care to accept the statement, which may be proved as follows.

$$I_{av} = \frac{\omega}{\pi} \int_0^{\frac{\pi}{\omega}} i \,.\, dt$$

$$= \frac{\omega}{\pi} \int_0^{\frac{\pi}{\omega}} I_m \cdot \sin \omega t \,.\, dt$$

$$= \frac{I_m \omega}{\pi} \left[\frac{-\cos \omega t}{\omega} \right]_0^{\frac{\pi}{\omega}}$$

$$= \frac{I_m \omega}{\pi} \cdot \left(\frac{1+1}{\omega} \right)$$

$$I_{av} = \frac{2}{\pi} \cdot I_m = 0{\cdot}636 I_m \tag{3.5}$$

Similarly for a sinusoidal voltage,

$$V_{av} = \frac{2}{\pi} \cdot V_m = 0{\cdot}636 V_m$$

The r.m.s. value can be related to the average value by a factor termed the form factor k_f.

$$k_f = \frac{\text{r.m.s. value}}{\text{average value}}$$

$$= \frac{0{\cdot}707 I_m}{0{\cdot}636 I_m}$$

$$k_f = 1{\cdot}11 \qquad (3.6)$$

Thus the form factor for sine waves is 1·11. A square wave can be shown to have a form factor of 1·0 whilst a triangular wave has a form factor of 1·15. Many modern control devices produce waveforms with form factors with very high values, say 4 or 5. In this respect, the form factors give an indication of the peakiness of the waveform.

For non-sinusoidal waves, the average values of an alternating quantity may be found over half a cycle by dividing the area under the waveform by the length of the base. Similarly the r.m.s. value is found by first drawing the waveform for the instantaneous squared values of the alternating quantity. The mean square value is then found in the manner already noted above. The root mean square value is then given by the square root of the mean square value.

Example 3.1 The equation relating the current in a circuit with time is

$$i = 141{\cdot}4 \sin 377t$$

where the current is measured in amperes and the time is measured in seconds. Find the values of

(*a*) the r.m.s. current,
(*b*) the frequency,
(*c*) the instantaneous value of the current when t is 3·0 ms.

$$i = I_m \sin \omega t = 141{\cdot}4 \sin 377t$$

$$I_m = 141{\cdot}4 \text{ A}$$

$$I = \frac{I_m}{\sqrt{2}} = \frac{141{\cdot}4}{\sqrt{2}} = \underline{100{\cdot}0 \text{ A}}$$

$$2\pi f = \omega = 377 \text{ rad/s}$$

$$f = \frac{377}{2\pi} = \underline{60 \text{ Hz}}$$

At t = 3·0 ms

$$i = 141{\cdot}4 \,.\, \sin(377 \times 3 \times 10^{-3}) = 141{\cdot}4 \sin 1{\cdot}135$$

$$= 141{\cdot}4 \sin 64{\cdot}9°$$

$$= \underline{128{\cdot}9 \text{ A}}$$

Example 3.2 A current has the following steady values in amperes for equal intervals of time changing instantaneously from one value to the next:

0, 10, 20, 30, 20, 10, 0, −10, −20, −30, −20, −10, 0, etc.

Calculate the r.m.s. value of the current and its form factor.

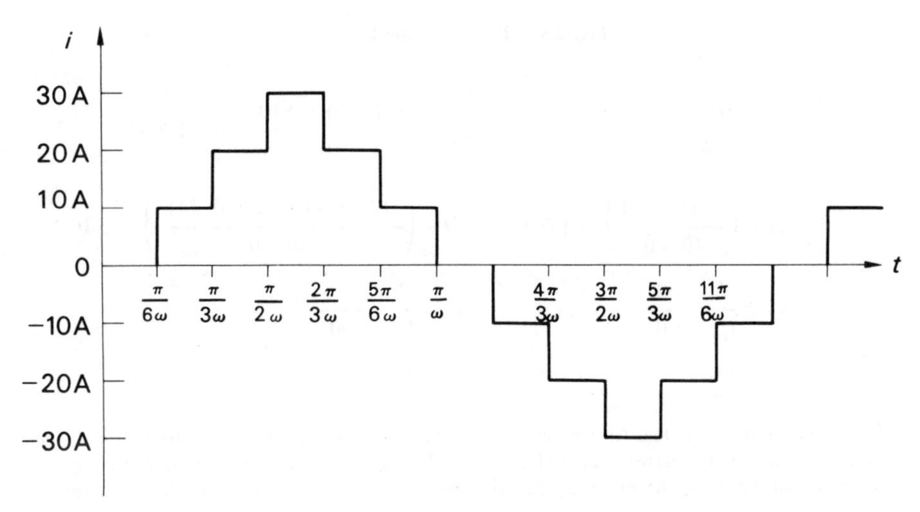

Fig. 3.4. Part of Example 3.2

Because of the symmetry of the waveform, it is only necessary to calculate the values over the first half cycle.

$$I_{av} = \frac{\text{area under curve}}{\text{length of base}}$$

$$= \frac{0\left(\frac{\pi}{6\omega} - 0\right) + 10\left(\frac{2\pi}{6\omega} - \frac{\pi}{6\omega}\right) + 20\left(\frac{3\pi}{6\omega} - \frac{2\pi}{6\omega}\right) + 30\left(\frac{4\pi}{6\omega} - \frac{3\pi}{6\omega}\right) + 20\left(\frac{5\pi}{6\omega} - \frac{4\pi}{6\omega}\right) + 10\left(\frac{6\pi}{6\omega} - \frac{5\pi}{6\omega}\right)}{\frac{\pi}{\omega} - 0}$$

$$= 15{\cdot}0 \text{ A}$$

$$I^2 = \frac{0^2\left(\frac{\pi}{6\omega} - 0\right) + 10^2\left(\frac{2\pi}{6\omega} - \frac{\pi}{6\omega}\right) + 20^2\left(\frac{3\pi}{6\omega} - \frac{2\pi}{6\omega}\right) + 30^2\left(\frac{4\pi}{6\omega} - \frac{3\pi}{6\omega}\right) + 20^2\left(\frac{5\pi}{6\omega} - \frac{4\pi}{6\omega}\right) + 10^2\left(\frac{6\pi}{6\omega} - \frac{5\pi}{6\omega}\right)}{\frac{\pi}{\omega} - 0}$$

$$= 316$$

$$I = \sqrt{316} = \underline{17{\cdot}8 \text{ A}}$$

$$k_f = \frac{I}{I_{av}} = \frac{17{\cdot}8}{15{\cdot}0} = \underline{1{\cdot}19}$$

Example 3.3 Calculate the form factor for each of the following waveforms.

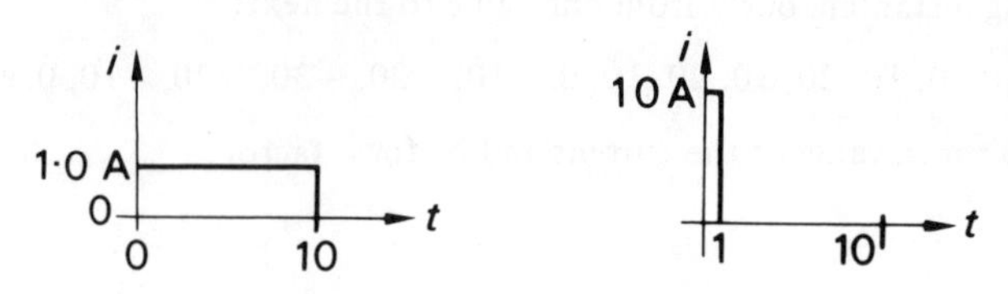

Fig. 3.5. Part of Example 3.3

$$I_{av} = \frac{1(10-0)}{10-0} = 1{\cdot}0\text{ A} \qquad I_{av} = \frac{10(1-0)+0(10-1)}{10-0} = 1{\cdot}0\text{ A}$$

$$I = \left(\frac{1^2(10-0)}{10-0}\right)^{\frac{1}{2}} = 1{\cdot}0\text{ A} \qquad I = \left(\frac{10^2(1-0)+0^2(10-1)}{10-0}\right)^{\frac{1}{2}} = 3{\cdot}16\text{ A}$$

$$k_f = \frac{I}{I_{av}} = \frac{1{\cdot}0}{1{\cdot}0} \qquad k_f = \frac{I}{I_{av}} = \frac{3{\cdot}16}{1{\cdot}0}$$

$$= \underline{1{\cdot}0} \qquad = \underline{3{\cdot}16}$$

It will be noted that the first waveform is that of direct current in which the r.m.s. current and the mean current have the same value. It is for this reason that the r.m.s. value of an alternating current may be equated to the mean value of a direct current.

3.4 Vector, complexor and phasor diagrams

A vector quantity is one that has magnitude and direction at a point in space. It can be represented in a diagram by a line drawn in the appropriate direction and of a length appropriate to the magnitude of the quantity. Vectors are taught in most elementary mathematics courses and it is therefore assumed that the reader will be familiar with vectors. Those who have not been introduced to them in mathematics will probably have used vectors to solve mechanical force problems.

Any electrical quantity that has direction may be expressed vectorially. However, none of the electrical quantities so far introduced are vectors; they are generally scalars in that they have magnitude only.

One of the most useful analytical devices available to the electrical engineer is the complexor. This is a non-vector quantity that may be illustrated in a manner similar to the vector but with the important difference that the complexor is drawn with angular or co-ordinate reference to some given step, i.e. a datum. In chapter 8, it will be shown that a complexor can be defined by complex numbers hence its name.

If a complexor is made to rotate about one end in the plane of the diagram with constant angular velocity, its geometric projection upon a datum varies sinusoidally. This process may be reversed so that a complexor may represent sinusoidally alternating voltages and current more concisely than the appropriate waveform diagram.

Consider a complexor OA rotating with constant angular velocity ω as shown in Fig. 3.6. If time is measured from the instant when OA is horizontal and about to rotate into the first quadrant, let OA_1 be some random position of OA some θ/ω

seconds later. It follows that OA_1 makes an angle θ with the horizontal axis. From the constructed figure,

$$\frac{A_1 P_1}{OA_1} = \sin\theta_1$$

where P_1 is the projection of A_1 on the horizontal.

If this process is repeated at regular intervals and the resulting values corresponding to A_1P_1 are plotted to a base of time, the resulting graph is a sine wave.

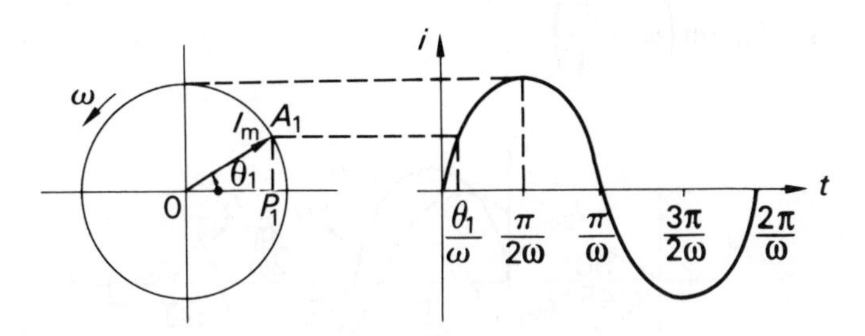

Fig. 3.6. Production of a sine wave from a complexor

The angle θ swept out by the complexor OA is given by

$$\theta = \omega t$$

When $\theta = 2\pi$, one complete sine wave will have been produced on the graph. Also it will be seen that the length OA represents the maximum value of the sinusoidal quantity.

It will be noted that the rotation of the complexor in the above case is time dependent. A complexor (such as voltage or current) that is derived from a time-varying sinusoidal quantity is termed a phasor, thus OA in Fig. 3.6 is a phasor.

An a.c. circuit is chiefly analysed in terms of the relationship between phasors denoting the various physical quantities, such as voltage and current. These have the same angular frequency and there is no relative motion between the phasors hence they can be displayed in a stationary instantaneous diagram, the common angular rotation being disregarded. Such diagrams are referred to as phasor diagrams.

A phasor is denoted in print by italic bold-faced or Clarendon type, e.g. $\boldsymbol{I}$. In handwriting, an over-dotted symbol is used, e.g. $\dot{I}$. Note that $\boldsymbol{I}$ represents a current phasor whilst I represents the magnitude only of the current.

It should be clearly understood that vectors and complexors and phasors are different in that the vector has space co-ordinates whilst the phasor is derived from time-varying sinusoids. A phasor diagram representing voltage, say, is therefore not a vector diagram.

In earlier analyses, it has been assumed that any alternating quantity is of the form $i = I_m \sin \omega t$. This infers that $i = 0$ when $t = 0$. However, it is quite possible that the quantity will have some value when $t = 0$. Suppose that $i = I_m \sin \phi$ when $t = 0$. The general expression for i must be modified to

$$i = I_m \sin(\omega t + \phi) \tag{3.7}$$

ϕ is termed the phase angle and represents the rotation of the phasor from the horizontal axis when $t = 0$.

To illustrate the relation between expression (3.7) and the corresponding waveforms, consider the following currents:

$$i_1 = I_{1\mathrm{m}} \sin(\omega t)$$

$$i_2 = I_{2\mathrm{m}} \sin\left(\omega t + \frac{\pi}{6}\right)$$

$$i_3 = I_{3\mathrm{m}} \sin\left(\omega t - \frac{\pi}{4}\right)$$

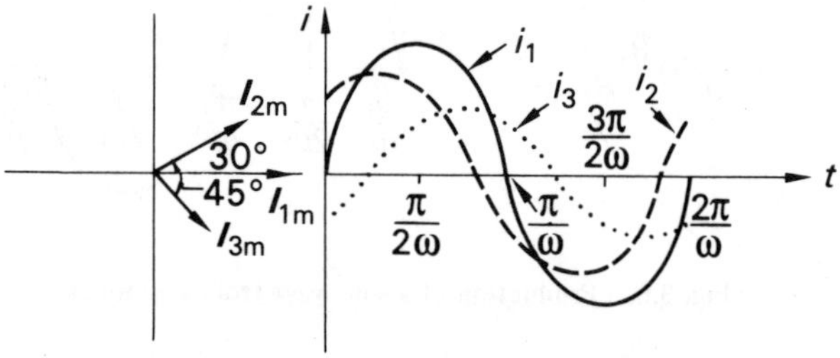

Fig. 3.7. Sinusoidal currents of the same frequency

The instantaneous phasor diagram for these currents and the corresponding waveforms are shown in Fig. 3.7. The diagram is drawn for the instant $t = 0$. It is admissible to draw the three phasors on one diagram since each as the same angular frequency.

Instead of drawing the phasor diagram to a scale appropriate to the maximum values, it could have been drawn to a scale appropriate to the r.m.s. values, because there is a fixed relationship between maximum and r.m.s. values. The diagram showing r.m.s. values is usually drawn in the same position as the instantaneous phasor diagram corresponding to $t = 0$. The r.m.s. phasor and instantaneous phasor diagrams are compared in Fig. 3.8.

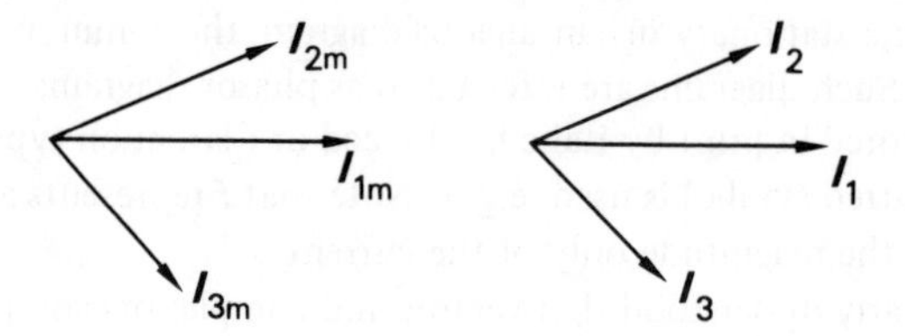

Fig. 3.8. Instantaneous phasor and r.m.s. phasor diagrams

Because most phasor diagrams show r.m.s. values, the term r.m.s. may be dropped for convenience; thus it may be assumed that any further reference to phasor diagrams infers the use of r.m.s. values. Instantaneous phasor diagrams retain their former interpretation.

In the phasor diagram of Fig. 3.8, the phasors are denoted $\boldsymbol{I}_1, \boldsymbol{I}_2$ and $\boldsymbol{I}_3$. This corresponds to the previous remarks on notation.

It should be noted that only the instantaneous phasor diagrams can be made to rotate and produce corresponding waveform diagrams. The r.m.s. diagrams must not be made to rotate with this object in mind, since they merely denote the effective values of alternating quantities. R.M.S. diagrams are really complexor diagrams only but, because an r.m.s. value is so closely linked to the maximum value, it has become general practice to retain the term phasor diagram in this connection.

This close relationship is responsible for the following:

1. The phase angles between the phasors in the phasor diagram denote the order in which the various quantities reach their maximum values. In Fig. 3.6, this order is I_2, I_1, I_3.

2. Since I_2 reaches its peak before I_1 and I_3, it is said to lead each of the others. Similarly I_1 leads I_3. Note that I_2 leads I_1 by 30° and I_2 leads I_3 by 75°. Similarly I_1 leads I_3 by 45°.

3. Since I_3 reaches its peak after I_1 and I_2, it is said to lag each of the others. Similarly I_1 lags I_2. Note that I_3 lags I_2 by 75° and I_3 lags I_1 by 45°. Also I_1 lags I_2 by 30°.

4. The phase angles permit the time intervals between the instants of peak value to be determined. Thus I_2 leads I_3 by

$$\frac{\frac{\pi}{6}+\frac{\pi}{4}}{\omega}=\frac{5\pi}{12\omega}\text{ seconds}$$

In each case, the information is derived from either diagram although the figures above are those of the r.m.s. phasors. Although currents have been used in the above discussion, it could equally well have been applied to any other alternating quantity. In electrical network analysis, the phase angle ϕ is generally the phase difference between voltage and current. It then only remains to state whether the current leads or lags the voltage.

Example 3.4 A sinusoidal alternating voltage of peak value 24 V, when applied to a certain circuit, gives rise to a current of maximum value 16 A. This current is also sinusoidal and lags the voltage by 45°. Represent the voltage and current waveforms in proper phase relationship over one complete cycle.

Taking the voltage as reference, $v = V_m \sin \omega t = 24 \sin \omega t$.

Similarly $i = I_m \sin(\omega t + \phi) = 16 \sin(\omega t - 45°)$

From these expressions, the instantaneous phasor diagram can be drawn and hence the waveform diagram shown in Fig. 3.9.

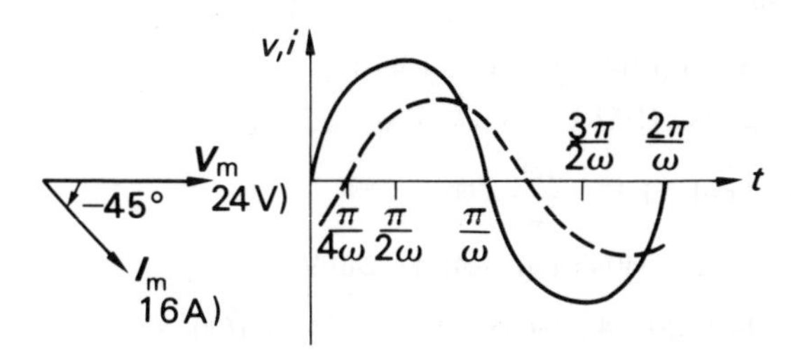

Fig. 3.9.

Example 3.5 In an a.c. circuit, the voltage is found to have a peak value of 340 V whilst the current has a peak value of 14·1 A. At some instant, taken as $t = 0$, the voltage has an instantaneous value of 240 V and the current has a corresponding value of 5·0 A. If the supply frequency is 50 Hz, derive relations for the instantaneous values of voltage and current.

Calculate the instantaneous values of the voltage and current at the instant $t = 5{\cdot}0$ ms and find the phase angle between the voltage and the current.

Since there is no statement to the contrary, it may be assumed that the alternating quantities are sinusoidal.

$$f = 50 \text{ Hz}$$

$$\omega = 2\pi f = 2\pi \times 50 = 314 \text{ rad/s}$$

The voltage is of the form

$$v = V_m \sin(\omega t + \theta)$$

but $\quad V_m = 340$ V and $\omega = 314$ rad/s

hence $\quad v = 340 . \sin(314t + \theta)$

At $t = 0$, $v = 240$ V

thus $\quad 200 = 340 . \sin\theta$

$$\sin\theta = \frac{240}{340} = 0{\cdot}707$$

$$\theta = 45^\circ$$

Only the first available solution need be taken.

$$\underline{v = 340 . \sin(314t + 45^\circ)}$$

It will be seen that this solution provides a trigonometrical solution in which radian measurement is mixed with measurement in degrees.

Similarly

$$i = I_m \sin(\omega t + \beta) = 14{\cdot}1 \sin(314t + \beta)$$

When $t = 0$, $i = 5{\cdot}0$ A, thus

$$5{\cdot}0 = 14{\cdot}1 \sin\beta$$

$$\sin\beta = \frac{5{\cdot}0}{14{\cdot}1}$$

$$\beta = 20{\cdot}6^\circ$$

$$\underline{i = 14{\cdot}1 \sin(314t + 20{\cdot}6^\circ)}$$

At $t = 5{\cdot}0$ ms

$$v = 340 \sin(314 \times 5 \times 10^{-3} + 45^\circ)$$

$$= 340 \sin(1{\cdot}57 + 45^\circ)$$

But $\quad 1{\cdot}57 \text{ rad} = 1{\cdot}57 \times \dfrac{180}{\pi} = 90^\circ$

$$v = 340 \sin(90^\circ + 45^\circ) = \underline{240 \text{ V}}$$

Also $\quad i = 14{\cdot}1 \sin(314 \times 0{\cdot}005 + 20{\cdot}6^\circ) = 14{\cdot}1 \sin(90^\circ + 20{\cdot}6^\circ)$

$$= \underline{13{\cdot}3 \text{ A}}$$

The phase angle between the voltage and current is given by

$$\phi = \theta - \beta = 45° - 20{\cdot}6° = 24{\cdot}4°$$

It can also be seen that the voltage leads the current. However the current is usually referred to the voltage thus the phase angle is defined as

$$\phi = \underline{24{\cdot}4° \text{ lag}}$$

3.5 Polar notation

One form of notation used to define a complexor is polar notation. This takes the general form $A\angle\theta \,.\, \mathbf{1}$, where **1** is a unit reference step. A unit reference step is one of unit scale length in some fixed axis. The value A serves to change the magnitude whilst the angle θ is the angle through which the complexor has been rotated. Since the unit reference step is taken to lie on the horizontal axis acting from left to right, it may be omitted, but its existence must not be forgotten. The effect of the unit reference step will become apparent in chapter 8.

In general, therefore, complex quantities can be expressed in the form $A\angle\theta$. A is the magnitude or tensor of the quantity whilst θ is the argument or versor. A diagram illustrating these remarks is given in Fig. 3.10.

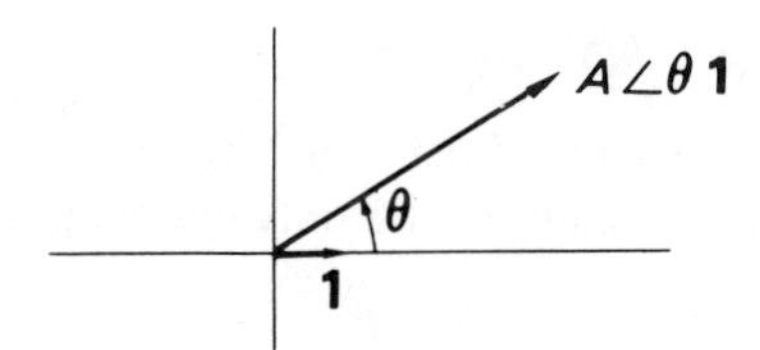

Fig. 3.10. Polar diagram

Any diagram using co-ordinate axes of the form shown in Fig. 3.10 is termed an Argand diagram or a complexor diagram. Thus any quantity expressed on such a diagram is in general terms a complexor. However, if its position represents time with respect to some datum, the complexors are termed phasors; in the given case, the unit reference step would represent some instant in time. In the case of the r.m.s. phasor diagram, this condition is not fulfilled but the phase relationship remains.

Phasors can be multiplied and divided. Consider

$$A\angle\theta_1 \times B\angle\theta_2$$

This gives rise to a new quantity, the magnitude of which is AB. The versors are additive, since each determines the angle through which the magnitude should be rotated. In this case, it is first rotated through θ_1 and then θ_2. Hence

$$A\angle\theta_1 \times B\angle\theta_2 = AB\angle(\theta_1 + \theta_2)$$

Similarly:

$$\frac{A\angle\theta_1}{B\angle\theta_2} = \frac{A}{B}\angle(\theta_1 - \theta_2)$$

In such operations, the unit step has been omitted. If it is now included, the above relations become

$$A\angle\theta_1 .1 \times B\angle\theta_2 .1 = AB\angle(\theta_1+\theta_2)(1)^2$$

$$\frac{A\angle\theta_1 .1}{B\angle\theta_2 .1} = \frac{A}{B}\angle(\theta_1-\theta_2)$$

In either case, the initial conception of the unit step has been destroyed–in the second case it has disappeared–and thus it can be seen that multiplication or division of phasors produces terms that are complex but no longer phasors. The importance of this statement will become apparent in chapter 7.

3.6 Kirchhoff's Laws applied to a.c. circuits

Kirchhoff's 1st law states that the sum of the currents flowing into a junction in an electric circuit is equal to the sum of the currents flowing out from the junction. In the junction shown in Fig. 3.11.

$$i_1 = i_2 + i_3$$

Let $\quad i_2 = I_{2m}\sin(\omega t + \phi_2)$

and $\quad i_3 = I_{3m}\sin(\omega t + \phi_3)$.

These are two sinusoidal currents of the same frequency; this is a reasonable assumption since the majority of a.c. circuits involve only one frequency.

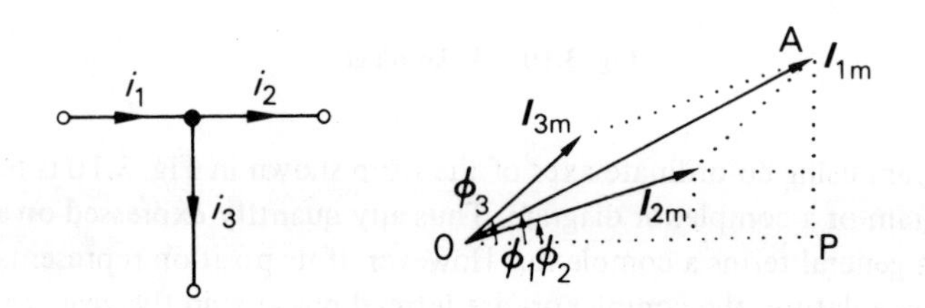

Fig. 3.11. Kirchhoff's 1st law applied to an a.c. circuit

$$\begin{aligned}
i_1 &= I_{2m}\sin(\omega t + \phi_2) + I_{3m}\sin(\omega t + \phi_3)\\
&= I_{2m}\sin\omega t\,.\cos\phi_2 + I_{2m}\cos\omega t\,.\sin\phi_2 + I_{3m}\sin\omega t\,.\cos\phi_3\\
&\quad + I_{3m}\cos\omega t\,.\sin\phi_3\\
&= (I_{2m}\cos\phi_2 + I_{3m}\cos\phi_3)\sin\omega t + (I_{2m}\sin\phi_2 + I_{3m}\sin\phi_3)\cos\omega t\\
&= \sqrt{[(I_{2m}\cos\phi_2 + I_{3m}\cos\phi_3)^2}\\
&\quad \overline{+ (I_{2m}\sin\phi_2 + I_{3m}\sin\phi_3)^2]}\ \sin(\omega t + \phi_1) \qquad (3.8)
\end{aligned}$$

where $$\phi_1 = \tan^{-1}\left(\frac{I_{2m}\sin\phi_2 + I_{3m}\sin\phi_3}{I_{2m}\cos\phi_2 + I_{3m}\cos\phi_3}\right) \qquad (3.9)$$

The instantaneous phasors I_{2m} and I_{3m} representing i_2 and i_3 are drawn in Fig. 3.11. I_{1m} is the complexor sum of I_{2m} and I_{3m}. It can be seen that, from Fig. 3.11,

$$\begin{aligned} OA^2 &= OP^2 + PA^2 \\ &= (I_{2m}\cos\phi_2 + I_{3m}\cos\phi_3)^2 + (I_{2m}\sin\phi_2 + I_{3m}\sin\phi_3)^2 \\ &= I_{1m}{}^2 \end{aligned}$$

$$I_{1m} = \sqrt{[(I_{2m}\cos\phi_2 + I_{3m}\cos\phi_3)^2 + (I_{2m}\sin\phi_2 + I_{3m}\sin\phi_3)^2]}$$

Thus the magnitude OA represents I_{1m} and similarly ϕ_1 represents the phase angle of I_{1m}. These results from the diagram compare with relations (3.8) and (3.9). It follows that OA is the complexor sum of I_{2m} and I_{3m} and further that the diagram is the representation of

$$I_{1m} = I_{2m} + I_{3m}$$

Thus a method for adding instantaneous phasors has been obtained. The result would apply to any number of branches at the junction point. Also the result can be extended to cover not only instantaneous phasors but also r.m.s. phasors. Thus

$$I_1 = I_2 + I_3$$

Kirchhoff's 2nd law states that at any instant, the sum of the e.m.f.'s round any closed loop is equal to the sum of the volt drops. In the circuit shown in Fig. 3.12, it follows that

$$e_1 = v_1 + v_2$$

By the same method described above, an instantaneous phasor diagram can be constructed as shown in Fig. 3.12. Again the diagram interprets the relation

$$E_{1m} = V_{1m} + V_{2m}$$

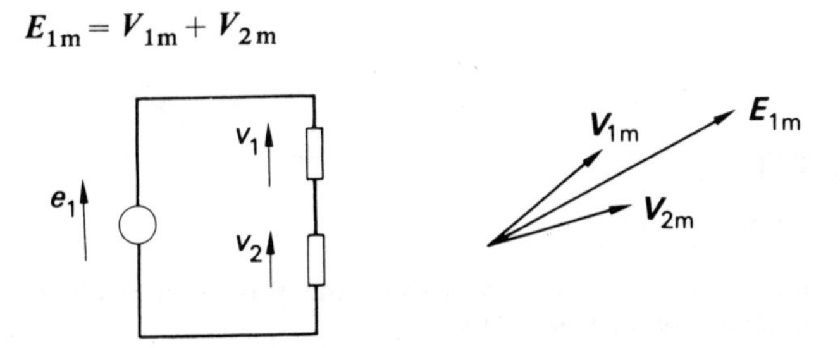

Fig. 3.12. **Kirchhoff's 2nd law applied to an a.c. circuit**

Finally instead of using instantaneous phasor values, r.m.s. values could have been used and a diagram drawn to illustrate

$$E_1 = V_1 + V_2$$

It should be noted that phasors can only be added complexorially provided they represent quantities which are both sinusoidal and of the same frequency.

Example 3.6 A circuit comprises four loads connected in series; the voltages across these loads are given by the following relations measured in volts:

$$v_1 = 50 \sin\omega t$$
$$v_2 = 25 \sin(\omega t + 60°)$$
$$v_3 = 40 \cos\omega t$$
$$v_4 = 30 \sin(\omega t - 45°)$$

Calculate the supply voltage, giving the relation in similar form.

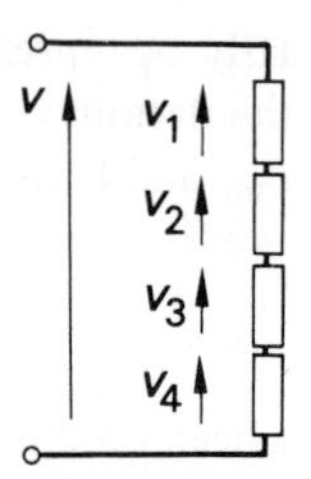

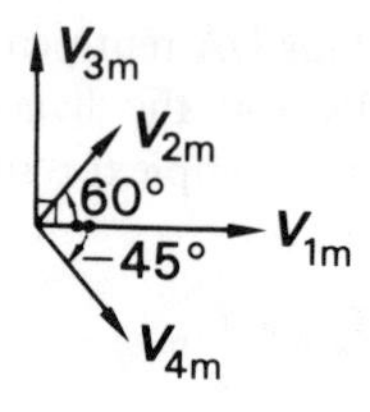

Fig. 3.13

The instantaneous phasor diagram is shown above in Fig. 3.13. The sum of the voltages may be carried out complexorially as follows. Note that $v_3 = 40 \cos \omega t = 40 \sin(\omega t + 90°)$

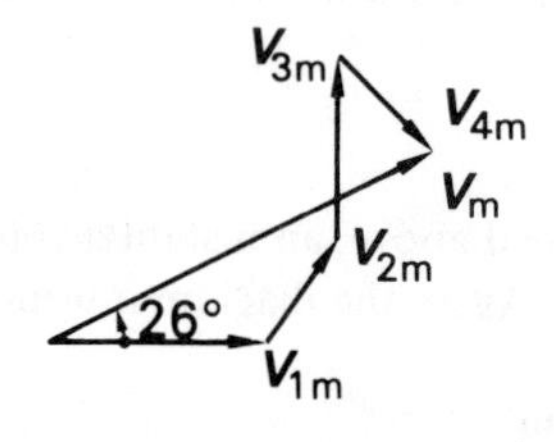

Fig. 3.14

From Fig. 3.14,

$$\underline{v = 93 \sin(\omega t + 26°)}$$

The solution can also be obtained by analysis, the phasors are resolved into their vertical and horizontal components, thus

$$V_m \cos\phi = V_{1m}\cos\phi_1 + V_{2m}\cos\phi_2 + V_{3m}\cos\phi_3 + V_{4m}\cos\phi_4$$
$$= 50 \times \cos 0° + 25 \times \cos 60° + 40 \times \cos 90° + 30 \times \cos -45°$$
$$= 83{\cdot}7 \text{ V}$$

$$V_m \sin\phi = V_{1m}\sin\phi_1 + V_{2m}\sin\phi_2 + V_{3m}\sin\phi_3 + V_{4m}\sin\phi_4$$
$$= 50 \times \sin 0° + 25 \times \sin 60° + 40 \times \sin 90° + 30 \times \sin -45°$$
$$= 40{\cdot}4 \text{ V}$$

$$V_m = (V_m{}^2 \cos^2\phi + V_m{}^2 \sin^2\phi)^{\frac{1}{2}} = (83{\cdot}7^2 + 40{\cdot}4^2)^{\frac{1}{2}}$$
$$= 93{\cdot}0 \text{ V}$$

$$\sin\phi = \frac{V_m \sin\phi}{V_m} = \frac{40{\cdot}4}{93{\cdot}0} = 0{\cdot}435$$

$$\phi = 25{\cdot}6°$$

Hence $$\underline{v = 93\sin(\omega t + 25{\cdot}6°)}$$

PROBLEMS ON ALTERNATING QUANTITIES

1 An alternating current of sinusoidal waveform has an r.m.s. value of 10·0 A. What are the maximum values of this current during the period of one waveform?
14·14 A; – 14·14 A

2 An alternating voltage wave is represented by v = 353·5 sin ωt. What are the maximum and r.m.s. values of this voltage? 353·5 V; 250 V

3 A 50-Hz sinusoidal voltage has an r.m.s. value of 200 V. The initial instantaneous voltage is zero and rising positively; find the time taken for the voltage to reach a value of 141·4 V for the first time. 1·67 ms

4 The equation relation current (in amperes) with time (in seconds) for a circuit is i = 141·4 sin 377t. What are the values of

(*a*) r.m.s. current,
(*b*) frequency,
(*c*) instantaneous value of the current when t is 3·0 ms?
100 A; 60 Hz; 127·8 A

5 Find the time taken for a sinusoidal alternating current of maximum value 20·0 A to reach 15·0 A for the first time after being instantaneously zero, the supply frequency being 40 Hz. Graphically or otherwise find the rate of change of current at the stated condition 3·375 ms; 3324 A/s

6 A sinusoidal alternating voltage of peak value 24·0 V is applied to a circuit resulting in a current of maximum value 16·0 A. This current is also sinusoidal and lags the voltage by 45°. Draw the voltage and current waves in proper phase relationship over one complete cycle.

7 A current has the following steady values (in amperes) for equal intervals of time, changing instantaneously from one value to the next:

0, 10, 20, 30, 20, 10, 0, – 10, –20, – 30, – 20, – 10, etc.

Calculate the r.m.s. value of the current and its form factor. 17·8 A; 1·18

8 Plot the half wave of current corresponding to the values in the table below. Draw a smooth curve through the points and use the mid-ordinate method to determine

(*a*) the average current,
(*b*) the r.m.s. current.

Time base in degrees	0	20	40	60	80	100	110	120	140	160	180
Current in amperes	0	15	24	35	54	68	70	64	35	12	0

33·5 A; 39·9 A

9 The positive half cycle of a symmetrical alternating current waveform varies as follows:

Time (ms)	0-2	2-6	6-8
Current	Uniform increase from zero to 1 A	Constant at 1 A	Uniform decrease from 1 A to zero

Plot to convenient scales the current waveform over one cycle and determine the r.m.s. value of the current. 0·79 A

10 A resistor takes 14·14 A from a 240-V supply, the voltage and current being sinusoidal. To a base of one cycle, plot graphs of current, voltage and power.
Determine the average power from the power waveform. 3·39 kW

11 An inductive load takes a current of 70·7 A at a power factor of 0·5 lagging from a 120-V sinusoidal supply. Plot graphs of the voltage, current and power to a base of one cycle and hence determine the average power by analysing the power waveform.
4·24 kW

12 Two volt drops (in volts) in a series circuit are represented by $v_1 = 100 \sin \omega t$ and $v_2 = 100 \cos \omega t$. Add these voltages complexorially and express the resultant voltage in the form $v = V_m \sin (\omega t \pm \phi)$. $141{\cdot}4 \sin (\omega t + 45°)$ V

13 The instantaneous voltages across each of four series coils are given by:

$$v_1 = 100 \sin 471t$$
$$v_2 = 250 \cos 471t$$
$$v_3 = 150 \sin (471t + \pi\sqrt{6})$$
$$v_4 = 200 \sin (471t - \pi\sqrt{4})$$

Determine the total potential difference expressing the value in similar form.
What will be the resultant potential difference if the polarity of v_2 is reversed?
$414 \sin (471t + 26{\cdot}5°)$; $486 \sin (471t - 40{\cdot}0°)$

14 Add complexorially the following voltages:

$$v_a = 100 \sin (\omega t - 45°)$$
$$v_b = 50 \sin (\omega t + 30°)$$
$$v_c = 60 \cos \omega t$$

$115 \sin (\omega t + 7°)$

15 By means of a suitable complexor construction to a suitable scale, add the following voltages:

$$v_1 = 50 \sin \omega t$$
$$v_2 = 25 \sin (\omega t + 60°)$$
$$v_3 = 40 \cos \omega t$$
$$v_4 = 30 \sin (\omega t - 45°)$$

$94 \sin (\omega t + 25°)$

4

Elements of electromagnetism

Electromagnetism is the study of magnetic fields set up by the passage of electric currents through a system of conductors. This study leads, in the first instance, to an understanding of most electrical machines and it is also important when considering electronic circuitry particularly with respect to communications.

This chapter has the objective of defining the terms of electromagnetic theory required for a basic understanding of electrical engineering. The application of this work to practical magnetic circuits and the consequent results, that affect an engineer, are left to chapter 5.

4.1 Magnetic field displays and polarity

If an electric current can be passed through a long straight conductor and a small compass needle is placed close to that conductor, it is found that, when the current is switched on, the compass needle deflects. It is also found that the direction of the deflection depends on the direction of the current flow. This is illustrated in Fig. 4.1.

The space surrounding the conductor in which the effect can be observed is termed

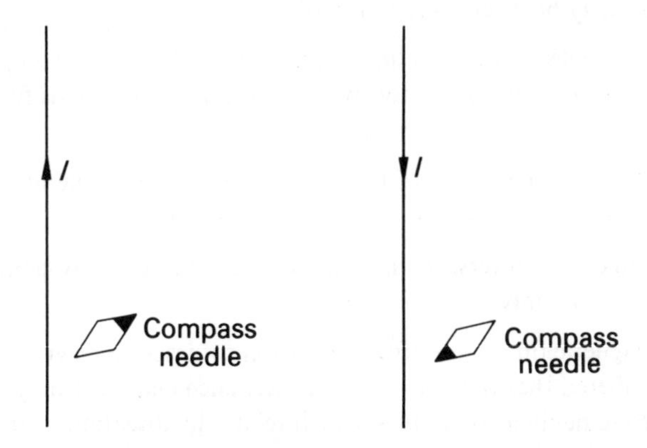

Fig. 4.1. Deflection of compass needle due to passage of electric current

the magnetic field of the conductor and the force on the compass needle causing it to deflect is termed the magnetic force.

A picture of the magnetic field can be obtained if the compass needle is moved progressively in the direction of its north pole. If this is done, without deviation, it is found that a complete path is traced out; the path is therefore a complete loop surrounding the conductor. The path is said to link the conductor.

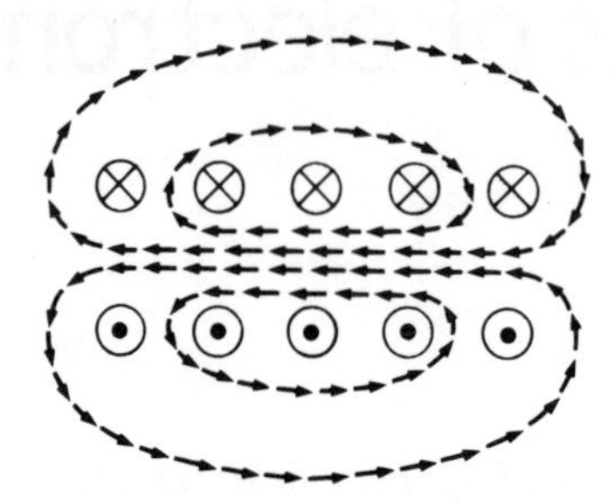

Fig. 4.2. Compass positions in an axial plane of a simple coil

Rather than investigate the field of a straight conductor, consider the field about the axis of a coil. The cross-section of a coil is shown in Fig. 4.2. The direction of the current is shown by the dot and cross convention. The dot signifies a current coming out from the plane of the paper whilst the cross signifies a current entering the plane of the paper. The dot represents the point of the approaching current 'arrow' whilst the cross represents the tail feathers of the departing current 'arrow'. The needle indications have also been mapped out. It will be noted that the resulting paths are closest at the centre of the coil in which area the force on the compass needle is strongest. In future these paths shall be shown more simply by lines, the direction of which gives the direction in which the north pole of the compass needle points, and the spacing of which gives an indication of the strength of the force on the compass.

These paths or lines are termed lines of flux. Typical field displays are shown in Fig. 4.3 for different arrangements of current-carrying conductors. From the diagrams, certain properties may be given to the lines of flux:

1. In an electromagnetic field, each line of magnetic flux forms a complete loop round at least one current-carrying conductor, which it is said to link. This forms a flux linkage.

2. The direction of the line is that of the force experienced by the north pole of a compass needle placed at that point in the electromagnetic field.

3. The lines of flux never intersect since the resultant force at any point in an electromagnetic field can have only one direction.

The line of magnetic flux is a useful convention to describe magnetic fields. However it must be remembered that they have no real existence and are purely imaginary.

It is helpful to remember some rules which relate the direction of the field to the direction of the current flow. One such rule is the right-hand rule. If a conductor is gripped by the right hand in such a manner that the thumb indicates the direction of flow of the current, then the fingers indicate the direction of the lines of flux about the conductor. This is illustrated in Fig. 4.3a.

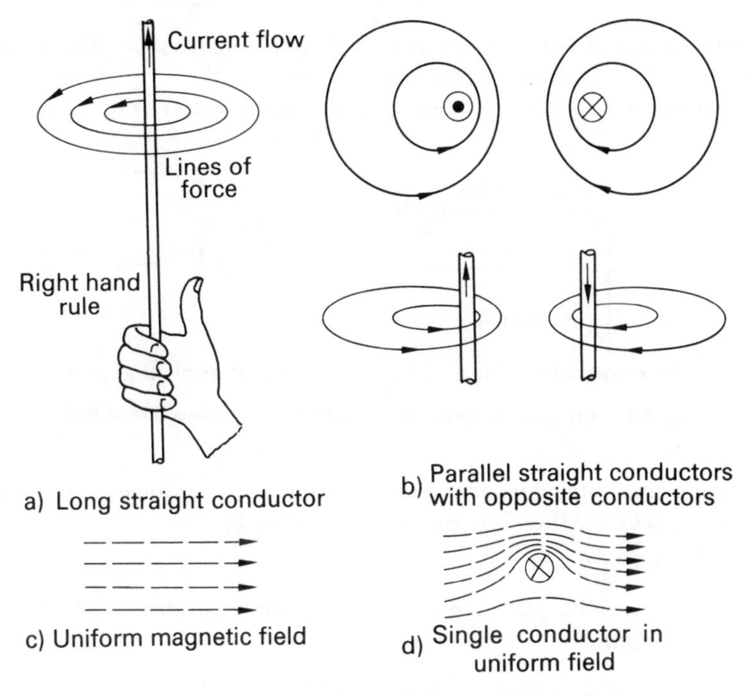

Fig. 4.3. Typical lines of force in electromagnetic fields

4.2 Magnetic flux and flux density

The total magnetic effect of a system as described by the lines of magnetic flux is termed the magnetic flux linking the system. The unit of magnetic flux is the weber.

Magnetic Flux	Symbol: Φ	Unit: weber (Wb)

When a conductor moves through a magnetic field, an e.m.f. is induced in it. This important concept is further discussed in para. 4.4, but it is used to define the weber as that flux which, when cut by a conductor in one second causes an average e.m.f. of one volt to be induced in the conductor.

Nevertheless the flux tells little of the field concentration. It is of greater importance, for reasons that will become apparent at a later stage, to signify the concentration of the field, this being termed the flux density. The flux density B is given by the flux Φ passing uniformly and normally, i.e. at right angles, through a surface of area A; at each point on the surface, the flux density is given by

$$B = \frac{\Phi}{A} \tag{4.1}$$

$$\Phi = BA \tag{4.2}$$

Flux density	Symbol: B	Unit: tesla (T)

Formerly the unit of flux density was the weber per square metre (Wb/m^2) and this unit is still to be found in older publications.

A plane cutting the flux at right angles is shown in Fig. 4.4(a). If the plane is not

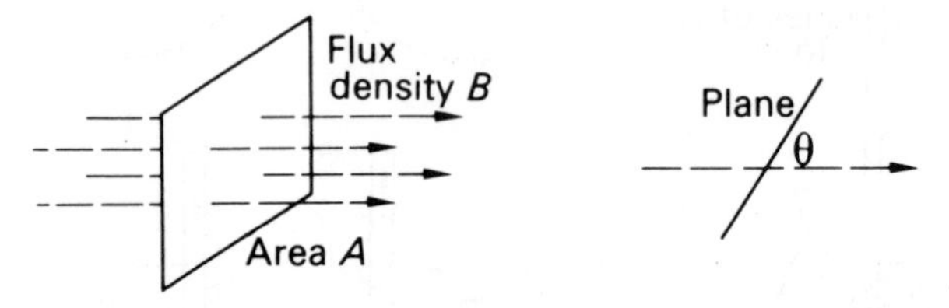

Fig. 4.4. Effect of the angle between the plane and direction of flux

taken at right angles to the direction of the field, then, by the geometry of the diagram shown in Fig. 4.4(b), the flux through the plane is given by

$$\Phi = BA \sin \theta \tag{4.2.1}$$

It follows that when the plane is parallel to the direction of the flux, no flux will pass through the plane.

Example 4.1 A rectangular coil measuring 200 mm by 100 mm is mounted such that it can be rotated about the midpoints of the 100-mm sides. The axis of rotation is at right angles to a magnetic field of uniform flux density 0·05 T. Calculate the flux in the coil for the following conditions:

1. The maximum flux through the coil and the position at which it occurs.
2. The flux through the coil when the 100-mm sides are inclined at 45° to the direction of the flux.

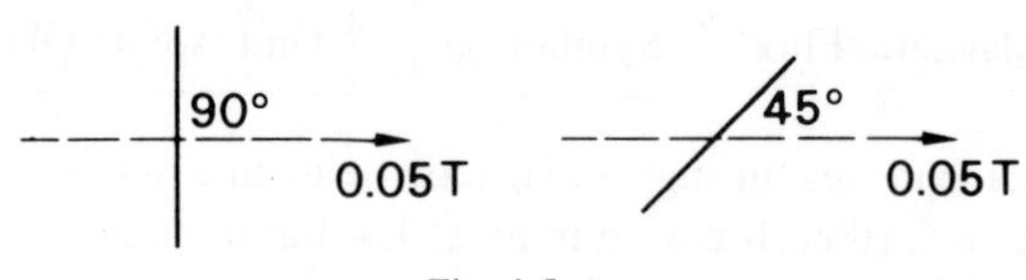

Fig. 4.5.

(1) The maximum flux will pass through the coil when the plane of the coil is at right angles to the direction of the flux.

$$\Phi = BA = 0{\cdot}05 \times 200 \times 10^{-3} \times 100 \times 10^{-3} = 1{\cdot}0 \times 10^{-3}\ \text{Wb}$$
$$= \underline{1{\cdot}0\ \text{mWb}}$$

(2)
$$\Phi = BA \sin\theta = 1{\cdot}0 \times 10^{-3} \times \sin 45° = 0{\cdot}71 \times 10^{-3}\ \text{Wb}$$
$$= \underline{0{\cdot}71\ \text{Wb}}$$

4.3 Force on a current-carrying conductor

The definition of the ampere depends on the principle that a current-carrying conductor situated in a magnetic field experiences a mechanical force.

Consider a straight uniform conductor placed in and normally to a magnetic field of uniform flux density B. The arrangement is shown in Fig. 4.6. The conductor carries a current I and has an effective length l, i.e. that length of the conductor that lies within the field.

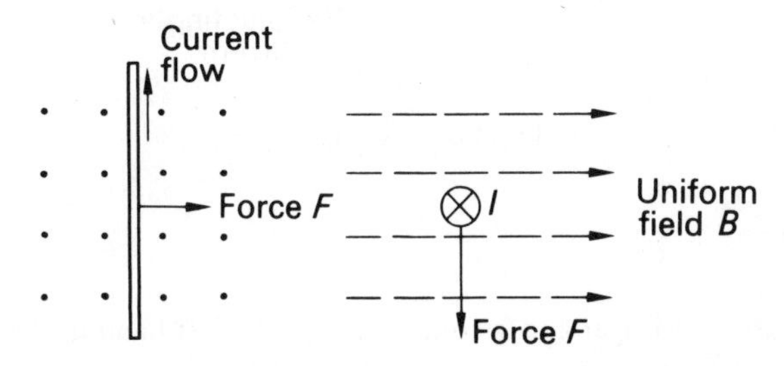

Fig. 4.6. Current-carrying conductor in a uniform magnetic field

It can be shown by experiment that the resulting force is proportional to the flux density, the effective length and the current.

$$F \propto BlI$$

However the flux density is measured in units such that

$$F = BlI \tag{4.3}$$

Whilst it is from this relation that the unit of flux measurement is derived, it will be shown that this principle can be related to the resulting induced e.m.f. which forms a simpler definition for the unit of flux–the weber. The definition of the weber will be derived in para. 4.6.

Nevertheless relation (4.3) is important as the basis of all magnetic measurements. The force BlI, which is measured in newtons, is only obtained when the magnetic field and the conductor are at right angles. If the conductor and the field make an angle θ with one another then the effective length of the conductor in the field, i.e. the length presented normally to the field, is $l \,.\, \sin\theta$ and the force is given by

$$F = BlI \sin\theta \tag{4.3.1}$$

The corresponding arrangement is shown in Fig. 4.7.

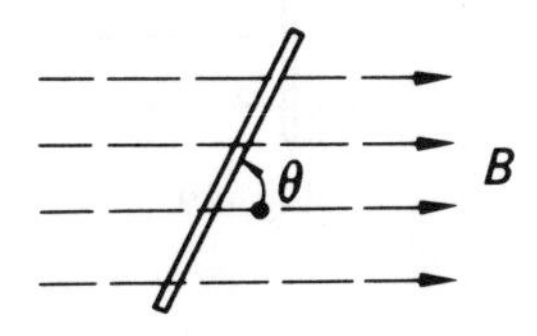

Fig. 4.7. Current-carrying conductor in a uniform magnetic field at an angle to the field

By experiment it can be shown that the mechanical force exerted by the conductor always acts in a direction perpendicular to the plane of the conductor and the magnetic field directions. The direction of the force is given by the Left-hand Rule, which is illustrated in Fig. 4.8.

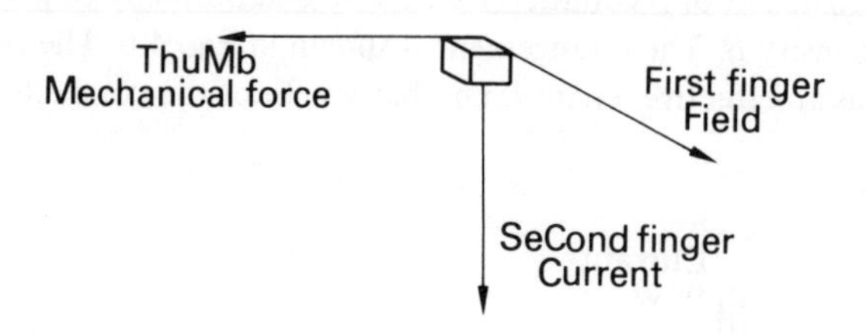

Fig. 4.8. Left-hand rule

Left-hand rule:

1. Hold the thumb, first finger and second finger of the left hand in the manner indicated by Fig. 4.6 whereby they are mutually at right angles.

2. Point the *F*irst finger in the *F*ield direction.

3. Point the se*C*ond finger in the *C*urrent direction.

4. The thu*M*b then indicates the direction of the *M*echanical force exerted by the conductor.

It follows that if either the current or the direction of the field is reversed, then the direction of the force is also reversed. If both current and field are reversed, the direction of the force remains unchanged.

Example 4.2 A conductor of length 0·5 m is situated in and at right angles to a uniform magnetic field of flux density 1·0 T and carries a current of 100 A. Calculate the force exerted by the conductor when

(*a*) it lies in the given position,
(*b*) it lies in a position such that it is inclined at 30° to the direction of the field.

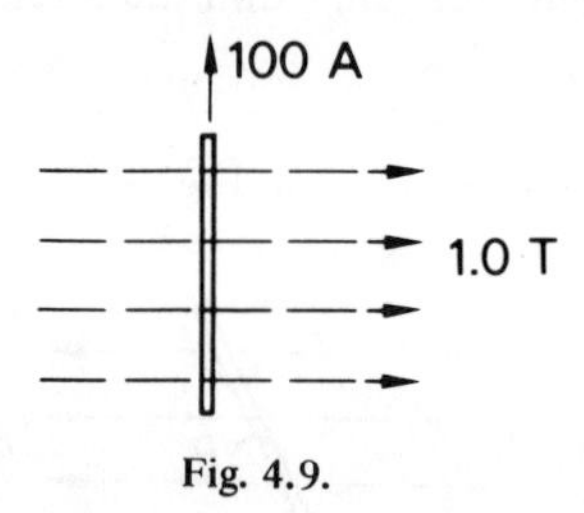

Fig. 4.9.

$$F = BIl = 1 \times 0{\cdot}5 \times 100 = \underline{50 \text{ N}}$$

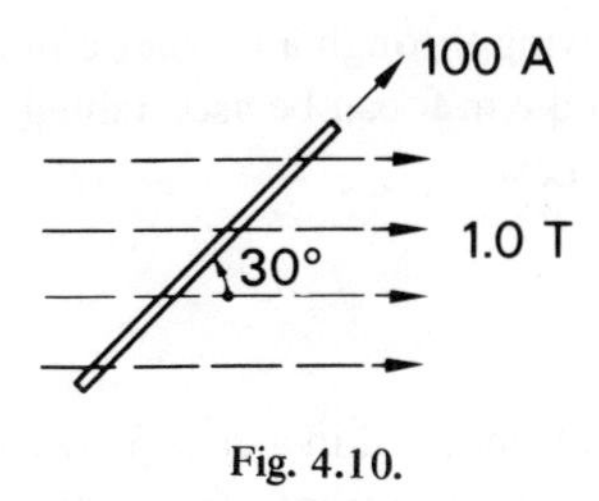

Fig. 4.10.

$$F = BIl \sin\theta = 50 \times \sin 30° = \underline{25 \text{ N}}$$

4.4 Induced e.m.f.

It has already been stated that when a conductor is moved through a magnetic field, an e.m.f. is induced in it. Faraday discovered this principle, which may be expressed in more general terms—when the magnetic flux linking a circuit is changing, an e.m.f. is induced in the circuit. This statement is the more appropriate since only the e.m.f. induced in a closed loop can be measured. For instance, if a single conductor is connected to an appropriate voltmeter so that the e.m.f. may be measured, the voltmeter connections complete a loop or circuit.

The manner in which the flux linking a circuit is varied may be achieved in one of three ways:

1. The coil may be moved in and relative to a fixed magnetic field so that the flux established in the coil varies in magnitude.

2. The field may be varied by changing the current which gives rise to it, the coil remaining stationary. Again the flux in the coil will vary.

3. The above conditions may occur simultaneously.

An e.m.f. induced by method 1 is termed a motional e.m.f. whilst an e.m.f. induced by method 2 is termed a transformer e.m.f. Possible arrangements for inducing these e.m.f.'s are shown in Fig. 4.11.

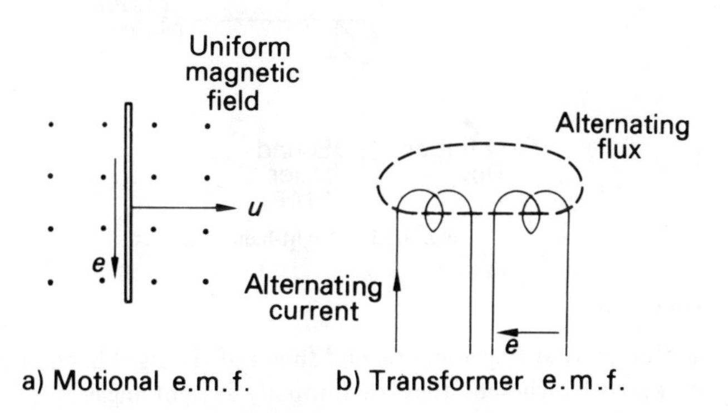

Fig. 4.11. Arrangements for inducing e.m.f.'s in coils

The instance of the coil moving through a magnetic field introduces the problem of how the polarity of the induced e.m.f. can be ascertained. To solve this problem, it is necessary to introduce Lenz's Law.

4.5 Lenz's Law

Lenz's Law states that the e.m.f. induced in a circuit by a change of flux linkage will be of a polarity that tends to set up a current which will oppose the change of flux linkage.

As an instance of this law, consider the two coils mounted in close proximity as shown in Fig. 4. 12. Coil B has its circuit completed by a resistor. If the current in coil A is increased, the flux in coil B is increased thus the flux linkage in coil B increase causing an e.m.f. to be induced in coil B. This e.m.f. will cause a current to flow such that the current will create a flux which will oppose the increase of initial flux. To satisfy this requirement terminal 2 must be driven positive with respect to terminal 1.

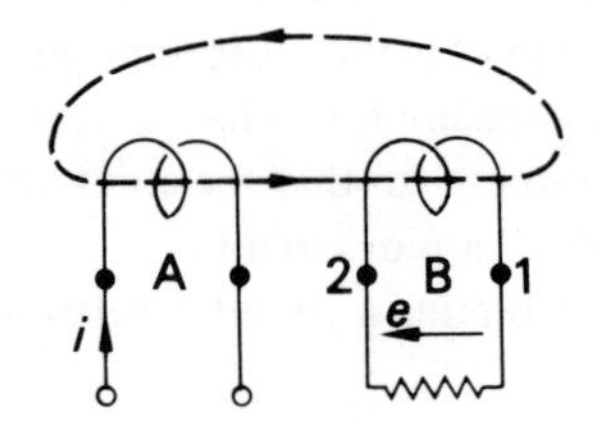

Fig. 4.12. Arrangment of two coils to illustrate Lenz's Law

Lenz's Law is effectively an extension of the Law of Conservation of Energy. It is an electrical application of this principle that for every action, there is an equal and opposite reaction. In the case of the two coils, the resistance of circuit B prevents the current attaining a sufficient value to prevent the flux linkages from changing.

From Lenz's Law, it can be shown that the polarity of the induced e.m.f. in a moving conductor is given by the Right-hand Rule. This can be illustrated using Fig. 4.13.

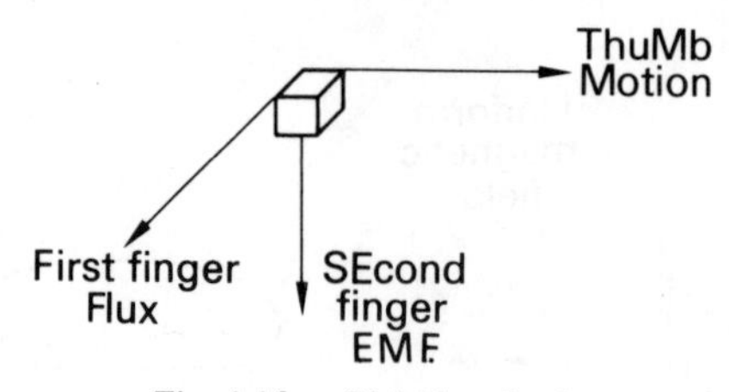

Fig. 4.13. Right-hand rule

Right-hand Rule:

1. Hold the thumb, first finger and second finger of the right hand in the manner indicated by Fig. 4.13 whereby they are mutually at right angles.
2. Point the thu*M*b in the direction of the conductor *M*ovement normal to the field.
3. Point the *F*irst finger in the direction of the *F*ield.

4. The s*E*cond finger indicates the polarity of the resulting induced *E*.m.f. with respect to the conductor.

4.6 Faraday's Law

Consider a single conductor moving normally across a uniform magnetic field as shown in Fig. 4.14.

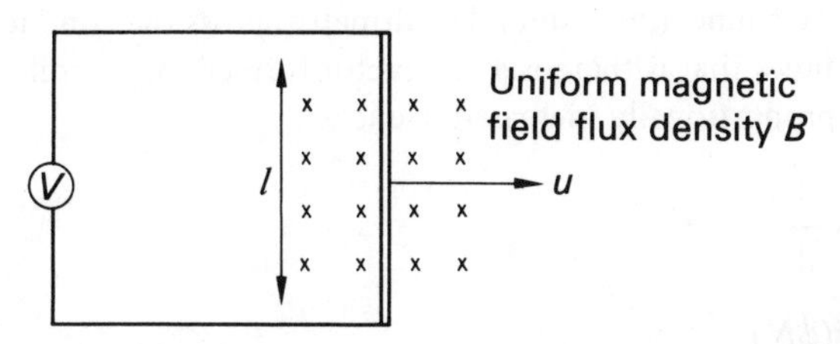

Fig. 4.14. Current-carrying conductor moving across a uniform magnetic field

Since the conductor is carrying a current I, it experiences a force given by

$$F = BlI$$

This is the force which is exerted by the conductor on the mounting. It is also the force which is exerted on the conductor by the mechanical mounting and which will be equal in magnitude but opposite in sense.

If the conductor moves a distance Δx in a time Δt, where $u = \dfrac{\Delta x}{\Delta t}$, then the average power to move the conductor is

$$P = Fu$$

$$= BlI \cdot \frac{\Delta x}{\Delta t}$$

However, by the law of the conservation of energy, the power input from the mechanical source must equal the electrical power output from the conductor. If the average induced e.m.f. is E, then

$$P = EI$$

$$= BlI \cdot \frac{\Delta x}{\Delta t}$$

$$\therefore \quad E = \frac{Bl \, . \, \Delta x}{\Delta t}$$

$$= \frac{B \, . \, \Delta A}{\Delta t}$$

where ΔA is the area traced out by the movement of the conductor. Since $\Phi = BA$, then $\Delta\Phi = B \, . \, \Delta A$ hence

$$E = \frac{\Delta \Phi}{\Delta t}$$

In the notation of calculus,

$$e = \frac{d\phi}{dt} \tag{4.4}$$

By experiment, it can be shown that the induced e.m.f. exists regardless of the direction of the motion of the conductor relative to the field and regardless of whether the flux linked is increasing or decreasing. These e.m.f.'s induced in the conductor can be demonstrated by connecting a suitable voltmeter across the conductor ends. This experiment also shows that if there is a conductor formed into a coil of N turns, the induced e.m.f. is proportionally increased. Hence

$$e = N\frac{d\phi}{dt} \tag{4.4.1}$$

$$e = \frac{d(\phi N)}{dt} \tag{4.5}$$

The product ϕN is termed the flux linkage of the coil or circuit.

Flux linkage	Symbol: Ψ	Unit: weber turn (Wb t)

$$\Psi = \Phi N \tag{4.6}$$

$$e = \frac{d(\phi N)}{dt} = \frac{d\psi}{dt} \tag{4.7}$$

This is a mathematical expression for Faraday's 2nd Law of Electrodynamics, but which is usually termed Faraday's Law. The definition of the weber is obtained from relation (4.7).

Example 4.3 A coil of 100 turns is linked by a flux of 20 mWb. If this flux is reversed in a time of 2 ms, calculate the average e.m.f. induced in the coil.

$$\Delta\Psi = N.\Delta\Phi = 100 \times (20 \times 10^{-3} - (-20 \times 10^{-3}))$$

$$= 4 \text{ Wbt}$$

$$E = \frac{\Delta\Psi}{\Delta t} = \frac{4}{2 \times 10^{-3}} = \underline{2000 \text{ V}}$$

Relation (4.7) may be modified to relate the induced e.m.f. to the velocity at which the conductor is moving in a magnetic field. A conductor is used in this context because of its practical implications in electrical machines rather than a loop or coil. Nevertheless the conductor will eventually form part of a circuit and thus a conductor can be said to have flux linkages. Consider again the conductor of Fig. 4.10. It will be noted that only the conductor lies within the magnetic field.

The conductor is moving with a velocity u. The flux cut by the conductor and therefore entering the circuit loop in a time Δt is given by

$$\Delta\Phi = Bl.\Delta x = Blu.\Delta t$$

It follows that the average e.m.f. is given by

$$E = \frac{\Delta\Phi}{\Delta t} = Blu$$

Using the notation of calculus,

$$e = \frac{d\phi}{dt} = Blu \tag{4.8}$$

As before, this e.m.f. can be measured using a voltmeter. Provided the connecting wires from the voltmeter to the conductor play no part in the cutting of the flux, then relation (4.8) remains valid. This factor applies to all similar cases in this chapter. If the connections cut the field then an e.m.f. is induced in them which will contribute to the voltmeter reading.

Note the case of a rectangular coil moving through a uniform magnetic field. The rate at which flux enters the loop at one side is equal to the rate at which the flux leaves the loop at the other side. There is therefore no change in the flux linkages of the loop and hence there is no e.m.f. induced in the loop. However, if the loop sides are considered to be separate conductors, then each will have an e.m.f. Blu induced in it. These e.m.f.'s act in opposite directions around the loop and so cancel one another out–the total effective e.m.f. is therefore zero as stated before.

The conductor does not actually require to move across the field. The field may instead sweep past the coil which is then held stationary. In the relation $e = Blu$, the velocity u is that of the conductor relative to the field.

The motion of the conductor must be across the field if the conductor is to cut the field. Any motion parallel with the field therefore will not produce any change in flux linkages and hence will fail to induce an e.m.f. If the conductor is moving at an angle α to the field then the velocity at which the conductor moves across the field is $u \,.\, \sin\alpha$ hence

$$e = Blu \,.\, \sin\alpha \tag{4.8.1}$$

Example 4.3 A conductor of length 0·5 m is situated in and at right angles to a uniform magnetic field of flux density 1·0 T moves with a velocity of 40 m/s. Calculate the e.m.f. induced in the conductor when the direction of motion is:

(*a*) at right angles to the direction of the field;
(*b*) inclined at 30° to the direction of the field;
(*c*) as in (*b*) but with the conductor inclined at 45° to the field.

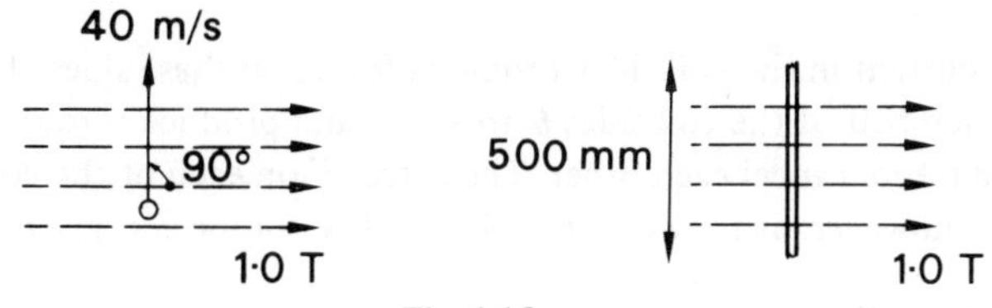

Fig. 4.15.

$$E = Blu \,.\, \sin\alpha = 1 \times 0{\cdot}5 \times 40 \times \sin 90^\circ$$
$$= \underline{20\ \text{V}}$$

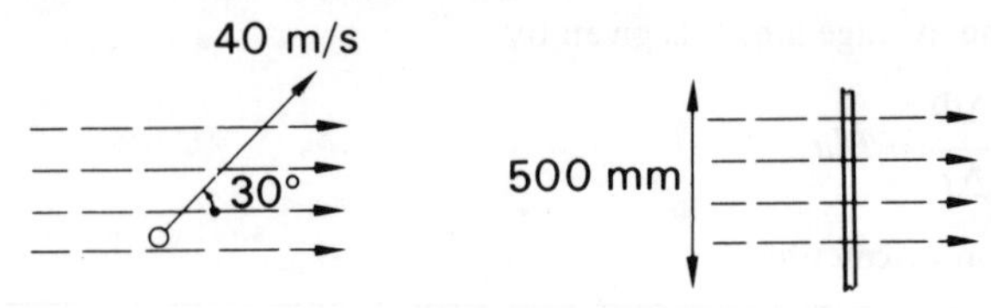

Fig. 4.16.

$$E = Blu\,.\sin\alpha = 1 \times 0{\cdot}5 \times 40 \times \sin 30^\circ$$
$$= \underline{10\text{ V}}$$

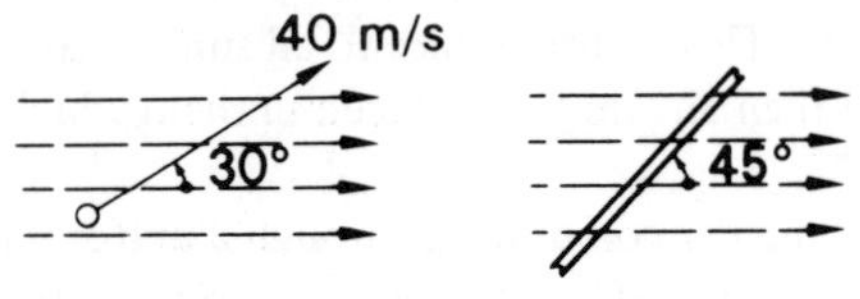

Fig. 4.17.

$$E = B(l\,.\sin\theta)\,u\,.\sin\alpha = 10 \times \sin 45^\circ$$
$$= \underline{7{\cdot}1\text{ V}}$$

4.7 Magnetic field strength and magnetomotive force

Consider a rectangular coil of dimensions $l \times b$ as shown in Fig. 4.18 in which the plane of the coil is at an angle θ to the direction of a uniform magnetic field of density B. The coil is pivotted about the midpoints of the sides b and carries a current I. The

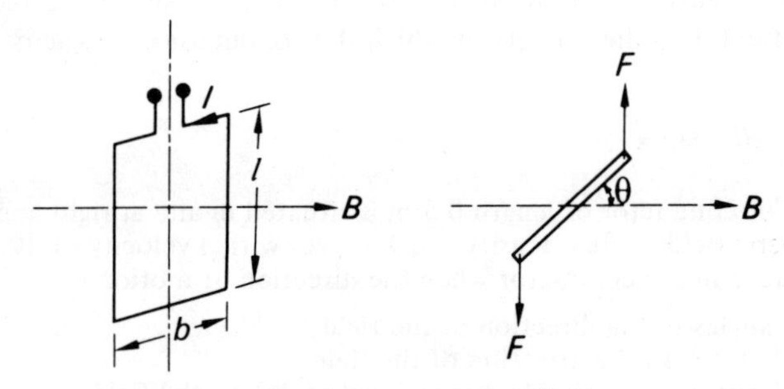

Fig. 4.18. Coil in a magnetic field

passage of the current in the coil sides produces forces on these sides. The forces developed in each half of the coil sides b are equal and produce torques of opposing sense. They therefore cancel each other. The forces F on each of the coil sides l give rise to a torque however, as shown in Fig. 4.18*b*. The torque T is given by

$$T = 2F \cdot \frac{b}{2}\cos\theta$$
$$= BIlb\,.\cos\theta$$
$$= BIA\,.\cos\theta \qquad (4.9)$$

For the purposes of investigating the properties of a magnetic field, it is convenient to consider a magnet which is equivalent to an electromagnetic coil of the type shown in Fig. 4.18. The magnet used is termed a magnetic dipole, which is the ultimate individual unit of magnetism. To fulfil the requirement of equivalence, it might be associated with a permanent magnet which has a pair of equal and opposite magnetic poles or it might equally well consist of an electric current in a small coil.

The strength of a magnetic pole is measured by the magnetic flux Φ which emanates from it. If the dipole is of length $2l$ and it has a flux Φ which emanates from the north pole and returns to the south pole, the magnetic moment m is defined by

$$m = 2\Phi l \tag{4.10}$$

where m is measured in weber metres.

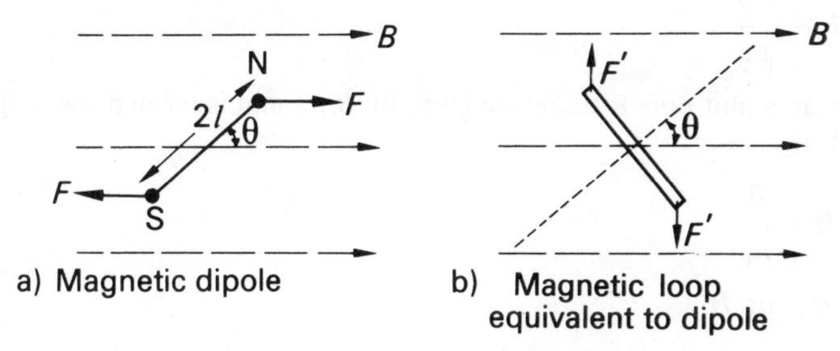

Fig. 4.19. Magnetic dipole arrangements

If the dipole is situated in a uniform magnetic field of density B and which exists in a vacuum or free space as shown in Fig. 4.19*a*, the force of each pole of the dipole can be shown to be proportional to the flux density B and also to the pole flux Φ. Hence

$$T \propto B\Phi 2l \,.\, \sin\theta$$

$$\propto mB \,.\, \sin\theta$$

$$T = \text{constant} \times mB \,.\, \sin\theta$$

Let the constant be $\dfrac{1}{\mu_0}$

$$T = \frac{mB}{\mu_0} \cdot \sin\theta$$

The torque of the equivalent 1-turn coil shown in Fig. 4.19 is

$$T = BIA \,.\, \cos(90° - \theta)$$

$$= BIA \,.\, \sin\theta$$

For equivalence of dipole in either form:

$$BIA \,.\, \sin\theta = \frac{mB}{\mu_0} \cdot \sin\theta$$

$$IA = \frac{m}{\mu_0}$$

μ_0 is the magnetic constant known as the permeability of free space which will be further discussed in para. 4.8. The torque on the dipole is thus given by

$$T = BIA \cdot \sin\theta$$
$$= \frac{mB}{\mu_0} \cdot \sin\theta$$
$$= \frac{2\Phi lB \cdot \sin\theta}{\mu_0}$$

but $\quad T = 2Fl \cdot \sin\theta$

where F is the force on each pole of the dipole.

$$F = \frac{\Phi B}{\mu_0}$$

The force on a unit pole is therefore given by B/μ_0 and is termed the magnetic field strength H.

$$H = \frac{B}{\mu_0}$$

$$B = \mu_0 H \tag{4.11}$$

The constant μ_0 in relation (4.11) is applicable when the dipoles are considered to exist in free space. This was given as an introductory condition.

The magnetising force at any point in an electromagnetic field is therefore the force experienced by a unit magnetic pole placed at that point. If the unit magnetic pole is made to move in a complete path round N current-carrying conductors then work is done provided the movement is in opposition to the lines of force. Conversely, if the movement is in the direction of the magnetic field, work will again be done by the magnetic force on whatever force is restraining the movement of the pole. In either case, the unit pole makes one complete loop around the N conductors. The work done is given by the Work Law which states that the net work done on or by a unit pole in moving once round any complete path is equal numerically to the product of the current and the number of turns linked within the path. It follows that if there are N conductors each carrying a current I then the work done is IN measured in joules.

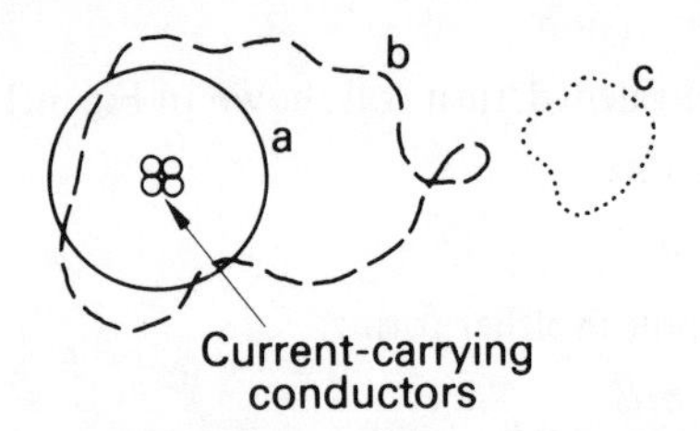

Fig. 4.20. Work law

A simple regular path (a) is shown in Fig. 4.20 around which a unit pole may be moved but any irregular path, such as path (b) would yield the same result. Path (c)

on the other hand fails to link any conductors and therefore no total work is done in moving a unit pole round such a path.

The law is applicable to all magnetic fields regardless of the dimensions of the field or any magnetic properties that it may possess. The Work Law can be proved by considering the effect of a unit magnetic pole, but the mathematics involved are too complex for an introductory chapter on electromagnetism.

To demonstrate the application of the Work Law, consider the movement of a unit pole in a circular path about a current-carrying conductor, as shown in Fig. 4.21.

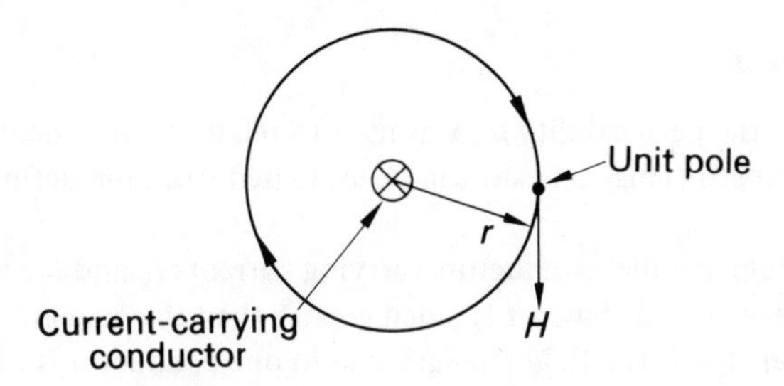

Fig. 4.21. Magnetising field strength about a current-carrying conductor

If a unit pole is placed on the circle, it experiences a force H. When the unit pole is moved around the circle once, the work done is

$$2\pi rH$$

By the Work Law, this is numerically equal to the current linked, hence

$$I = 2\pi rH$$

$$H = \frac{I}{2\pi r}$$

If there had been N conductors linked then

$$H = \frac{NI}{2\pi r} \tag{4.12}$$

By observation of this expression, it can be seen that the force termed the magnetic field strength is measured in amperes per metre or ampere turns per metre. Since the number of turns serves as a dimensionless factor, it may either be included or omitted from the unit of measurement without prejudice. The magnetic field strength is a vector quantity since it has magnitude and direction.

Magnetic field strength	Symbol: H	Unit: ampere turn per metre (At/m)

The work done in moving a unit pole around a magnetic circuit is given by the product IN. This is termed the magnetomotive force or m.m.f. Here again the number of turns is a dimensionless quantity.

Magnetomotive force	Symbol: F	Unit: ampere turn (At)

Since the magnetic field strength and the m.m.f. can be measured in ampere turns per metre and ampere turns respectively, it is no longer necessary to consider the unit magnetic pole, which is purely an imaginary device used to derive the concept of magnetic field strength.

The m.m.f. is analogous to e.m.f. since each represents the energy introduced into the circuits. From the definition of the magnetomotive force, it follows that, in a uniform arrangement,

$$F = Hl \tag{4.13}$$

4.8 Permeability

In relation (4.11), the permeability μ_0 was used to relate the flux density to the magnetic field strength. This constant can be evaluated from the definition of the ampere.

Consider two long parallel conductors carrying currents I_1 and I_2 spaced a distance d apart in free space. The distance d is much greater than the diameter of either conductor in order that if the field strength due to one conductor is known at the centre of the second conductor, this value will be approximately valid for the remainder of the cross section of the conductor. The arrangement is shown in Fig. 4.22.

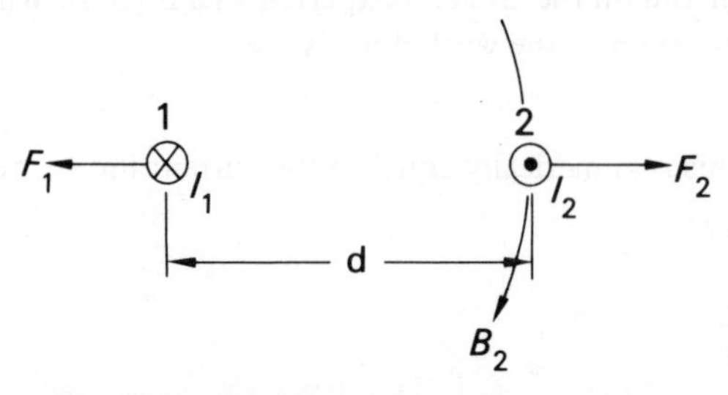

Fig. 4.22. Parallel current-carrying conductors

Let B_2 be the average flux density at the second conductor due to the current I_1 in the first conductor.

$$B_2 = \frac{\mu_0 I_1}{2\pi d}$$

Therefore the force on length l_2 of conductor 2 is given by

$$F_2 = B_2 l_2 I_2$$

Per unit length of conductor 2, the force is given by

$$F_2 = B_2 I_2$$

$$= \frac{\mu_0 I_1 I_2}{2\pi d} \tag{4.14}$$

In the definition of the ampere, it is stated that if such a system of conductors is placed one metre apart and a current of one ampere is passed through each of the conductors, then the force experienced by the conductors is 2×10^{-7} newtons per metre run.

$$\mu_0 = \frac{2\pi d F_2}{I_1 I_2} = \frac{2\pi \times 1 \times 2 \times 10^{-7}}{1 \times 1}$$

$$\mu_0 = 4\pi \times 10^{-7}$$

If the flux density due to a magnetic field strength exists in a medium other than free space, the permeability is found to change in value. This is the absolute permeability μ.

Absolute permeability	Symbol: μ	Unit: henry per metre (H/m)

The absolute permeability may be expressed as a multiple of that for free space μ_0.

Absolute permeability of free space	Symbol: μ_0	Unit: henry per metre (H/m)

The unit of absolute permeability will be justified in para. 5.2.

$$\mu_0 = 4\pi \times 10^{-7} \text{ H/m}$$

The ratio of the absolute permeability to that for free space is termed the relative permeability μ_r.

Relative permeability	Symbol: μ_r	Unit: none

$$\mu_r = \frac{\mu}{\mu_0}$$

$$\mu = \mu_0 \mu_r$$

It follows that the general case of relation (4.11) is

$$B = \mu_0 \mu_r H \tag{4.15}$$

For air, μ_r may be taken as unity; most other materials except the ferromagnetic materials have similar values of μ_r. Ferromagnetic materials such as iron, cobalt and nickel can have very much higher values of relative permeability.

Example 4.4 Two parallel conductors A and B are placed 100 mm apart in air. Conductor A carries a current of 100 A in one direction whilst conductor B carries a current of 60 A in the other direction. Calculate the magnetic field strength at a point C which is 60 mm from conductor A and 40 mm from conductor B.

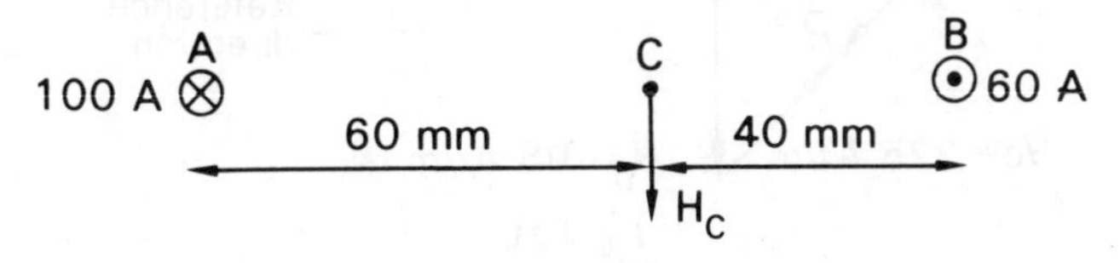

Fig. 4.23.

$$H_{CA} = \frac{I_A N_A}{2\pi r_{CA}} = \frac{100 \times 1}{2\pi \times 60 \times 10^{-3}} = 265 \text{ At/m}$$

$$H_{CB} = \frac{I_B N_B}{2\pi r_{CB}} = \frac{60 \times 1}{2\pi \times 40 \times 10^{-3}} = 239 \text{ At/m}$$

Both magnetic field strengths act vertically downwards. Therefore the resultant magnetic field strength at C is given by

$$H_C = H_{CA} + H_{CB} = 265 + 239 = \underline{504 \text{ At/m}}$$

This addition is really a vector addition. The next example involves a complex vector addition.

Example 4.5 Four conductors A, B, C and D viewed end-on are situated at the corners of a square of side 100 mm. The current in each conductor is 200 A having the directions shown in the Fig. 4.24.
Determine:

(*a*) The magnetic field stength at conductor A due to the currents in conductors B, C and D.
(*b*) The magnitude and direction of the force acting on conductor A.

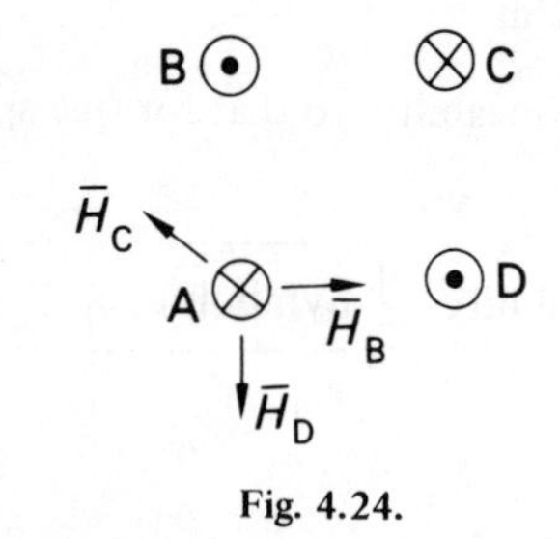

Fig. 4.24.

(*a*) Calculate the magnetic field strength at A due to the other conductors.

$$H_B = \frac{I_B N_B}{2\pi r_B} = \frac{200 \times 1}{2\pi \times 100 \times 10^{-3}} = 318 \text{ At/m}$$

$$H_C = \frac{I_C N_C}{2\pi r_C} = \frac{200 \times 1}{2\pi \times \sqrt{2} \times 100 \times 10^{-3}} = 225 \text{ At/m}$$

$$H_D = \frac{I_D N_D}{2\pi r_D} = \frac{200 \times 1}{2\pi \times 100 \times 10^{-3}} = 318 \text{ At/m}$$

The resultant magnetic field strength at A can be derived from a vector diagram drawn to scale–see Fig. 4.25.

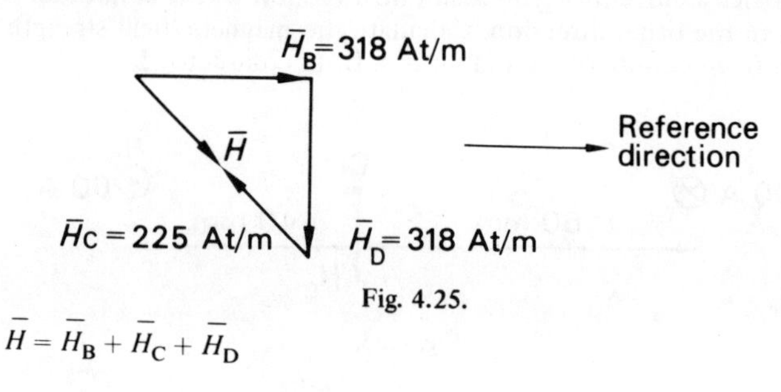

Fig. 4.25.

$$\bar{H} = \bar{H}_B + \bar{H}_C + \bar{H}_D$$

$\bar{H} = 225 \angle -45°$ At/m relative to the reference direction.

$H = 225$ At/m

The magnetic field strength could also have been obtained by resolving the vectors into horizontal and vertical components.

Horizontal components:

$$H_B \cos 0° + H_C \cos 135° + H_D \cos 270° = 159 \text{ At/m}$$

Vertical components:

$$H_B \sin 0° + H_C \sin 135° + H_D \sin 270° = -159 \text{ At/m}$$

The resultant magnetic field strength is therefore given by

$$H = (159^2 + 159^2)^{\frac{1}{2}} = \underline{225 \text{ At/m}}$$

The angle of action is given by

$$\theta = \tan^{-1} \frac{-156}{156} = -45°$$

$$B = \mu_0 H = 4\pi \times 10^{-7} \times 220 = 2{\cdot}76 \times 10^{-4} \; T$$

$$F = BI_A = 2{\cdot}76 \times 10^{-4} \times 200 = 55{\cdot}2 \times 10^{-3} \text{ N/m}$$

$$= \underline{55{\cdot}2 \text{ mN/m}}$$

By the left-hand rule, the angle of action relative to the reference direction is $-135°$.

4.9 Ferromagnetism and domain theory

The reader will be familiar with the concept that when a permanent magnet is cut in two, it forms two smaller magnets. It is now necessary to take this idea to the ultimate step whereby only one atom is left. The magnetic field of such an atom is derived from the movement of its electrons. These electrons move in two distinct ways:

1. Round the nucleus of the atom
2. Round their own axes.

The resultant magnetic moment of an atom is the resultant of the moments due to electron spins and the orbital motions, and it can be shown that the electron spin is the more important of the two factors. The calculation of the resultant moment is simplified by the fact that completed electron shells have no magnetic moment on account of their symmetry.

Many atoms have no magnetic moment except when in the presence of a magnetic field. This is because their symmetry is deformed by the field. In such a case, the magnetic moment is reduced and the atom is said to be diamagnetic. Its relative permeability will fall by the order of 10^{-6}. Mercury and silver exhibit diamagnetic effects.

Certain groups of atoms have incomplete inner shells to their atoms. This causes them to have permanent resultant magnetic moments. They tend to align themselves with any external field and hence reinforce it. These atoms are said to be paramagnetic. In these atoms, diamagnetism is always present but is overwhelmed by the greater paramagnetism. A paramagnetic material has a relative permeability greater than unity by the order of 10^{-3}. Platinum and tungsten exhibit paramagnetic effects.

The most important group are the ferromagnetics–principally iron, nickel and cobalt. In these cases, the atoms do not act singly but in groups called domains, each containing between 10^9 and 10^{15} atoms. The domains are smaller than the grains of the materials. Each atom or ion has a permanent magnetic moment and in this respect the ions of ferromagnetics are similar to those of paramagnetics but the ferromagnetic properties are characteristically not due to the properties of single ions. All the ions in each domain have their permanent magnetic axes pointing in the same direction, being aligned by a permanent intramolecular field which extends over the complete domain. This intramolecular field is the characteristic of ferromagnetism.

It should be noted that this domain structure depends on temperature. With increase in temperature, thermal agitation tends to break up the domains and, in the case of iron, finally succeeds at about 750°C.

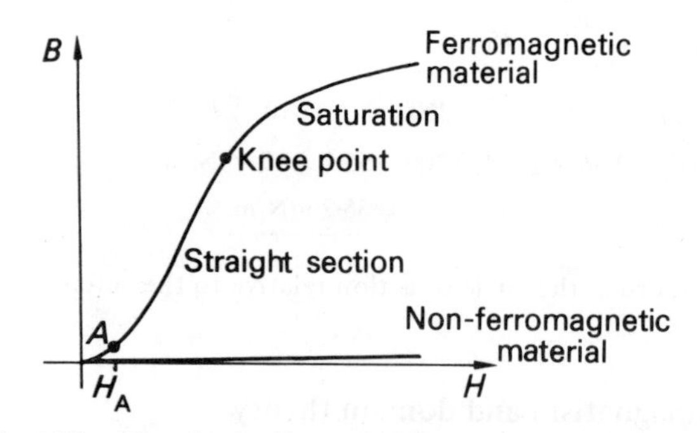

Fig. 4.26. Flux density/magnetic field strength characteristic

The reason that iron, cobalt and nickel form domains is complex. The third electron ring from the nucleus is incomplete, which is the condition for paramagnetism. However, the spacing of the ions is such that the orbits of these magnetic electrons can interpenetrate one another in adjacent ions. Intramolecular fields occur in other substances, but only in ferromagnetic materials does the field have the right direction to cause alignment of the magnetic axes of the ions.

When a specimen of ferromagnetic material is placed in a magnetic field, the domains tend to turn into line. It follows that their magnetic fields add to the external field resulting in a stronger total field. This effect can be observed from the characteristic shown in Fig. 4.26 relating the resultant flux density to the magnetic field strength.

Initially the specimen is magnetically neutral. This is caused by the domains having random orientation and the resultant magnetic moment is zero. Any small magnetic field strength up to H_A produces a flux density which is greater than that due to the magnetic field strength acting on the equivalent space alone. This extra flux is due to the actions of the domains. Up to this value, it is the boundaries of certain domains which change, these domains being those which are more or less parallel to the external field growing at the expense of their immediate neighbours. This growth is reversible and if the magnetic field strength is removed, the flux will also disappear.

As the magnetic field strength increases further, the domains not only continue to grow but many of them are turned round into alignment. This movement will not be reversed by the removal of the magnetic field strength, hence, if the magnetic field strength is removed, a field will still remain due to the alignment of the domains which have been rotated.

The process of building the domains up into an aligned condition continues at a reasonably steady rate with increase in magnetic field strength until most of the domains have been aligned. Because of this steady rate, the appropriate part of the characteristic is termed the straight part of the characteristic. When most of the domains are aligned, the material is said to be saturated. Strictly speaking saturation is never achieved under normal conditions. Even if saturation were completed, the flux density will still increase with magnetic field strength but at a rate appropriate to unity relative permeability. The relative domain positions are indicated in the schematic diagrams shown in Fig. 4.27.

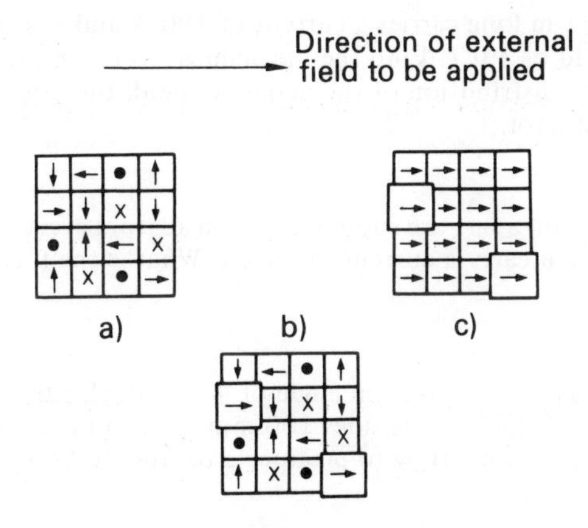

Fig. 4.27. Domain arrangements (*a*) Initial random arrangement (*b*) Domain growth (*c*) 'Saturation'

It should be noted that, as the magnetising effect is increased, the domains will not turn uniformly. For instance, if a coil is placed round the specimen, the domain movements will induce e.m.f.'s in it, which can be amplified and heard on a loudspeaker. The resultant noise either takes the form of a rustling or a series of clicks, thus indicating the discontinuity of movement. This is known as the Barkhausen Effect.

Using any magnetisation characteristic, it is possible to derive the values of the relative permeability corresponding to the values of magnetic field strength or flux density from the relation $B = \mu_0 \mu_r H$. The resulting characteristics are shown in Fig. 4.28.

It should be noted that the early part of the B/H characteristic has been exaggerated, i.e. that part during which the domains grow but do not rotate. In practice, it is very small: the smaller it is then the closer does the domain-notation section of the characteristics become linear. It is for this reason that in many magnetic circuit calculations, the relative permeability may be considered constant.

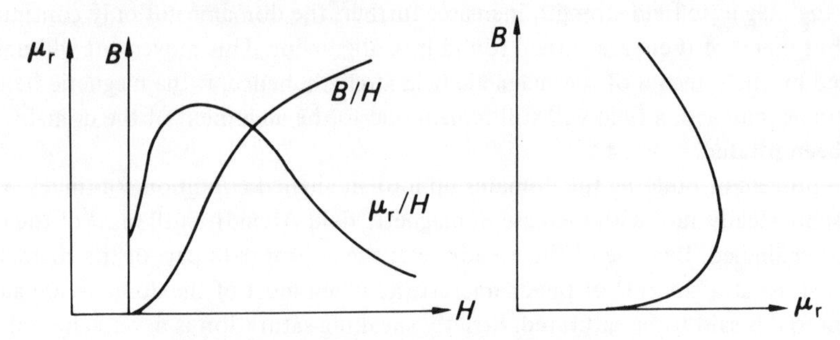

Fig. 4.28. Relative permeability characteristics

PROBLEMS ON ELEMENTS OF ELECTROMAGNETISM

1 A straight wire 0·1 m long carries a current of 100 A and lies at right angles to a uniform magnetic field of 1·0 T. Find the mechanical force on the conductor and give a diagram showing the distribution of the magnetic field, the direction of current and the force on the conductor.

10 N

2 Two busbars 100 mm apart are supported by insulators every metre along their length. The busbars each carry a current of 15 kA. What is the force acting on each insulator?

450 N

3 The plane of a square coil makes an angle of 45° with the direction of a uniform field of density 0·4 T. If the coil has sides 100 mm long and is wound with 500 turns, calculate the current that must flow to produce a torque of 3·0 N m about the neutral axis of the coil.

2·12 A

4 An e.m.f. of 1·5 V is induced in a straight conductor which moves with a velocity of 5·0 m/s perpendicular to a magnetic field of density 0·75 T. Calculate the effective length of the conductor.

0·4 m

5 The magnetic flux linking 1800 turns of an electromagnet changes uniformly from 0·6 mWb to 0·5 mWb in 50 ms. Find the average value of the induced e.m.f.

3·6 V

6 Each field pole of a 4-pole, d.c. generator produces a magnetic flux of 20 mWb. The armature conductors rotate at 1 200 rev/min and may be assumed to cut the flux at right angles. What is the average value of the induced e.m.f. in each armature conductor?

1·6 V

7 A conductor of effective length 150 mm moves with a velocity of 5 m/s perpendicular to a magnetic field of uniform flux density 0·4 T. Calculate:

(*a*) The e.m.f. induced in the conductor.
(*b*) The force acting on the conductor when it carries a current of 20 A.
(*c*) The power required to drive the conductor.

Give a diagram of the relative directions of the field, the motion of the conductor and the induced e.m.f.

(SANCAD)

0·3 V; 1·2 N; 6·0 W

8 A straight wire 1·0 m long carries a current of 100 A and lies at right angles to a uniform magnetic field of 1·5 T. Find the mechanical force on the conductor and also calculate the power to move the wire at a speed of 10 m/s in a plane at right angles to the field.

150 N; 1·5 kW

5

Applied electromagnetic theory

Having introduced the physical background to electromagnetism, it is now necessary to consider the salient points that concern an engineer. Due to the number of methods of implementing the theory, this chapter splinters into a number of separate topics as opposed to the progressive study apparent in other chapters.

5.1 Self inductance

When a current flows in a winding, it sets up a magnetic field and hence a system of flux linkage. It has been noted in para. 4.6 that if this flux linkage varies, an e.m.f. is induced in the circuit, the e.m.f. being given by

$$e = \frac{\mathrm{d}\psi}{\mathrm{d}t} = N \cdot \frac{\mathrm{d}\phi}{\mathrm{d}t}$$

The effect of the e.m.f. can be observed from measurements in the electric circuit and since the flux is due to the current, it follows that the e.m.f. is proportional to the rate of change of current, i.e.

$$e \propto \frac{\mathrm{d}\phi}{\mathrm{d}t} \propto \frac{\mathrm{d}i}{\mathrm{d}t}$$

Let

$$e = L . \frac{\mathrm{d}i}{\mathrm{d}t} \tag{5.1}$$

where L is a suitable constant termed the inductance.

Inductance	Symbol: L	Unit: henry (H)

The winding e.m.f. is entirely self-induced, hence the inductance is termed the self inductance. The unit of inductance called the henry is the inductance of a closed circuit in which an e.m.f. of 1 volt is produced when the electric current in the circuit varies uniformly at the rate of 1 ampere per second.

It is not the case however that the inductance of a winding is an absolute constant, thus relation (5.1) may be more generally stated in the form

$$e = \frac{d(Li)}{dt} = \frac{d\psi}{dt}$$

$$\psi = Li$$

$$L = \frac{\psi}{i} = \frac{N\phi}{i} \tag{5.2}$$

It can thus be stated that a circuit has an inductance of 1 henry if 1 weber turn of flux linkage is set up by a current of 1 ampere in the circuit. Regardless of how inductance is defined, however, it has the important role of relating magnetic and electric systems in terms of the effects on the electric system.

Relation (5.2) is given in instantaneous terms because the inductance of a circuit is not necessarily constant. It depends on the shape of the magnetisation characteristic, which, in the case of ferromagnetic materials, gives rise to a varying value of relative permeability. In turn, this gives rise to a varying value of the ratio ϕ/i. However, provided the magnetisation characteristic is reasonably linear up to the maximum flux density experienced, the inductance is almost constant.

In cases where there is no ferromagnetic material present, the magnetisation characteristic is a straight line and the self inductance is constant.

The inductance of a circuit depends on the construction of the circuit, the number of turns and the relative permeability of the constituent materials.

The effects of inductance on electric networks will be considered in later chapters. However, it is necessary to develop the relation of inductance to magnetic and electric quantities before this can satisfactorily be undertaken.

5.2 Reluctance and permeance

In a complete magnetic circuit, the m.m.f. around the circuit can be related to the flux in the circuit by a constant termed the reluctance of the magnetic circuit. This is comparable to the resistance of an electric circuit.

Just as part of an electric circuit can be considered apart, so can part of a magnetic circuit be considered. In such a section, only the change in magnetic potential is relevant as opposed to the m.m.f. Magnetic p.d. is comparable with electric p.d. in the same manner that m.m.f. is comparable with e.m.f.

Magnetic potential difference	Symbol: F	Unit: ampere turn

The magnetic potential difference is the energy required to move a unit magnetic pole between two points. The sum of the magnetic p.d.'s around a circuit is equal to the m.m.f.

The reluctance, like resistance, is dependent on the physical dimensions of the magnetic circuit and on the material from which it is made.

Consider part of a uniform magnetic field as shown in Fig. 5.1.

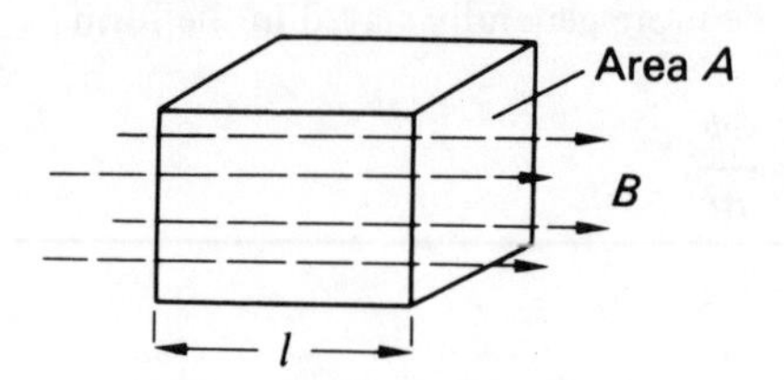

Fig. 5.1. Derivation of reluctance

For the section of the field indicated in the diagram:

$$F = Hl$$

$$= \frac{Bl}{\mu_0 \mu_r}$$

$$= \frac{\Phi l}{\mu_0 \mu_r A}$$

$$F = \Phi S \tag{5.3}$$

where S is termed the reluctance.

Reluctance	Symbol: S	Unit: reciprocal henry (/H) *or* ampere turn per weber (At/Wb)

$$S = \frac{l}{\mu_0 \mu_r A} \tag{5.4}$$

Alternatively:

$$\Phi = F\Lambda \tag{5.5}$$

where $$\Lambda = \frac{1}{S} = \frac{\mu_0 \mu_r A}{l} \tag{5.6}$$

is termed the permeance.

Permeance	Symbol: Λ	Unit: henry (H)

Example 5.1 A ring of cross-sectional area 500 mm² and mean diameter 300 mm is wound with 100 turns of conductor. If the relative permeability of the ring is 560 and a current of 1·5 A is passed through the coil, calculate the flux in the ring.

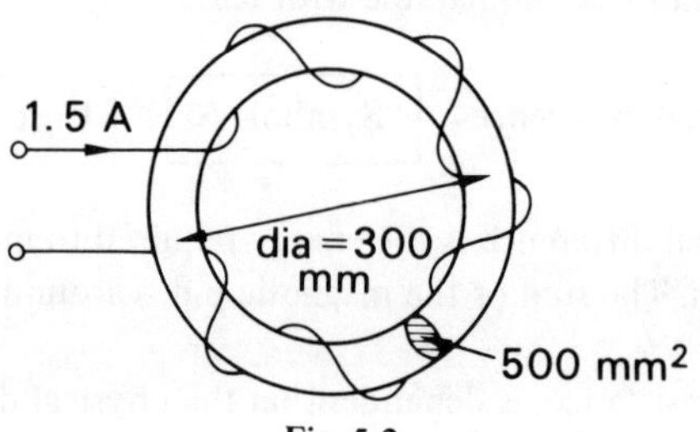

Fig. 5.2

$$F = NI = 100 \times 1{\cdot}5 = 150 \text{ At}$$

$$\Lambda = \frac{\mu_0 \mu_r A}{l} = \frac{4\pi \times 10^{-7} \times 560 \times 500 \times 10^{-6}}{300\pi \times 10^{-3}}$$

$$= 373 \times 10^{-9} \text{ H}$$

$$\Phi = F\Lambda = 150 \times 373 \times 10^{-9} = 56 \times 10^{-6} \text{ Wb}$$

$$= \underline{56 \ \mu\text{Wb}}$$

The self-inductance of a magnetic circuit can be derived in terms of the reluctance as follows:

$$B = \mu_0 \mu_r H$$

$$= \mu_0 \mu_r \cdot \frac{NI}{l}$$

$$\Phi = BA$$

$$= \mu_0 \mu_r \cdot \frac{NIA}{l}$$

$$L = \frac{N\Phi}{I}$$

$$= \frac{N}{I} \cdot \mu_0 \mu_r \cdot \frac{NIA}{l}$$

$$= \frac{N^2}{S} \tag{5.7}$$

$$= N^2 \Lambda \tag{5.7.1}$$

Finally it is of interest that consideration should be given to some of the units of measurement, which have been presented, by definition

$$L = \frac{N\phi}{i}$$

It has already been noted that the turn is not a unit of measurement but is a dimensionless factor. Consequently the henry is effectively the weber per ampere.

$$S = \frac{F}{\phi} = \frac{Ni}{\phi}$$

In the expression for reluctance, it can be seen that the reluctance is equated to current per unit of flux, i.e. it will be measured in amperes per weber. By comparison with the above argument, this unit of measurement is abbreviated to the reciprocal henry.

$$S = \frac{l}{\mu_0 \mu_r A}$$

$$\mu_0 = \frac{l}{S \mu_r A}$$

Dimensionally, the absolute permeability is hence measured in metres per square metre per reciprocal henry, which may be reduced to henrys per metre.

5.3 Self inductance of a long solenoid

A long solenoid is a coil, the cross-section of which is shown in Fig. 5.3. It has a wound length l, which must be much greater than the breadth d, say 10 times greater. It also has N turns. The following assumptions are then permissible:

1. The field strength external to the solenoid is effectively zero.
2. The field strength inside the solenoid is uniform.

These assumptions compare with the field indicated in Fig. 5.2 by a few lines of force. They are crowded together within the solenoid suggesting a relatively high field strength but they are spread out outwith the solenoid suggesting a correspondingly weak magnetising force.

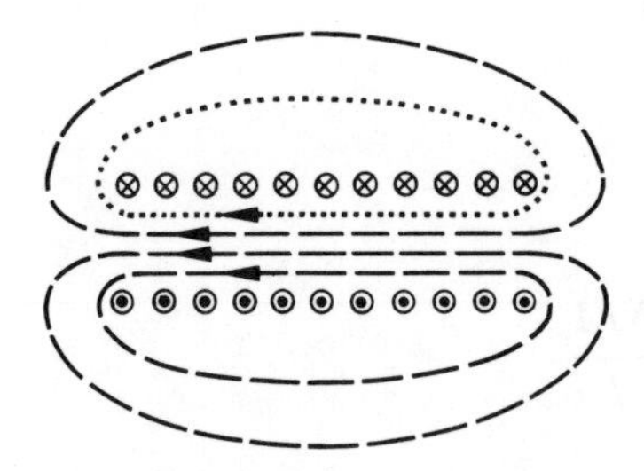

Fig. 5.3 Magnetic field around a long solenoid

If the field strength is assumed to be uniform within the solenoid, let it have a value H and let it be created by a winding current I. Apply the Work Law to any closed path, say the dotted one shown.

Total work done round closed path = Ampere turns linked

Since there is negligible field strength outside the solenoid, the only work done will be in travelling the length l within the solenoid.

$$Hl = IN$$

$$H = \frac{NI}{l} \tag{5.8}$$

Also

$$B = \mu_0 \mu_r H$$

$$= \mu_0 \mu_r \cdot \frac{NI}{l}$$

If A is the cross-sectional area of the solenoid through which the flux passes.

$$\Phi = BA$$

$$= \mu_0 \mu_r \cdot \frac{NIA}{l}$$

Provided all the flux links all the turns then

$$L = \frac{\Phi N}{I}$$

$$= \mu_0 \mu_r \cdot \frac{NIA}{l} \cdot \frac{N}{I}$$

$$= \frac{\mu_0 \mu_r N^2 A}{l} \qquad (5.9)$$

$$= \frac{N^2}{S} \text{ as before.}$$

This solution only approximates toward the correct value of the self inductance of a solenoid. More accurate methods for the determination of the self inductance are available but they involve more advanced mathematics.

The solenoid is an important winding arrangement being simple to manufacture. It is found in relays, inductors, small transformers, etc. in the form considered above.

Example 5.2 A solenoid 800 mm long and 20 mm in diameter is uniformly wound with a coil of 1 000 turns. Determine the self inductance of the coil assuming that it is air-cored. Also determine the flux density within the solenoid, when the coil current is 1·0 A.

$$S = \frac{l}{\mu_0 \mu_r A} = \frac{800 \times 10^{-3} \times 4}{4\pi \times 10^{-7} \times 1 \times \pi \times 2^2 \times 10^{-4}}$$

$$= 20{\cdot}3 \times 10^8 \text{ At/Wb}$$

$$L = \frac{N^2}{S} = \frac{1000^2}{20{\cdot}3 \times 10^8} = 4{\cdot}94 \times 10^{-4} \text{ H} = \underline{494\ \mu\text{H}}$$

$$H = \frac{NI}{l} = \frac{1000 \times 1}{800 \times 10^{-3}} = 1250 \text{ At/m}$$

$$B = \mu_0 \mu_r H = 4\pi \times 10^{-7} \times 1 \times 1250 = 1{\cdot}57 \times 10^{-3} \text{ T}$$

$$= \underline{1{\cdot}57 \text{ mT}}$$

5.4 Self inductance of a uniformly-wound toroid

A toroid is a ring and hence a toroidal winding is one which incorporates a ring as a former. Although it is virtually impossible to produce a uniformly-wound toroid, it is a convenient arrangement to consider when introducing magnetism. Also it is significant in the testing of magnetic materials. Such a toroidal winding with N turns carrying a current I is shown in Fig. 5.4.

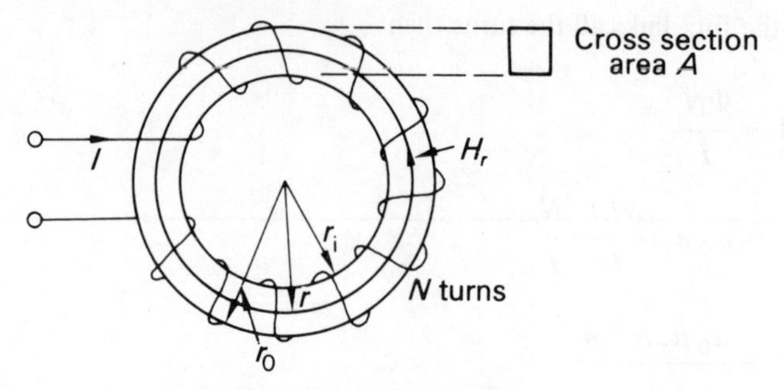

Fig. 5.4. Toroidal winding

Apply the Work Law to a path having a radius r and lying within the core. At every point on the path, the unit pole will experience a force H_r and hence:

Total work done round closed path = Ampere turns linked

$$2\pi r H_r = NI$$

$$H_r = \frac{NI}{2\pi r} \tag{5.10}$$

At the inner edge, where $r = r_i$,

$$H_i = \frac{NI}{2\pi r_i}$$

At the outer edge, where $r = r_o$,

$$H_0 = \frac{NI}{2\pi r_0}$$

Hence the value of the magnetic field strength at any point within the core can be ascertained. The magnetic field strength, however, is seen to decrease as the radius increases. If the toroid has a mean radius much greater than its cross-sectional thickness $(r_o \rightarrow r_i)$, then H_i is approximately equal to H_o and it may be assumed that at each point in the cross-section, the magnetic field strength is given by H where H is the magnetic field strength at the mean radius.

$$H = \frac{NI}{2\pi r_{av}} \tag{5.10.1}$$

where r_{av} is the mean radius

Also $$B = \mu_0 \mu_r H$$

$$= \frac{\mu_0 \mu_r NI}{2\pi r_{av}}$$

If A is the cross-sectional area of the toroid,

$$\Phi = BA$$

$$= \frac{\mu_0 \mu_r NIA}{2\pi r_{av}}$$

Provided all the flux links all the turns:

$$L = \frac{\Phi N}{I}$$

$$= \frac{\mu_0 \mu_r N^2 A}{2\pi r_{av}} \tag{5.11}$$

$$= \frac{N^2}{S} \text{ as before.}$$

Example 5.3 A wooden toroid of mean diameter 400 mm and cross-sectional area 400 mm² is uniformly wound with a coil of 1 000 turns, which carries a current of 2·0 A. Determine the self inductance of the coil and the e.m.f. induced in it when the current is uniformly reduced to zero is 10 ms.

$$L = \frac{N^2}{S} = \frac{N^2 \mu_0 \mu_r A}{l} = \frac{1000^2 \times 4\pi \times 10^{-7} \times 1 \times 400 \times 10^{-6}}{0{\cdot}4\pi}$$

$$= 4 \times 10^{-4} \text{ H}$$

$$= \underline{0{\cdot}4 \text{ mH}}$$

$$E = L \cdot \frac{\Delta I}{\Delta t} = 0{\cdot}4 \times 10^{-3} \times \frac{2}{10 \times 10^{-3}} = \underline{0{\cdot}08 \text{ V}}$$

5.5 Mutual inductance

If two circuits are positioned in such a manner that the flux set up by one links the other circuit, then that flux which is common to both circuits is termed the mutual flux. If the current in the first circuit is varied, the mutual flux also varies and hence induces an e.m.f. in the second circuit. This effect is termed mutual inductance.

Two circuits have a mutual inductance of 1 henry if an e.m.f. of 1 volt is induced in one when the current in the other is changing at a rate of 1 ampere per second.

Mutual inductance	Symbol: M	Unit: henry (H)

It follows that the induced e.m.f. is given by the following relation provided that the mutual inductance is constant.

$$e = M \cdot \frac{di}{dt} \tag{5.12}$$

The mutual inductance can be related to the self inductances of the constituent circuits. Consider two concentrated coils mounted in the manner shown in Fig. 5.5.

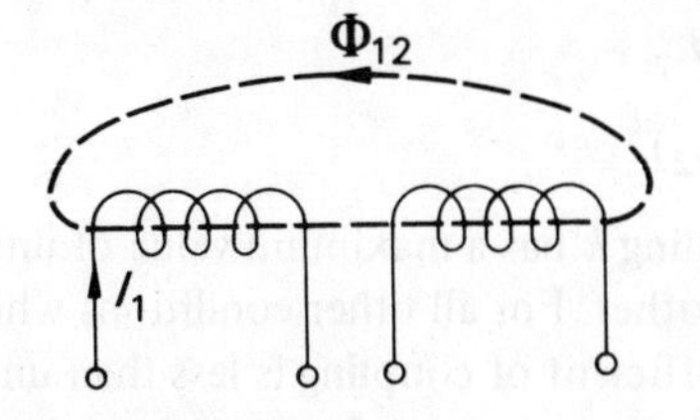

Fig. 5.5. Two coils linked by mutual flux

The flux is created by a current I_1 in coil 1 of N_1 turns. Of the total resulting flux Φ_1, $k_{12}\Phi_1$ links coil 2. Hence by definition:

$$L_1 = \frac{\Phi_1 N_1}{I_1}$$

The mutual inductance can be similarly defined hence:

$$M_{12} = \frac{k_{12}\Phi_1 N_2}{I_1}$$

but $$\Phi = \frac{F}{S} = \frac{NI}{S}$$

$$\frac{\Phi}{I} = \frac{N}{S}$$

hence $$M_{12} = \frac{k_{12}\Phi_1 N_2}{I_1}$$

$$= \frac{N_1 N_2}{S}$$

where S is the reluctance of the common flux path.

In the case of a current I_2 in coil 2 producing the same mutual flux linkages as in the above situation then:

$$L_2 = \frac{\Phi_2 N_2}{I_2}$$

and $$M_{21} = \frac{k_{21}\Phi_2 N_1}{I_2}$$

$$= \frac{N_2 N_1}{S}$$

Therefore $M_{12} = M_{21} = M$

Also $$M^2 = M_{12} M_{21}$$

$$= \frac{k_{12}\Phi_1 N_2}{I_1}\frac{k_{21}\Phi_2 N_1}{I_2}$$

$$= k_{12}k_{21}L_1 L_2$$

$$= k^2 L_1 L_2$$

$$M = k(L_1 L_2)^{\frac{1}{2}} \tag{5.13}$$

The coefficient of coupling k has a maximum value of unity when the entire self flux of each coil links the other. For all other conditions where less than the entire self flux is mutual, the coefficient of coupling is less than unity. Ferromagnetic-cored coils represent the nearest approach to perfect or unity coupling. It has been assumed that the ratio of coil flux to coil current is a constant hence instantaneous values were not used in the above theory. If the ratio is not a constant, the relationships still hold for instantaneous values.

Example 5.4 A ferromagnetic ring of cross-sectional area 800 mm^2 and of mean radius 170 mm has two windings connected in series, one of 500 turns and one of 700 turns. If the relative permeability is 1 200, calculate the self inductance of each coil and the mutual inductance of each assuming that there is no flux leakage.

$$S = \frac{l}{\mu_0 \mu_r A} = \frac{2\pi \times 170 \times 10^{-3}}{4\pi \times 10^{-7} \times 1\,200 \times 800 \times 10^{-6}}$$

$$= 8{\cdot}85 \times 10^5 \text{/H}$$

$$L_1 = \frac{N_1^{\,2}}{S} = \frac{500^2}{8{\cdot}85 \times 10^5} = 0{\cdot}283 \text{ H}$$

$$L_2 = \frac{N_2^{\,2}}{S} = \frac{700^2}{8{\cdot}85 \times 10^5} = 0{\cdot}552 \text{ H}$$

$$M = k(L_1 L_2)^{\frac{1}{2}} = 1 \times (0{\cdot}283 \times 0{\cdot}552)^{\frac{1}{2}} = 0{\cdot}395 \text{ H}$$

In example 5.4, the coils were connected in series. However, the manner in which one coil was wound with respect to the other was not stated. Depending on the physical winding of the second coil, the direction of any flux it may create, may add to or subtract from the flux of the first coil thus for two circuits with mutual inductive coupling, the relative directions of the e.m.f.'s in each winding depend on the relative directions of the turns and the relative position of the windings.

One system of defining the directions is the dot notation. At any instant, it may be considered that current is entering the first coil at one particular end. This end is indicated by placing a dot beside it. A second dot is placed at the end of the second coil at which current would have to enter at the same instant to give additive flux. These dots are indicated on a diagram as shown in Fig. 5.6.

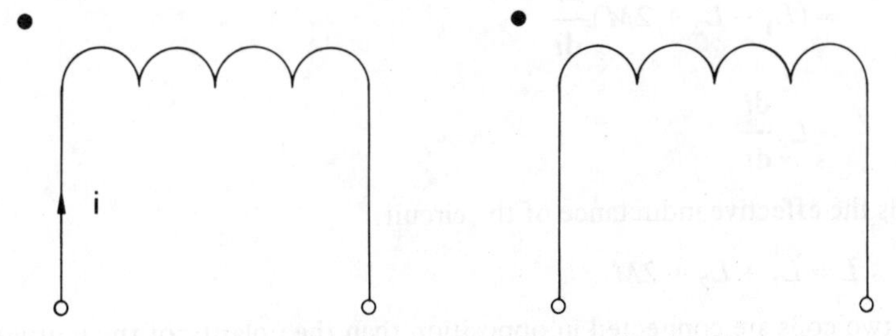

Fig. 5.6. Dot notation

The relative e.m.f. directions can now be derived. Suppose a voltage to be applied to the first coil. If this makes the dotted end positive with respect to the undotted end, then current enters at the dotted end and increases in magnitude due to the increase in potential difference across the coil. This change of current causes e.m.f.'s to be set up in the first coil due to the self flux linkages increasing and in the second coil due to the mutual flux linkages increasing. The e.m.f. in the first coil opposes the current and so makes the dotted end positive with respect to the undotted end, i.e. it opposes the change of flux linkages. Also the e.m.f. induced in the second coil will make current leave at the dotted end of this coil. Hence in both coils, both dotted ends are driven simultaneously positive with respect to the undotted ends.

It follows that it would have been possible to connect the two coils in series in two ways such that they would be mutually linked. These ways are so that the common current enters on leaves both coils at the dotted ends, i.e. both produce flux in the same direction, or so that, while the current enters the dotted end of one coil, it leaves the dotted end of the other. These arrangements are termed series aiding and series opposing respectively. Circuit connection diagrams are shown in Fig. 5.7.

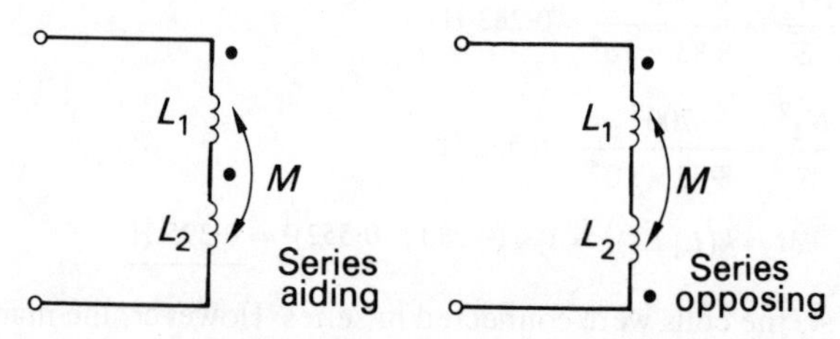

Fig. 5.7. Series connections of mutually coupled coils

In the first case, suppose that the current is changing at the rate di/dt. The total induced e.m.f. in the circuit will equal the self-induced e.m.f.'s plus the mutually-induced e.m.f.'s in magnitude. In the aiding connection, the e.m.f.'s are all driving the dotted ends positive with respect to the undotted ends.

$$e = L_1 \cdot \frac{di}{dt} + L_2 \cdot \frac{di}{dt} + M \cdot \frac{di}{dt} + M \cdot \frac{di}{dt}$$

$$= (L_1 + L_2 + 2M)\frac{di}{dt}$$

$$= L . \frac{di}{dt}$$

where L is the effective inductance of the circuit.

$$L = L_1 + L_2 + 2M \tag{5.14}$$

If the two coils are connected in opposition then the polarity of the mutually-induced e.m.f.'s is reversed, hence:

$$e = L_1 \cdot \frac{di}{dt} + L_2 \cdot \frac{di}{dt} - M \cdot \frac{di}{dt} - M \cdot \frac{di}{dt}$$

$$= (L_1 + L_2 - 2M)\frac{di}{dt}$$

$$= L \cdot \frac{di}{dt}$$

$$L = L_1 + L_2 - 2M \qquad (5.15)$$

Example 5.5 When two coils are connected in series, their effective inductance is found to be 10·0 H. However, when the connections to one coil are reversed, the effective inductance is 6·0 H. If the coefficient of coupling is 0·6, calculate the self inductance of each coil and the mutual inductance.

$$L = L_1 + L_2 \pm 2M = L_1 + L_2 \pm 2k(L_1 L_2)^{\frac{1}{2}}$$

$$10 = L_1 + L_2 + 2k(L_1 L_2)^{\frac{1}{2}}$$

$$6 = L_1 + L_2 - 2k(L_1 L_2)^{\frac{1}{2}}$$

$$8 = L_1 + L_2$$

$$10 = 8 - L_2 + L_2 + 1{\cdot}2(8L_2 - L_2{}^2)^{\frac{1}{2}}$$

$$0 = L_2{}^2 - 8L_2 + 2{\cdot}78$$

$$\therefore \quad \underline{L_2 = 7{\cdot}63 \text{ H or } 0{\cdot}37 \text{ H}}$$

$$\therefore \quad \underline{L_1 = 0{\cdot}37 \text{ H or } 7{\cdot}63 \text{ H}}$$

$$2M = 10 - 7{\cdot}63 - 0{\cdot}37$$

$$\underline{M = 1{\cdot}0 \text{ H}}$$

5.6 Induced e.m.f. in a circuit

It has been shown that

$$e = L \cdot \frac{di}{dt}$$

Whilst this term gives the magnitude of the e.m.f., there remains the problem of polarity. When a force is applied to a mechanical system, the system reacts by deforming, or mass-accelerating, or dissipating or absorbing energy. A comparable state exists when a force (voltage) is applied to an electric system, which accelerates (accepts magnetic energy in an inductor) or dissipates energy in heat (in a resistor). The comparable state to deformation is the acceptance of potential energy in a capacitor which is dealt with in Chapter 6. In the case of a series circuit containing resistance and inductance then

$$v = Ri + L \cdot \frac{di}{dt}$$

There are now two schools of thought as to how to proceed. One says

$$v = v_R + e_L$$

The other says

$$v = v_R - e_L$$

$$= v_R + v_L$$

This requires that

$$e = -L \cdot \frac{di}{dt}$$

The idea behind this second interpretation is a wish to identify active circuit components. These include batteries, generators and (because they store energy) inductors and capacitors. A charged capacitor can act like a battery for a short time and certainly a battery possesses an e.m.f. which, when the battery is part of a circuit fed from a source of voltage, acts against the passage of current through it from the positive terminal to the negative terminal. Similarly an inductor opposes the increase of current in it by presenting the applied voltage by an opposing e.m.f.

The first method suggests that the only voltage to be measured is a component of the voltage applied and this is

$$v_L = +L \cdot \frac{di}{dt}$$

Both arguments are acceptable and the reader will find both systems have wide application. Although the International Electrotechnical Commission prefer the second method, for the purposes of this text, the positive version will be used.

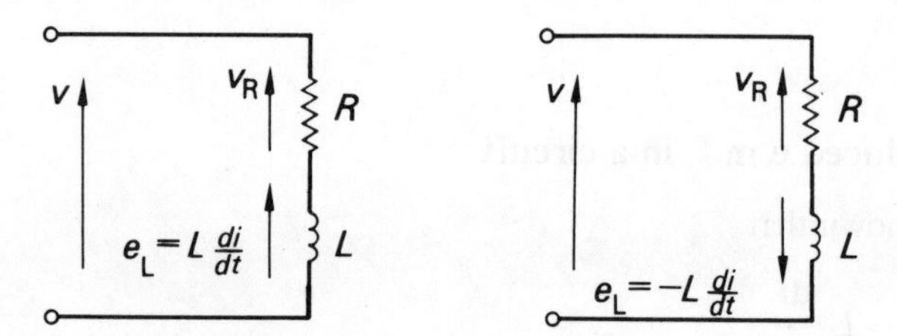

Fig. 5.8. Polarity of e.m.f. in a circuit diagram

It will be noted that the positive version of the relation appears to be more logical in a circuit diagram. In Fig. 5.8, both versions are considered. If the induced e.m.f. is taken as an effective volt drop, it may be represented by an arrow pointing upwards. If the negative version is used, the arrow must point in the direction of the current flow.

Example 5.6 Two mutually-coupled coils, A and B, are connected in series to a 360-V, d.c. supply. Coil A has a resistance 6 Ω and inductance 4 H. Coil B has a resistance 11 Ω and inductance 9 H. At a certain instant after the circuit is energised, the current is 10 A and increasing at the rate 10 A/s. Calculate:

(*a*) The mutual inductance between the coils.
(*b*) The coefficient of coupling.

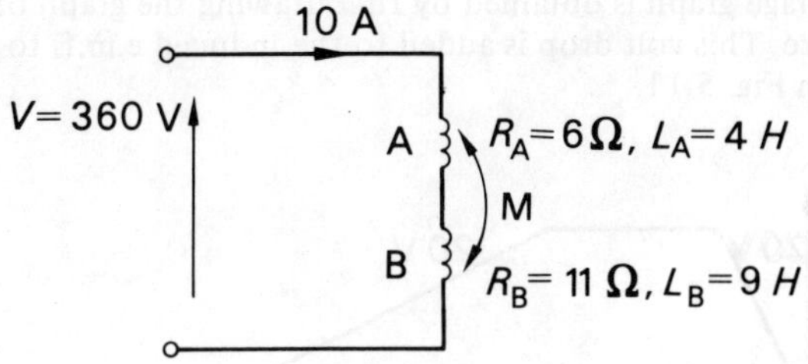

Fig. 5.9

$$V = iR + L \cdot \frac{di}{dt}$$

$$360 = 10(6 + 11) + 10L$$

$$L = 19$$

$$= L_A + L_B + 2M = 4 + 9 + 2M$$

$$\underline{M = 3\text{ H}}$$

$$M = k(L_A L_B)^{\frac{1}{2}}$$

$$3 = k(4 \times 9)^{\frac{1}{2}}$$

$$\underline{k = 0{\cdot}5}$$

Example 5.7 A coil having a resistance of 4 Ω and an inductance of 0·8 H has a current passed through it which varies as follows:

(*a*) Uniform increase from 0 to 5·0 A in 0·25 s.
(*b*) Constant at 5·0 A for 0·5 s.
(*c*) Uniform decrease from 5·0 A to 0 in 1 s.

Plot graphs representing the current, the induced e.m.f. and the applied voltage.

$$E_a = L \cdot \frac{\Delta I}{\Delta t} = 0{\cdot}8 \times \frac{5{\cdot}0}{0{\cdot}25} = 16\text{ V}$$

$$V_b = IR = 5 \times 4 = 20\text{ V}$$

$$E_c = L \cdot \frac{\Delta I}{\Delta t} = 0{\cdot}8 \times \frac{-5{\cdot}0}{1{\cdot}0} = -4\text{ V}$$

The graphs of the current and induced e.m.f. can now be drawn as shown in Fig. 5.10

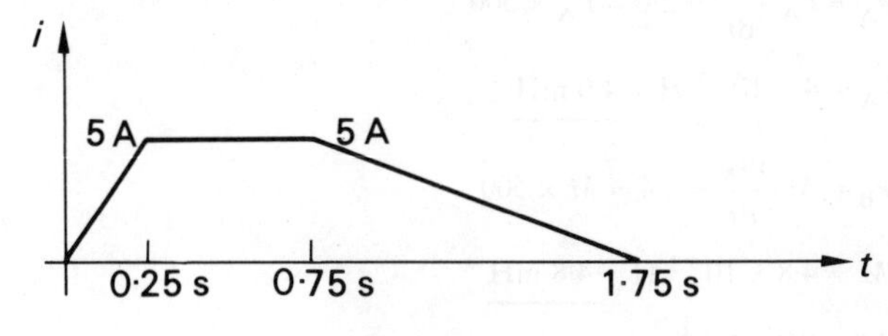

Fig. 5.10

The applied voltage graph is obtained by first drawing the graph of the volt drop across the resistance. This volt drop is added to the induced e.m.f. to obtain the applied voltage as shown in Fig. 5.11.

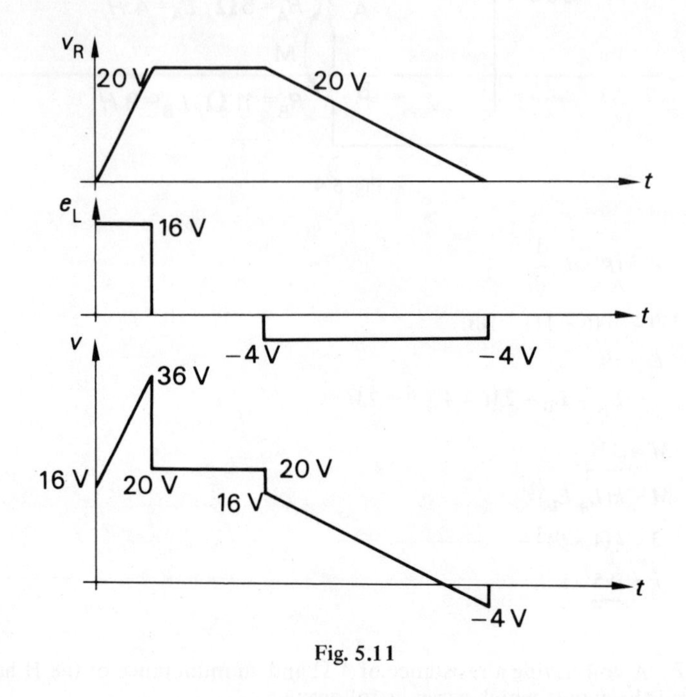

Fig. 5.11

Example 5.8 Two coils, A and B, are so placed that 80% of the total flux produced by one links the other. A has 2 000 turns and B has 3 000 turns. When the current in A is changing at the rate 500 A/s, the flux linking it is changing at the rate 1 mWb/s. Determine:

(*a*) The corresponding e.m.f. induced in each coil.
(*b*) The self inductance of each coil.
(*c*) The mutual inductance of each coil.
(*d*) The effective inductance of the two coils in series.

$$e_A = N_A \cdot \frac{d\phi_A}{dt} = 2000 \times 1 \times 10^{-3} = \underline{2{\cdot}0 \text{ V}}$$

$$e_B = N_B \cdot \frac{d\phi_B}{dt} = 3000 \times 0{\cdot}8 \times 1 \times 10^{-3} = \underline{2{\cdot}4 \text{ V}}$$

$$e_A = L_A \cdot \frac{di_A}{dt} = 2{\cdot}0 = L_A \times 500$$

$$L_A = 4 \times 10^{-3} \text{ H} = \underline{4{\cdot}0 \text{ mH}}$$

$$e_B = M \cdot \frac{di_A}{dt} = 2{\cdot}4 = M \times 500$$

$$M = 4{\cdot}8 \times 10^{-3} \text{ H} = \underline{4{\cdot}8 \text{ mH}}$$

$$M = k(L_A L_B)^{\frac{1}{2}}$$

$$4{\cdot}8 \times 10^{-3} = 0{\cdot}8(L_B \times 4{\cdot}0 \times 10^{-3})^{\frac{1}{2}}$$

$$L_B = 9{\cdot}0 \times 10^{-3}\ \text{H} = 9{\cdot}0\ \text{mH}$$

$$L = L_A + L_B + 2M \qquad or \qquad L = L_A + L_B - 2M$$
$$= 4{\cdot}0 + 9{\cdot}0 + 2 \times 4{\cdot}8 \qquad = 4{\cdot}0 + 9{\cdot}0 - 2 \times 4{\cdot}8$$
$$= 22{\cdot}6\ \text{mH} \qquad = 3{\cdot}4\ \text{mH}$$

5.7 Energy storage in inductors

In the circuit shown in Fig. 5.12, the closing of the switch causes a current to flow. In course of time, the current achieves a steady value given by

$$I = \frac{V}{R}$$

During the period of change, while the current is increasing, an e.m.f. is induced in the inductor due to the change in the flux linkage. This e.m.f. opposes the current,

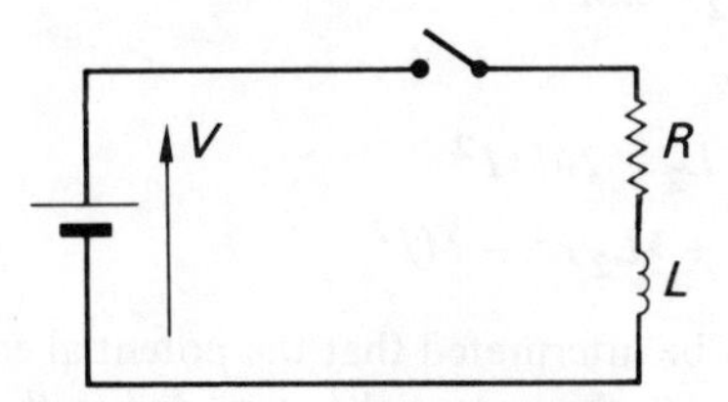

Fig. 5.12. Energy storage

which therefore is flowing against the e.m.f. As the current increases, the rate of change falls in value so in turn the back e.m.f. falls thus permitting further increase in current.

Potential energy is stored by a magnetic field so that when a current grows to a steady value, a definite amount of potential energy is stored in the magnetic field. The power dissipated into the magnetic field at any instant is given by ei. The e.m.f. will equal the applied voltage unless there is a further volt drop due to resistance in the coil of the inductor. For simplicity, assume that the current rises steadily from zero to a final value I in a time t.

$$\frac{\Delta I}{\Delta t} = \frac{I}{t}$$

$$E = L.\frac{\Delta I}{\Delta t} = \frac{LI}{t}$$

The mean current during the changing period is $\frac{1}{2}I$ since the current increases at a steady rate. Thus the energy supplied to the field W_f is given by

$$W_f = E.\tfrac{1}{2}I.t$$
$$= \tfrac{1}{2}LI^2 \qquad (5.16)$$

This proof requires the assumption that the current increases steadily. A more accurate proof is to integrate the power over the time involved as follows:

$$ei = L \cdot \frac{\mathrm{d}i}{\mathrm{d}t} \cdot i$$

$$\frac{\mathrm{d}W_f}{\mathrm{d}t} = Li . \frac{\mathrm{d}i}{\mathrm{d}t}$$

$$W_f = \int_0^I Li . \mathrm{d}i$$

$$= \tfrac{1}{2}LI^2 \qquad (5.16.1)$$

The use of integration may not be acceptable to the reader at this stage but if the answer is differentiated, the derivation may be 'proved'.

If two inductors are connected in series and can act mutually then the potential energy stored is again $\frac{1}{2}LI^2$ where L is the effective inductance of the system. In the case of the series-opposed coils:

$$L = L_1 + L_2 - 2M$$

$$W_f = \tfrac{1}{2}LI^2$$

$$= \tfrac{1}{2}(L_1 + L_2 - 2M)I^2$$

$$= \tfrac{1}{2}L_1 I^2 + \tfrac{1}{2}L_2 I^2 - MI^2 \qquad (5.18)$$

The negative sign may be interpreted that the potential energy due to the mutual inductance is in oppostion to the potential energy due to the self inductance. Similarly the expression for series-aiding coils:

$$W_f = \tfrac{1}{2}L_1 I^2 + \tfrac{1}{2}L_2 I^2 + MI^2 \qquad (5.19)$$

This paragraph commenced by considering the effect of suddenly applying a direct potential difference to a coil. It remains to consider the effect of suddenly removing this potential difference. The circuit remains that shown in Fig. 5.8.

When the switch is opened, the current falls to zero. This means that there will be a rapid collapse of the magnetic field in the coil and hence there will be a rapid change

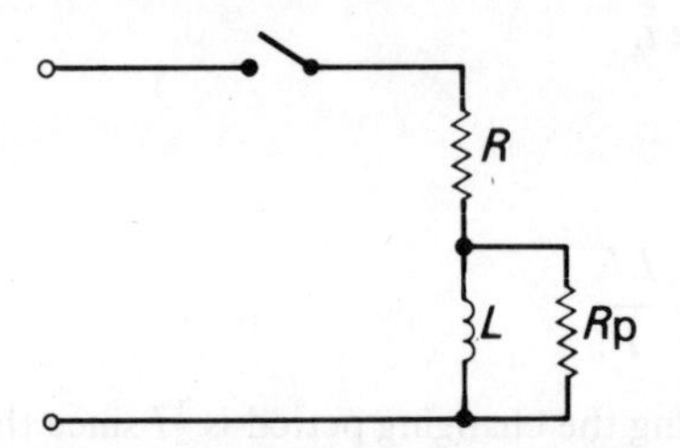

Fig. 5.13. Dissipation of stored energy

in the flux linkages. This effect produces a large self-induced e.m.f., which by Lenz's Law, tends to maintain the current flow. The induced e.m.f. is usually relatively large compared with the circuit voltages and, since the circuit is broken, it appears across the switch. Such an e.m.f. is sufficiently high to cause a breakdown of the air and hence cause a spark to jump across the gap between the switch contacts. This can be avoided to some extent by connecting a resistor in parallel with the inductor as shown in Fig. 5.13. The parallel resistor permits the self-induced e.m.f. to create a current round the closed loop thus dissipating the energy in the magnetic field.

5.8 Time constant

The growth of current in the circuit shown in Fig. 5.12 is shown graphically to a base of time in Fig. 5.14.

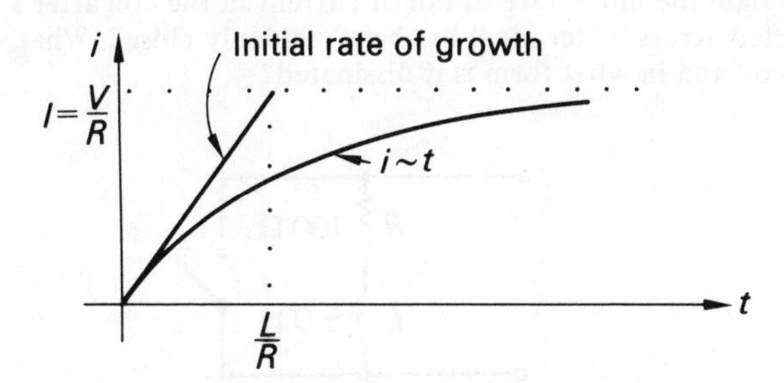

Fig. 5.14. Growth of current in inductive circuit

It can be shown that this curve rises exponentially towards a value given by

$$i = \frac{V}{R}$$

Such a curve takes the mathematical form

$$i = \frac{V}{R}\left(1 - e^{-\frac{R}{L}t}\right) \tag{5.20}$$

where e is the natural logarithmic base and is 2·718.

Whilst it is not generally necessary to be able to apply this relation in calculations at this introductory stage, the relation nevertheless merits some attention. In particular, consider the index of e. L/R is termed the time constant of the circuit. Let T be the time constant.

$$T = \frac{L}{R} \tag{5.21}$$

The time constant is the time required for the voltage and current to reach their steady-state values if they continued to change at their initial rates of decay and growth. Decay in this context refers to the decrease in induced e.m.f. The larger the time constant then the longer does the current take to reach its final steady-state value. The converse action takes place when the energy is released by an inductor.

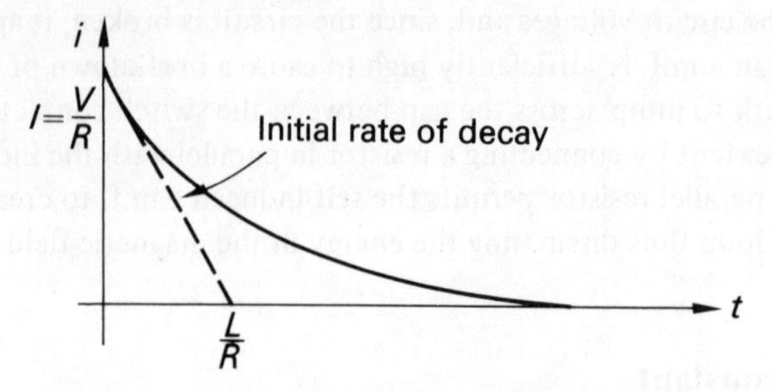

Fig. 5.15. Decay of current in inductive circuit

Example 5.9 A coil of inductance 5·0 H and resistance 100 Ω carries a steady current of 2·0 A. Calculate the initial rate of fall of current in the coil after a short-circuiting switch connected across its terminal has been suddenly closed. What was the energy stored in the coil and in what form is it dissipated?

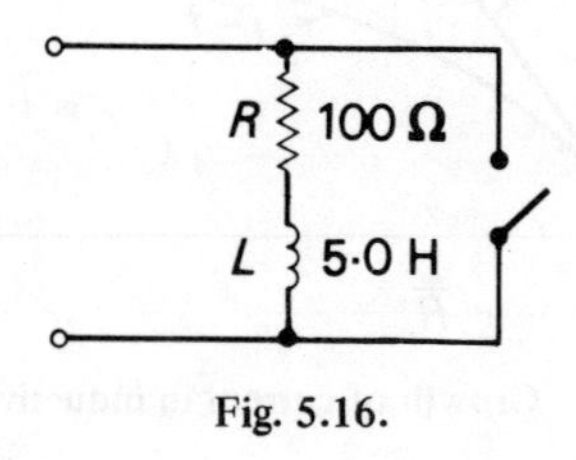

Fig. 5.16.

$$v_R = iR = 2 \times 100 = 200 \text{ V}$$

$$V = v_R + L \cdot \frac{di}{dt}$$

$$0 = 200 + 5 \times \frac{di}{dt}$$

$$\frac{di}{dt} = \underline{-40 \text{ A/s}}$$

$$W_f = \tfrac{1}{2}Li^2 = \tfrac{1}{2} \times 5 \times 2^2 = \underline{10 \text{ J}}$$

The energy is dissipated in the form of heat.

Example 5.10 A coil has a time constant of 1·0 s and an inductance of 10·0 H. What is the value of the current 0·1 s after switching on to a steady potential difference of 100 V? Find also the time taken for the current to reach half its steady-state value.

$$T = 1 = \frac{L}{R}$$

$$R = \frac{10}{1} = 10 \ \Omega$$

$$i = \frac{V}{R}\left(1 - e^{\frac{-Rt}{L}}\right) = \frac{100}{10}(1 - e^{-0 \cdot 1}) = \underline{0 \cdot 952 \text{ A}}$$

When the current is half its steady-state value:

$$i = \frac{V}{2R} = \frac{V}{R}(1 - e^{-t})$$

$$e^{-t} = 0{\cdot}5$$

$$\underline{t = 0{\cdot}697\ \text{s}}$$

5.9 Rotating coil in a uniform field

Consider the case of a rectangular coil rotating in a uniform field with a constant angular velocity. This arrangement constitutes the most simple electrodynamic machine. It is shown in Fig. 5.17.

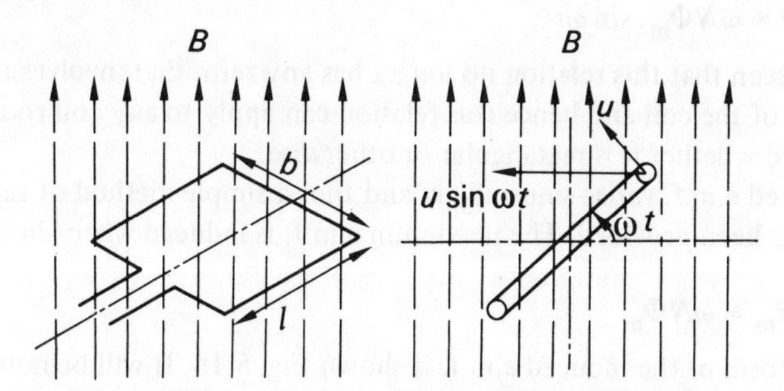

Fig. 5.17. E.M.F. induced in a rotating coil in a uniform field

The induced e.m.f.'s are motional. If the axis of the coil is normal to the field, no e.m.f.'s are induced in the coil ends hence only the coil sides need be considered. Let the coil sides be of length l and let the peripheral velocity of these sides be u. If the coil rotates with a constant angular velocity ω, then, at any instant t, the coil will have rotated through an angle ωt. For the arrangement shown in Fig. 5.17, the velocity of the coil sides at this instant perpendicular to the lines of force is

$$u\,.\sin\omega t$$

It follows from relation (4.8) that the e.m.f. induced in one side is given by

$$Blu\,.\sin\omega t$$

A similar e.m.f. is induced in the other side of the coil, hence the e.m.f. induced in a 1-turn coil is

$$e = 2Blu\,.\sin\omega t$$

The peripheral velocity can be related to the angular velocity since

$$n = \frac{\omega}{2\pi}$$

Because the diameter of the coil is b, then

$$u = \pi b n$$

$$= \frac{b\omega}{2}$$

Substituting this in the above expression for the e.m.f. induced in a 1-turn coil:

$$e = 2Bl \cdot \frac{b\omega}{2} \cdot \sin \omega t$$

$$= B \, . \, bl \, . \, \omega \, . \sin \omega t$$

The product bl represents the area of the coil. Also B is the flux density of the field. Therefore $B \, . \, bl$ must represent the flux linking the coil when the plane of the coil is normal to the field. This is the maximum flux which can link the coil and is designated Φ_m. Thus

$$e = \omega \Phi_m \, . \sin \omega t$$

Finally if the coil has N turns then the induced e.m.f. is given by

$$e = \omega N \Phi_m \, . \sin \omega t \qquad (5.22)$$

It can be seen that this relation no longer has any term that involves the physical arrangement of the coil and hence this relation can apply to any coil rotating in a magnetic field whether it is rectangular or otherwise.

The induced e.m.f. varies sinusoidally and thus a simple method of generating a.c. quantities has been provided. The maximum e.m.f. is induced when $\sin \omega t$ is unity hence

$$E_m = \omega N \Phi_m \qquad (5.22.1)$$

The waveform of the induced e.m.f. is shown Fig. 5.18. It will be noted that the expression is in terms of the maximum flux since it is the related maximum flux density that is the limiting factor of the magnetic system.

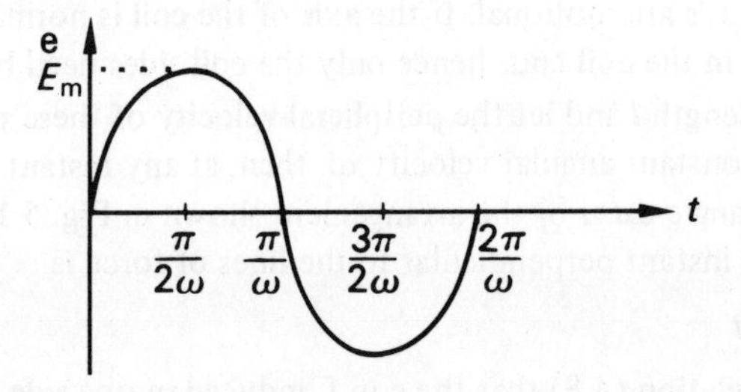

Fig. 5.18. Waveform of e.m.f. induced in rotating coil

An alternative derivation of relation (5.22) is worth noting. The flux linking the coil can be expressed as

$$\phi = \Phi_m \, . \cos \omega t$$

$$e = \frac{d\psi}{dt}$$

$$= N \cdot \frac{d\phi}{dt}$$

$$= N \cdot \frac{d(\Phi_m \, . \cos \omega t)}{dt}$$

$$= \omega N \Phi_m \, . \sin \omega t$$

Example 5.11 A circular coil of diameter 200 mm and of 1 000 turns is rotated at a uniform speed of 3 000 rev/min in a uniform field of density 0·3 T. Derive an expression for the e.m.f. induced in the coil.

$$A = \pi r^2 = \pi \times \left(\frac{200}{2}\right)^2 \times 10^{-6} = 3{\cdot}14 \times 10^{-2}\ m^2$$

$$\Phi_m = BA = 0{\cdot}3 \times 3{\cdot}14 \times 10^{-2} = 9{\cdot}42 \times 10^{-3}\ \text{Wb}$$

$$n = \frac{3000}{60} = 50\ \text{rev/s}$$

$$\omega = 2\pi n = 2\pi 50 = 314\ \text{rad/s}$$

$$e = \omega N\Phi_m . \sin \omega t = 314 \times 1000 \times 9{\cdot}42 \times 10^{-3} \sin 314t$$

$$= \underline{2960 . \sin 314t\ \text{V}}$$

5.10 Hysteresis loop

In para. 4.8, the magnetisation curve was obtained by considering the application of a magnetic field strength to a ferromagnetic material. The process commenced with an unmagnetised specimen and the direction of the applied magnetic field strength was of no consequence.

Consider now the action of reversing the direction of the applied magnetic field strength provided the specimen has been sufficiently magnetised to cause some of the domains to turn, say to the point 1 in Fig. 5.19. When the magnetic field strength is

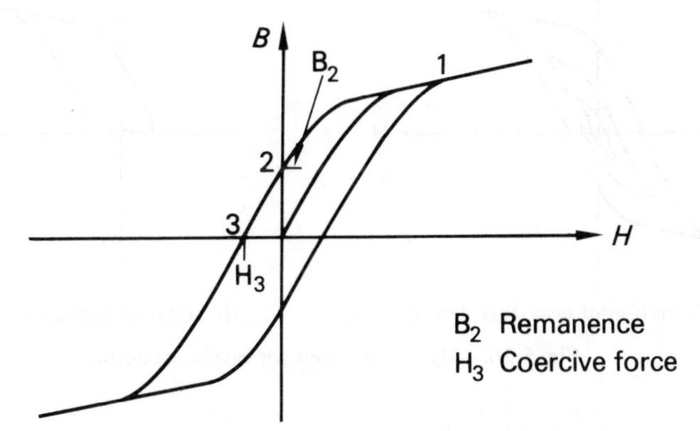

Fig. 5.19. Magnetisation curve and hysteresis loop

reduced to zero, a flux still remains due to the reorientation of the domains. The reduction in flux is due to a considerable extent to the reduction in the size of the domains. The resulting flux density B_2 is termed the remanence.

If a force of the opposite polarity to that formerly used is applied to the specimen, the domains will progressively turn until eventually there is no effective flux. The magnetic field strength H_3 to give this situation is termed the coercive force. Further increase of the magnetic field strength will produce a flux acting in the opposite direction to that formerly noted. Eventually saturation will be achieved in this opposite sense.

If this sequence of events is reversed, a similar magnetic effect is produced and eventually the original point of saturation 1 is achieved. The resulting characteristic therefore forms a loop termed the hysteresis loop. It follows that when alternating current gives rise to a varying flux in a ferromagnetic material, the corresponding instantaneous values of flux density and magnetic field strength are shown by the hysteresis loop.

The hysteresis loop shown in Fig. 5.19 is only one of many possible loops. There are three factors which affect the shape and size:

The material. This may magnetise easily in which case the loop will appear narrow whilst conversely if the material does not magnetise easily, the loop will be wide. Also different materials will saturate at different values of flux density thus affecting the height of the loop.

The maximum value of flux density. This can be shown most easily in a diagram. Figure 5.20(*a*) shows how the loop area increases as the alternating magnetic field strength has progressively greater peak values.

The initial magnetic state of the specimen. Figure 5.20(*b*) shows a hysteresis loop in which the specimen has been saturated, the flux density has been reduced to zero and finally returned to the saturated condition.

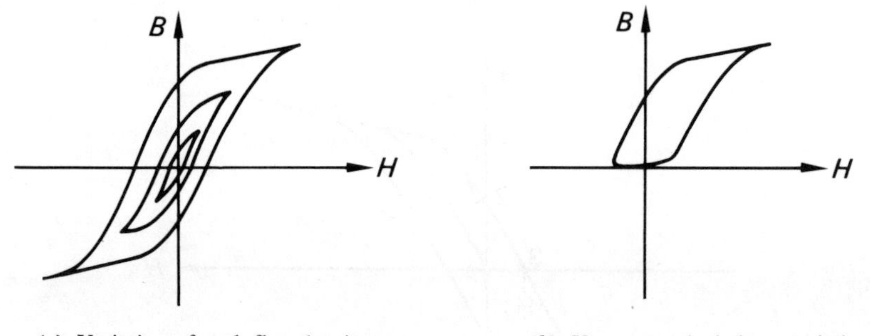

(*a*) Variation of peak flux density (*b*) Unsymmetrical characteristic

Fig. 5.20. Hysteresis loops for varying conditions

Finally, it should be noted that different materials can be compared if the loops are drawn to the same scale and each loop is taken to the saturated condition. One feature that is of particular importance is the area of the loop. The area is a measure of the work that is done in taking the material through a cycle of magnetisation. This work results in a loss of energy termed the hysteresis loss; this loss becomes apparent in the form of heat.

5.11 Hysteresis loss

Consider a magnetic circuit, e.g. a toroid of large diameter, of length l, uniform cross-sectional area A and linking a uniform winding of N turns. At some given instant let the current be i. The corresponding magnetic field strength is given by

$$H = \frac{Ni}{l}$$

The instantaneous conditions are shown in Fig. 5.21.

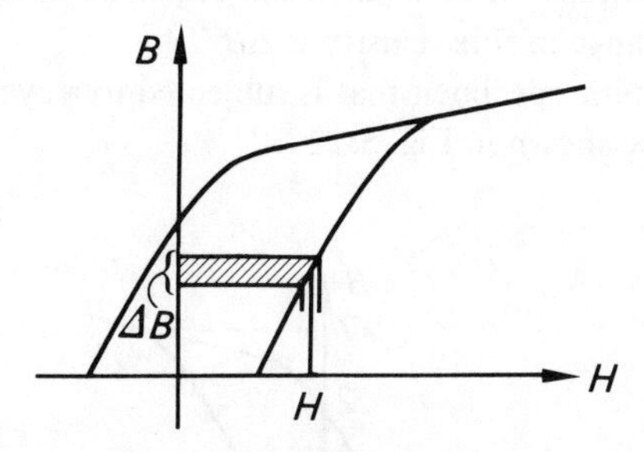

Fig. 5.21. Hysteresis loop

Suppose the current to increase by ΔI in a time Δt. This will cause an increase in the flux density ΔB and hence an increase in flux $\Delta\Phi$ $(= A \,.\, \Delta B)$. This causes an e.m.f. to be induced in the winding:

$$E = \frac{\Delta\Psi}{\Delta t}$$

$$= N \cdot \frac{\Delta\Phi}{\Delta t}$$

$$= AN \cdot \frac{\Delta B}{\Delta t}$$

By Lenz's Law, this e.m.f. opposes the current i so that energy ΔW is supplied to the system. Assuming the current i to be reasonably unchanged, i.e. ΔI is small, then approximately:

$$\Delta W = Ei \,.\, \Delta t$$

$$= AN \cdot \frac{\Delta B}{\Delta t} \cdot i \,.\, \Delta t$$

$$= A \,.\, N \,.\, \Delta B \cdot \frac{Hl}{N}$$

$$= H \,.\, \Delta B \times Al$$

$$= H \,.\, \Delta B \times V \qquad (5.23)$$

where $V = Al$ and is the volume of the magnetic specimen.

Provided that the increase in magnetic field strength ΔH is small, then $H \,.\, \Delta B$ is represented by the shaded area in the diagram. Thus for unit volume of the ferromagnetic specimen, the energy absorbed is $H \,.\, \Delta B$ measured in joules.

Relation (5.23) was derived on the basis that the magnetic field strength and the flux density were uniform for all parts in the specimen. If ΔB had represented a decrease in flux density due to a decrease in the current i, $H \,.\, \Delta B$ would be the energy transferred to the electric circuit from a unit volume of the ferromagnetic material. However, if the magnetic field strength and the flux density are not uniform throughout the specimen, $H \,.\, \Delta B$ represents the change in energy density–in joules per cubic metre–absorbed by or released from a particular region at which the magnetic field strength is H and the change in flux density is ΔB.

Consider a ferromagnetic specimen that is subjected to a cyclic magnetisation characteristic of the type shown in Fig. 5.22.

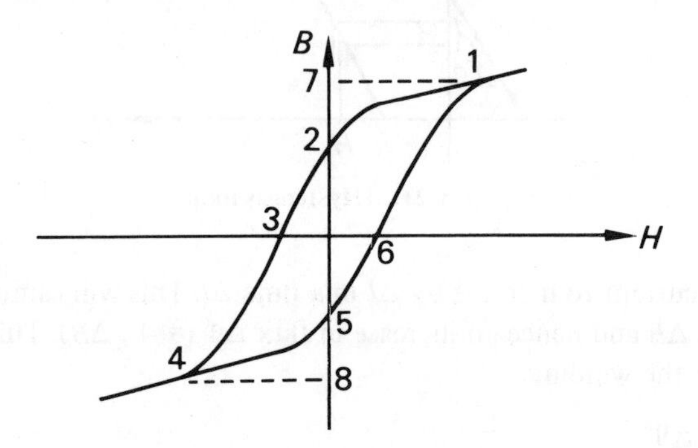

Fig. 5.22. Hysteresis loop showing energy losses

If the magnetic field strength is initially zero and the flux density has a corresponding value B_5, let the magnetic field strength be gradually increased in small positive steps such that the flux density is increased from B_5 to B_1. The steps should be such that each produces a change in flux density B, of which the step shown in Fig. 5.16 is typical. During such a step, $H \,.\, \Delta B$ joules of energy are supplied to unit volume of the specimen. It follows that, by summating the effects of all the steps, the energy absorbed by unit volume of the specimen, for a change of flux density from B_5 to B_1 is represented by the area 561725.

If the magnetic field strength is reduced from H_1 to zero again, it can be argued in a similar manner that the energy released by unit volume of the specimen is represented by the area 1721. Thus the energy absorbed exceeds the energy released. The net energy absorbed is represented by the area 56125. If the complete cycle of magnetisation is carried out, a similar sequence of events occurs in the opposite direction and so the energy absorbed over one complete cycle is represented by the area 5612345. Hence the net energy absorbed over one complete cycle is represented by the area of the hysteresis loop.

The area of the loop is measured as a product of flux density and magnetic field strength and is normally achieved by measuring the area in square metres and multiplying this by the appropriate scale factors–teslas per metre and ampere turns per metre per metre. The energy absorbed is therefore measured in joules per cubic metre per cycle.

Assuming that conditions are uniform throughout the specimen, if the energy absorbed is multiplied by the volume of the specimen, the total energy absorbtion per cycle of magnetisation W_h is found, hence

$$W_h = V \times (\text{energy density represented by area of loop}) \tag{5.24}$$

If the magnetising current is supplied from an a.c. source, the energy absorbtion per cycle can be multiplied by the source frequency. This product gives the hysteresis power loss P_h.

$$P_h = fW_h \tag{5.25}$$

The energy absorbed due to hysteresis is converted to thermal energy. The ferromagnetic material therefore experiences a rise in temperature.

It can be shown by experiment that the hysteresis power loss is proportional to the maximum flux density to which the specimen is being subjected. The relation is of the form

$$P_h \propto B_m{}^N$$

where N is a constant of value 1·6–2·5 depending on the material used. N is called the Steinmetz Index. Since the hysteresis power loss is also proportional to the frequency and to the volume, it follows that

$$P_h = k_h f B_m{}^N V \tag{5.26}$$

The value of the constant k_h also depends on the material; for instance, k_h for silicon steel is 100–200.

One important factor in the choice of ferromagnetic material for a particular application is the area and shape of the hysteresis loop of the material. Three points should be noted:

1. It is generally desirable to have high flux density resulting from a small magnetic field strength. This minimises the volume of material required to produce a desired value of flux and also minimises the current required to create the flux.

2. In consequence of (1), the hysteresis loop will be narrow and thus the loop area will be small. For machines experiencing cycling of the magnetisation, the loop area should be small to minimise the hysteresis power loss.

3. It will be seen in para. 5.15 that, for permanent magnets, a large loop is desirable to give both high remanence and high coercive force.

Example 5.12 A magnetic circuit core is made of silicon steel and has a volume of 1 000 000 mm³. Using the hysteresis loop shown in Fig. 5.23, calculate the hysteresis power loss when the flux is alternating at 50 Hz. (SANCAD)

$P_h = Vf\,(\text{area of loop})$

$= 1000000 \times 10^{-9} \times 50 \times (\text{area of loop})$

area of loop $=$ area in square units x scale factors

$= 7{\cdot}25 \times 4 \times 10$

$= 290\ \text{J/m}^3/\text{c}$

$P_h = 1000 \times 10^{-6} \times 50 \times 290 = \underline{14{\cdot}5\,\text{W}}$

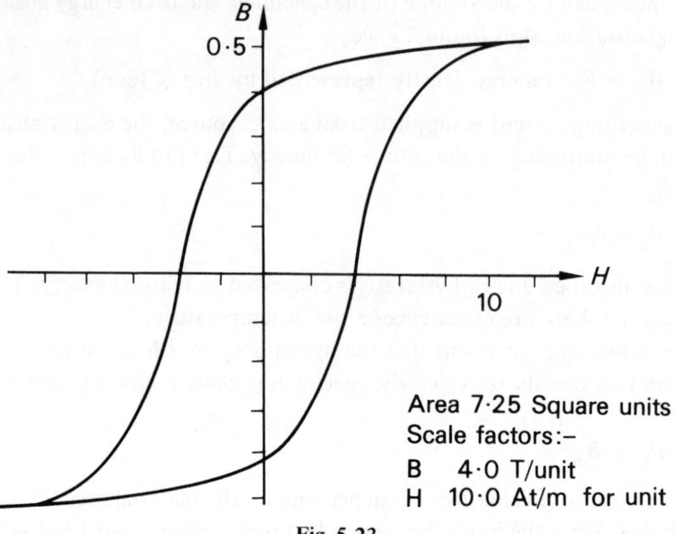

Fig. 5.23

Example 5.13 A sample of silicon steel has a hysteresis coefficient of 100 and a corresponding Steinmetz Index of 1·6. Calculate the hysteresis power loss in 1 000 000 mm³ when the flux is alternating at 50 Hz, such that the maximum flux density is 2·0 T

$$
\begin{aligned}
P_h &= k_h f V B_m{}^N \\
&= 100 \times 50 \times 1000000 \times 10^{-9} \times 2{\cdot}0^{1{\cdot}6} \\
&= 5 \times 2{\cdot}0^{1{\cdot}6} \\
&= 5 \times 3{\cdot}03 \\
&= \underline{15{\cdot}2\ \text{W}}
\end{aligned}
$$

This compares with the answer to Example 5.12.

5.12 Eddy-current loss

If a loop of conducting material is linked by a varying flux, an e.m.f. is induced in the loop and a circulating current will flow round the loop. Such an arrangement is shown in Fig. 5.24a. The current in the loop is termed an eddy current. The eddy current flows round a path of one turn which is effectively a short circuit.

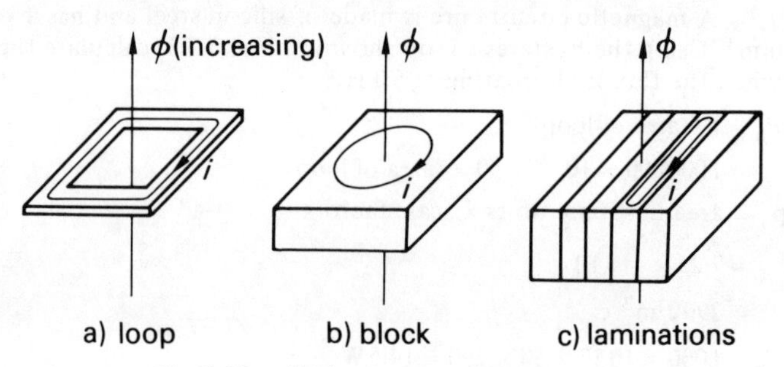

Fig. 5.24. Eddy-current arrangements

Had the loop been replaced by a block of conducting material, a similar system of eddy currents would result in the block. In either case, the eddy currents give rise to a power loss in the resistance of the eddy current path. The loss is called the eddy current power loss.

From Lenz's Law, the eddy currents flow in a direction such that their magnetic effects oppose the flux changes. At high frequencies, eddy currents may radically alter the flux distribution in the loop or block. However, for thin sheets at low frequencies, 50 Hz for example, there is little alteration to the flux: generally this effect may be neglected at low frequencies.

Suppose the flux to be varying sinusoidally in a specimen of cross-sectional area A normal to the direction of the field.

$$\phi = \Phi_m \sin \omega t$$
$$= AB_m \sin \omega t$$

The loop e.m.f. is proportional to the rate of change of the flux.

$$e \propto \frac{d(AB_m \sin \omega t)}{dt}$$
$$\propto \omega AB_m \cos \omega t$$

The r.m.s. value of this e.m.f is given by

$$E \propto \omega AB_m$$

$$I \propto \frac{\omega AB_m}{R}$$

$$\propto \frac{fAB_m}{R}$$

where R is the effective resistance to the eddy currents. The eddy-current power loss P_e is therefore given by

$$P_e = I^2 R$$

$$\propto \frac{A^2 B_m{}^2 f^2}{R}$$

$$\propto \frac{A^2 B_m{}^2 f^2}{\rho} \qquad \text{since } R = \rho \frac{l}{A}$$

$$P_e = \frac{k_e A^2 B_m{}^2 f^2}{\rho} \tag{5.27}$$

The eddy-current power loss is therefore proportional to the square of the cross-sectional area normal to the direction of the field, the square of the maximum flux density, the square of the frequency and is inversely proportional to the resistivity of the material form which the loop or block is made. To minimise the loss only the cross-sectional area and the resistivity can be varied.

The net area of the core of a magnetic circuit cannot be reduced since this is determined by the required flux and the maximum permissible flux density, but the area can be divided into smaller sections. This is achieved by making the core of a number of thin sheets, called laminations—shown in Fig. 5.24c—which are lightly insulated from one another, e.g. by an oxide. This reduces the area of each section and hence the induced e.m.f. It also increases the resistance of the eddy current paths since the area through which the currents can pass is smaller. Both effects combine to reduce the current and hence the power loss. The thickness of the laminations is therefore a factor in determining the eddy-current power loss. Further reductions can be obtained by using a material of high resistivity.

Lamination is only one possible method of splitting up the eddy currents. The thickness of the ferromagnetic components can be reduced in any of the following ways:

1. by the use of laminations already discussed,
2. by making the magnetic circuit from wire (unusual),
3. by using a ferromagnetic dust core,
4. by making the core from a suitable ferrite material.

The following precautions should therefore be taken to minimise the core loss in electromagnetic system:

(*a*) The flux density should be kept as economically low as possible.

(*b*) The core should be made from a material with a relatively small hysteresis loop.

(*c*) The core material should have high resistivity.

(*d*) The magnetic circuit should be laminated or similarly divided.

Finally it should be noted that eddy currents have many useful applications and need not be considered as imperfections of an electromagnetic system. Energy meters and eddy-current brakes depend on eddy currents for their principles of operation. Another application in moving-coil instrument movements is described in para. 17.4.

5.13 Magnetic circuits

In previous paragraphs, the magnetic circuits considered were simple ones, e.g. the toroid, in which the magnetic cores were uniform throughout. Many practical circuits are more complex and it is necessary to be able to analyse them with a view to optimising the use of core material by utilising the highest possible flux density.

First of all, it is of interest to compare a simple circuit with an equivalent electric circuit. This is illustrated by Fig. 5.25.

The magnetomotive force has already been defined as the work done in moving a unit magnetic pole once round the magnetic circuit. Thus the m.m.f. is analogous to the e.m.f. of the electric circuit, in that both are the source of energy to the circuit.

The e.m.f. E causes a current to flow given by

$$E = IR$$

The m.m.f. F causes a flux given by

$$F = \Phi S$$

It can be seen that the flux is analagous to current, but it would be incorrect to think of flux as a continuous flow.

Finally:

$$R = \rho \frac{l}{A}$$

and

$$S = \frac{1}{\mu_0 \mu_r} \cdot \frac{l}{A}$$

which are again similar.

If the electric circuit were complex, i.e. it were a series and/or parallel circuit, it would be necessary to consider the potential differences within the circuit. This would permit the introduction of Kirchhoff's Laws to the circuit analysis in their entirity.

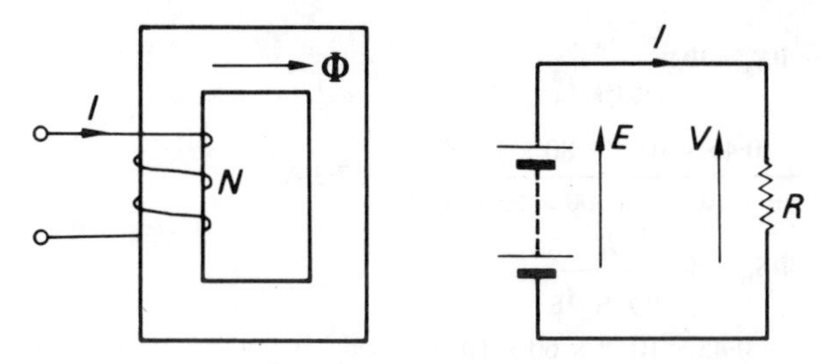

Fig. 5.25. Comparison between magnetic and electric circuits

Magnetic circuits require an equivalent to potential difference if only parts of the circuit are to be considered. This is the magnetic potential difference, which facilitates the application of Kirchhoff's Laws to magnetic circuit analysis. Hence the flux 'arriving' at any junction in a magnetic circuit is equal to that 'leaving' the circuit junction, and the sum of the m.m.f.'s around any closed loop of a magnetic circuit is equal to the sum of the magnetic potential differences.

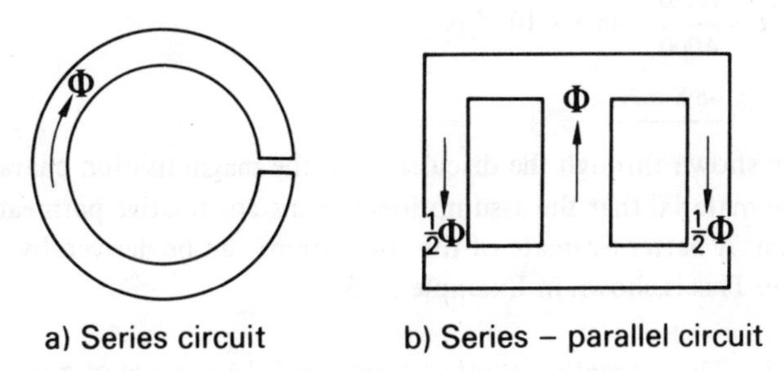

Fig. 5.26. Magnetic circuits

The reluctances of the various sections of a magnetic circuit can be manipulated in the same manner as resistances in an electric circuit. Figure 5.26 shows a series circuit and a series-parallel circuit.

The analysis of magnetic circuits is best illustrated by a series of worked examples. The simplest analysis is that of a series magnetic circuit. An approximate estimate of the circuit conditions can be obtained from the assumption that the relative permeability is constant.

Example 5.14 A magnetic circuit comprises three parts in series, each of uniform cross-sectional area (c.s.a.). They are:

(*a*) a length of 80 mm and c.s.a. 50 mm^2,
(*b*) a length of 60 mm and c.s.a. 90 mm^2,
(*c*) an air-gap of length 0·5 mm and c.s.a. 150 mm^2.

A coil of 4 000 turns is wound on part (*b*), and the flux density in the air-gap is 0·3 T. Assuming that all the flux passes through the given circuit, and that the relative permeability μ_r is 1 300, estimate the coil current to produce such a flux density.

$$\Phi = B_c A_c = 0{\cdot}3 \times 1{\cdot}5 \times 10^{-4} = 0{\cdot}45 \times 10^{-4}\,\mathrm{T}$$

$$F_a = \Phi S_a = \Phi \,.\, \frac{l_a}{\mu_0 \mu_r A_a}$$

$$= \frac{0{\cdot}45 \times 10^{-4} \times 80 \times 10^{-3}}{4\pi \times 10^{-7} \times 1\,300 \times 50 \times 10^{-6}} = 57{\cdot}3\ \mathrm{At}$$

$$F_b = \Phi S_b = \Phi \,.\, \frac{l_b}{\mu_0 \mu_r A_b}$$

$$= \frac{0{\cdot}45 \times 10^{-4} \times 60 \times 10^{-3}}{4\pi \times 10^{-7} \times 1\,300 \times 90 \times 10^{-6}} = 18{\cdot}4\ \mathrm{At}$$

$$F_c = \Phi S_c = \Phi \cdot \frac{l_c}{\mu_0 \mu_r A_c}$$

$$= \frac{0{\cdot}45 \times 10^{-4} \times 0{\cdot}5 \times 10^{-3}}{4\pi \times 10^{-7} \times 1 \times 150 \times 10^{-6}} = 119{\cdot}3\ \mathrm{At}$$

$$F = F_a + F_b + F_c = 57{\cdot}3 + 18{\cdot}4 + 119{\cdot}3 = 195{\cdot}0\ \mathrm{At}$$

$$= IN$$

$$I = \frac{195{\cdot}0}{4000} = 48{\cdot}8 \times 10^{-3}\ \mathrm{A}$$

$$= \underline{48{\cdot}8\ \mathrm{mA}}$$

It has been shown through the discussion on the magnetisation characteristic of a ferromagnetic material that the assumption of constant relative permeability is an approximation. A better estimate of the coil current can be derived by application of the *B*/*H* curve. This is shown in Example 5.15.

Example 5.15 The magnetic circuit of Example 5.14 is made of a material which has the following magnetic characteristic:

Flux density	T	0·2	0·4	0·6	0·8	1·0	1·2
Magnetic field strength	At/m	100	210	340	500	800	1 500

Using this characteristic, estimate the current which would now be required to produce the flux density of 0·3 T in the air-gap.

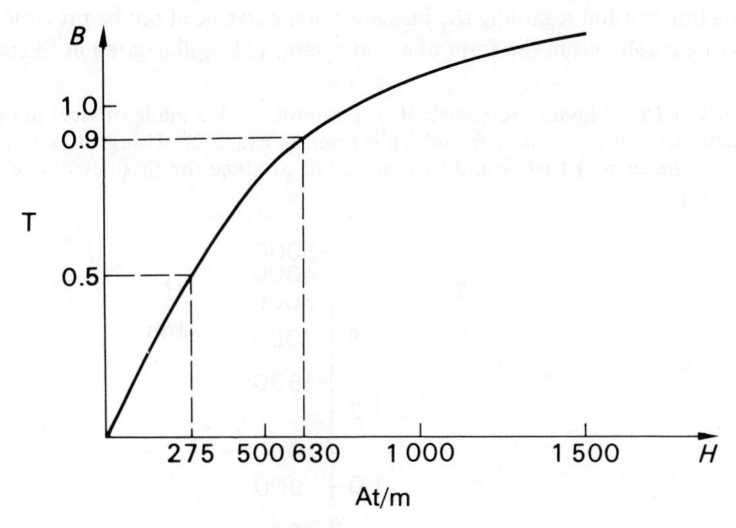

Fig. 5.27

Again $\Phi = B_c A_c = 0{\cdot}3 \times 150 \times 10^{-6} = 0{\cdot}45 \times 10^{-4}$ Wb

$$B_a = \frac{\Phi}{A_a} = \frac{0{\cdot}45 \times 10^{-4}}{50 \times 10^{-6}} = 0{\cdot}9 \text{ T}$$

From the magnetisation characteristic, the corresponding value of magnetic field strength can be obtained as

$$H_a = 630 \text{ At/m}$$

$$F_a = H_a l_a = 630 \times 80 \times 10^{-3} = 50{\cdot}4 \text{ At}$$

Also $B_b = \frac{\Phi}{A_b} = \frac{0{\cdot}45 \times 10^{-4}}{90 \times 10^{-6}} = 0{\cdot}5$ T

From the characteristic:

$$H_b = 275 \text{ At/m}$$

$$F_b = H_b l_b = 275 \times 60 \times 10^{-3} = 16{\cdot}5 \text{ At}$$

The magnetic potential difference across section c is estimated in the same manner that was used in Example 5.14. Thus

$$F_c = 119{\cdot}3 \text{ At}$$

$$F = F_a + F_b + F_c = 50{\cdot}4 + 16{\cdot}5 + 119{\cdot}3 = 186{\cdot}2 \text{ At}$$

$$= IN$$

$$I = \frac{F}{N}$$

$$= \frac{186{\cdot}2}{4000} = 46{\cdot}6 \times 10^{-3} \text{ A}$$

$$= \underline{46{\cdot}6 \text{ mA}}$$

The information regarding the magnetisation curve need not be presented in the form of a graph, but in the form of a nomogram. This is illustrated in Example 5.16.

Example 5.16 The magnetic circuit of Example 5.14 is made of steel plate, the magnetisation characteristic for which is given in Fig. 5.28. Using this characteristic, estimate the current that would be required to produce the flux density of 0·3 T in the air gap.

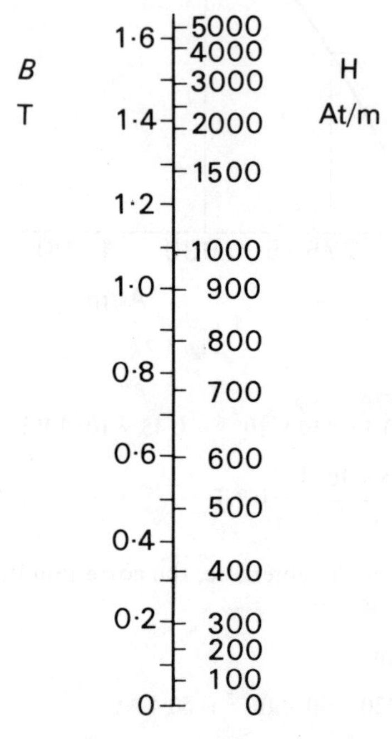

Fig. 5.28 (SANCAD)

It has been shown that $B_a = 0{\cdot}9$ T.
From the chart, $H_a = 810$ At/m
Hence $F_a = H_a l_a = 810 \times 80 \times 10^{-2} = 64{\cdot}8$ At

Similarly $B_b = 0{\cdot}5$ T

$H_b = 30{\cdot}6$ At/m

$F_b = H_b l_b = 510 \times 60 \times 10^{-3} = 30{\cdot}6$ At

Also $F_c = 119{\cdot}3$ At as before.

$$F = F_a + F_b + F_c = 64{\cdot}8 + 30{\cdot}6 + 119{\cdot}3$$

$$= 214{\cdot}7 \text{ At}$$

$$I = \frac{F}{N} = \frac{214{\cdot}7}{4000} = 53{\cdot}7 \times 10^{-3} \text{ A}$$

$$= 53{\cdot}7 \text{ mA}$$

Before leaving this consideration of series magnetic circuits, one further case merits attention–that of estimating the flux density due to a given magnetising current. In this case, the magnetic potential differences across the various sections of the magnetic circuit are not known. Due to the magnetisation characteristic the values of reluctance are not directly obtainable. Thus a solution of the circuit cannot be obtained in the manner previously used in the case of electric circuits. However, a solution can be estimated using a graph. This is illustrated in Example 5.17.

Example 5.17 The magnetic circuit of Example 5.16 is again made of steel plate, for which the characteristic remains unchanged, and a current of 50 mA is passed through the coil. Estimate the flux density in the air-gap.

$$F = NI = 4000 \times 50 \times 10^{-3} = 200 \text{ At}$$

Since it is not known how the m.m.f. is made up by the p.d.'s around the circuit, a preliminary assumption will be made that half of the m.m.f. ampere-turns are required by the air-gap.

$$\Phi = \frac{F_c}{S_c} = \frac{F_c \mu_0 \mu_r A_c}{l_c} = \frac{100 \times 4\pi \times 10^{-7} \times 1 \times 150 \times 10^{-6}}{0{\cdot}5 \times 10^{-3}}$$

$$= 0{\cdot}378 \times 10^{-4} \text{ Wb}$$

This is only an estimate which can be taken as $0{\cdot}4 \times 10^{-4}$ Wb. To produce the required graphical solution, the various magnetic p.d.'s can be calculated for flux values of $0{\cdot}35 \times 10^{-4}$ Wb, $0{\cdot}4 \times 10^{-4}$ Wb and $0{\cdot}45 \times 10^{-4}$ Wb. This is summarised below.

Φ (Wb)	$0{\cdot}35 \times 10^{-4}$	$0{\cdot}40 \times 10^{-4}$	$0{\cdot}45 \times 10^{-4}$
B_a (T)	0·7	0·8	0·9
H_a (At/m)	660	720	810
F_a (At)	52·9	57·6	64·8
B_b (T)	0·39	0·45	0·50
H_b (At/m)	430	480	510
F_b (At)	25·8	28·8	30·6
F_c (At)	92·8	106·0	119·3
F (At)	171·5	192·4	214·7

A graph is now drawn (Fig. 5.29) of the flux against m.m.f.

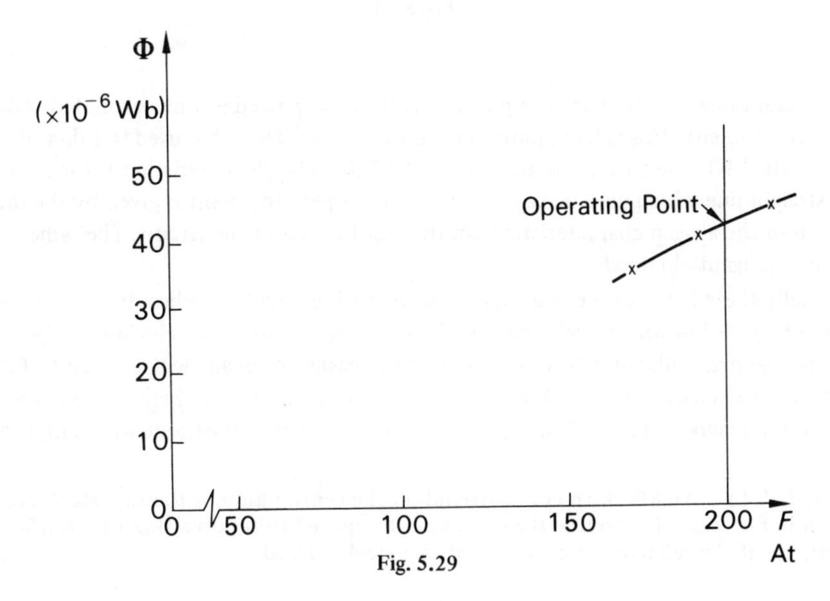

Fig. 5.29

From the graph, the flux corresponding to an m.m.f. of 200 At is found to be $0{\cdot}43 \times 10^{-4}$ Wb. This is therefore the total flux in the circuit.

$$B_c = \frac{\Phi}{A} = \frac{0{\cdot}43 \times 10^{-4}}{1{\cdot}5 \times 10^{-4}} = \underline{0{\cdot}277\ \text{T}}$$

This solution depended on an initial guess as to the order of the flux value. Had it not been as close to the correct value as proved to be the case, the graph would not have provided a solution. The graph would, however, have shown what extra values would require to be plotted.

Where an air-gap is involved, the solution may be simplified. Only the sum of the magnetic p.d.'s need be calculated in the same way that was used above. However,

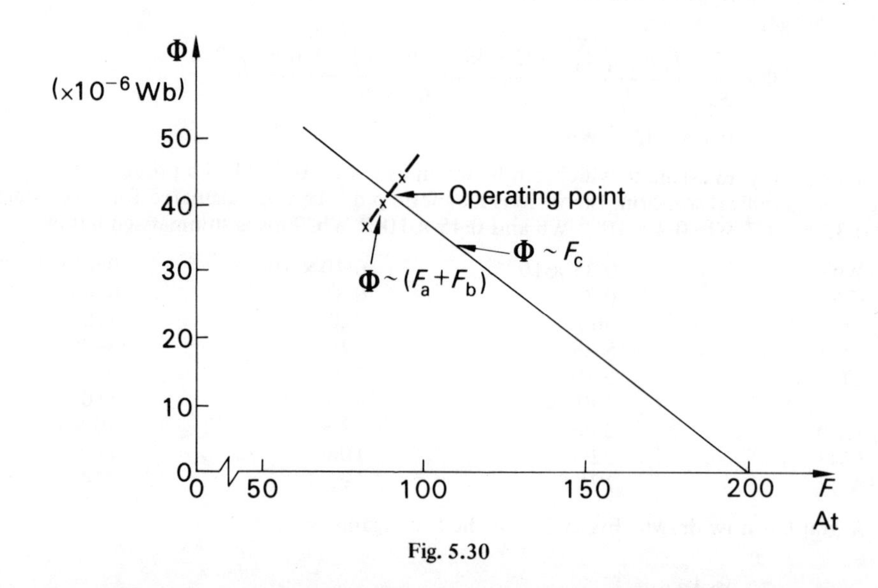

Fig. 5.30

the several values of the magnetic p.d. across the air-gap need not be calculated—one value is sufficient. This is back-plotted as shown above. The value used is a flux of $0{\cdot}45 \times 10^{-4}$ Wb due to a magnetic p.d. of 119·3 At. The characteristic for the air-gap is a straight line which may be easily plotted. The operating point is given by the intersection of the air-gap characteristic with that for the rest of the circuit. The same solution is again obtained.

Finally there is the case of magnetic circuits with parallel branches. If the relative permeability is known, the reluctances of the various sections are calculated. They can then be manipulated in a similar manner to resistances in an electric circuit. If the magnetisation characteristic is known instead then the m.m.f. can be balanced against the section magnetic p.d.'s. The first method of approach is illustrated in Example 5.18.

Example 5.18 An 800-turn coil is wound on the central limb of the cast-steel frame shown in Fig. 5.31. Determine the coil current required to set up a flux of 1·0 mWb in the air-gap if the relative permeability of cast-steel is 1 000.

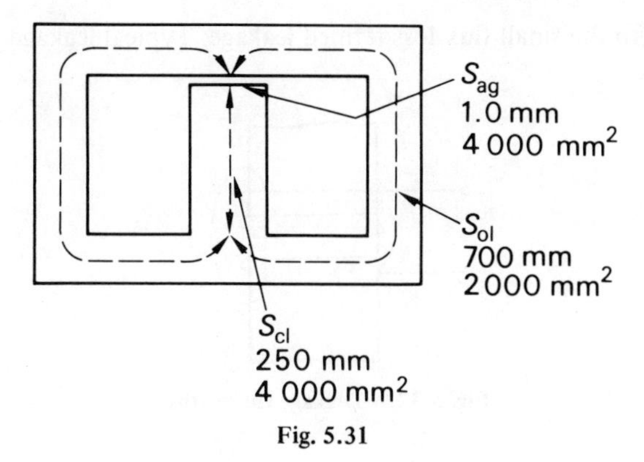

Fig. 5.31

For the centre limb:

$$S_{cl} = \frac{l_{cl}}{\mu_0 \mu_r A_{cl}} = \frac{250 \times 10^{-3}}{4\pi \times 10^{-7} \times 1000 \times 4000 \times 10^{-6}}$$
$$= 4{\cdot}97 \times 10^4/\text{H}$$

For one outer limb:

$$S_{ol} = \frac{l_{ol}}{\mu_0 \mu_r A_{ol}} = \frac{700 \times 10^{-3}}{4\pi \times 10^{-7} \times 1000 \times 2000 \times 10^{-6}}$$
$$= 27{\cdot}85 \times 10^4\ /\text{H}$$

Since the two outer limbs are similar and are in parallel, the total effective reluctance is halved in the same way that equal resistances in parallel have an effective resistance that is half the value of either constituent resistance. Hence for the outer limbs, the effective reluctance is

$$S_p = \tfrac{1}{2}S_{ol} = 13{\cdot}92 \times 10^4\ /\text{H}$$

For the air-gap:

$$S_{ag} = \frac{l_{ag}}{\mu_0 \mu_r A_{ag}} = \frac{1 \times 10^{-3}}{4\pi \times 10^{-7} \times 1 \times 4000 \times 10^{-6}}$$
$$= 19{\cdot}9 \times 10^4\ /\text{H}$$

The total reluctance is there given by

$$S = S_{cl} + S_p + S_{ag} = (4{\cdot}97 + 13{\cdot}92 + 19{\cdot}9) \times 10^4$$
$$= 38{\cdot}79 \times 10^4\ /\text{H}$$
$$F = \Phi S = 1 \times 10^{-3} \times 38{\cdot}79 \times 10^4 = 387{\cdot}9\ \text{At}$$
$$I = \frac{F}{N} = \frac{387{\cdot}9}{800}$$
$$= \underline{0{\cdot}485\ \text{A}}$$

5.14 Magnetic circuit factors

In practice, the total flux set up by a current in a magnetising coil wound on a magnetic circuit core is greater than the useful flux in the core, the difference being due in the first place to the small flux loss termed leakage. Typical leakage paths are shown in Fig. 5.32.

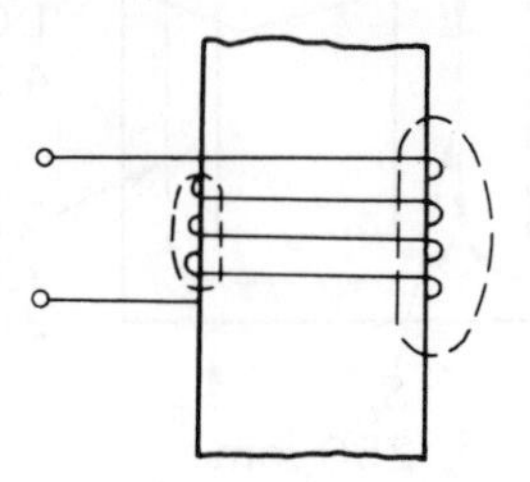

Fig. 5.32. Leakage flux paths

The ratio of total flux to useful flux is termed the leakage coefficient, i.e.

$$\sigma = \text{Leakage coefficient} = \frac{\text{Total flux}}{\text{Useful flux}} \tag{5.27}$$

If the magnetic circuit contains an air-gap, the flux tends to 'fringe' at the boundaries as illustrated in Fig. 5.33.

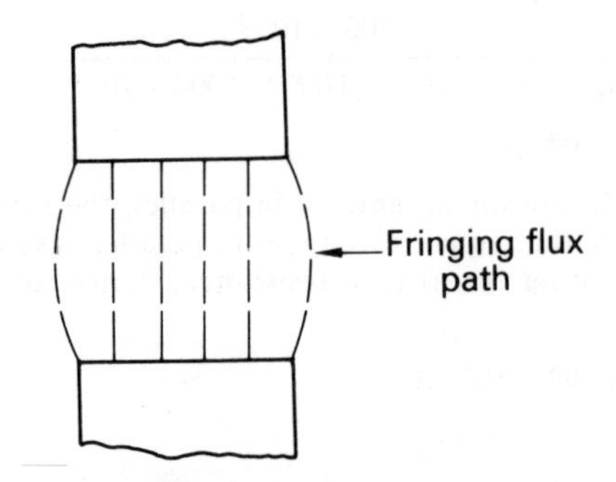

Fig. 5.33. Fringing flux

The effect of fringing is to effectively increase the cross-sectional area of the gap and hence reduce its effective reluctance. The effect can normally be neglected in cases when the air-gap is short.

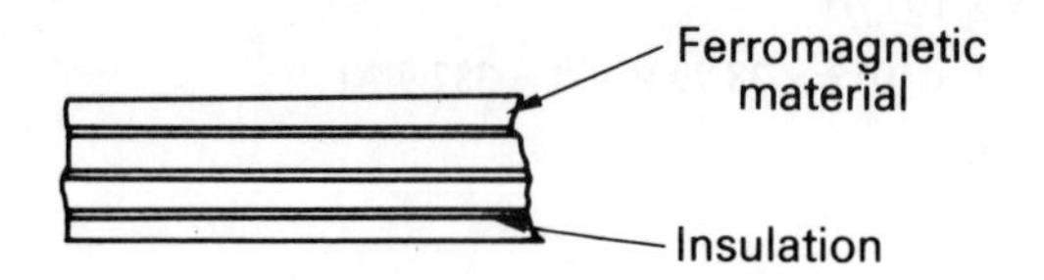

Fig. 5.34. Effect of lamination

When the flux through a ferromagnetic core varies, it has been shown that this gives rise to eddy-current power loss. In many magnetic circuits, this is minimised by lamination of the core; a cross-section of such a laminated core is shown in Fig. 5.34.

The cross-sectional area of the iron is now less than the apparent area due to the area taken up by the insulation. The ratio of iron area to apparent area is termed the space factor or stacking factor. This factor should be taken into account when doing calculations on laminated cores.

$$\beta = \text{Stacking factor} = \frac{\text{Useful area}}{\text{Total area}} \tag{5.28}$$

Example 5.19 A laminated soft-iron ring has a mean circumference of 600 mm, cross-sectional area of 500 mm² and has a radial air-gap of 1·0 mm cut through it. It is wound with a coil of 1 000 turns. Estimate the current in the coil to produce a flux of 0·5 Wb in the air-gap assuming:

(*a*) The relative permeability of the soft iron is 1 000.
(*b*) The leakage factor is 1·2.
(*c*) Fringing is negligible.
(*d*) The space factor is 0·9.

$$F = \Phi S_{ag} + \Phi S_i$$

$$= \frac{\Phi l_{ag}}{\mu_o \mu_r A_{ag}} + \frac{\Phi l_i \sigma}{\mu_o \mu_r A_i B}$$

$$= \frac{5 \times 10^{-4} \times 10^{-3}}{4\pi \times 10^{-7} \times 500 \times 10^{-6}} + \frac{5 \times 10^{-4} \times 600 \times 10^{-3} \times 1{\cdot}2}{4\pi \times 10^{-7} \times 10^{3} \times 500 \times 10^{-6} \times 0{\cdot}9}$$

$$= 1433 \text{ At}$$

$$I = \frac{F}{N} = \frac{1433}{1000}$$

$$= \underline{1{\cdot}44 \text{ A (say)}}$$

Note the answer is only an estimate hence the approximation of the numerical value.

PROBLEMS ON APPLIED ELECTROMAGNETIC THEORY

1 A 350-turn coil carries a current of 0·6 A and is wound round a magnetic circuit 0·5 m long. Calculate the magnetic field strength in the magnetic circuit. 420 At/m

2 A current of 4 A produces a flux of 4·8 μWb in an air-cored coil of 1 600 turns. What is the coil inductance? 1·92 mH

3 An air-cored coil of inductance 4·5 mH has an axial length of 314 mm and a cross-sectional area of 500 mm². How many turns are there in the coil? 1 500 turns

4 A current of 2·5 A flows through a 1 000-turn coil that is air-cored. The coil inductance is 0·6 H. What magnetic flux is set up? 1·5 mWb

5 A ferromagnetic circuit has a uniform cross-sectional area of 500 mm² and a mean diameter of 200 mm. The relative permeability of the ferromagnetic material is 2000 and the flux density in it is 1·5 T. Calculate the permeance of the magnetic circuit and the m.m.f. 2 μH; 375 At

6 A toroid of mean diameter 200 mm has a uniformly wound conductor passed ten times through its centre. The cross-sectional area of the ring is 100 mm^2 and the relative permeability is 850. A current of 1·5 A is passed through the conductor. Calculate the flux in the ring. 2·55 μWb

7 A coil of insulated wire of 400 turns and of resistance 2·5 Ω is wound tightly round a steel ring and is connected to a d.c. supply of 4·0 V. The steel ring is of uniform cross-sectional area 600 mm^2 and of mean diameter 150 mm. The relative permeability of the steel is 450. Calculate the total flux in the ring. 0·46 mWb

8 A steel ring has a circular cross-sectional area of 150 mm^2. The ring has a mean diameter of 85 mm and is wound with 250 turns of wire. The steel has a relative permeability of 500 and the total flux in the steel is 0·35 Wb. What current is required in the coil to produce this flux? 3·9 A

9 A 2 000-turn coil is uniformly wound on an ebonite ring of mean diameter 320 mm and cross-sectional area 400 mm^2. Calculate the inductance of the toroid so formed.
2 mH

10 An air-cored coil is 1·0 m long, 60 mm in diameter and has 5 000 turns of wire. Calculate the inductance of the coil.

A 500-turn coil of 40 mm diameter is placed axially in the centre of the previous coil. A current of 3·0 A in the 5 000-turn is reversed at a uniform rate in 10 ms. Calculate the e.m.f. induced in the 500-turn coil.
88·8 mH; 2·37 V

11 A 100-turn coil is uniformly wound on a perspex ring of mean diameter 200 mm and of cross-sectional area 800 mm^2. Determine:

(*a*) The magnetising force at the mean circumference of the ring when the coil is 2·0 A.
(*b*) The current required to produce a flux of 1·0 μWb.
(*c*) The self-inductance of the coil.

318 At/m; 6·24 A; 16 μH

12 A coil is wound uniformly with 300 turns over a steel ring of relative permeability 900 having a mean circumference of 400 mm and a cross-sectional area of 500 mm^2. If the coil has a resistance of 8·0 Ω and is connected across a 20-V d.c. supply, calculate:
(*a*) The coil m.m.f.
(*b*) The magnetic field strength.
(*c*) The total flux.
(*d*) The reluctance of the ring.
(*e*) The permeance of the ring.
750 At; 1875 At/m; 1·06 mWb; 707 000 /H; 1·41 μH

13 A non-magnetic former of a toroid has a mean circumference of 450 mm and a uniform cross-sectional area of 500 mm^2. The coil consists of 900 turns of wire and carries a current of 3·0 A. Calculate:
(*a*) The m.m.f.
(*b*) The magnetic field strength.
(*c*) The reluctance.
(*d*) The flux.
(*e*) The flux density.
2700 At; 6000 At/m; 716 MAt/Wb; 3·77 μWb; 7·54 mT

14 A ferromagnetic-cored coil is in two sections. One section has an inductance of 0·9 H and the other an inductance of 0·1 H. The coefficient of coupling is 0·5.
Calculate:
(*a*) The mutual inductance.
(*b*) The total inductance, when the sections are connected first in series aiding and then in series opposing. 0·15 H; 1·3 H; 0·7 H

15 A mutual inductor is constructed by winding a primary coil of 723 turns uniformly on a cylindrical former over a length of 840 mm and placing a second coil of 1 000 turns wound on a diameter of 31·8 mm coaxially within the main coil. Assuming the field within the main coil is uniform and the reluctance of the path of the return flux is negligible, calculate the mutual inductance between the two coils. 861 μH

16 A coil having a resistance of 5 Ω and an inductance of 0·3 H has a current passing through it which varies as follows:
(*a*) Increasing uniformly from 8 A to 12 A in 0·2 s.
(*b*) Constant at 12 A for 0·5 s.
(*c*) Falling uniformly to 0 A in 0·8 s.
Plot graphs representing the current, the induced e.m.f. and the applied voltage.

17 A coil has a resistance of 20 Ω and an inductance of 0·1 H. It is connected to a source of current which rises uniformly from zero to 2 A in 10 ms, remains constant at 2 A for 20 ms and falls uniformly to zero in 5 ms. Plot to scale a graph showing the variation with time of:
(*a*) The volt drop across the resistance.
(*b*) The induced e.m.f.
(*c*) The potential difference across the circuit.

18 A constant-voltage d.c. supply is applied through a switch to a circuit comprising resistance in series with inductance. The resistor is 100 Ω and the inductor is 10 H. The applied voltage is 100 V. Calculate:
(*a*) The initial rate of current rise.
(*b*) The final steady current. 10 A/s; 1·0 A

19 The resistance and inductance of a series circuit are 5 Ω and 20 H respectively. At the instant of closing the supply switch, the current increases at the rate of 4 A/s. Calculate:
(*a*) The applied voltage.
(*b*) The rate of growth of current when the current rises to 5 A.
(*c*) The stored energy under both conditions. –80 V; 2·75 A/s; 0; 0·25 J

20 A battery of e.m.f. E is supplying a steady current to a series circuit of total resistance R and inductance L. A part R_1 of the total resistance is suddenly short-circuited. Derive an expression for the current in the circuit subsequent to this operation. If $E = 100$ V, $R = 20$ Ω, $R_1 = 10$ Ω and $L = 2$ H, plot the current/time curve and determine the current (*a*) 0·1 s, (*b*) 0·5 s after short circuit. 6·97 A; 9·59 A

21 The field windings of a machine consist of eight coils connected in series each containing 1 200 turns. When the current is 3·0 A, the flux linked with each coil is 20 mWb. Calculate:
(*a*) The inductance of the circuit.
(*b*) The energy stored therein.
(*c*) The average value of the induced e.m.f. if the circuit is broken in 0·1 s.
64 H; 288 J; 1920 V

22 A current of 15 A flows through a coil of inductance 60 mH. If the circuit is to be opened in 15 ms, calculate the average power dissipation. 450 W

23 A magnetic circuit consists of a toroid of mean diameter 150 mm. The c.s.a. of the toroid is 500 mm^2. The toroid is made from silicon steel and is subjected to an alternating m.m.f. which produces the hysteresis effect shown on chart 5.1. Calculate the power loss in the steel due to hysteresis. 3·4 W

Chart 5.1. Hysteresis loops

MATERIAL	SCALE FACTORS per unit value of	
	B (T)	H (At/m)
Silicon Steel	4	10
Mumetal	1	1

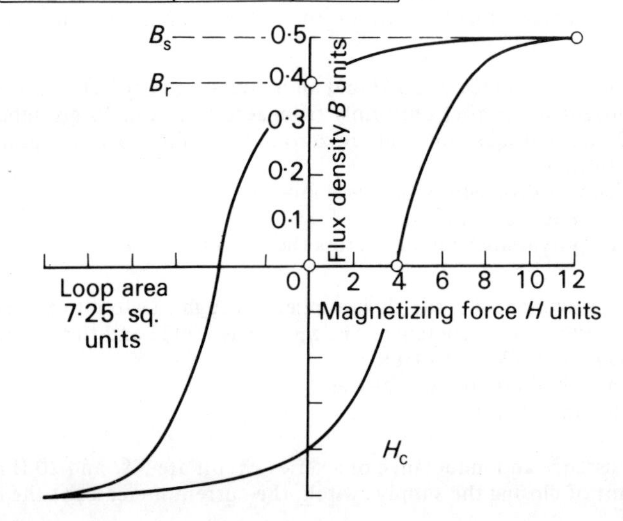

24 A magnetic core is made from mumetal, the hysteresis loop for which is shown on Chart 5.1. The core measures 50 mm long and has an average c.s.a. of 1·0 cm^2. The hysteresis loss is 1·0 W. Calculate the frequency of the alternating flux. 27·6 kHz

25 Prove that for a specimen of steel, the hysteresis loss in watts per kilogram per cycle is given by the area of the B/H loop divided by a constant. The hysteresis loop for a specimen of mass 12 kg is equivalent to 300 W/mm^3. Find the loss of energy per hour at 50 Hz. The relative density of the steel is 7500 kg/m^3. 0·024 kWh

26 The measured total core losses occurring in transformer stampings 14 mm thick when tested at 50 Hz at a maximum flux density of 1·0 T amounted to 1·4 W/kg. For the B/H loop between the same flux density limits, the hysteresis loop at 50 Hz was calculated to be 1·0 W/kg. Estimate the probable core loss in watts per kilogram when the same material 20 mm thick is operated with a maximum induction of 1·3 T at 60 Hz. Take $N = 1{\cdot}6$. 3·82 W/kg

27 What m.m.f. is required to produce a flux density of 1·28 T in a closed steel ring of mean length 700 mm; the relative permeability for the steel at this flux density is 540.

What extra m.m.f. is required to maintain a flux density of 1·28 T when a radial air-gap of 1·59 mm is introduced to the ring? 1 320 At; 1622 At

28 A steel ring 300 mm in diameter has a cross-sectional area of 400 mm^2 for one third of its radial axis, 800 mm^2 for the second third and 1 200 mm^2 for the remainder The relative permeability of the steel is 3 000; calculate the m.m.f. required to produce a flux of 1·0 mWb. 382 At

29 An electromagnet has a magnetic circuit that can be regraded as comprising three parts in series, each of uniform cross-sectional area.

Part a is of length 80 mm and cross-sectional area 50 mm^2.
Part b is of length 60 mm and cross-sectional area 90 mm^2.
Part c is an air-gap of length 0·5 mm and cross-sectional area 150 mm^2.
Parts a and b are made from a material having the following corresponding magnetic characteristic values:

$H = 620$ At/m; $B = 0{\cdot}9$ T $\qquad H = 275$ At/m; $B = 0{\cdot}5$ T

Determine the current necessary in a coil of 4 000 turns wound on part b to produce an air-gap flux density of 0·3 T. 46·4 mA

30 A series magnetic circuit has a steel path of length 500 mm and an air-gap of length 1·0 mm. The c.s.a of the steel is 600 mm^2 and the exciting coil has 400 t. Determine the current required to produce a flux of 0·9 mWb in the circuit, if the following points were taken on the magnetisation curve for the steel:

Flux density	(T)	1·2	1·35	1·45	1·55
Magnetising force	(At/m)	500	1 000	2 000	4 500

6·35 A

31 A transformer core is made from core steel. The outer limbs are effectively 500 mm long with a c.s.a. of 2 000 mm^2. The inner limb is 200 mm long and has a c.s.a of 3 000 mm^2. The air-gap is 1·0 mm long. Using chart 5.2, calculate the flux density in the centre limb when a current of 1·33 A is passed through a 1 000-turn coil wound on the same limb. 1·33 T

32 Figure 5.35 represents an inductor built of transformer steel laminations having the B/H characteristic given in chart 5.2. The centre limb carries a 1 000-turn exciting coil. Neglecting leakage and fringing, estimate the current required in the coil to produce a flux of 6 mWb in the air-gap. 2·5 A

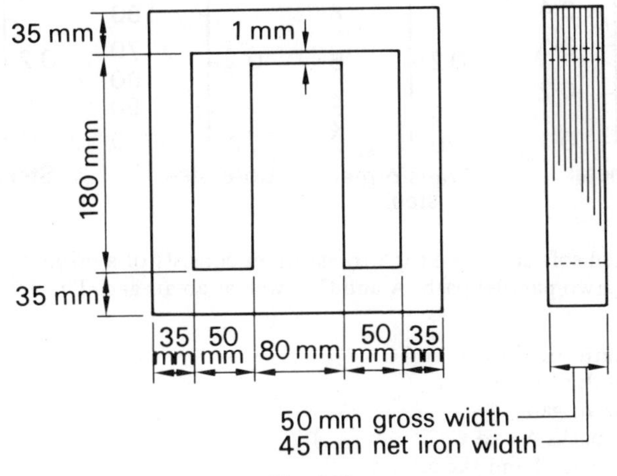

Fig. 5.35

33 The core of a magnetic circuit consists of three parts:

Part A—c.s.a. 500 mm^2, length 125 mm.
Part B—c.s.a. 250 mm^2, length 100 mm.
Part C—c.s.a. 500 mm^2, length 0·5 mm.

Parts A and B are made from mumetal, the characteristic of which is given on chart 5.2. Part C is an air-gap. A 1 000-turn coil is wound on Part B. Calculate the coil current to produce a flux density of 0·5 T in Part A. 0·2 A

Chart 5.2. Magnetization characteristics giving B in teslas and H in ampere-turns per metre

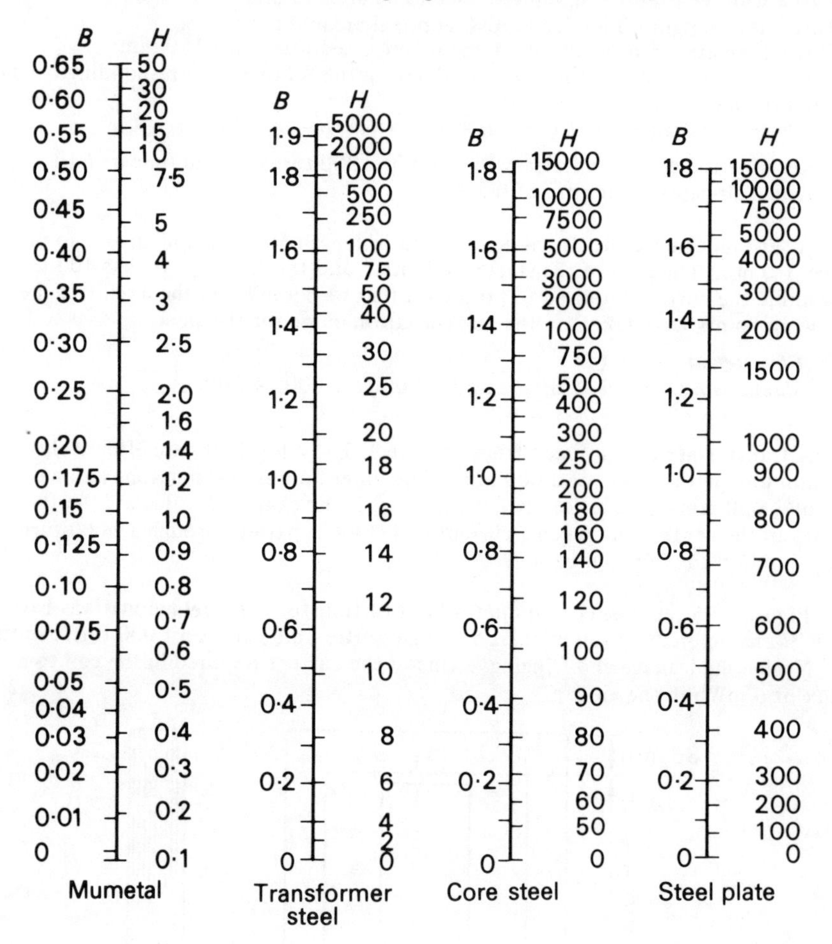

34 A laminated soft-steel core for a transformer consists of a magnetic circuit path B in series with two parallel paths A and C as well as an air-gap. The details are as follows:

Mean lengths of paths A and C	500 mm
Mean length of path B	200 mm
Length of air-gap	2 mm
C.S.A. of paths A and C	1 000 mm^2
C.S.A. of path B and the air-gap	2 000 mm^2
Space factor	0·9

A 1 000-turn coil is wound on the central limb–path B. Assuming that the relative permeability of the iron is 1 000, the flux density in the air-gap is 1·0 T and neglecting the effects of leakage and fringing, determine:

(*a*) The coil current.
(*b*) The energy stored in the magnetic field. 2·21 A; 2·21 J

35 The transformer core shown in Fig. 5.36 is 100 mm thick and made from transformer steel, the characteristic for which is shown in chart 5.2.

A 2 000-turn coil is wound on the centre limb. Calculate the current in the coil to produce a flux density in the centre limb of 1·4 T. 0·292 A

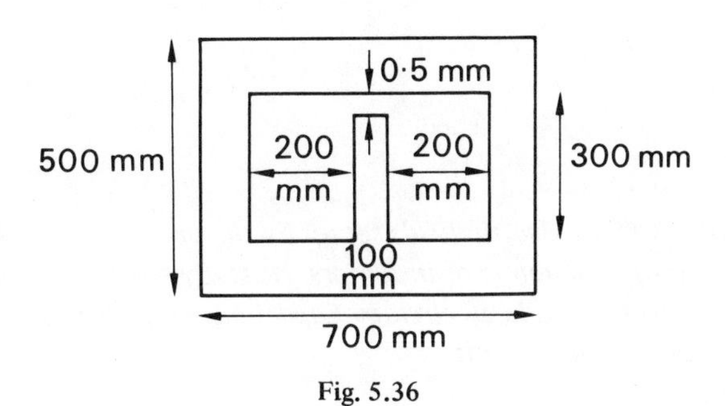

Fig. 5.36

36 A magnetic circuit consists of a toroid with a radial air-gap. The toroid, which is made of steel plate, has an effective annular length of 800 mm and its cross-section is 100 mm x 50 mm. The air-gap is 1·0 mm long and of the same effective cross-section as the steel. A 2 000-turn coil is wound on the toroid. Using the magnetisation characteristic from steel plate given on chart 5.2, estimate the coil current necessary to produce a flux of 6 mWb in the magnetic circuit.

What flux would be produced by a coil current of 0·75 A?

0·98 A; 5·0 mWb

37 A stamping is made from 0·5-mm thick steel and measures 300 mm x 240 mm. A window is cut from the middle measuring. 200 mm x 200 mm. Two hundred such stampings are clamped together and coils of 60 turn are wound on each of the longer sides and are connected in series. The magnetisation characteristic for the steel is given below:

H (At/m)	30	50	70	90	120	160	200	280	400	500	600	700
B (T)	0·10	0·20	0·33	0·45	0·63	0·83	0·98	1·20	1·37	1·45	1·51	1·55

Determine:

(*a*) The coil currents necessary to establish fluxes of 1 mWb and 3 mWb.
(*b*) The flux when the current is 2·0 A.
(*c*) The flux this current will produce if the stampings are made up of four straight pieces butt-jointed at each corner with an effective overall air-gap of 0·4 mm length, i.e. 0·1 mm per joint.

0·58 A; 2·83 A; 2·21 mWb; 0·93 mWb

6

Electrostatics and capacitance

Electrostatics is the study of electric fields set up by the presence of charges of electricity in systems of conductors or insulators. Its study leads to an understanding of the capacitance effect found in all circuits. Capacitance is particularly important in electronic and communication circuits.

6.7 Properties of an electric field

When a current flows at the rate of one ampere, the charge that passes through a cross-section of the conductor during a period of one second is one coulomb. The coulomb is the unit of electrical charge.

If a current can be arranged to flow into a body during a finite period of time, the body will acquire a charge. Since conventional current is a flow of positive charge, the body will acquire a positive charge. Conversely if the current had been arranged to flow from the body, it would become deficient in positive charge, i.e. it would acquire a negative charge. Prior to external interference, the body would have had equal positive and negative charges due to the balance of its atomic constitution. After it has had its balance upset, it will try to regain its original condition.

This quest for balance takes the form of positive and negative charges searching for one another. These unlike charges attract one another. Similarly it may be observed that similar or like charges repel one another. These forces of attraction and repulsion may be very strong.

The space surrounding a charge can be investigated using a small charged body. This investigation is similar to that applied to the magnetic field surrounding a current-carrying conductor. However in this case the charged body is either attracted or repelled by the charge under investigation. The space in which this effect can be observed is termed the electric field of the charge. The force on the charged body is termed the electric field of the charge. The force on the charged body is termed the electric force.

As in the magnetic case, line of force can be traced out. These lines are again given certain properties:

1. In an electric field, each line of force will emanate from or terminate in a charge. The conventional direction is from a positive charge to a negative charge.

2 The direction of the line is that of the force experienced by a positive charge placed at a point in the electric field. It is assumed that the search charge has no effect on the field distribution.

3. The lines of force never intersect since the resultant force at any point in an electric field can have only one direction.

It should be noted that whilst it is possible to observe the electric force acting on a small charged body in principle, it is extremely difficult to obtain experimental

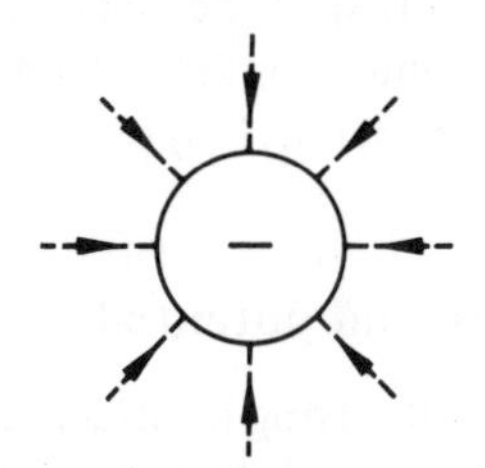

Fig. 6.1. Electric field about an isolated spherical charge

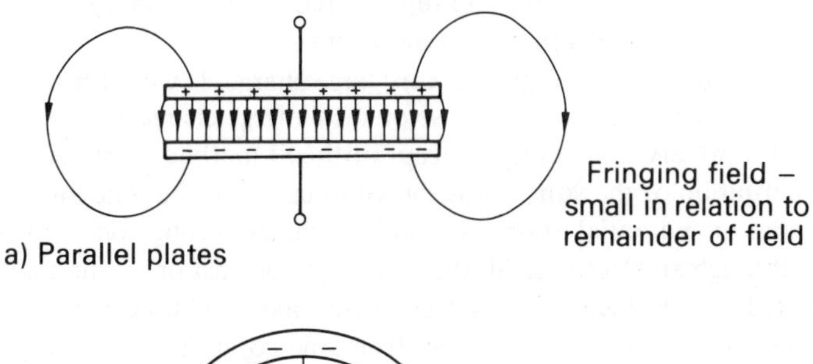

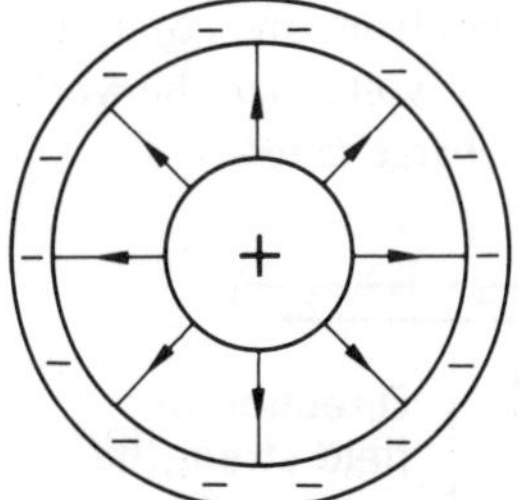

b) Concentric cylinders (cable)

Fig. 6.2. Electric fields between oppositely charged surfaces

verification of the field distribution and indirect methods have to be used. The concept can be furthered by considering the forces that would act on an imaginary positive charge as it is moved about in the field.

The force of attraction or repulsion acts directly between two adjacent charges. The lines of force therefore always radiate out from equipotential surfaces at right angles. A surface which is equipotential has the same potential at all points on it. A conductor will have an equipotential surface and therefore all lines of force will be normal to its surfaces. The most simple case is that of the isolated spherical charge shown in Fig. 6.1.

Most electric fields exist between two conductors having equal but opposite charges. The two important arrangements of conductors are the parallel plates and the concentric cylinders. The former appears in practical form in many capacitors whilst the latter is the cross-section of a coaxial cable. The appropriate fields are shown in Fig. 6.2.

In the arrangements shown in Fig. 6.2, it is important that the charges be kept apart. They must therefore be insulated from one another although they are held in conducting materials. The intervening space between the charges is filled with any insulating material which, in this context, is termed a dielectric.

Suitable dielectrics include air, paper, nylon, etc.

6.2 Electric field strength and potential

An electric field is investigated by observing its effect on a charge. Let the charge be a unit charge–a coulomb. The magnitude of the force experienced at a point in a field by this unit charge is termed the electric field strength at that point. It may be measured in newtons per unit charge and represented by the symbol E. It can also be termed the electric stress or electrostatic field strength.

It should be noted that a coulomb is a very large charge. It would therefore disrupt the field being investigated. The use of such a large charge is therefore pure supposition but it preserves the unity concept of the SI method of measurement.

From the definition of the volt, it was shown in para. 1.8 that one joule of work is necessary to raise the potential of one coulomb of charge through one volt. When a charge moves through an electric field, the work done against or by the electric field forces is reflected by the change in potential of the charge. This neglects any mechanical resistance to motion. If a unit charge moves from one point in an electric field to another thereby changing its potential by V volts then the work done is V joules. This holds true no matter what path the charge takes.

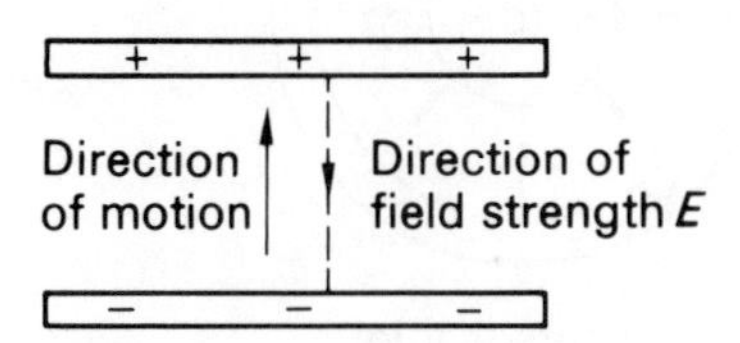

Fig. 6.3. Work done by moving a charge within the field between parallel charged plates

The most simple field arrangement to investigate is that between parallel charged plates. This is shown in Fig. 6.3. The field in the centre of the system is essentially uniform.

Let the p.d. between the plates be V volts. The work done in transferring a unit charge from one plate to the other through the electric field will be V joules. Consider part of the transfer process as indicated in Fig. 6.4. Here the unit charge is moved a distance Δl in a direction at an angle θ to the field direction. Since the field is uniform then the work done is

$$-E.\Delta l.\cos\theta$$

The negative sign is included because the motion of the charge is in the opposite direction to that of the electric field strength. Alternatively it can be stated that the electric field strength has a negative direction with respect to the movement.

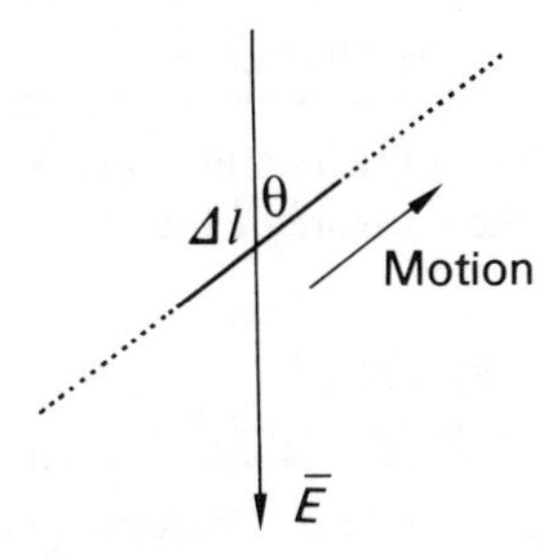

Fig. 6.4. Movement of a charge in an electric field

In the parallel-plate arrangement, the usual charge movement is considered parallel to the lines of force. This action may be assumed in any electric field unless otherwise stated. In consequence of this assumption, $\theta = 0$ and $\cos\theta = 1$. The work done is therefore given by

$$-\Delta l.E$$

Let ΔV be the increase in potential due to the movement Δl of the charge.

$$\Delta V = -\Delta l.E$$

$$E = -\frac{\Delta V}{\Delta l}$$

As $\Delta l \to 0$, then the expression may be given in calculus form, i.e.

$$\frac{\Delta V}{\Delta l} \to \frac{dV}{dl}$$

$$E = -\frac{dV}{dl} \qquad (6.1)$$

It can be seen that the electric field strength measured in the direction of the field is given by the rate of change of potential measured in the opposite direction at any point in the field. Although the electric field strength can be measured in newtons per unit charge, it follows from relation 6.1 that it can also be measured in volts per metre.

Electric field strength	Symbol: E	Unit: Volt per metre (V/m)

Note that the electric field strength at a point in the field is a vector quantity since it has both magnitude and direction. It also has the direction of the greatest rate of change of voltage. The voltage or potential at a point is not a vector quantity.

6.3 Electric flux and flux density

The total electric effect of a system as described by the lines of electric force is termed the electric flux linking the system. The unit of electric flux is the coulomb.

Electric flux	Symbol: Q	Unit: Coulomb (C)

It should be noted that a flux of Q coulombs is created by a charge of Q coulombs.

The electric flux density is the measure of the electric flux passing normally through unit area.

Electric flux density	Symbol: D	Unit: Coulomb per square metre (C/m^2)

In a uniform field, if a flux Q passes normally through an area A, then

$$D = \frac{Q}{A} \tag{6.2}$$

Again the electric flux density is a vector quantity having both magnitude and direction. The direction is that of the lines of force.

6.4 Permittivity

The flux density may be considered to result from the electric field strength. For any given value of electric field strength E, the value of the resulting flux density D depends on the medium in which the flux is produced. The ratio of D to E is termed the absolute permittivity of the medium.

Absolute permittivity	Symbol: ϵ	Unit: Farad per metre (F/m)

It follows therefore that

$$\frac{D}{E} = \epsilon$$

$$D = \epsilon E \tag{6.3}$$

This relation may be compared with its magnetic equivalent, $B = \mu H$. Again a basic value of permittivity is set by taking that value corresponding to free space. If the flux density due to an electric field strength exists in a medium other than free space, the permittivity is found to change in value. The absolute permittivity ϵ is then expressed as a multiple of that for free space ϵ_0, where ϵ_0 is the absolute permittivity for free space.

Permittivity of free space	Symbol: ϵ_0	Unit: Farad per metre (F/m)

The unit of measurement will be discussed in para. 6.6.

$$\epsilon_0 = \frac{1}{36\pi} \times 10^{-9} \text{ F/m}$$

$$= 8{\cdot}854 \times 10^{-12} \text{ F/m}$$

Theory more advanced than that covered by this present text, shows that there is a relation between absolute permittivity, absolute permeability and the speed of propagation of electromagnetic waves in free space. This relation is

$$\frac{1}{\mu_0 \epsilon_0} = c_0^2 \tag{6.4}$$

where c_0 is the velocity of propagation. This velocity is an absolute constant whilst the value of absolute permeability has already been determined in line with the primary unit of electricity. It follows that the value of absolute permittivity can therefore be determined from relation 6.4. Taking $c_0 = 3 \times 10^8$ m/s, then ϵ_0 has the value indicated above. This value is slightly high but is sufficient for most purposes.

The ratio of the permittivity to the absolute permittivity is termed the relative permittivity ϵ_r.

Relative permittivity	Symbol: ϵ_r	Unit: none

$$\epsilon_r = \frac{\epsilon}{\epsilon_0}$$

$$\therefore \quad \epsilon = \epsilon_0 \epsilon_r$$

It follows that the general case of relation 6.3 is

$$D = \epsilon_0 \epsilon_r E \tag{6.5}$$

For air, ϵ_r may be taken as unity; most other materials have values between 1 and 10. Only water and titanate compounds exceed this range, the latter giving values of several hundreds.

Example 6.1 Two parallel plates are charged to have a potential difference of 100 V. Each has an area of 0·05 m² and they are separated by 1·0 mm of air. Assuming that all the flux is contained between the plates, calculate the electric charge on each plate.

$$E = \frac{\Delta V}{\Delta l} \text{ (the sign of no consequence since the direction is not required)}$$

$$= \frac{100}{1 \times 10^{-3}}$$

$$= 100 \times 10^3 \text{ V/m}$$

$$D = \epsilon_0 \epsilon_r E$$

$$= 1 \times 8{\cdot}854 \times 10^{-12} \times 100 \times 10^3$$

$$= 8{\cdot}854 \times 10^{-7} \text{ C/m}^2$$

$$Q = DA$$
$$= 8{\cdot}854 \times 10^{-7} \times 0{\cdot}05$$
$$= 4{\cdot}427 \times 10^{-8}\,\mathrm{C}$$

It will be noted that Q represents both electric flux and also the charge on each plate.

6.5 Capacitance

This chapter commenced with the charging of a body by causing a flow of current to enter or leave the body. In the analysis that followed, specific arrangements such as the pair of parallel plates were used to retain a charge. Devices created specifically for this purpose are termed capacitors.

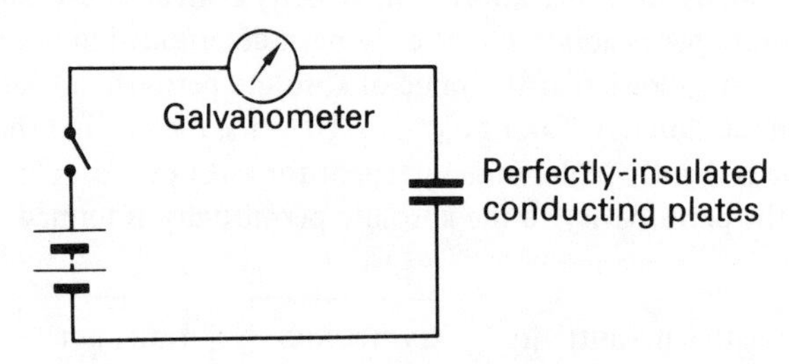

Fig. 6.5. Charging a capacitor

Consider again the charging process. One possible circuit is that shown in Fig. 6.5. When the switch is closed, the galvanometer deflects momentarily showing that a current is passed for a short period of time and has then ceased to flow. This burst of current is the flow of charge to or from the capacitor plates.

From the circuit diagram, it can be seen that the top plate of the capacitor now has a deficiency of electrons, these having been attracted to the positive plate of the battery. The transfer of these electrons is equivalent to a conventional current flowing into the top plate of the capacitor. However the plates are insulated from one another by the dielectric and the circuit is therefore incomplete. The current flow can only accumulate on the top plate making it positively charged. An equal but opposite action takes place at the lower plate making it negatively charged.

The process of transferring charge cannot take place indefinitely because the accumulated plate charges repel further charge movements within the circuit. For instance the first quantity of charge to arrive at a plate of the capacitor does so unimpeded. However, the second quantity to arrive does so against the repelling electrostatic force of the first charge. This difficulty increases with the arrival of each charge until it is impossible for any further quantity of charge to be transferred.

This effect is better described in terms of the potential difference that appears between the plates due to the charges on the plates. The magnitude of the p.d. depends on the quantity of charge that has accumulated and is equal to the work done in moving a unit charge from one plate to the other.

The cause of the charge transfer is the e.m.f. of the battery. This e.m.f. acts against the p.d. between the capacitor plates and so long as it exceeds the capacitor p.d., charge continues to move around the circuit. As the difference decreases, the rate of charge decreases until it ceases altogether. The capacitor is then charged such that the p.d. between the plates is equal to the e.m.f. of the battery. If the e.m.f. of the battery is V, it follows that the p.d. between the capacitor plates is V.

It should be noted that the charging process takes a very short period of time unless the resistance of the circuit is high relative to the size of the capacitor.

The quantity of charge that is transferred from one plate to the other can be measured if the galvanometer is replaced by a ballistic galvanometer (see para 17.16). By noting the charge transferred due to the application of different e.m.f.'s, then, provided that the capacitor arrangement remains unchanged, it can be shown that the p.d. V is proportional to the transferred charge Q.

$$V \propto Q$$

Let $$Q = CV \tag{6.6}$$

C is a constant termed the capacitance of the capacitor. The capacitance of a capacitor is that property whereby a capacitor accumulates charge when a potential difference is applied to it. Capacitance is measured in farads–a capacitor that accumulates a charge of one coulomb when a potential difference of one volt is applied to it has a capacitance of one farad.

Capacitance	Symbol: C	Unit: Farad (F)

The farad is rather an unfortunate unit because it is immense. Most capacitors are therefore rated either in microfarads or picofarads.

So long as the e.m.f. of the battery in Fig. 6.5 is maintained, the charge will remain on the capacitor. If it is removed and then replaced by a resistor or even a short-circuit, the plates' charges will move round the circuit to recombine and neutralise one another. This effect can be observed from the deflection experienced by the galvanometer in Fig. 6.6 when the switch is moved from the charge position 1 to the discharge position 2.

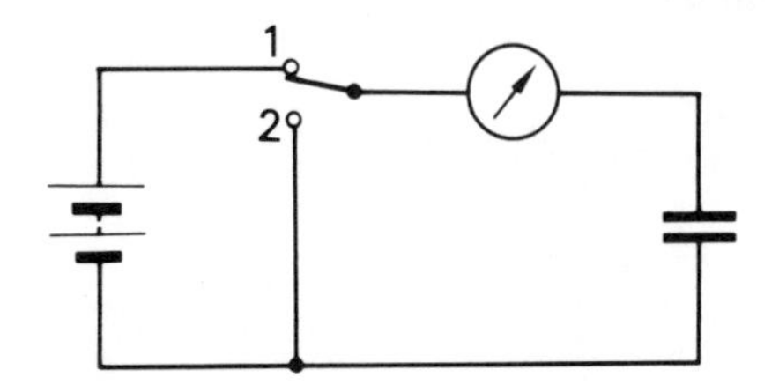

Fig. 6.6. Discharge of a capacitor

It will be noted that the charged capacitor gives rise to a current flow when it is discharged. This will serve to dissipate heat in the circuit resistance. The p.d. between the plates is therefore related to a source of energy and is more appropriately termed an e.m.f.

It should not be thought that if a capacitor is charged and then disconnected from its source of e.m.f., it will retain its charge indefinitely. Instead the charge will slowly leak away. Because no insulator is perfect, there is always a leakage path between one plate and the other through which the charges can move and neutralise one another. Most capacitors will retain their charges for some considerable period of time. When handling large capacitors, it is as well to check that they are discharged.

Finally it should be noted that capacitance is not a property of capacitors alone. Capacitance exists between any two parts of a circuit that are at different potentials. However the effect of capacitance is generally so small that it need only be taken into account either in the case of capacitance of very long transmission lines or of operation at higher frequencies.

6.6 Simple parallel-plate capacitor

Although there are many forms of capacitor, the most important arrangement is the parallel-plate capacitor. Most capacitors are either formed by a stack of such plates or by a pair of such plates rolled up like a swiss roll. It is therefore fortunate that the parallel-plate capacitor is also the most suitable to analyse due to the uniformity of its field.

Consider two parallel plates each of area A and spaced a distance d apart. Because the area involved is large whilst the spacing is very small, the fringing field at the edges of the plates may be neglected. It may also be assumed that the field between the plates is uniform.

Let the plates be given a charge of Q coulombs hence giving rise to an electric flux of Q coulombs and a potential difference of V volts between the plates.

Since $$D = \frac{Q}{A}$$

then $$Q = DA$$

Also the work in transferring a unit charge from one plate to the other through the electric field is

$$Ed = V$$

However, from relation 6.6,

$$C = \frac{Q}{V}$$

$$= \frac{DA}{Ed}$$

$$= \frac{D}{E} \cdot \frac{A}{d}$$

But $$D = \epsilon_0 \epsilon_r E$$

where ϵ_r is the relative permittivity of the dielectric separating the plates of the capacitor.

$$\frac{D}{E} = \epsilon_0 \epsilon_r$$

$$C = \frac{\epsilon_0 \epsilon_r A}{d} \tag{6.7}$$

Re-arranging this relation

$$\epsilon = \frac{Cd}{A}$$

Inspection of this relation shows that the permittivity must be measured in farads per metre as noted in para. 6.4.

Example 6.2 Calculate the capacitance of two metal plates of area 30 m² and separated by a dielectric 2·0 mm thick and of relative permittivity 6. If the electric field strength in the dielectric is 500 V/mm, calculate the total charge on each plate.

$$C = \frac{\epsilon_0 \epsilon_r A}{d} = \frac{8{\cdot}854 \times 10^{-12} \times 6 \times 30}{2 \times 10^{-3}}$$

$$= 0{\cdot}798 \times 10^{-6}\ \text{F}$$

$$= \underline{0{\cdot}798\ \mu\text{F}}$$

$$V = Ed = 500 \times 10^{3} \times 2 \times 10^{-3} = 1000\ \text{V}$$

$$Q = CV = 0{\cdot}798 \times 10^{-6} \times 1000$$

$$= 0{\cdot}798 \times 10^{-3}\ \text{C}$$

$$= \underline{0{\cdot}798\ \text{mC}}$$

6.7 Charging current and energy storage

During the charging process of a capacitor, let ΔQ be the small charge accumulated during a period Δt. The average current during this period is given by the rate of arrival of charge, i.e.

$$\frac{\Delta Q}{\Delta t}$$

As Δt tends to zero, the current is given at any instant by the following relation in calculus form,

$$i = \frac{\mathrm{d}q}{\mathrm{d}t} \tag{6.8}$$

However, at any instant,

$$q = Cv$$

hence $$i = \frac{\mathrm{d}}{\mathrm{d}t}(Cv)$$

It is usual to find that the capacitance C is a constant unless there is movement between the plates. Assuming it constant,

$$i = C \cdot \frac{\mathrm{d}v}{\mathrm{d}t} \tag{6.9}$$

Although the above relations were derived on the basis of an accumulation of charge taking place in a particular time, it should be remembered that this can either be an increase or a decrease.

Example 6.3 A direct potential of 200 V is suddenly applied to a circuit comprising an uncharged 100-μF capacitor in series with a 1 000-Ω resistor. Calculate the initial rate of rise of voltage across the capacitor.

Once the capacitor is completely charged, the source of e.m.f. is short-circuited. Calculate the initial rate of decrease of voltage across the capacitor.

Because the capacitor is initially uncharged, there will be no initial p.d. between the plates. When the supply e.m.f. is applied, the corresponding volt drop must appear across the 1 000-Ω resistor. Let the initial current be i_0.

$$i_0 = \frac{V}{R} = \frac{200}{1000} = 0{\cdot}2 \text{ A}$$

$$= C.\frac{dv}{dt} = 100 \times 10^{-6} \times \frac{dv}{dt}$$

$$\frac{dv}{dt} = \frac{0{\cdot}2}{100 \times 10^{-6}} = 2000 \text{ V/s}$$

$$= \underline{2 \text{ kV/s}}$$

When the capacitor is discharged, the full p.d. is applied to the resistor. This gives the same current as before but the direction of flow is reversed and it follows that the initial rate of voltage decrease is

$$\underline{-2 \text{ kV/s}}$$

Returning to the charging process, consider a capacitor of capacitance C being charged at a constant rate I for a period of time t. Provided that the capacitor was initially discharged, the charge at the end of time t is

$$Q = It$$

The p.d. between the capacitor plates will have steadily increased during this time from zero to a final value V. The average p.d. between the plates during charging is therefore $\frac{1}{2}V$. The average power to the capacitor during charging is $\frac{1}{2}VI$ and therefore the total energy stored in the capacitor is

$$W_f = \tfrac{1}{2}VIt$$

$$= \tfrac{1}{2}VQ$$

$$= \tfrac{1}{2}CV^2 \qquad (6.10)$$

A more general proof is as follows. Let the p.d. between the plates at any instant be v and the corresponding charging current be i. During a period dt, the energy supplied to the capacitor is $vi\,.\,dt$. But

$$i = C.\frac{dv}{dt}$$

$$i.dt = C.dv$$

The energy supplied therefore is $Cv\,.\,dv$. If the capacitor is charged to a p.d. V, then the energy stored is given by

$$W_f = \int_0^V Cv\,.\,dv = \tfrac{1}{2}CV^2 \text{ as before}$$

It should be noted that the energy is stored in the electric field within the dielectric. This energy can be returned to the source provided there is no resistance in the circuit. However any circuit will have some resistance and the passage of the charging or discharging current through it will dissipate some energy in the form of heat.

The total energy acquired by the capacitor is $\tfrac{1}{2}CV^2$ whilst the total energy delivered from the source during the charging process is

$$W_E = \int_0^\infty iv_R\,.\,dt + \tfrac{1}{2}CV^2$$

where v_R is the volt drop across the circuit resistance at any instant. It can be reasoned by comparison with the above derivation of the stored energy that

$$iv_R\,.\,dt = \tfrac{1}{2}CV^2$$

This means that the energy dissipated in the resistor is equal to the energy stored in the capacitor. It should be noted that this relation holds true regardless of the value of the circuit resistance.

Apart from the energy dissipated in the circuit resistance, it is not possible to return all the stored energy to the circuit. Apart from the loss due to leakage, there is a further loss termed the dielectric loss. This loss compares somewhat with the hysteresis loss in ferromagnetic materials. In most capacitors, this loss is so small that it may be neglected.

Example 6.4 In the circuit shown below, the battery e.m.f. is 100 V and the capacitor has a capacitance of 1 μF. The switch is operated 100 times every second, i.e. to operational cycles per second. Calculate the average current through the switch between switching operations and also calculate the average power dissipation in the resistor. It may be assumed that the capacitor is ideal and that the capacitor is fully charged or discharged before the subsequent switching.

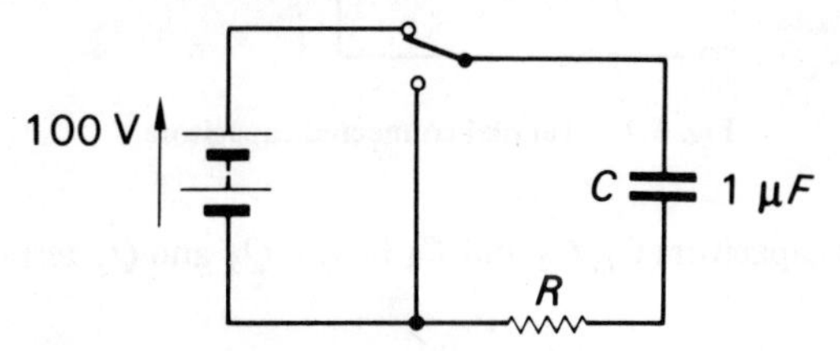

The maximum charge accumulated by the capacitor is

$$Q = CV = 1 \times 10^{-6} \times 100 = 100 \times 10^{-6}\text{ C}$$

The time taken to acquire this charge (or to lose it) is

$$T = \frac{1}{f} = \frac{1}{100} = 0{\cdot}01\text{ s}$$

hence $$I_{av} = \frac{\Delta Q}{\Delta t} = \frac{100 \times 10^{-6}}{0{\cdot}01} = 0{\cdot}01 \text{ A}$$

$$= 10 \text{ mA}$$

The maximum energy stored during charging is

$$W_f = \tfrac{1}{2}CV^2 = \tfrac{1}{2} \times 1 \times 10^{-6} \times 100^2 = 0{\cdot}005 \text{J}$$

During the charging period, a similar quantity of energy must be dissipated in the resistor. In the subsequent discharging period, the stored energy in the capacitor is dissipated in the resistor. Hence for every switching action, 0·005 J are dissipated in the resistor. For 100 switching operations, the energy dissipated is

$$W = 100 \times 0{\cdot}005 = 0{\cdot}5 \text{ J}$$

Hence the average power is given by

$$P = \frac{\Delta W}{\Delta t} = \frac{0{\cdot}5}{1}$$

$$= 0{\cdot}5 \text{ W}$$

6.8 Parallel- and series-connected capacitors

Consider three capacitors connected in parallel as shown in Fig. 6.7. When the switch is closed, all three capacitors are charged to have a p.d. V between their plates. Let the

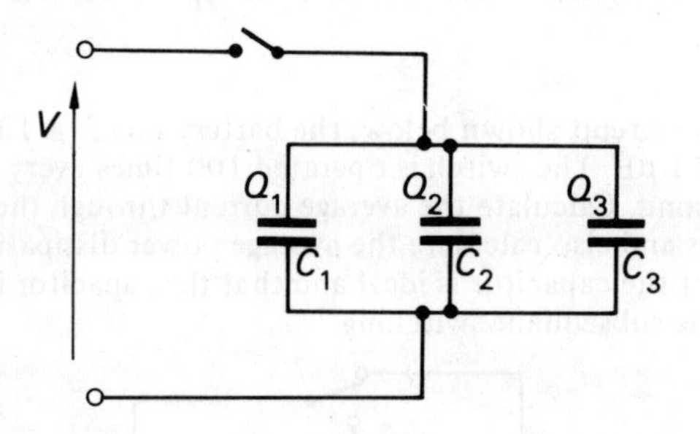

Fig. 6.7. Parallel-connected capacitors

charges acquired by capacitors C_1, C_2 and C_3 be Q_1, Q_2 and Q_3 respectively. It follows that

$$Q_1 = C_1 V$$

$$Q_2 = C_2 V$$

and $$Q_3 = C_3 V$$

Let the total charge be Q and the total effect capacitance be C; therefore

$$Q = CV$$

The total charge acquired by the capacitors is

$$Q = Q_1 + Q_2 + Q_3$$
$$= C_1 V + C_2 V + C_3 V$$
$$= (C_1 + C_2 + C_3)V$$
$$= CV$$
$$C = C_1 + C_2 + C_3 \qquad (6.11)$$

Although the above analysis was carried out using three capacitors, the number is of no significance.

Example 6.5 A 16-μF capacitor is charged to 100 V and is then disconnected from the source of e.m.f. It is then connected across a 4-μF capacitor which is initially uncharged. Determine:

(*a*) The original energy stored in the 16-μF capacitor.
(*b*) The voltage across the combined circuit.
(*c*) The total stored energy.

$$W_f = \tfrac{1}{2}CV^2 = \tfrac{1}{2} \times 16 \times 10^{-6} \times 100^2 = \underline{0{\cdot}08\,\text{J}}$$

The original charge on the 16-μF capacitor is

$$Q = C_1 V = 16 \times 10^{-6} \times 100 = 1{\cdot}6 \times 10^{-3}\ \text{C}$$

This charge is now distributed between the two capacitors. The total capacitance of the arrangement is

$$C = C_1 + C_2 = (16 + 4) \times 10^{-6} = 20 \times 10^{-6}\ \text{F}$$

$$V = \frac{Q}{C} = \frac{1{\cdot}6 \times 10^{-3}}{20 \times 10^{-6}} = \underline{80\ \text{V}}$$

The energy stored in the capacitors is now

$$W_f = \tfrac{1}{2}CV^2 = \tfrac{1}{2} \times 20 \times 10^{-6} \times 80^2 = \underline{0{\cdot}064\,\text{J}}$$

It will be noted that there is a loss in energy. This is due to heat dissipated by the resistance of the circuit. This resistance cannot be ignored otherwise the charge transfer would comprise an infinite current for zero time which is impossible.

Before progressing to consider the effect of connecting capacitors in series, it is necessary to give some consideration to induced charges. If an uncharged body, which is a conductor, is brought towards a charged body, a movement of charge will take

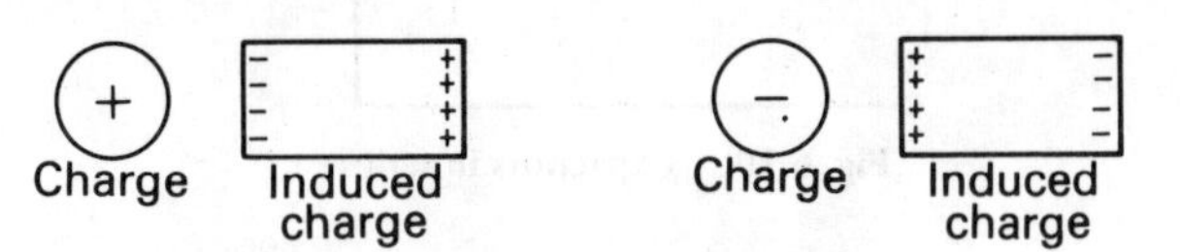

Fig. 6.8. Induced charges

place within the conductor. The unlike charges within the conductor will be attracted towards the charged body and an equivalent like charge will be repelled in the opposite direction. These separated charges within the conductor are induced charges; this is indicated by Fig. 6.8.

The most simple method of arranging the introduction of a conductor into the influence of charged bodies is to place a conductor slab between the plates of a parallel-plate capacitor. It may be assumed that the entire electric field passes through the conductor slab. The charges on the capacitor plates attract the unlike charges of the slab. This movement within the slab continues until the induced charges on each surface of the slab are equal to the charges on the corresponding capacitor plates. This corresponds with the force per unit charge within the conductor being zero. The arrangement is shown in Fig. 6.9.

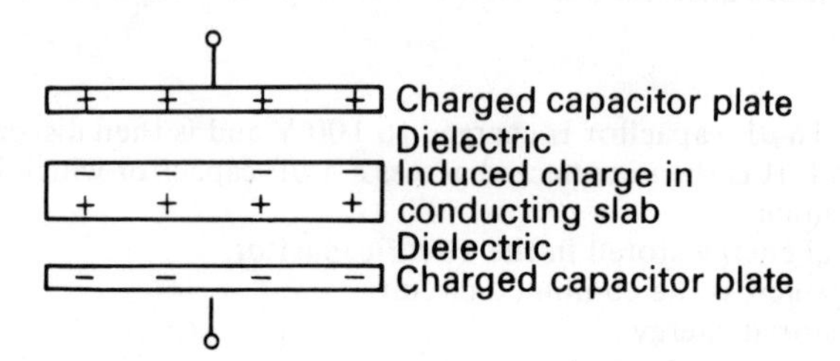

Fig. 6.9. Effect of a conductor slab in between the plates of a parallel-plate capacitor

It should be noted that the number of charges which move is only a small part of the total charges available in the conductor.

When two capacitors are connected in series, the applied e.m.f. causes charge to accumulate on the plates connected to the external supply. The intermediate plates are charged due to induced charge moving within that part of the system so that balance of charge is maintained on each plate. This compares with the movement within the slab described above except that the surfaces of the slab have been separated leaving only a connecting wire between them. It should be noted that in the series connection, again the charge on each plate is numerically the same.

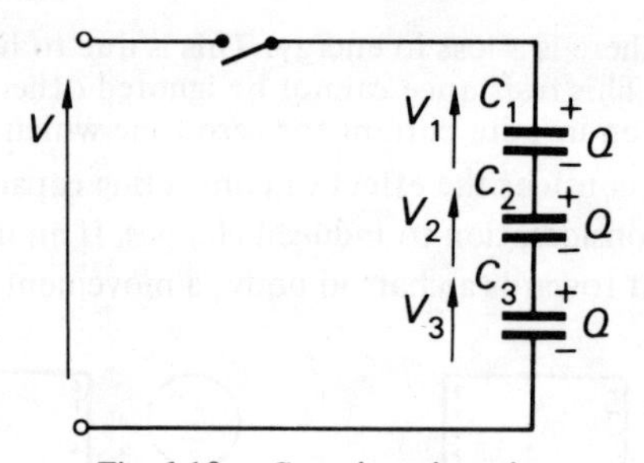

Fig. 6.10. Capacitors in series

Consider three capacitors connected in series as shown in Fig. 6.10. When the switch is closed, a charge Q moves round the circuit and thus each capacitor acquires a charge Q as shown.

Let the capacitors C_1, C_2 and C_3 acquire p.d.'s V_1, V_2 and V_3 respectively. Therefore

$$Q = C_1 V_1 = C_2 V_2 = C_3 V_3$$

Let the total applied voltage be V and the effective capacitance of the series capacitors be C. Hence

$$Q = CV$$

By Kirchhoff's Second Law,

$$V = V_1 + V_2 + V_3$$

$$= \frac{Q}{C_1} + \frac{Q}{C_2} + \frac{Q}{C_3}$$

$$= Q\left(\frac{1}{C_1} + \frac{1}{C_2} + \frac{1}{C_3}\right)$$

$$= Q \cdot \frac{1}{C}$$

$$\frac{1}{C} = \frac{1}{C_1} + \frac{1}{C_2} + \frac{1}{C_3} \qquad (6.12)$$

Although this analysis is based on three capacitors only, the relation holds true for any number. In particular, note the case of two series-connected capacitors.

$$\frac{1}{C} = \frac{1}{C_1} + \frac{1}{C_2}$$

$$C = \frac{C_1 \cdot C_2}{C_1 + C_2} \qquad (6.13)$$

Example 6.6 Two ideal capacitors of 4 μF and 1 μF are connected in series across a 100-V, d.c. supply. Find the p.d. across each capacitor.

$$C = \frac{C_1 \cdot C_2}{C_1 + C_2}$$

$$Q = CV = \frac{VC_1 C_2}{C_1 + C_2}$$

P.D. across 4-μF capacitor (C_1) is given by

$$V_1 = \frac{Q}{C_1} = \frac{VC_2}{C_1 + C_2} = \frac{100 \times 1 \times 10^{-6}}{4 \times 10^{-6} + 1 \times 10^{-6}}$$

$$= \underline{20 \text{ V}}$$

P.D. across 1-μF capacitor is given by

$$V_2 = \frac{Q}{C_2} = \frac{VC_1}{C_1 + C_2} = \frac{100 \times 4 \times 10^{-6}}{4 \times 10^{-6} + 1 \times 10^{-6}}$$

$$= \underline{80 \text{ V}}$$

Alternatively:

$$V_2 = V - V_1$$
$$= 100 - 20$$
$$= \underline{80 \text{ V}}$$

Example 6.7 Find the equivalent capacitance of the circuit shown in Fig. 6.11.

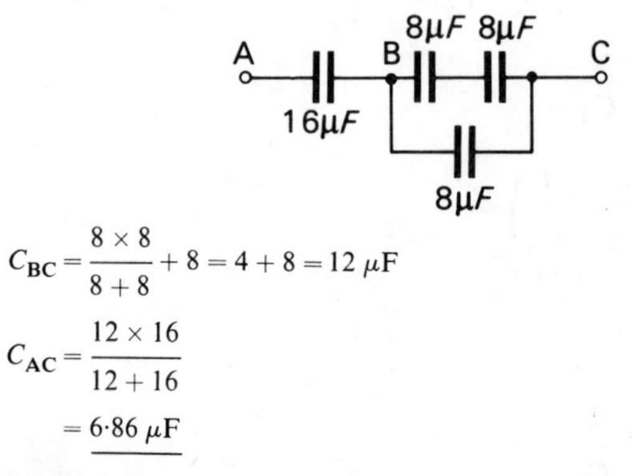

$$C_{BC} = \frac{8 \times 8}{8 + 8} + 8 = 4 + 8 = 12 \ \mu\text{F}$$

$$C_{AC} = \frac{12 \times 16}{12 + 16}$$
$$= \underline{6{\cdot}86 \ \mu\text{F}}$$

6.9 Parallel-plate capacitor with two dielectrics

A parallel-plate capacitor may have two dielectrics and be constructed in the form shown in Fig. 6.12. The electric stress will be directed from one plate straight across to the other plate at all points in the dielectrics. The p.d. between the plates is therefore given by

$$V = E_1 d_1 + E_2 d_2$$

Dielectric no 1 ... ϵ_{r_1} ... d_1

Dielectric no 2 ... ϵ_{r_2} ... d_2

Area A

Fig. 6.12. Parallel-plate capacitor with two dielectrics

Let the charge on the capacitor plates be Q. Consequently the electric flux density at any point in the dielectrics is

$$D = \frac{Q}{A}$$

If ϵ_{r_1} and ϵ_{r_2} are relative permittivities of the dielectrics, then

$$V = E_1 d_1 + E_2 d_2$$
$$= \frac{D d_1}{\epsilon_{r_1} \epsilon_0} + \frac{D d_2}{\epsilon_{r_2} \epsilon_0}$$
$$= \frac{Q}{A \epsilon_0} \left(\frac{d_1}{\epsilon_{r_1}} + \frac{d_2}{\epsilon_{r_2}} \right)$$

Note that it follows that

$$V = Q\left(\frac{1}{C_1} + \frac{1}{C_2}\right)$$

where C_1 and C_2 are the effective capacitances of the two dielectrics. The dielectrics therefore appear to act separately as series capacitors.

The capacitance of the total capacitor is therefore

$$C = \frac{Q}{V}$$

$$= \frac{Q}{\dfrac{Q}{A\epsilon_0}\left(\dfrac{d_1}{\epsilon_{r_1}} + \dfrac{d_2}{\epsilon_{r_2}}\right)}$$

$$= \frac{\epsilon_0 A}{\dfrac{d_1}{\epsilon_{r_1}} + \dfrac{d_2}{\epsilon_{r_2}}} \quad (6.14)$$

It will be shown in the following example (6.8) that the electric stress is greatest in the material of lesser permittivity. The lowest permittivity generally experienced is that of air. It follows that air should be excluded from dielectric materials. Any small voids in the dielectric will tend to have a high electric stress that may lead to a breakdown. The resulting burning will then lead to a complete breakdown of the capacitor dielectrics in the course of time.

Example 6.8 A capacitor of area 100 m² has two dielectrics 1 mm and 2 mm thick. The relative permittivities of these dielectrics is 2 and 5 respectively. Calculate the capacitance of the capacitor and the electric stresses in the dielectrics, if a p.d. of 1000 V is applied between the plates.

$$C = \frac{\epsilon_0 A}{\dfrac{d_1}{\epsilon_{r_1}} + \dfrac{d_2}{\epsilon_{r_2}}}$$

$$= \frac{8{\cdot}854 \times 10^{-12} \times 100}{\left(\dfrac{1}{2} + \dfrac{2}{5}\right)10^{-3}}$$

$$= 0{\cdot}985 \times 10^{-6}\ \text{F}$$

$$= \underline{0{\cdot}985\ \mu\text{F}}$$

$$Q = CV = 0{\cdot}985 \times 10^{-6} \times 1000 = 0{\cdot}985 \times 10^{-3}\ \text{C}$$

$$D = \frac{Q}{A} = \frac{0{\cdot}985 \times 10^{-3}}{100}$$

$$= 9{\cdot}85 \times 10^{-6}\ \text{C/m}^2$$

$$E_1 = \frac{D}{\epsilon_0 \epsilon_{r_1}} = \frac{9{\cdot}85 \times 10^{-6}}{8{\cdot}854 \times 10^{-12} \times 2}$$

$$= \underline{560000 \text{ V/m}}$$

$$E_2 = \frac{D}{\epsilon_0 \epsilon_{r_2}} = \frac{9{\cdot}85 \times 10^{-6}}{8{\cdot}854 \times 10^{-12} \times 5}$$

$$= \underline{224000 \text{ V/m}}$$

6.10 Force on charged plates

When two parallel plates carry opposite charges, a force of attraction exists between the plates. Thus the plates of a capacitor are attracted towards one another.

If one of the plates can be moved towards or away from the other, let the plate separation be increased by Δx as indicated in Fig. 6.13. Assuming the plate charges to

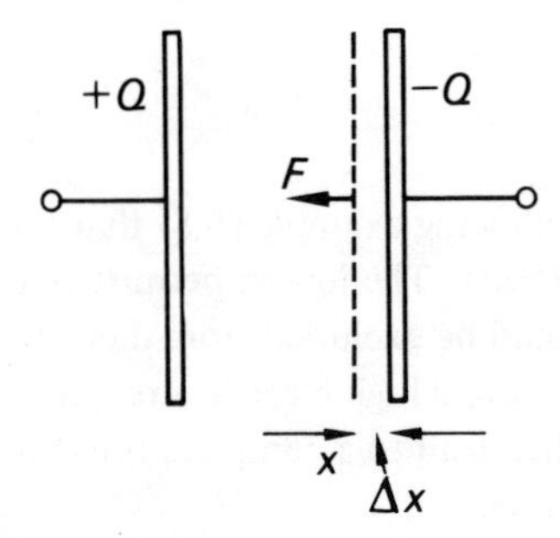

Fig. 6.13. Force between charged parallel plates.

remain unchanged, no electrical energy can enter or leave the system. Nevertheless the stored energy in the gap changes. This change is due to the transfer of mechanical energy. For the small movement, the work done is

$$F . \Delta x$$

where F is the force of attraction between the plates.

The initial energy stored in the capacitor is given by

$$W_f = \tfrac{1}{2}CV^2$$

where C is the initial capacitance and V is the initial p.d. between the plates due to the charge Q. However, $V = Q/C$ whence

$$W_f = \tfrac{1}{2}C\left(\frac{Q}{C}\right)^2$$

$$= \frac{Q^2}{2C}$$

This is a preferable expression since it defines the initial energy stored in terms of the given charge. Because the separation of the plates has been increased, the capacitance will decrease by ΔC. The final capacitance is therefore $(C - \Delta C)$ and the final stored energy is given by

$$W_f + \Delta W_f = \frac{1}{2} \cdot \frac{Q^2}{(C - \Delta C)}$$

$$= \frac{Q^2(C + \Delta C)}{2(C^2 - \Delta C^2)}$$

Since Δx is small, ΔC is small relative to C and therefore ΔC^2 can be neglected,

$$W_f + \Delta W_f = \frac{Q^2(C + \Delta C)}{2C^2}$$

$$= \frac{Q^2}{2C} + \frac{Q^2}{2C^2} \cdot \Delta C$$

$$\Delta W_f = \frac{Q^2}{2C^2} \cdot \Delta C$$

$$= F . \Delta x$$

$$F = \frac{Q^2}{2C^2} \frac{\Delta C}{\Delta x}$$

In the notation of calculus, as $x \to 0$, then

$$\frac{\Delta C}{\Delta x} \to \frac{dC}{dx}$$

$$F = \frac{Q^2}{2C^2} \cdot \frac{dC}{dx}$$

$$= \tfrac{1}{2} V^2 . \frac{dC}{dx} \qquad (6.15)$$

It follows that

$$F = \frac{d}{dx}(\tfrac{1}{2} CV^2)$$

$$= \frac{dW_f}{dx}$$

The mechanical force is therefore equal to the rate of change of stored energy with deformation of the arrangement. This is the fundamental principle on which all electro-mechanical conversion devices are based. Its electromagnetic applications are studied in chapter 12.

Considering relation (6.15) again, for a parallel-plate arrangement,

$$C = \frac{\epsilon_0 \epsilon_r A}{x}$$

$$\frac{dC}{dx} = -\frac{\epsilon_0 \epsilon_r A}{x^2}$$

Hence the force of attraction is given by

$$F = -\tfrac{1}{2}V^2 . \frac{\epsilon_0 \epsilon_r A}{x^2} \tag{6.16}$$

The negative sign indicates that the force acts in the opposite direction to that in which the plate separation was measured. This corresponds with the concept that the force is one of attraction. In relation (6.15) no negative sign appeared but when dC/dx is evaluated, it is found to have a negative value since the capacitance decreases with increase of plate separation.

The above relations could also have been derived if constant voltage instead of constant charge had been assumed. However, this does not mean that the energy transfer processes are the same in each case. In the constant-charge case, the mechanical energy was transferred to energy stored in the field, there being no change in the electrical energy.

In the constant-voltage case, there must be a transfer of charge to or from the plates to maintain the voltage. Let the plates be drawn apart by a distance Δx as before. With constant voltage and the consequent decrease in capacitance, charge must leave the plates. Also the stored energy must decrease. The mechanical energy fed into the system to separate the plates has not gone to the stored field energy since this has decreased. Both these energy changes are transferred to the electrical system.

The mechanical work done is

$$W_M^f = F . \Delta x$$

$$= \tfrac{1}{2}V^2 . \frac{dC}{dx} . \Delta x$$

The change in stored energy is given by the rate of change of stored energy with change of plate separation times the change Δx.

$$W_f^M = \frac{d}{dx}(\tfrac{1}{2}CV^2) . \Delta x$$

$$= \tfrac{1}{2}V^2 . \frac{dC}{dx} . \Delta x$$

Thus the field and mechanical changes in energy are equal as before. It is for this reason that the force expression remains unchanged although the above does not prove this to be so.

On the electrical system, the change in the charge is given by the rate of change of charge with plate separation times the change Δx.

$$\Delta Q = \frac{\mathrm{d}}{\mathrm{d}x}(CV).\Delta x$$

$$= V\cdot\frac{\mathrm{d}C}{\mathrm{d}x}\cdot\Delta x$$

Since the voltage is constant, the change in electrical energy is given by

$$W_E^f = V^2\cdot\frac{\mathrm{d}C}{\mathrm{d}x}\cdot\Delta x$$

This is the result already suggested, i.e. that the change of electrical energy is equal to the sum of the changes of mechanical and stored energies.

The most important point to note is the equal division of the energy. Again this will be discussed further in chapter 12 and is a further example of the 50-50 rule discussed therein.

Example 6.9 A parallel-plate capacitor has its plates separated by 0·5 mm of air. The area of the plates is 1 m² and they are charged to have a p.d. of 100 V. The plates are pulled apart until they are separated by 1·0 mm of air. Assuming the p.d. to remain unchanged, what is the average mechanical force experienced in separating the plates.

When the plate separation is 0·5 mm, the capacitance of the arrangement is

$$C_1 = \frac{\epsilon_0 A}{d_1} = \frac{8{\cdot}854\times10^{-12}\times1}{0{\cdot}5\times10^{-3}}$$

$$= 17{\cdot}7\times10^{-9}\,\mathrm{F}$$

The initial energy stored is

$$W_{f_1} = \tfrac{1}{2}CV^2 = 0{\cdot}5\times17{\cdot}7\times10^{-9}\times100^2$$

$$= 8{\cdot}854\times10^{-5}\,\mathrm{J}$$

After the movement has taken place,

$$C_2 = \tfrac{1}{2}.C_1 = 8{\cdot}854\times10^{-9}\,\mathrm{F}$$

and $$W_{f_2} = \tfrac{1}{2}W_{f_1} = 4{\cdot}427\times10^{-5}\,\mathrm{J}$$

The change in energy is

$$W_{f_1} - W_{f_2} = (8{\cdot}854 - 4{\cdot}427)\times10^{-5}$$

$$= 4{\cdot}427\times10^{-5}\,\mathrm{J}$$

$$F = \frac{\Delta W_f}{\Delta x}$$

$$= \frac{4{\cdot}427\times10^{-5}}{0{\cdot}5\times10^{-3}}$$

$$= 8{\cdot}854\times10^{-2}\,\mathrm{N}$$

$$= \underline{88{\cdot}54\,\mathrm{mN}}$$

6.11 Time constant

When a direct potential difference is applied to a capacitor, the growth of p.d. between the plates is shown graphically to a base of time in Fig. 6.14.

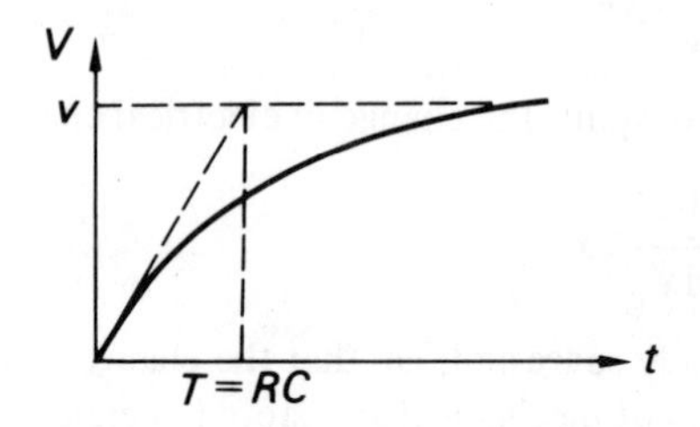

Fig. 6.14. Growth of p.d. across a capacitor

It can be shown that this curve rises exponentially towards a value equal to that of the supply voltage V. Such a curve takes the mathematical form

$$v = V(1 - e^{-(1/RC)t}) \tag{6.17}$$

where e is the natural logarithmic base and is 2·718.

Whilst is it not necessary to be able to apply this relation in calculations at this introductory stage, the relation nevertheless merits some attention. In particular, consider the index of e. RC is termed the time constant of the circuit. Let T be the time constant.

$$T = RC \tag{6.18}$$

The time constant is the time required for the voltage and current to reach their steady-state values if they continued to change at their initial rates of growth and

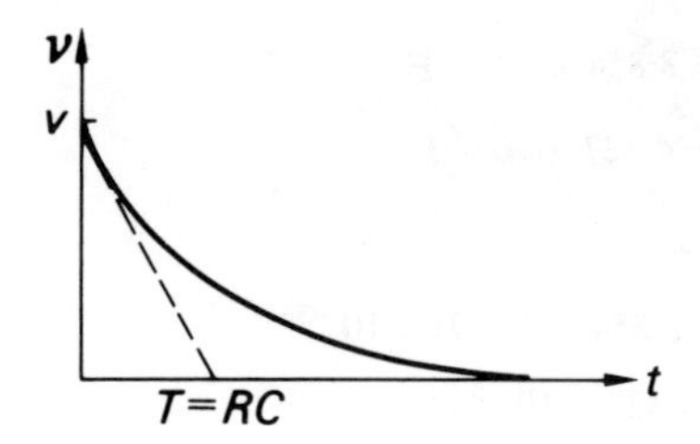

Fig. 6.15. Decay of p.d. in a capacitive circuit

decay. Growth in this context refers to the increase in p.d. between the capacitor plates. The larger the time constant then the longer does the capacitor p.d. take to reach its final steady-state value. The converse takes place when the energy is released by the capacitor on discharge.

PROBLEMS ON ELECTROSTATICS AND CAPACITANCE

1 A direct potential of 200 V is applied across the plates of a two-plate capacitor of capacitance 0·05 μF. The effective area of each plate is 0·05 m^2 and the absolute permittivity of the dielectric is 20 pF/m. Calculate:
(*a*) The electric flux density.
(*b*) The electric field strength.

0·2 mC/m^2; 10 MV/m

2 A parallel-plate capacitor has 500 μF capacitance and an effective plate area of 0·025 m^2. The relative permittivity of the dielectric is 2·5. If a constant voltage of 500 V is maintained across the plates, calculate:
(*a*) The electric flux.
(*b*) The electric flux density.
(*c*) The electric field strength.

0·25 μC; 0·01 mC/m^2; 45·3 MV/m

3 An 8-μF capacitor is charged by a constant current. If the voltage across the capacitor rises from zero to 240 V in 40 ms, find the value of the current.

48 mA

4 The potential difference of 10 kV is applied to the terminals of a capacitor consisting of two circular plates each having an area of 0·01 m^2 separated by a dielectric 1·0 mm thick. The resulting capacitance of the arrangement is 300 pF. Calculate:
(*a*) The total electric flux.
(*b*) The electric flux density.
(*c*) The relative permittivity of the dielectric.

3 μC; 0·3 mC/m^2; 3·39

5 A fully-charged capacitor is discharged through a resistor. The voltage across the capacitor falls from 250 V to 50 V in 5 s and during this time the average discharge current is 2·0 mA. Calculate the capacitance of the capacitor. 50 μF

6 A parallel-plate capacitor is built up from two metal plates, each of area 0·02 m^2, separated 0·5 mm by a sheet of mica. The relative permittivity of mica is 6. Calculate the capacitance of this capacitor.

The capacitor is charged to a potential difference of 100 V between the plates and is then isolated from the supply. The sheet of mica is removed without altering the distance between the plates and without allowing any charge to leak away. What is the new value of the potential difference between the plates?

2 120 pF, 600 V (SANCAD)

7 A parallel-plate capacitor with a distance of 2 mm between the plates has a dielectric with a relative permittivity of 2·5. It is charged to a p.d. of 1 kV.

Calculate the flux density between the plates. 11·07 $\mu C/m^2$

8 Circuit ABCD is made up as follows. AB consists of a 5-μF capacitor in parallel with a 7-μF capacitor. BC consists of a 2-μF capacitor. CD is a 4-μF capacitor shunted by two series capacitors of 20 μF and 5 μF. Further a 2-μF capacitor is connected between A and D. If a supply of 100 V direct current is applied at A and D, determine:
(*a*) The effective capacitance between A and D.
(*b*) The charge held by and the energy stored in the 20-μF capacitor.

3·41 μF; $7{\cdot}05 \times 10^{-5}$ C; $12{\cdot}4 \times 10^{-5}$ J

9 Circuit ABCD is made up as follows. AB consists of a 10-μF capacitor. BC has three parallel branches, two being 8-μF capacitors whilst the third consists of two 8-μF capacitors in series. CD consists of a 12-μF capacitor shunted by one of 8 μF. The circuit is supplied from a d.c. source of 400 V.

Determine:

(*a*) The effective capacitance between A and D.
(*b*) The charge held by and the energy stored in the 12-μF capacitor.
(*c*) The potential difference between A and the junction between the series branch capacitors.

5 μF; 1·2 mC; 60 mJ; 250 V (SANCAD)

10 A capacitor is formed by two flat metal plates each 5 000 mm^2 in area and separated by a dielectric 1·0 mm thick. The capacitance of this arrangement is 0·000 2 μF and a p.d. of 10 kV is applied to the terminals.

Calculate:

(*a*) The charge on the plates.
(*b*) The relative permittivity of the dielectric.
(*c*) The electric flux density.
(*d*) The energy stored.

2 μC; 4·52; 400 μC/m^2; 10 mJ

11 A 20-pF capacitor is to be constructed using two parallel metal plates, the width of which is fixed at 50 mm. The maximum potential gradient in the air dielectric is not to exceed 10 kV/m when a p.d. of 100 V is applied across the plates.

Determine the length of the plates.

If the capacitor is connected in series with a second capacitor of 50 pF and if a p.d. of 100 V is applied across the combination, find the voltage across each.

Calculate the energy stored in the electric fields, assuming that the capacitors are perfect.

45·2 cm; 71·5 V; 28·5 V; 7×10^{-8} J

12 A 5-μF capacitor having a paper dielectric has to operate on a peak voltage of 500 V. This capacitor is to be charged from a d.c. source. Determine:

(*a*) The plate area and spacing if the maximum permissible stress in the paper is 250 kV/m and the relative permittivity is 6.
(*b*) The p.d. across the capacitor plates after 7·5 s if the charging current is maintained constant at 0·1 mA.

188·5 m^2; 2·0 mm; 150 V

13 A perfectly-insulated parallel-plate capacitor is formed by two flat plates each having an area 0·05 m^2 separated by a solid dielectric of relative permittivity 4 and thickness 0·2 mm. The p.d. across the plates is 50 V. Calculate:

(*a*) The capacitance, and;
(*b*) The charge held by as well as the energy stored in the capacitor.

The above capacitor is connected to a d.c. supply such that the p.d. across the plates is increased from 50 V to 200 V at the uniform rate 20 V/s. Calculate the charging current and the change in the stored energy.

8 850 pF; 0·4 425 μC; 11·07 μJ; 0·177 μA; 166 μJ

14 Two parallel-plate capacitors A and B are connected in series. A has a plate area of 5 000 mm^2, an air dielectric and the distance between the plates is 1·0 mm. B has a plate area of 2 000 m^2, a solid dielectric of relative permittivity 4 and of thickness 0·5 mm.

Find:

(*a*) The voltage across the combination if the potential gradient associated with the capacitor A is 100 kV/m.

(*b*) The charge on each capacitor.
(*c*) The energy stored in the electrostatic fields.

131 V; 4 400 pC; 0·288 μJ

15 A uniform sheet of dielectric material of thickness t rests on the lower horizontal plate of a parallel-plate capacitor and partially fills the space between the plates; the dielectric is then removed and the plate separation decreased by an amount d in order to restore the capacitance of the system to its original value. Neglecting fringing, show that for material

$$e_r = \frac{t}{t-d}$$

16 A parallel-plate capacitor is made from two plates 254 mm in diameter and mounted 2·0 mm apart in air. A 1·0-mm thick sheet of nylon is placed between the plates and the resulting capacitance of the arrangement is 360 pF. Find the relative permittivity of the nylon.

4.1 (SANCAD)

17 Explain how a capacitor may be used to reduce the sparking at the contacts of a d.c. switch on opening.

A direct potential difference of 100 V is applied suddenly to a circuit comprising an uncharged 100-μF capacitor in series with a 500-Ω resistor. Calculate the initial rate of rise of voltage across the capacitor. What is the final energy stored in the capacitor?

2 kV/s; 0·5 J (SANCAD)

7

Single-phase alternating current circuits

In chapter 2, only the effect of resistance had to be taken into account when analysing direct current circuits. Alternating current circuits have other factors which must be considered; these factors are inductance and capacitance. In order to limit this introductory study, only those circuits in which both the voltage and current waveforms are sinusoidal will be considered. This is not an unreasonable assumption since it applies to the majority of a.c. circuits; its validity, however, will be further discussed at the end of the chapter.

7.1 Resistance

All circuits incorporate resistance to some degree. However, just as in the d.c. circuit case, only the relatively large values of resistance in a circuit need be considered. In the purely resistive circuit shown in Fig. 7.1, suppose that the source voltage may be represented by $v = V_m \sin \omega t$ and that the circuit resistance may be represented by R. At any instant, the resulting current is given by

$$i = \frac{v}{R} = \frac{V_m}{R} \sin \omega t \qquad (7.1)$$

The maximum value of this expression occurs when $\sin \omega t = 1$, hence

$$I_m = \frac{V_m}{R}$$

Substituting in (7.1)

$$i = I_m \sin \omega t \qquad (7.2)$$

Also

$$V_m = I_m R$$

$$\frac{V_m}{\sqrt{2}} = \frac{I_m}{\sqrt{2}} \cdot R$$

Hence in r.m.s. values

$$V = IR \qquad (7.3)$$

The voltage and current waveforms for such an arrangement are shown in Fig. 7.1. Both waveforms are sinusoidal and they 'rise and fall' together. When two sine waves behave in this manner, they are said to be 'in phase'. In such a case, they pass through zero at the same instants, i.e. when $t = 0$, π/ω, $2\pi/\omega$, etc., pass through their positive maximum values at the same instants, i.e. when $t = \pi/2\omega$, etc. and pass through their negative maximum values at the same instants, i.e. when $t = 3\pi/2\omega$, etc. Thus the voltage and the current in a circuit, which is purely resistive, are in phase.

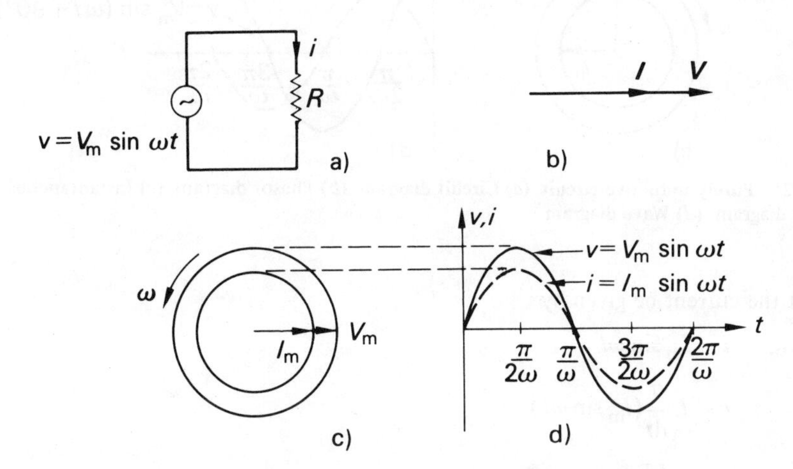

Fig. 7.1. Purely resistive circuit (*a*) Circuit diagram (*b*) Phasor diagram (*c*) Instantaneous phasor diagram (*d*) Wave diagram

The instantaneous phasor diagram has been drawn for the instant $t = 0$. When the instantaneous values of the phasors are plotted against the vertical datum, the instantaneous voltages and currents are evaluated thereby. The phasor diagram, on the other hand, represents the r.m.s. values and the relative directions of the phasors have time significance only. It is generally convenient to take the voltage as reference and the phasor is drawn horizontally.

7.2 Inductance

A circuit must complete at least one turn and consequently the circuit will have inductance. Under the steady conditions of a d.c. circuit, the inductance has no effect and therefore does not require to be included in the steady-state d.c. circuit analysis of chapter 2. In an a.c. circuit, the fluctuating current gives rise to a fluctuating flux, which induces an e.m.f. in the circuit. This appears as an effective volt-drop.

The effect of inductance only in a circuit, as shown in Fig. 7.2., will now be analysed. From the definition of inductance (see para. 5.6)

$$v = L\frac{di}{dt}$$

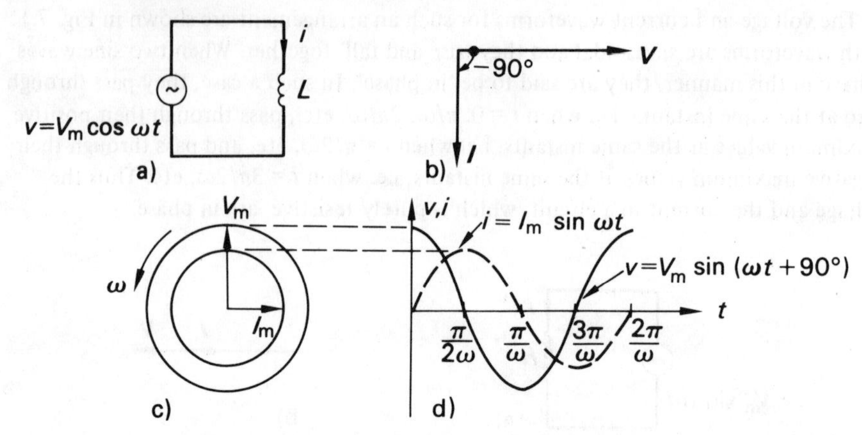

Fig. 7.2. Purely inductive circuit (*a*) Circuit diagram (*b*) Phasor diagram (*c*) Instantaneous phasor diagram (*d*) Wave diagram

Let the current be given by

$$i = I_m \sin \omega t$$

$$v = L\frac{d}{dt}(I_m \sin \omega t)$$

$$= \omega L I_m \cos \omega t$$

$$= \omega L I_m \sin(\omega t + \pi/2) \quad (7.4)$$

Hence $\quad V_m = \omega L I_m$

$$\frac{V_m}{\sqrt{2}} = \frac{I_m}{\sqrt{2}}\omega L$$

In r.m.s. values

$$V = I\omega L$$

$$= I(2\pi f L) \quad (7.5)$$

ωL ($= 2\pi f L$) is termed the inductive reactance of the circuit and is represented by the symbol X_L. Thus

$$V = IX_L \quad (7.5.1)$$

Inductive reactance	Symbol: X_L	Unit: ohm (Ω)

Also $\quad X_L = 2\pi f L = \omega L \quad (7.5.2)$

Hence it will be seen that the inductive reactance depends on the construction of the circuit and the frequency of the alternating quantities.

The waveform diagram (Fig. 7.2(*d*)) shows the phase relationship between the voltage and the current as derived from relation (7.4). Clearly the voltage and current are no longer in phase but are instead $\pi/2\omega$ seconds apart, i.e. they are quarter of a cycle apart, the voltage leading the current. Such waveforms are 'in quadrature'.

In the phasor diagram, the voltage is taken as reference and now the current is seen to lag the voltage by 90°. It is a characteristic of inductive circuits that the current lags the voltage. It should be noted that since both the voltage and the current have the same angular frequency, they can again be shown in the same phasor diagram which only indicates their relative phase displacement.

The instantaneous phasor diagram is drawn for the instant $t = 0$ in relation (7.4). This compares with Fig. 7.1*c*.

7.3 Resistance and inductance in series

Having considered the effects of resistance and inductance separately in a circuit, it is now necessary to consider their combined effects. This can be most simply achieved by connecting the resistance and inductance in series, as shown in Fig. 7.3(*a*).

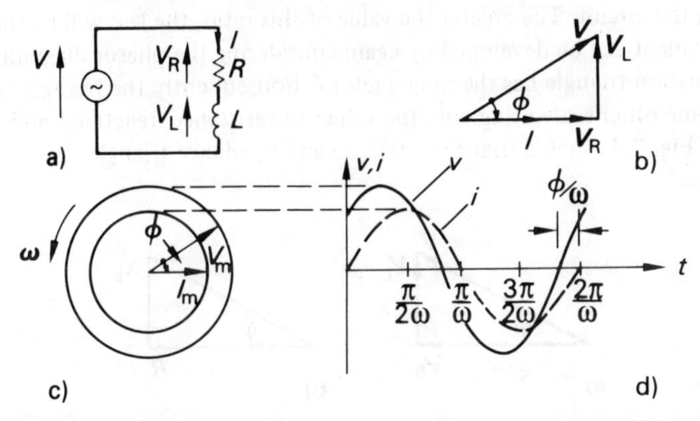

Fig. 7.3. Resistance and inductance in series (*a*) Circuit diagram (*b*) Phasor diagram (*c*) Instantaneous phasor diagram (*d*) Wave diagram

The phasor diagram results from an application of Kirchhoff's 2nd Law. For convenience, the current is taken as reference since it is common to all the elements of a series circuit. The circuit voltage may then be derived from the following relations:

$V_R = IR$, where V_R is in phase with I.

$V_L = IX_L$, where V_L leads I by 90°

$$V = V_R + V_L \tag{7.6}$$

In the phasor diagram, shown in Fig. 7.3(*b*), the total voltage is thus obtained from relation (7.6), which is a complexor summation. The arithmetical sum of V_R and V_L is incorrect, giving too large a value for the total voltage V.

The angle of phase difference between V and I is termed the phase angle and is represented by ϕ.

Also $$V = (V_R^2 + V_L^2)^{\frac{1}{2}}$$
$$= (I^2 R^2 + I^2 X_L^2)^{\frac{1}{2}}$$
$$= I(R^2 + X_L^2)^{\frac{1}{2}}$$

Hence $V = IZ$ (7.7)

where $Z = (R^2 + X_L{}^2)^{\frac{1}{2}}$

$= (R^2 + \omega^2 L^2)^{\frac{1}{2}}$ (7.8)

Z is termed the impedance of the circuit. Relation (7.7) will be seen to be a development of the relation $V = IR$ used in d.c. circuit analysis. However, for any given frequency, the impedance is constant and hence Ohm's Law also applies to a.c. circuit analysis.

Impedance	Symbol: Z	Unit: ohm (Ω)

The instantaneous phasor diagram, and the resulting wave diagram, shows that the current lags the voltage by a phase angle greater than 0° but less than 90°. The phase angle between voltage and current is determined by the ratio of resistance to inductive reactance in the circuit. The greater the value of this ratio, the less will be the angle ϕ.

This statement can be developed by again considering the phasor diagram. Each side of the summation triangle has the same factor I. Consequently the triangle can be drawn to some other scale using only the values of resistance, reactance and impedance, as shown in Fig. 7.4. Such a triangle is termed an impedance triangle.

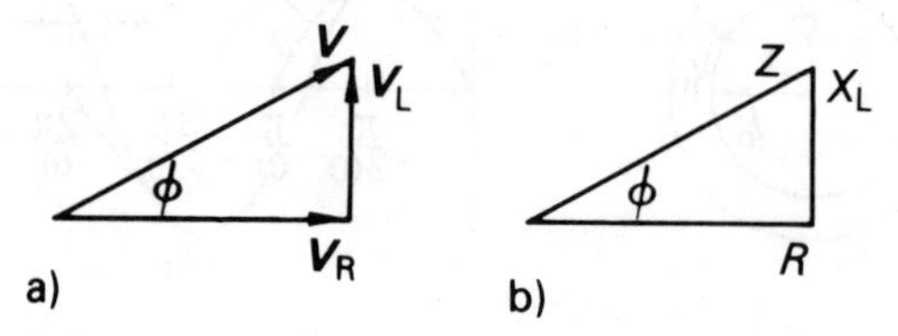

Fig. 7.4. Voltage and impedance triangles (*a*) **Voltage diagram** (*b*) **Impedance diagram**

Just as in Fig. 7.3(*b*), the triangle is again right-angled. This compares with relation (7.8). By the geometry of the diagram:

$$\phi = \tan^{-1}\frac{V_L}{V_R} = \tan^{-1}\frac{IX_L}{IR}$$

$$= \tan^{-1}\frac{X_L}{R} \qquad (7.8.1)$$

To emphasise that the current lags the voltage, it is usual to give either the resulting angle as a negative value or else to use the word 'lag' after the angle. This is illustrated in Example 7.1.

The phase angle may also be derived as follows:

$$\phi = \cos^{-1}\frac{V_R}{V} = \cos^{-1}\frac{R}{Z} \qquad (7.8.2)$$

hence $$\phi = \cos^{-1}\frac{R}{(R^2 + \omega^2 L^2)^{\frac{1}{2}}}$$

Example 7.1 A resistance of 7·0 Ω is connected in series with a pure inductance of 31·4 mH and the circuit is connected to a 100-V, 50-Hz, sinusoidal supply. Calculate:

(*a*) The circuit current.
(*b*) The phase angle.

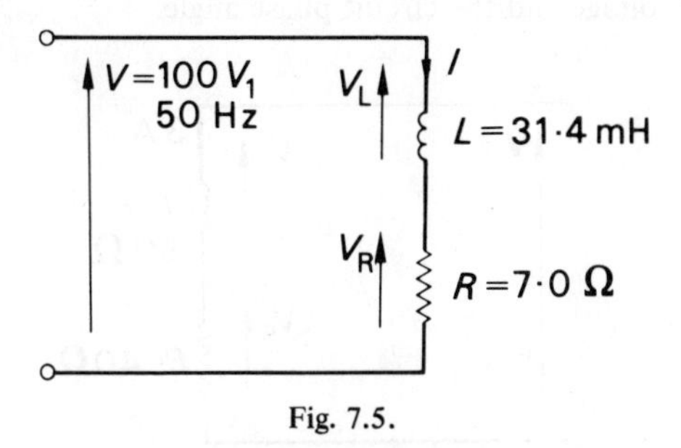

Fig. 7.5.

$$X_L = 2\pi f L = 2\pi 50 \times 31{\cdot}4 \times 10^{-3} = 10{\cdot}0\ \Omega$$

$$Z = (R^2 + X_L{}^2)^{\frac{1}{2}} = (7{\cdot}0^2 + 10{\cdot}0^2)^{\frac{1}{2}} = 12{\cdot}2\ \Omega$$

$$I = \frac{V}{Z} = \frac{100}{12{\cdot}2} = \underline{8{\cdot}2\ \text{A}}$$

$$\phi = \tan^{-1}\frac{X_L}{R} = \tan^{-1}\frac{10{\cdot}0}{7{\cdot}0} = \underline{55{\cdot}1^\circ\ \text{lag or} -55{\cdot}1^\circ}$$

Example 7.2 A pure inductance of 318 mH is connected in series with a pure resistance of 75 Ω. The circuit is supplied from a 50-Hz sinusoidal source and the voltage across the 75-Ω resistor is found to be 150-V.
Calculate the supply voltage.

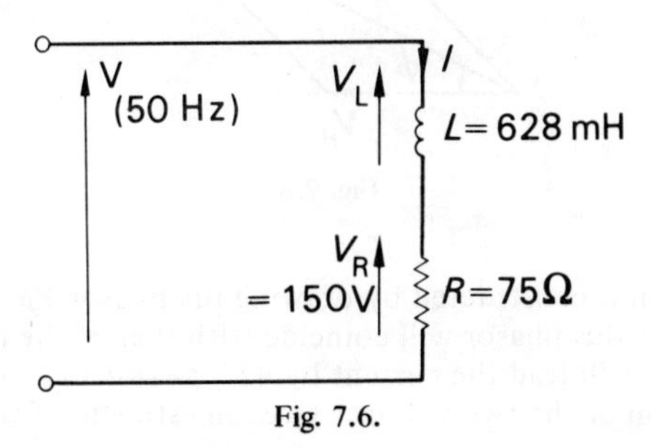

Fig. 7.6.

$$V_R = 150\ \text{V}$$

$$I = \frac{V}{R} = \frac{150}{75} = 2\ \text{A}$$

$$X_L = 2\pi f L = 2\pi 50 \times 318 \times 10^{-3} = 100\ \Omega$$

$$V_L = IX_L = 2 \times 100 = 200\ \text{V}$$

$$V = (V_R{}^2 + V_L{}^2)^{\frac{1}{2}} = (150^2 + 200^2)^{\frac{1}{2}} = \underline{250\ \text{V}}$$

Alternatively $Z = (R^2 + X_L{}^2)^{\frac{1}{2}} = (75^2 + 100^2)^{\frac{1}{2}} = 125\ \Omega$

$$V = IZ = 2 \times 125 = \underline{250\ \text{V}}$$

Example 7.3 A coil, having both resistance and inductance, has a total effective impedance of 50 Ω and the phase angle of the current through it with respect to the voltage across it is 45° lag. The coil is connected in series with a 40-Ω resistor across a sinusoidal supply. The circuit current is 3·0 A; by constructing a phasor diagram, estimate the supply voltage and the circuit phase angle.

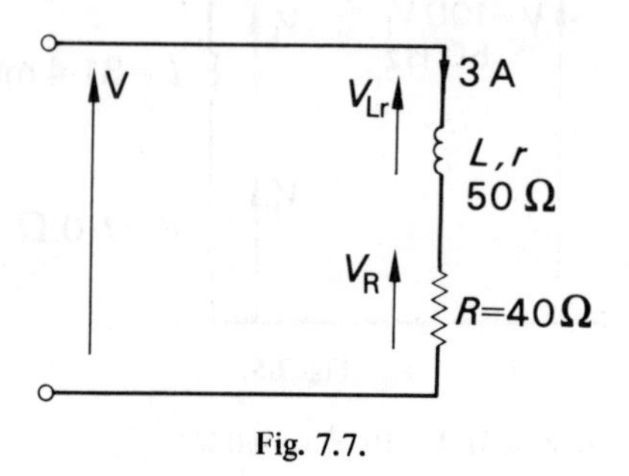

Fig. 7.7.

$$V_R = IR = 3 \times 40 = 120 \text{ V}$$

$$V_{Lr} = IZ_{Lr}\, 3 \times 50 = 150 \text{ V}$$

The use of subscript notation should be noted in the previous line. It would have been incorrect to write that $V_{Lr} = IZ$, since Z is used to represent the total circuit impedance. In more complex problems, numbers can be used, i.e. Z_1, Z_2, Z_3, etc. In this example, such a procedure would be tedious.

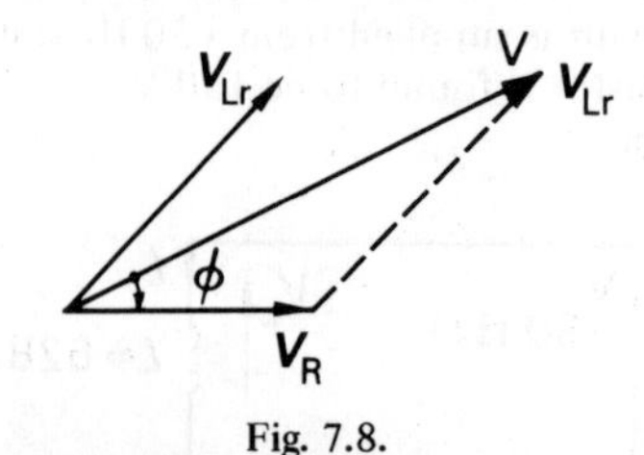

Fig. 7.8.

The phasor diagram is constructed by drawing the phasor V_R to some appropriate scale. The direction of this phasor will coincide with that of the current I. Since the voltage across the coil will lead the current by 45°, phasor V_{Lr} can also be drawn. Complexor summation of the two voltages gives an estimate of the total voltage.

From the diagram:

$$V = 250 \text{ V}$$

$$\phi = 26° \text{ lag}$$

7.4 Capacitance

Just as the physical layout of a circuit incorporates inductance into a circuit, so there must also be capacitance due to the proximity of conductors at different potentials. Consider the simple circuit shown in Fig. 7.9(*a*); this circuit comprises pure capacitance. In the circuit, a voltage v (= $V_m \sin \omega t$) is applied to a capacitor C.

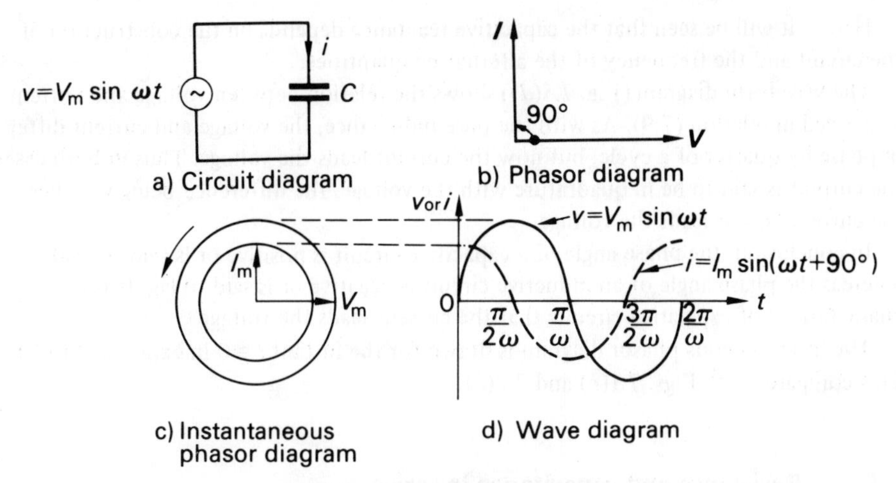

Fig. 7.9. **Purely capacitive circuit**

Following from the definition of capacitance (see para. 6.4)

$$q = Cv = CV_m \sin \omega t$$

$$i = \frac{dq}{dt} = C\frac{dv}{dt}$$

$$= C\frac{d}{dt}(V_m \sin \omega t)$$

$$= \omega C V_m \cos \omega t$$

$$= \omega C V_m \sin(\omega t + 90°) \qquad (7.9)$$

Hence $\quad I_m = \omega C V_m$

$$\frac{I_m}{\sqrt{2}} = \frac{\omega C V_m}{\sqrt{2}}$$

In r.m.s. values

$$I = \omega C V$$

$$V = I.\frac{1}{\omega C} = I.\frac{1}{2\pi f C} \qquad (7.10)$$

$1/\omega C$ (= $1/2\pi fC$) is termed the capacitive reactance of the circuit and is represented by the symbol X_C. Thus

$$V = IX_C \qquad (7.10.1)$$

Capacitive reactance	Symbol: X_C	Unit: Ohm (Ω)

Also $\quad X_C = \dfrac{1}{\omega C} = \dfrac{1}{2\pi f C} \qquad (7.10.2)$

Hence, it will be seen that the capacitive reactance depends on the construction of the circuit and the frequency of the alternating quantities.

The waveform diagram (Fig. 7.5(d)) shows the relation between voltage and current as derived in relation (7.9). As with the pure inductance, the voltage and current differ in phase by quarter of a cycle, but now the current leads the voltage. Thus in both cases, the current is said to be in quadrature with the voltage, the difference being whether the current lags or leads the voltage.

In conclusion, the phase angle of a capacitive circuit is positive or is said to lead, whereas the phase angle of an inductive circuit is negative or is said to lag. It is a characteristic of capacitive circuits that the current leads the voltage.

The instantaneous phasor diagram is drawn for the instant $t = 0$ in expression (7.9). This compares with Figs. 7.1(c) and 7.2(c).

7.5 Resistance and capacitance in series

The effect of connecting resistance and capacitance in series is illustrated in Fig. 7.10. The current is again taken as reference.

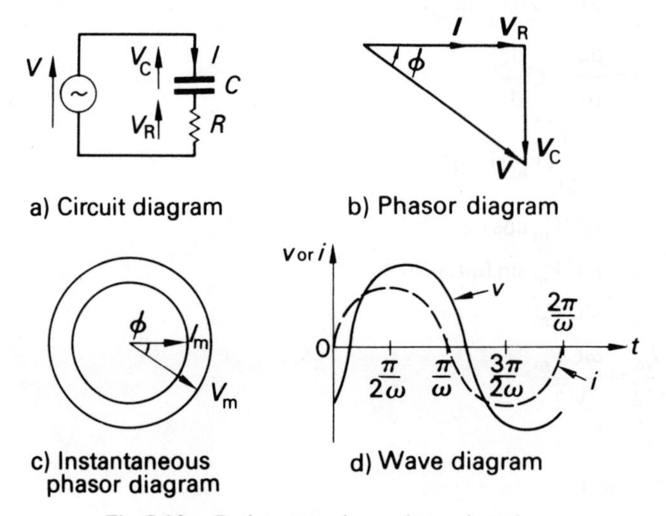

Fig. 7.10. Resistance and capacitance in series

The circuit voltage is derived from the following relations.

$V_R = IR$, where V_R is in phase with I.

$V_C = IX_C$, where V_C lags I by 90°.

$$V = V_R + V_C$$

Also

$$V = (V_R^2 + V_C^2)^{\frac{1}{2}}$$
$$= (I^2 R^2 + I^2 X_C^2)^{\frac{1}{2}}$$
$$= I(R^2 + X_C^2)^{\frac{1}{2}}$$

Hence $V = IZ$

where $Z = (R^2 + X_C{}^2)^{\frac{1}{2}}$

$$= \left(R^2 + \frac{1}{\omega^2 C^2}\right)^{\frac{1}{2}} \tag{7.11}$$

Again Z is the impedance of the circuit. For any given frequency, the impedance remains constant and is thus the constant used in Ohm's Law, i.e. the impedance is the ratio of the voltage across the circuit to the current flowing through it, other conditions remaining unchanged.

The instantaneous phasor diagram, and the resulting wave diagram, show that the current leads the applied voltage by a phase angle greater than 0° but less than 90°. The phase angle between voltage and current is determined by the ratio of resistance

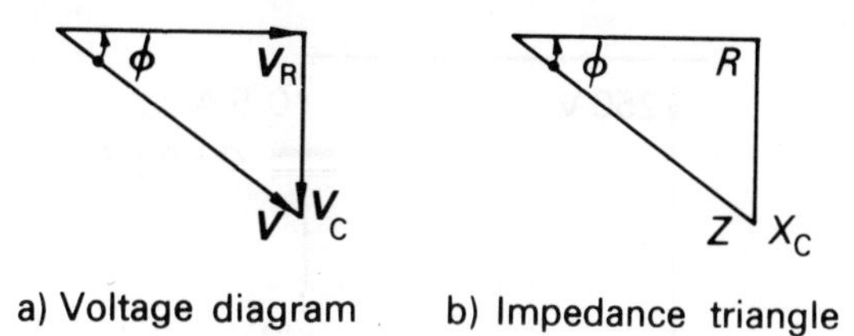

Fig. 7.11. Voltage and impedance diagrams

to capacitive reactance in the circuit. The greater the value of this ratio, the less will be the angle ϕ. This can be illustrated by drawing the impedance triangle for the circuit, as shown in Fig. 7.11.

By the geometry of the diagram:

$$\phi = \tan^{-1}\frac{V_C}{V_R} = \tan^{-1}\frac{IX_C}{IR}$$

$$= \tan^{-1}\frac{X_C}{R} \tag{7.11.1}$$

To emphasise that the current leads the voltage, it is usual to either give the resulting angle as a positive value or else to use the word 'lead' after the angle. This is illustrated in Example 7.4.

The phase angle can also be derived as follows:

$$\phi = \cos^{-1}\frac{V_R}{V} = \cos^{-1}\frac{R}{Z} \tag{7.11.2}$$

$$= \cos^{-1}\frac{R}{\left(R^2 + \dfrac{1}{\omega^2 C^2}\right)^{\frac{1}{2}}}$$

Example 7.4 A capacitor of 8 μF takes a current of 1·0 A when the alternating voltage applied across it is 250 V. Calculate:

(*a*) The frequency of the applied voltage.
(*b*) The resistance to be connected in series with the capacitor to reduce the current in the circuit to 0·5 A at the same frequency.
(*c*) The phase angle of the resulting circuit.

$$X_C = \frac{V}{I} = \frac{250}{1 \cdot 0} = 250\ \Omega$$

$$= \frac{1}{2\pi f C}$$

$$\therefore\ f = \frac{1}{2\pi C X_C} = \frac{1}{2\pi \times 8 \times 10^{-6} \times 250} = \underline{79 \cdot 5\ \text{Hz}}$$

When a resistance is connected in series with the capacitor, the circuit is now as follows:

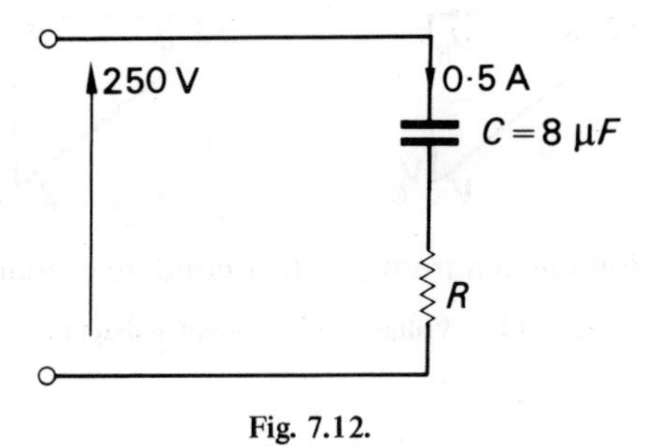

Fig. 7.12.

$$Z = \frac{V}{I} = \frac{250}{0 \cdot 5} = 500\ \Omega$$

$$= (R^2 + X_C^2)^{\frac{1}{2}}$$

but $X_C = 250\ \Omega$

hence $\underline{R = 433\ \Omega}$

$$\phi = \cos^{-1}\frac{R}{Z} = \cos^{-1}\frac{433}{500} = \underline{+30^\circ} \text{ or } \underline{30^\circ \text{ lead}}$$

7.6 General a.c. series circuit

Having considered the various possible components separately, it now remains to analyse the general a.c. series circuit containing resistance, inductance and capacitance. As before, the complexor diagram, shown in Fig. 7.13(*b*), is drawn by taking the current as reference. However when the voltage complexors are added together, it is seen that voltage across the inductance V_L and the voltage across the capacitance V_C are in phase opposition. It follows that the circuit can either be effectively inductive or capacitive depending on which voltage is predominant.

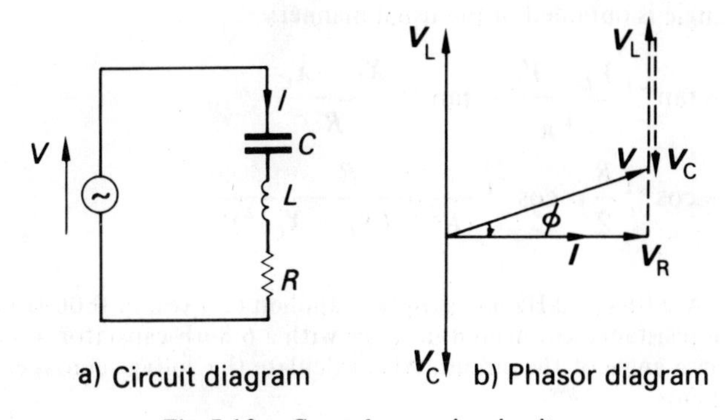

Fig. 7.13. General a.c. series circuit

$$V = (V_R{}^2 + (V_L - V_C)^2)^{\frac{1}{2}}$$
$$= (I^2 R^2 + (IX_L - IX_C)^2)^{\frac{1}{2}}$$
$$= I(R^2 + (X_L - X_C)^2)^{\frac{1}{2}}$$

Hence $\quad V = IZ$

where $\quad Z = (R^2 + (X_L - X_C)^2)^{\frac{1}{2}} \qquad (7.12)$

Solutions to this relation lie in one of the following groups:

(*a*) $X_L - X_C$ is positive ($X_L > X_C$) and the circuit is inductive,

(*b*) $X_L - X_C$ is negative ($X_C > X_L$) and the circuit is capacitive,

(*c*) $X_L - X_C$ is zero ($X_L = X_C$) and the circuit is purely resistive; this solution is further discussed in para. 7.12.

The impedance triangle corresponding to solution (*a*) is shown in Fig. 7.14. It should be noted that the negative sign used in conjunction with the value of the capacitive

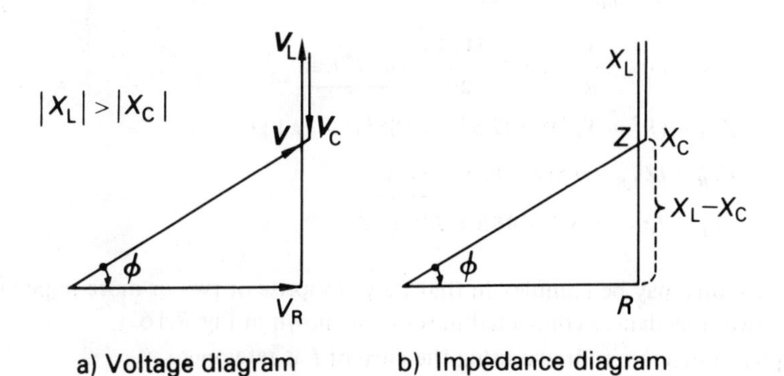

Fig. 7.14. Voltage and impedance triangles for series circuit

reactance X_C does not effect the value of the derived impedance since it is squared values which are being used; the negative sign therefore disappears. The negative sign arises from the convention of complexor diagram notation.

The phase angle is obtained in the usual manner:

$$\phi = \tan^{-1}\frac{V_L - V_C}{V_R} = \tan^{-1}\frac{X_L - X_C}{R} \tag{7.12.1}$$

Also

$$\phi = \cos^{-1}\frac{R}{Z} = \cos^{-1}\frac{R}{(R^2 + (X_L - X_C)^2)^{\frac{1}{2}}} \tag{7.12.2}$$

Example 7.5 A 230-V, 50-Hz a.c. supply is applied to a coil of 0·06 H inductance and 2·5 Ω effective resistance connected in series with a 6·8-μF capacitor. Calculate the current and phase angle of the circuit. Also calculate the voltage across each of the circuit components.

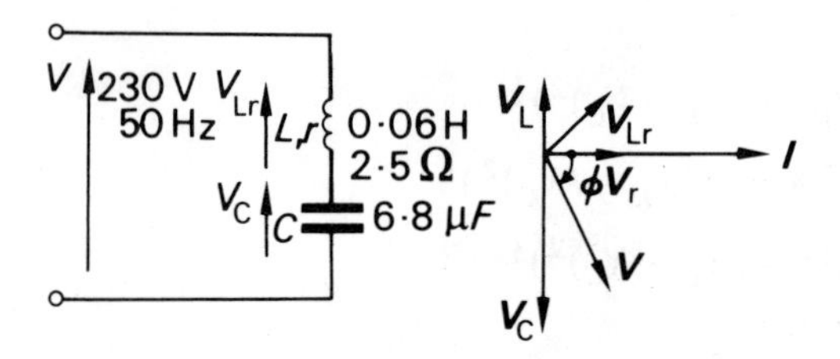

Fig. 7.15.

$$X_L = 2\pi fL = 2\pi 50 \times 0{\cdot}06 = 18{\cdot}85\ \Omega$$

$$X_C = \frac{1}{2\pi fC} = \frac{1}{2\pi 50 \times 6{\cdot}8 \times 10^{-6}} = 468\ \Omega$$

$$X = X_L - X_C = 18{\cdot}85 - 468 = -449{\cdot}15\ \Omega$$

$$Z = (R^2 + X^2)^{\frac{1}{2}} = (2{\cdot}5^2 + 449{\cdot}15^2)^{\frac{1}{2}} = 449{\cdot}2\ \Omega$$

$$I = \frac{V}{Z} = \frac{230}{449{\cdot}2} = \underline{0{\cdot}512\ \text{A}}$$

$$\phi = \tan^{-1}\frac{X}{R} = \tan^{-1}\frac{449{\cdot}15}{2{\cdot}5} = \underline{89{\cdot}7^\circ\ \text{lead}}$$

$$Z_{LR} = (R^2 + X_L{}^2)^{\frac{1}{2}} = (2{\cdot}5^2 + 18{\cdot}85^2)^{\frac{1}{2}} = 19{\cdot}1\ \Omega$$

$$V_{LR} = IZ_{LR} = 0{\cdot}512 \times 19{\cdot}1 = \underline{9{\cdot}8\ \text{V}}$$

$$V_C = IX_C = 0{\cdot}512 \times 468 = \underline{239{\cdot}6\ \text{V}}$$

Series circuits may be complex in that they comprise of two or more impedances. Consider two impedances connected in series as shown in Fig. 7.16.

The phasor diagram is drawn using the current $\boldsymbol{I}$ as reference.

It is completed by applying the following relations:

$$\boldsymbol{V}_1 = \boldsymbol{I}\boldsymbol{Z}_1$$

$$\boldsymbol{V}_2 = \boldsymbol{I}\boldsymbol{Z}_2$$

$$\boldsymbol{V} = \boldsymbol{V}_1 + \boldsymbol{V}_2$$

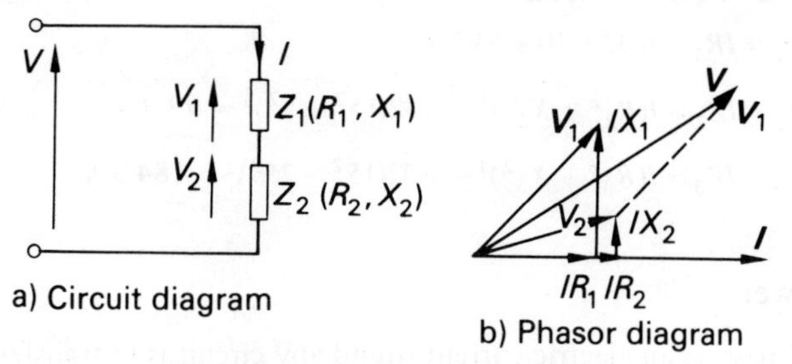

Fig. 7.16. Complex series circuit

By the geometry of the diagram:

$$V = [(IR_1 + IR_2)^2 + (IX_1 + IX_2)^2]^{\frac{1}{2}}$$
$$= I[(R_1 + R_2)^2 + (X_1 + X_2)^2]^{\frac{1}{2}}$$
$$= IZ$$

where $\quad Z = [(R_1 + R_2)^2 + (X_1 + X_2)^2]^{\frac{1}{2}}$

Thus the circuit impedance is found by collecting the separate resistances and reactances into like groups. It is thus simplified into one effective impedance.

Example 7.6 Three impedances are connected in series across a 200-V, 50-Hz a.c. supply. The first impedance is a 10-Ω resistor, the second is a coil of 15-Ω inductive reactance and 5-Ω resistance, and the third consists of a 15-Ω resistor in series with a 25-Ω capacitor. Calculate:

(*a*) The circuit current.
(*b*) The circuit phase angle.
(*c*) The impedance volt drops.

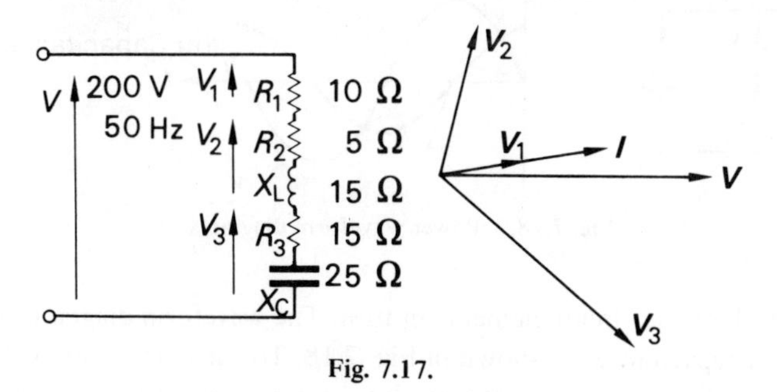

Fig. 7.17.

$$R_1 + R_2 + R_3 = 10 + 5 + 15 = 30\ \Omega$$
$$X_2 + X_3 = 15 - 25 = -10\ \Omega$$
$$Z = (R^2 + X^2)^{\frac{1}{2}} = (30^2 + 10^2)^{\frac{1}{2}} = 31{\cdot}6\ \Omega$$
$$I = \frac{V}{Z} = \frac{200}{31{\cdot}6} = \underline{6{\cdot}32\ \text{A}}$$
$$\phi = \cos^{-1}\frac{R}{Z} = \cos^{-1}\frac{30{\cdot}0}{31{\cdot}6} = \underline{18{\cdot}5^\circ\ \text{lead}}$$

$$V_1 = IR_1 = 6{\cdot}32 \times 10 = \underline{63{\cdot}2\ \text{V}}$$

$$V_2 = IZ_2 = I(R_2^2 + X_2^2)^{\frac{1}{2}} = 6{\cdot}32(15^2 + 5^2)^{\frac{1}{2}} = \underline{99{\cdot}9\ \text{V}}$$

$$V_3 = IZ_3 = I(R_3^2 + X_3^2)^{\frac{1}{2}} = 6{\cdot}32(15^2 + 25^2)^{\frac{1}{2}} = \underline{184{\cdot}3\ \text{V}}$$

7.7 Power

The object of passing an electric current round any circuit is to transfer energy from one place to another. Consequently the ultimate conclusion to any circuit analysis is to calculate the rate of energy transfer, i.e. the power developed or dissipated in the circuit.

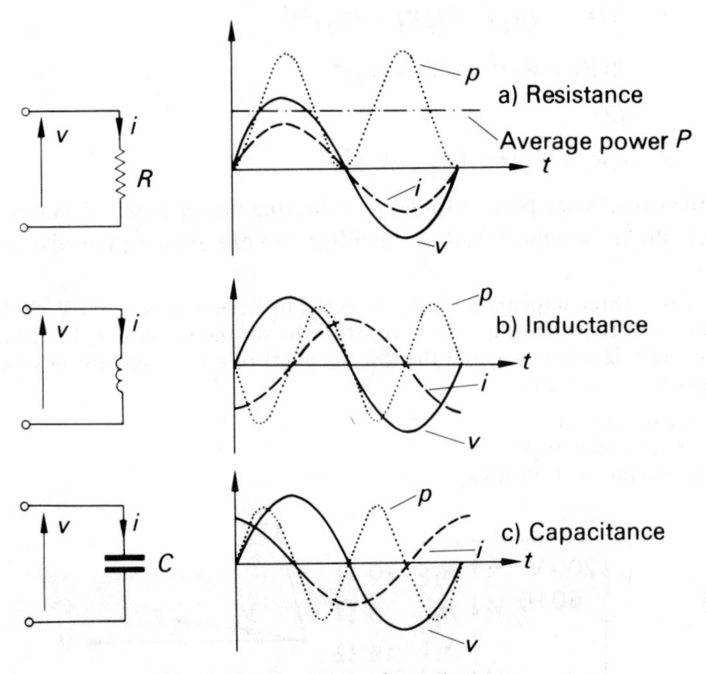

Fig. 7.18. Power waveform diagrams

Consider each of the circuit elements in turn. The waveform diagrams for resistance, inductance and capacitance are shown in Fig. 7.18. To the current and voltage waves, there has been added the waves of the product vi. Since the instantaneous values of vi represent the instantaneous power p, it follows that these waves are the power waves. Because the power is continually fluctuating, the power in an a.c. circuit is taken to be the average value of the wave.

In the case of the pure resistance, the average power can be most easily obtained from the definition of the r.m.s. current in the circuit, i.e.

$$P = I^2 R \qquad (7.13)$$

This relation can also be expressed as:

$$P = VI \qquad (7.13.1)$$

Alternatively, the average power can be derived from a formal analysis of the power waveform.

$$P = \frac{\omega}{2\pi}\int_0^{\frac{2\pi}{\omega}} (V_m \sin \omega t . I_m \sin \omega t)\,dt$$

$$= V_m I_m \frac{\omega}{2\pi}\int_0^{\frac{2\pi}{\omega}} (\sin^2 \omega t)\,dt$$

$$= V_m I_m \frac{\omega}{2\pi}\int_0^{\frac{2\pi}{\omega}} \left(\frac{1-\cos 2\omega t}{2}\right)dt$$

From this relation it can be seen that the wave has a frequency double that of the component voltage and current waves. This can be seen in Fig. 7.18; however, it also confirms that the wave is sinusoidal although it has been displaced from the horizontal axis.

$$P = V_m I_m \frac{\omega}{2\pi}\left[\frac{t}{2} - \frac{\sin 2\omega t}{4\omega}\right]_0^{\frac{2\pi}{\omega}}$$

$$= V_m I_m \frac{\omega}{2\pi} \cdot \frac{2\pi}{2\omega}$$

$$= \frac{V_m I_m}{2}$$

$$= VI \tag{7.13.1}$$

In the cases of inductance and capacitance, consideration of the power waveform, shows that as before its frequency is twice that of the supply frequency, but now the wave is symmetrical about the horizontal axis. Hence its average value is zero and thus pure inductors and capacitors do not dissipate electrical energy. They merely store energy for a short period of time and then release it back into the circuit.

That the inductive and capacitive circuits do not dissipate power can be proved by an analysis of the power wave. Consider the case of the capacitor.

$$P = \frac{\omega}{2\pi}\int_0^{\frac{2\pi}{\omega}} (V_m \sin \omega t . I_m \cos \omega t)\,dt$$

$$= V_m I_m \frac{\omega}{2\pi}\int_0^{\frac{2\pi}{\omega}} (\sin \omega t . \cos \omega t)\,dt$$

$$= V_m I_m \frac{\omega}{2\pi}\int_0^{\frac{2\pi}{\omega}} \frac{\sin 2\omega t}{2} \cdot dt$$

$$= V_m I_m \frac{\omega}{2\pi}\left[-\frac{\cos 2\omega t}{4\omega}\right]_0^{\frac{2\pi}{\omega}}$$

$$= 0 \tag{7.14}$$

In the case, shown in Fig. 7.19, where there are two or more elements present in the circuit, the power can still be derived from the r.m.s. current as follows:

$$P = I^2 R = \frac{V}{Z} \cdot IR = VI \cdot \frac{R}{Z}$$

$$= VI \cos\phi \qquad (7.15)$$

Thus the power developed in the circuit is somewhat less than that value given by the product of the voltage and the current, because $1 > \cos\phi > 0$. The wave diagram shows that the power waveform is again of a sinusoidal nature with double frequency but it has not been completely displaced from the horizontal axis.

The product of the voltage and the current in an a.c. circuit is termed the apparent power.

Apparent Power	Symbol: S	Unit: volt ampere (VA)

$$S = VI \qquad (7.16)$$

The factor $\cos\phi$ is termed the power factor, which can be defined as that function by which the apparent power must be multiplied to give the true power. $\cos\phi$ is the power factor for a sinusoidally-varying system.

$$P = S \cos\phi$$

$$= VI \cos\phi \qquad (7.16.1)$$

The significance of the power factor will be further discussed in para. 7.13.

Finally the formal analysis of the power wave shown in Fig. 7.19 is as follows:

$$P = \frac{\omega}{2\pi}\int_0^{\frac{2\pi}{\omega}} (V_m \sin \omega t . I_m \sin(\omega t + \phi))\mathrm{d}t$$

$$= V_m I_m \frac{\omega}{2\pi}\int_0^{\frac{2\pi}{\omega}} (\sin \omega t(\sin \omega t . \cos\phi + \cos\omega t . \sin\phi))\mathrm{d}t$$

$$= V_m I_m \frac{\omega}{2\pi}\int_0^{\frac{2\pi}{\omega}} (\sin^2 \omega t . \cos\phi + \sin\omega t . \cos\omega t . \sin\phi)\mathrm{d}t$$

$$= V_m I_m \frac{\omega}{2\pi}\int_0^{\frac{2\pi}{\omega}} \left(\frac{1 - \cos 2\omega t}{2} \cdot \cos\phi + \frac{\sin 2\omega t}{2} \cdot \sin\phi\right)\mathrm{d}t$$

By comparison with the previous proofs, it follows that

$$P = VI\cos\phi \qquad (7.16.2)$$

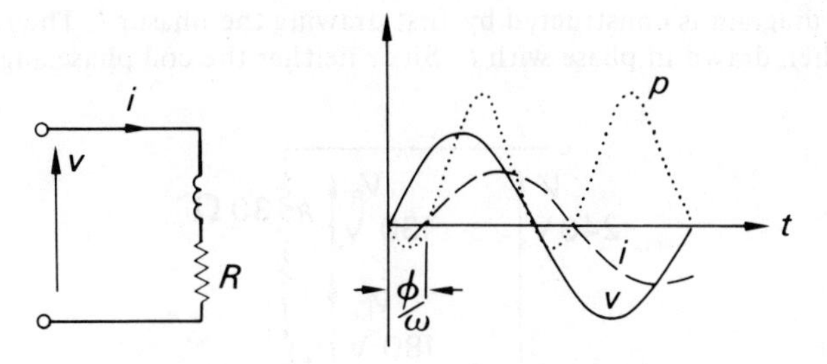

Fig. 7.19. Power in a complex series circuit

Example 7.7 An inductor coil is connected to a supply of 250 V at 50 Hz and takes a current of 5 A. The coil dissipates 750 W. Calculate:

(*a*) The resistance and the inductance of the coil.
(*b*) The power factor of the coil.

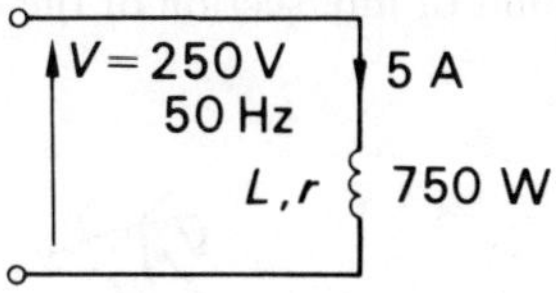

V = 250 V
50 Hz
5 A
L, r
750 W

Fig. 7.20.

In this example, the symbol r will be used to denote the resistance of the coil instead of R. This is done to draw attention to the fact that the resistance is not a separate component of the circuit but is an integral part of the inductor coil. This device was also used in Example 7.3.

$$Z = \frac{V}{I} = \frac{250}{5} = 50\ \Omega$$

$$r = \frac{P}{I^2} = \frac{750}{5^2} = \underline{30\ \Omega}$$

$$X_L = (Z^2 - r^2)^{\frac{1}{2}} = (50^2 - 30^2)^{\frac{1}{2}} = 40\ \Omega$$

$$L = \frac{X_L}{2\pi f} = \frac{40}{2\pi 50} = \frac{40}{314} = 0{\cdot}127\ \text{H}$$

$$= \underline{127\ \text{mH}}$$

$$\text{Power factor} = \cos\phi = \frac{P}{S} = \frac{I^2R}{VI}$$

$$= \frac{750}{250 \times 5} = \underline{0{\cdot}6\ \text{lag}}$$

Example 7.8 An inductor coil is connected in series with a pure resistor of 30 Ω across a 240-V, 50-Hz supply. The voltage measured across the coil is 180 V and the voltage measured across the resistor is 130 V. Calculate the power dissipated in the coil.

The phasor diagram is constructed by first drawing the phasor I. The resistor voltage phasor V_R is then drawn in phase with I. Since neither the coil phase angle nor the

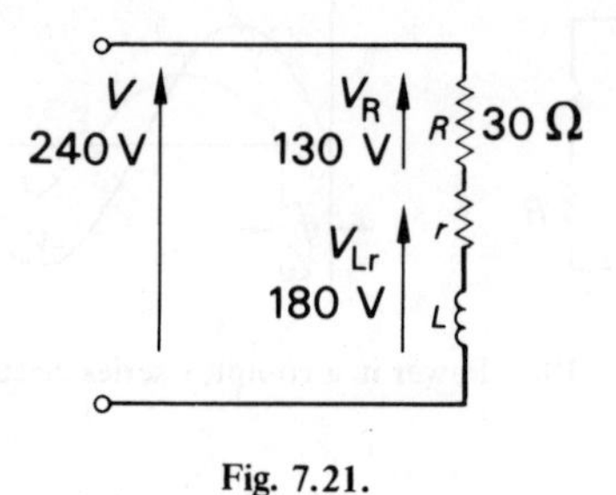

Fig. 7.21.

circuit phase angle are known, it is necessary to derive the remainder of the diagram by construction to scale. Circles of radius V and V_{Lr} are drawn radiating from the appropriate ends of V_R. The point of intersection of the circles satisfies the relation

$$V = V_R + V_{Lr}$$

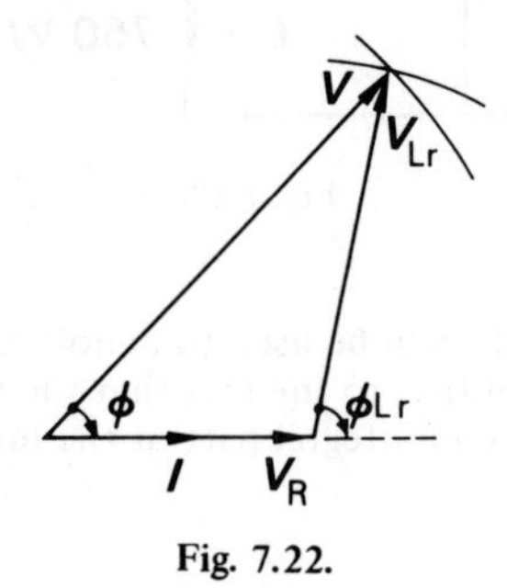

Fig. 7.22.

By the geometry of the diagram,

$$V^2 = V_R{}^2 + V_{Lr}{}^2 + 2V_R V_{Lr} \cos\phi_{Lr}$$

$$240^2 = 130^2 + 180^2 + 2 \times 130 \times 180 \times \cos\phi_{LR}$$

$$\cos\phi_{Lr} = 0{\cdot}177 \text{ lag}$$

$$I = \frac{V_R}{R} = \frac{130}{30} = 4{\cdot}33 \text{ A}$$

$$P_r = V_{Lr} I \cos\phi_{Lr} = 180 \times 4{\cdot}33 \times 0{\cdot}177 = \underline{138 \text{ W}}$$

Alternatively:

$$Z_{Lr} = \frac{V_{Lr}}{I} = \frac{180}{4{\cdot}33} = 41{\cdot}5\ \Omega$$

$$r = Z_{Lr} \cos\phi_{Lr} = 41{\cdot}5 \times 0{\cdot}177 = 7{\cdot}35\ \Omega$$

$$P_r = I^2 r = 4{\cdot}33^2 \times 7{\cdot}35 = \underline{138 \text{ W}}$$

7.8 Simple parallel circuits

Having completed an introductory study of series circuits, it is now possible to consider parallel circuits. Because the reader is familiar with a.c. circuit analysis and its notation, it will be in order to make the study less detailed.

There are two arrangements of simple parallel circuits which require analysis; these are resistance in parallel with inductance and resistance in parallel with capacitance. The effect of inductance in parallel with capacitance will be considered in para. 7.12.

When analysing a parallel circuit, it should be remembered that it consists of two or more series circuits connected in parallel. Therefore each branch of the circuit can be analysed separately as a series circuit and then the effect of the separate branches can be combined by applying Kirchhoff's 1st Law, i.e. the currents of the branches can be added complexorially, that is, by phasor diagram.

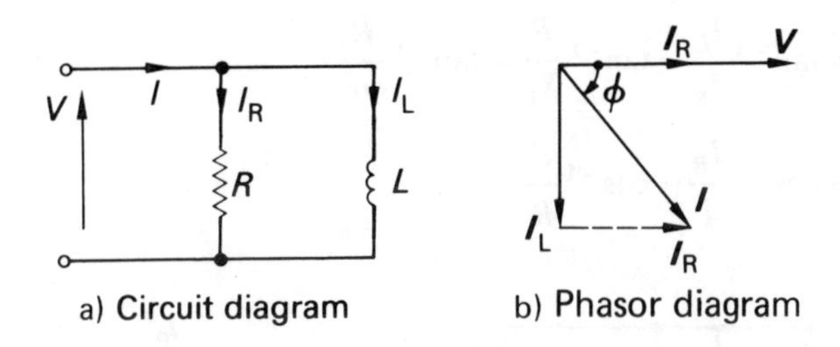

Fig. 7.23. Resistance and inductance in parallel

The circuit for resistance and inductance in parallel is shown in Fig. 7.23*a*. In the resistive branch, the current is given by

$$I_R = \frac{V}{R}, \text{ where } I_R \text{ and } V \text{ are in phase.}$$

In the inductive branch, the current is given by

$$I_L = \frac{V}{X_L}, \text{ where } I_L \text{ lags } V \text{ by } 90^\circ.$$

The resulting phasor diagram is shown in Fig. 7.23(*b*). The voltage which is common to both branches is taken as reference. Since parallel circuits are more common, this is one reason that it is usual to take the voltage as reference in circuit analysis. The total supply current I is obtained by adding the branch currents complexorially, i.e.

$$\mathbf{I} = \mathbf{I}_R + \mathbf{I}_L \qquad (7.17)$$

From the phasor diagram:

$$I = (I_R^2 + I_L^2)^{\frac{1}{2}}$$

$$= \left(\left(\frac{V}{R}\right)^2 + \left(\frac{V}{X_L}\right)^2\right)^{\frac{1}{2}}$$

$$= V\left(\frac{1}{R^2} + \frac{1}{X_L^2}\right)^{\frac{1}{2}}$$

$$\frac{V}{I} = Z = \frac{1}{\left(\frac{1}{R^2} + \frac{1}{X_L^2}\right)^{\frac{1}{2}}} \quad (7.18)$$

It can be seen from the phasor diagram that the phase angle ϕ is a lagging angle.

$$\phi = \tan^{-1}\frac{I_L}{I_R} = \tan^{-1}\frac{R}{X_L} = \tan^{-1}\frac{R}{\omega L} \quad (7.18.1)$$

Also $$\phi = \cos^{-1}\frac{I_R}{I} = \cos^{-1}\frac{Z}{R} \quad (7.18.2)$$

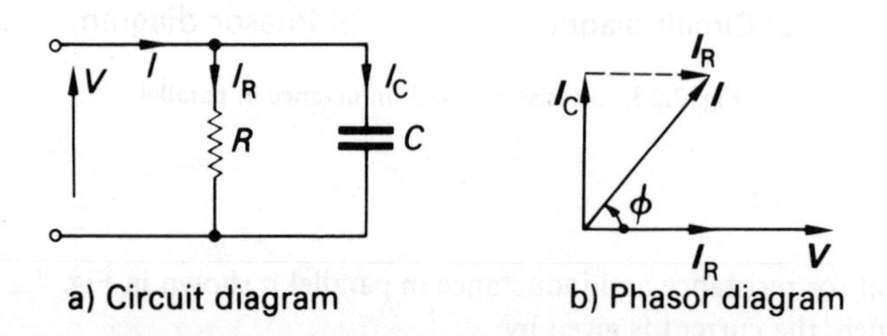

Fig. 7.24. Resistance and capacitance in parallel

In the case of resistance and capacitance connected in parallel, as shown in Fig. 7.24(*a*) the current in the resistive branch is again given by

$$I_R = \frac{V}{R}, \text{ where } I_R \text{ and } V \text{ are in phase.}$$

In the capacitive branch, the current is given by

$$I_C = \frac{V}{X_C}, \text{ where } I_C \text{ leads } V \text{ by } 90°.$$

The phasor diagram is constructed in the usual manner based on the relation

$$I = I_R + I_C \quad (7.19)$$

From the phasor diagram:

$$I = (I_R^2 + I_C^2)^{\frac{1}{2}}$$

$$= \left(\left(\frac{V}{R}\right)^2 + \left(\frac{V}{X_C}\right)^2\right)^{\frac{1}{2}}$$

$$= V\left(\frac{1}{R^2} + \frac{1}{X_C^2}\right)^{\frac{1}{2}}$$

$$\frac{V}{I} = Z = \frac{1}{\left(\frac{1}{R^2} + \frac{1}{X_C^2}\right)^{\frac{1}{2}}} \quad (7.20)$$

It can be seen from the phasor diagram that the phase angle ϕ is a leading angle. It follows that parallel circuits behave in a similar fashion to series circuits in that the combination of resistance with inductance produces a lagging circuit whilst the combination of resistance with capacitance gives rise to a leading circuit.

$$\phi = \tan^{-1}\frac{I_C}{I_R} = \tan^{-1}\frac{R}{X_C} = \tan^{-1} R\omega C \quad (7.20.1)$$

Also $$\phi = \cos^{-1}\frac{I_R}{I} = \cos^{-1}\frac{Z}{R} \quad (7.20.2)$$

Example 7.9 A circuit consists of a 120-Ω resistor in parallel with a 40-μF capacitor and is connected to a 240-V, 50-Hz supply. Calculate:

(*a*) The branch currents and the supply current.
(*b*) The circuit phase angle.
(*c*) The circuit impedance.

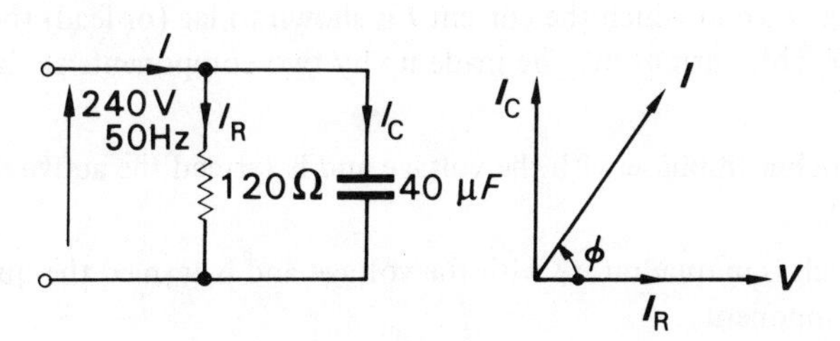

Fig. 7.25.

$$I_R = \frac{V}{R} = \frac{240}{120} = \underline{2{\cdot}0 \text{ A}}$$

$$X_C = \frac{1}{2\pi fC} = \frac{1}{2\pi 50 \times 40 \times 10^{-6}} = 80\ \Omega$$

$$I_C = \frac{V}{X_C} = \frac{240}{80} = \underline{3{\cdot}0 \text{ A}}$$

$$I = (I_R^2 + I_C^2) = (2{\cdot}0^2 + 3{\cdot}0^2)^{\frac{1}{2}} = \underline{3{\cdot}6 \text{ A}}$$

$$\phi = \cos^{-1}\frac{I_R}{I} = \cos^{-1}\frac{2{\cdot}0}{3{\cdot}6} = \underline{56{\cdot}3^\circ \text{ lead}}$$

$$Z = \frac{V}{I} = \frac{240}{3{\cdot}6} = \underline{66{\cdot}7\ \Omega}$$

7.9 Parallel impedance circuits

The analysis of impedances in parallel is similar to that of the previous paragraph in that the voltage is taken as reference and the branch currents are calculated with respect to the voltage. However the summation of the branch currents is now made more difficult since they do not necessarily remain either in phase or quadrature with one another. Thus before it is possible to analyse parallel impedance circuits, it is necessary to introduce a new analytical device–current components.

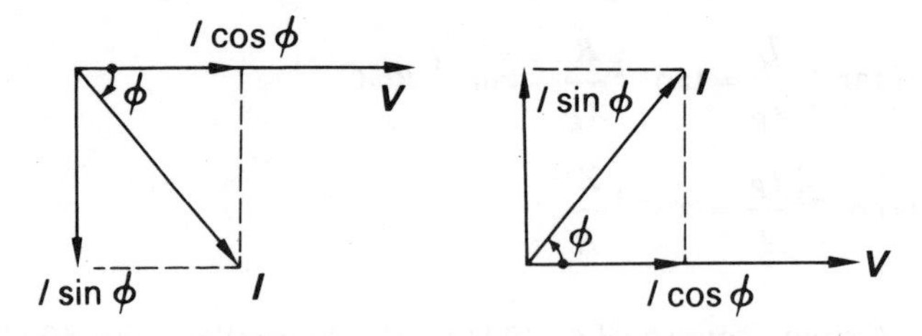

Fig. 7.26. Components of a current

Consider Fig. 7.26 in which the current I is shown to lag (or lead) the voltage V by a phase angle ϕ. This current may be made up by two components at right angles to one another:

(a) $I \cos \phi$, which is in phase with the voltage and is termed the active or power component.

(b) $I \sin \phi$, which is in quadrature with the voltage and is termed the quadrature or reactive component.

By the geometry of the diagram:

$$I^2 = (I \cos \phi)^2 + (I \sin \phi)^2$$

The term $I \cos \phi$ has already been met in relation 7.15 in which the power in an a.c. circuit is given by $VI \cos \phi$. It can be seen that the power absorbed by the load can be attributed to the current component $I \cos \phi$. This is the reason for it being called the power component.

The reactive component will either lag (or lead) the voltage by 90° depending on whether the current I lags (or leads) the voltage V.

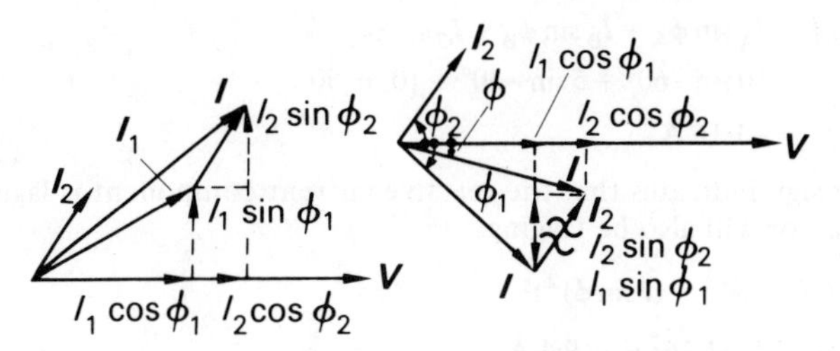

Fig. 7.27. Addition of current phasors

Consider the addition of the currents I_1 and I_2 as shown in Fig. 7.27, i.e.

$$I = I_1 + I_2$$

The value of I can be achieved by drawing a phasor diagram to scale but this is not generally practicable. It can, howerver, be calculated if the currents are resolved into components, then

$$I\cos\phi = I_1 \cos\phi_1 + I_2 \cos\phi_2$$

$$I\sin\phi = I_1 \sin\phi_1 + I_2 \sin\phi_2$$

But $$I^2 = (I\cos\phi)^2 + (I\sin\phi)^2$$

Hence $$I^2 = (I_1 \cos\phi_1 + I_2 \cos\phi_2)^2 + (I_1 \sin\phi_1 + I_2 \sin\phi_2)^2 \tag{7.21}$$

Also $$\phi = \tan^{-1}\frac{I_1 \sin\phi_1 + I_2 \sin\phi_2}{I_1 \cos\phi_1 + I_2 \cos\phi_2}$$

$$= \cos^{-1}\frac{I_1 \cos\phi_1 + I_2 \cos\phi_2}{I} \tag{7.21.1}$$

Example 7.10 A parallel circuit consists of branches A, B and C. If $I_A = 10\angle -60^\circ$ A, $I_B = 5\angle -30^\circ$ A and $I_C = 10\angle 90^\circ$ A, all phase angles being relative to the supply voltage, determine the total supply current and the overall power factor.

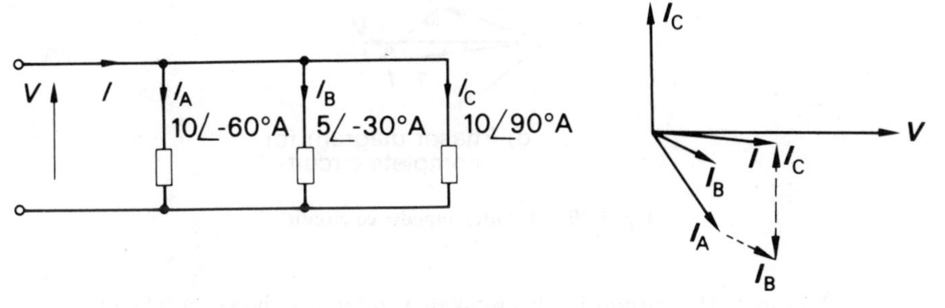

Fig. 7.28.

$$I = I_A + I_B + I_C$$

$$I\cos\phi = I_A \cos\phi_A + I_B \cos\phi_B + I_C \cos\phi_C$$

$$= 10\cos -60^\circ + 5\cos -30^\circ + 10\cos 90^\circ$$

$$= 9{\cdot}33 \text{ A}$$

$$I\sin\phi = I_A\sin\phi_A + I_B\sin\phi_B + I_C\sin\phi_C$$
$$= 10\sin -60° + 5\sin -30° + 10\sin 90°$$
$$= -1{\cdot}16\ \text{A}$$

The negative sign indicates that the reactive current component is lagging, so the overall power factor will also be lagging.

$$I = ((I\cos\phi)^2 + (I\sin\phi)^2)^{\frac{1}{2}}$$
$$= (9{\cdot}33^2 + 1{\cdot}16^2)^{\frac{1}{2}} = 9{\cdot}4\ \text{A}$$

$$\phi = \tan^{-1}\frac{I\sin\phi}{I\cos\phi} = \tan^{-1}\frac{1{\cdot}16}{9{\cdot}33} = 7{\cdot}1°\ \text{lag}$$

$$I = 9{\cdot}4\angle -7{\cdot}1°\ \text{A}$$

Power factor = $\cos\phi = \cos 7{\cdot}1° = 0{\cdot}993$ lag

Consider the circuit shown in Fig. 7.29 in which two series circuits are connected in parallel. To analyse the arrangement, the phasor diagrams for each branch have been drawn—Figs. 7.29(*b*) and 7.29(*c*)—in accordance with the principles of paras. 7.3 and

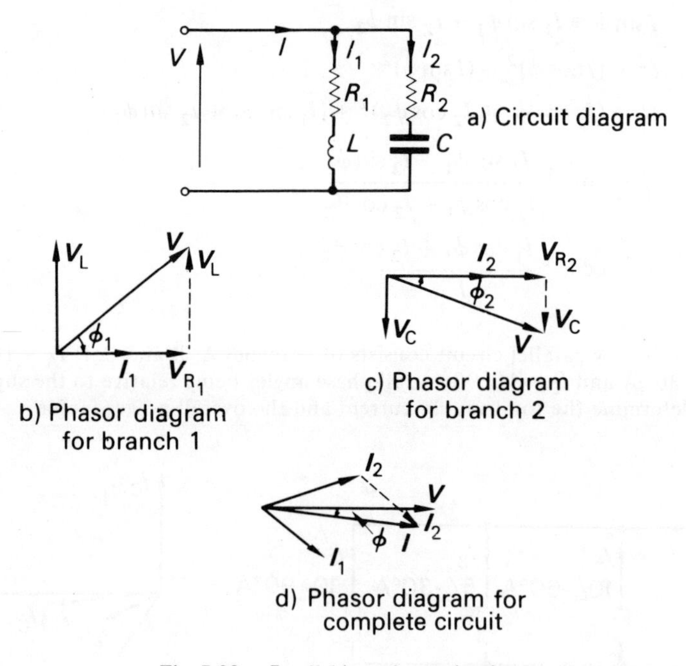

Fig. 7.29. Parallel-impedance circuit

7.5. In each branch, the current has been taken as reference; however when the branches are in parallel, it is easier to take the supply voltage as reference, hence Figs. 7.29(*b*) and 7.29(*c*) have been separately rotated and then superimposed on one another to give Fig. 7.29(*d*). The current phasors may then be added to give the total current in correct phase relation to the voltage. The analysis of the diagram is carried out in the manner noted above.

The phase angle for the circuit shown in Fig. 7.29(*a*) is a lagging angle if $I_1 \sin \phi_1 > I_2 \sin \phi_2$ and is a leading angle if $I_1 \sin \phi_1 < I_2 \sin \phi_2$.

It should be noted, however, that this was only an example of the method of analysis. Both circuit branches could have been inductive or capacitive. Alternatively, there could have been more than two branches. The main concern of this study has been to illustrate the underlying principles of the method of analysis.

Example 7.11 A coil of resistance 50 Ω and inductance 0·318 H is connected in parallel with a circuit comprising a 75-Ω resistor in series with a 159-μF capacitor. The resulting circuit is connected to a 240-V, 50-Hz a.c. supply. Calculate:

(*a*) The supply current.
(*b*) The circuit impedance, resistance and reactance.

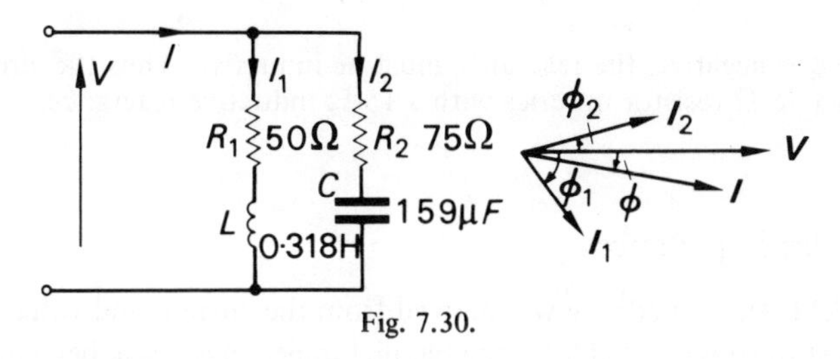

Fig. 7.30.

$$X_L = 2\pi f L = 2\pi 50 \times 0{\cdot}318 = 100\ \Omega$$

$$Z_1 = (R_1{}^2 + X_L{}^2)^{\frac{1}{2}} = (50^2 + 100^2)^{\frac{1}{2}} = 112\ \Omega$$

$$I_1 = \frac{V}{Z_1} = \frac{240}{112} = 2{\cdot}15\ \text{A}$$

$$\phi_1 = \cos^{-1}\frac{R_1}{Z_1} = \cos^{-1}\frac{50}{112} = \cos^{-1} 0{\cdot}447 = 63{\cdot}5^\circ \text{ lag}$$

$$\boldsymbol{I}_1 = 2{\cdot}15\angle{-63{\cdot}5^\circ}\ \text{A}$$

$$X_C = \frac{1}{2\pi f C} = \frac{1}{2\pi 50 \times 159 \times 10^6} = 20\ \Omega$$

$$Z_2 = (R_2{}^2 + X_C{}^2)^{\frac{1}{2}} = (75^2 + 20^2)^{\frac{1}{2}} = 77{\cdot}7\ \Omega$$

$$I_2 = \frac{V}{Z_2} = \frac{240}{77{\cdot}7} = 3{\cdot}09\ \text{A}$$

$$\phi_2 = \tan^{-1}\frac{X_C}{R_2} = \tan^{-1}\frac{20}{77{\cdot}7} = \tan^{-1} 0{\cdot}267 = 15^\circ \text{ lead}$$

NOTE: the solution incorporating the use of the tangent is used because ϕ_2 is relatively small.

$$\boldsymbol{I}_2 = 3{\cdot}09\angle 15^\circ\ \text{A}$$

$$\boldsymbol{I} = \boldsymbol{I}_1 + \boldsymbol{I}_2$$

$$I\cos\phi = I_1 \cos\phi_1 + I_2 \cos\phi_2$$

$$= 2{\cdot}15\cos{-63{\cdot}5^\circ} + 3{\cdot}09\cos 15^\circ = 3{\cdot}94\ \text{A}$$

$$I\sin\phi = I_1\sin\phi_1 + I_2\sin\phi_2$$

$$= 2{\cdot}15\sin -63{\cdot}5^\circ + 3{\cdot}09\sin 15^\circ = -1{\cdot}13\ \text{A}$$

$$I = ((I\cos\phi)^2 + (I\sin\phi)^2)^{\frac{1}{2}} = (3{\cdot}94^2 + 1{\cdot}13^2)^{\frac{1}{2}}$$

$$= \underline{4{\cdot}1\ \text{A}}$$

$$Z = \frac{V}{I} = \frac{240}{4{\cdot}1} = \underline{58{\cdot}5\ \Omega}$$

$$R = Z\cos\phi = Z\cdot\frac{I\cos\phi}{I} = 58{\cdot}5 \times \frac{3{\cdot}94}{4{\cdot}1} = \underline{56\ \Omega}$$

$$X = Z\sin\phi = Z\cdot\frac{I\sin\phi}{I} = 58{\cdot}5 \times \frac{1{\cdot}13}{4{\cdot}1} = \underline{15\ \Omega}$$

Since $I\sin\phi$ is negative, the reactance must be inductive. Thus the circuit is equivalent to a 56-Ω resistor in series with a 15-Ω inductive reactance.

7.10 Polar impedances

In Example 7.11, the impedance was derived from the current and voltage. However, it may be questioned why could not the parallel impedances have been handled in a similar manner to parallel resistors. Consider then three impedances connected in parallel as shown in Fig. 7.31.

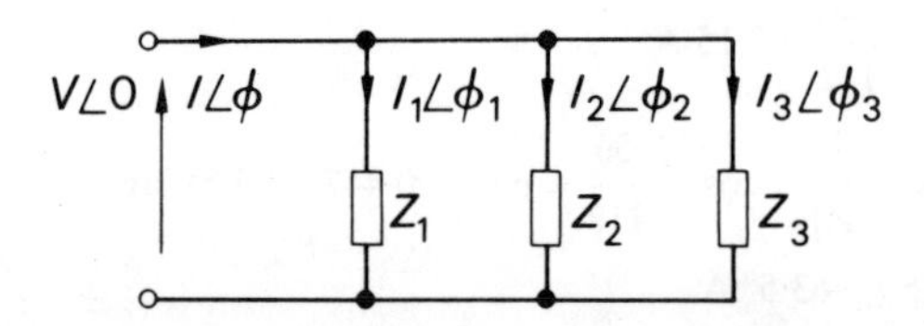

Fig. 7.31. Polar impedances in parallel

In the first branch:

$$V = I_1 Z_1$$

However, if consideration is given to the phase angles of V and I then to maintain balance, the impedance must also act like a complexor and have a phase angle, i.e.

$$V\angle 0 = I_1\angle\phi_1 . Z_1\angle -\phi_1$$

The impedance phase angle is the conjugate of the circuit phase angle. This compares with the impedance triangles previously (shown, apart from the reversal of the 'polarity'). In complexor notation:

$$\boldsymbol{I} = \boldsymbol{I}_1 + \boldsymbol{I}_2 + \boldsymbol{I}_3$$

In polar notation:

$$I\angle\phi = I_1\angle\phi_1 + I_2\angle\phi_2 + I_3\angle\phi_3$$

$$\therefore \quad \frac{V\angle 0}{Z\angle -\phi} = \frac{V\angle 0}{Z_1\angle -\phi_1} + \frac{V\angle 0}{Z_2\angle -\phi_2} + \frac{V\angle 0}{Z_3\angle -\phi_3}$$

$$\therefore \quad \frac{1}{Z\angle -\phi} = \frac{1}{Z_1\angle -\phi_1} + \frac{1}{Z_2\angle -\phi_2} + \frac{1}{Z_3\angle -\phi_3} \qquad (7.22)$$

This relation compares with that for parallel resistors but it has the complication of having to consider the phase angles. Because of this, it is not considered, at this introductory stage, prudent to use the polar approach to the analysis of parallel impedances; the method used in Example 7.11 is more suitable and less prone to error.

It is most important that the impedance phase angles are not ignored and it would be quite incorrect to state:

$$\frac{1}{Z} = \frac{1}{Z_1} + \frac{1}{Z_2} + \frac{1}{Z_3}$$

A similar situation occurs when impedances are connected in series. Consider the case shown in Fig. 7.32.

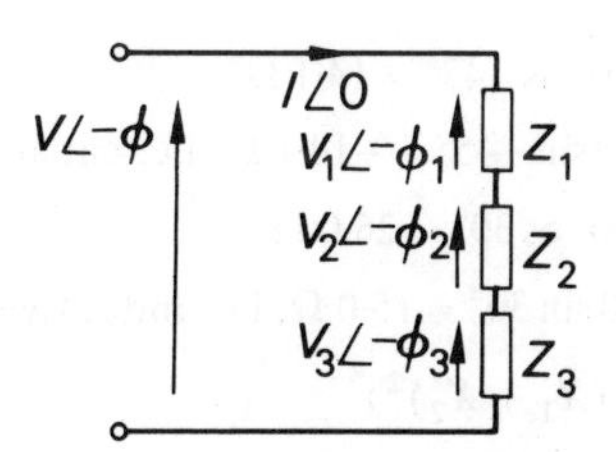

Fig. 7.32. Polar impedances in series

In complexor notation:

$$\boldsymbol{V} = \boldsymbol{V}_1 + \boldsymbol{V}_2 + \boldsymbol{V}_3$$

In polar notation:

$$V\angle -\phi = V_1\angle -\phi_1 + V_2\angle -\phi_2 + V_3\angle -\phi_3$$

$$I\angle 0 . Z\angle -\phi = I\angle 0 . Z_1\angle -\phi_1 + I\angle 0 . Z_2\angle -\phi_2 + I\angle 0 . Z_3\angle -\phi_3$$

$$Z\angle -\phi = Z_1\angle -\phi_1 + Z_2\angle -\phi_2 + Z_3\angle -\phi_3 \qquad (7.23)$$

However, it has been shown that in a series circuit:

$$Z\cos\phi = Z_1\cos\phi_1 + Z_2\cos\phi_2 + Z_3\cos\phi_3$$

hence $\quad R = R_1 + R_2 + R_3 \qquad (7.23.1)$

Similarly $\quad X = X_1 + X_2 + X_3 \qquad (7.23.2)$

Relation (7.23) has already been effectively used in Example 7.6.

As before, it would have been incorrect to state:

$$Z = Z_1 + Z_2 + Z_3$$

It may therefore be concluded that, whilst it is practical to deal with impedances in series using polar notation, it is not practical to deal with impedances in parallel in this manner. Parallel circuit calculations are better approached on the basis of analysing the branch currents.

Example 7.12 Two impedances of $20\angle -45°\ \Omega$ and $30\angle 30°\ \Omega$ are connected in series across a certain supply and the resulting current is found to be 10 A. If the supply voltage remains unchanged, calculate the supply current when the impedances are connected in parallel.

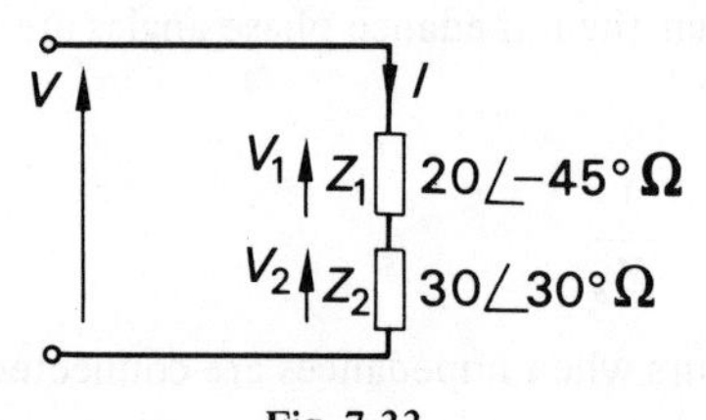

Fig. 7.33.

$$R_1 = Z_1 \cos -\phi_1 = 20 \cos -45° = 14{\cdot}1\ \Omega$$

$$X_1 = Z_1 \sin -\phi_1 = 20 \sin -45° = -14{\cdot}1\ \Omega, \text{ i.e. capacitive}$$

$$R_2 = Z_2 \cos -\phi_2 = 30 \cos 30° = 26{\cdot}0\ \Omega$$

$$X_2 = Z_2 \sin -\phi_2 = 30 \sin 30° = 15{\cdot}0\ \Omega, \text{ i.e. inductive}$$

$$Z = ((R_1 + R_2)^2 + (X_1 + X_2)^2)^{\frac{1}{2}}$$

$$= ((14{\cdot}1 + 26{\cdot}0)^2 + (-14{\cdot}1 + 15{\cdot}0)^2)^{\frac{1}{2}}$$

$$= 40{\cdot}1\ \Omega$$

$$V = IZ = 10 \times 40{\cdot}1 = 401\ \text{V}$$

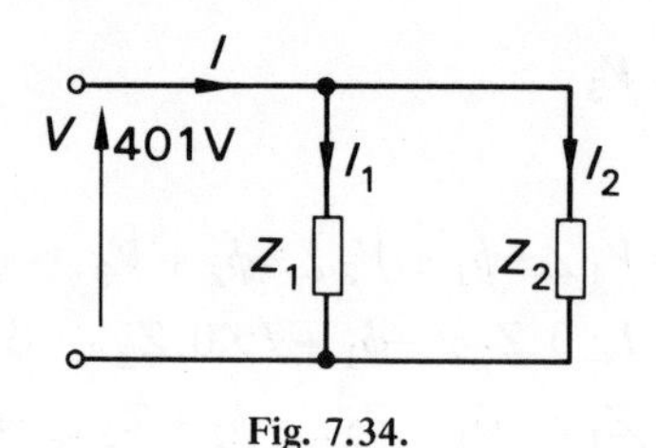

Fig. 7.34.

$$I_1 = \frac{V}{Z_1} = \frac{401}{20} = 20{\cdot}1\ \text{A}$$

$$\phi_1 = 45°$$

$$I_1 \cos \phi_1 = 20{\cdot}1 \times 0{\cdot}707 = 14{\cdot}2\ \text{A}$$

$$I_1 \sin \phi_1 = 20{\cdot}1 \times 0{\cdot}707 = 14{\cdot}2\ \text{A}$$

$$I_2 = \frac{V}{Z_2} = \frac{401}{30} = 13{\cdot}4 \text{ A}$$

$$\phi_2 = -30°$$

$$I_2 \cos\phi_2 = 13{\cdot}4 \times 0{\cdot}866 = 11{\cdot}6 \text{ A}$$

$$I_2 \sin\phi_2 = 13{\cdot}4 \times -0{\cdot}50 = -6{\cdot}7 \text{ A}$$

$$I\cos\phi = I_1 \cos\phi_1 + I_2 \cos\phi_2$$

$$= 14{\cdot}2 + 11{\cdot}6 = 25{\cdot}7 \text{ A}$$

$$I\sin\phi = I_1 \sin\phi_1 + I_2 \sin\phi_2$$

$$= 14{\cdot}2 - 6{\cdot}7 = 7{\cdot}5 \text{ A}$$

$$I = ((I\cos\phi)^2 + (I\sin\phi)^2)^{\frac{1}{2}}$$

$$= (25{\cdot}7^2 + 7{\cdot}5^2)^{\frac{1}{2}}$$

$$= 26{\cdot}8 \text{ A}$$

7.11 Polar admittances

An alternative approach to parallel a.c. circuits using polar notation can be made through admittance instead of impedance. The admittance is the inverse of the impedance in the same way that the conductance is the inverse of the resistance.

Admittance	Symbol: Y	Unit: siemens (S)

Thus in any branch of a parallel circuit,

$$\frac{V}{Z} = I = VY \tag{7.24}$$

When the phase angles are included in this relation, it becomes

$$I\angle\phi = V\angle 0 \,.\, Y\angle\phi = \frac{V\angle 0}{Z\angle -\phi}$$

$$Y\angle\phi = \frac{1}{Z\angle -\phi}$$

The resulting change in sign of the phase angle should be noted when the inversion takes place. Hence from relation 7.22:

$$Y\angle\phi = Y_1\angle\phi_1 + Y_2\angle\phi_2 + Y_3\angle\phi_3 \tag{7.25}$$

Hence $\quad Y\cos\phi = Y_1 \cos\phi_1 + Y_2 \cos\phi_2 + Y_3 \cos\phi_3$

$$G = G_1 + G_2 + G_3$$

G is the conductance of the circuit as in the d.c. circuit analysis. This must be the case since the current and voltage are in phase; this corresponds to the resistance of a circuit.

Also $\quad Y\sin\phi = Y_1 \sin\phi_1 + Y_2 \sin\phi_2 + Y_3 \sin\phi_3$

$$B = B_1 + B_2 + B_3$$

B is termed the susceptance of the circuit and is the reactive component of the admittance.

Susceptance	Symbol: B	Unit: siemens (S)

The power dissipated in an admittance may be derived as follows:

$$P = VI\cos\phi$$
$$= V^2 Y\cos\phi$$
$$= V^2 G \qquad (7.26)$$

From this expression, it can be seen that only the conductance of the circuit is responsible for power dissipation.

$$P = V^2 G$$
$$= I^2 Z^2 G$$
$$= I^2 R$$
$$G = \frac{R}{Z^2}$$

Similarly:

$$B = -\frac{X}{Z^2}$$

The negative sign in this expression is due to the change of sign of the phase angle noted above. Except in a purely resistive circuit, it must be remembered that $G \neq 1/R$.

Example 7.13 Three impedances $10\angle -30^\circ\ \Omega$, $20\angle 60^\circ\ \Omega$ and $40\angle 0^\circ\ \Omega$ are connected in parallel. Calculate their equivalent impedance.

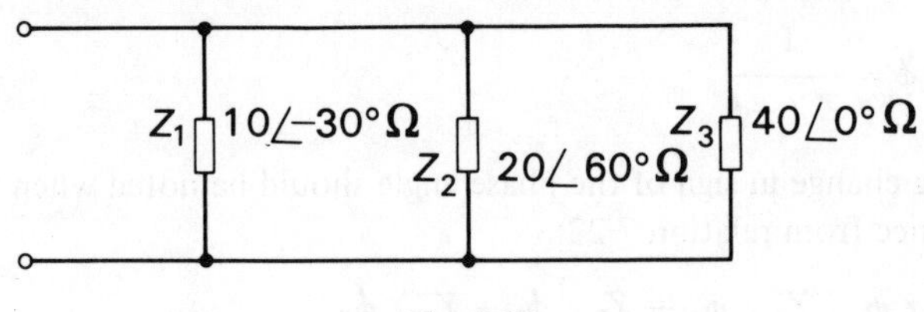

Fig. 7.35.

$$Y_1\angle\phi_1 = \frac{1}{Z_1\angle -\phi_1} = \frac{1}{10\angle -30^\circ} = 0{\cdot}1\angle 30^\circ\ \text{S}$$

Similarly $Y_2\angle\phi_2 = 0{\cdot}05\angle -60^\circ\ \text{S}$

$$Y_3\angle\phi_3 = 0{\cdot}025\angle 0^\circ\ \text{S}$$
$$G = G_1 + G_2 + G_3$$
$$= 0{\cdot}1\cos 30^\circ + 0{\cdot}05\cos -60^\circ + 0{\cdot}025\cos 0^\circ$$
$$= 0{\cdot}087 + 0{\cdot}025 + 0{\cdot}025 = 0{\cdot}137\ \text{S}$$

$$B = B_1 + B_2 + B_3$$
$$= 0{\cdot}1 \sin 30° + 0{\cdot}05 \sin -60° + 0{\cdot}025 \sin 0°$$
$$= 0{\cdot}05 - 0{\cdot}043 + 0{\cdot}0 = 0{\cdot}007 \text{ S}$$
$$Y = (G^2 + B^2)^{\frac{1}{2}} = (0{\cdot}137^2 + 0{\cdot}007^2)^{\frac{1}{2}}$$
$$= 0.137 \text{ S}$$
$$\phi = \tan^{-1}\frac{B}{G} = \tan^{-1}\frac{0{\cdot}007}{0{\cdot}137} = 3°$$
$$Z\angle{-\phi} = \frac{1}{Y\angle\phi} = \frac{1}{0{\cdot}137\angle 3°} = \underline{7{\cdot}32\angle{-3°}\ \Omega}$$

7.12 Resonance

In the analysis of the general series circuit, one case was not completely argued, namely that in which the inductive reactance is equal numerically to the capacitive reactance. When this occurs, $V_L = V_C$ and consequently V and I are in phase with one another. This is an instance of a condition that is termed resonance–in this case, series resonance because it concerns a series circuit. At resonance, the circuit is said to resonate. The phasor diagram for a series resonant circuit is shown in Fig. 7.36.

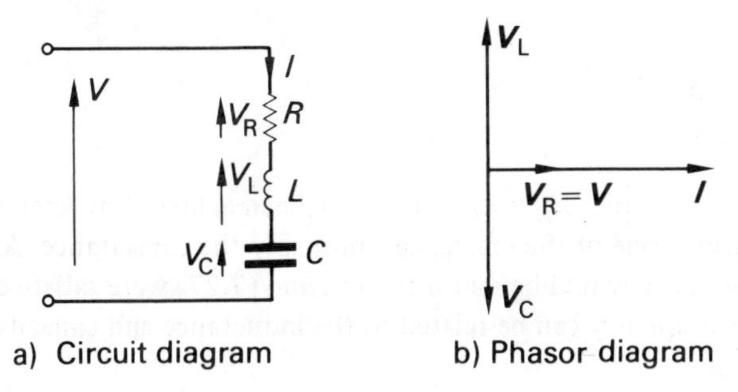

Fig. 7.36. Series resonant circuit

The condition of resonance is used extensively in electronic networks but, prior to explaining the phenomena of resonance, it is worth completing the investigation of the a.c. series circuit in terms of the reactances and their respective volt drops being equal.

The a.c. circuits discussed in this chapter have generally had the condition of constant supply frequency. However, many circuits–in electronic or communication networks particularly–are supplied from variable-frequency sources or sources supplying a number of frequencies. It is therefore pertinent to consider the effect of variation of supply frequency on the general series circuit.

When the frequency is zero, i.e. corresponding to a d.c. supply, $X_L = 2\pi fL = 0$ and $X_C = 1/2\pi fC = \infty$. As the frequency increases, X_L increases in direct proportion, whilst X_C decreases inversely. For resonance:

$$V_L = V_C$$
$$IX_L = IX_C$$
$$X_L = X_C \tag{7.27}$$

From the graph of reactance against frequency shown in Fig. 7.37 it is seen that this condition is satisfied when the frequency has a value f_r; f_r is termed the resonant frequency.

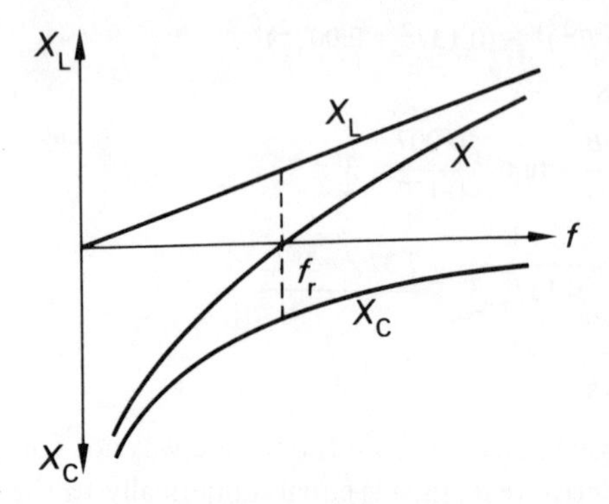

Fig. 7.37. Variation of reactance with frequency

For the general series circuit,

$$Z = (R^2 + (X_L - X_C)^2)^{\frac{1}{2}}$$

At resonance

$$X_L = X_C$$

hence
$$Z_r = R \tag{7.28}$$

The same condition that $X_L = X_C$ could have been achieved by keeping the frequency constant and varying one of the reactances, probably the capacitance. Again the condition of resonance would be met if the relation (7.27) were satisfied.

The resonant frequency can be related to the inductance and capacitance of the circuit as follows:

$$X_L = X_C$$

$$2\pi f_r L = \frac{1}{2\pi f_r C}$$

$$f_r^2 = \frac{1}{(2\pi)^2 LC}$$

$$f_r = \frac{1}{2\pi (LC)^{\frac{1}{2}}} \tag{7.29}$$

This is the resonant frequency of the general series circuit.

For the condition of the reactances of the inductor and the capacitor being equal, it has therefore been noted that V and I are in phase and that the circuit impedance is a minimum being equal to the resistance R. The resonant current is dependent on the resistance and the value of current, given by

$$I_r = \frac{V}{R}$$

may be considerably higher than that at other frequencies. A consequence of this is that the volt drops across the reactive components may reach very high values–many times that of the supply voltage. This was illustrated in Fig. 7.36. It may be added that the power dissipation in the circuit is consequently a maximum whilst the peak rates of energy storage for the reactors become equal and maximum. This variation follows from the current and impedance characteristics shown in Fig. 7.38.

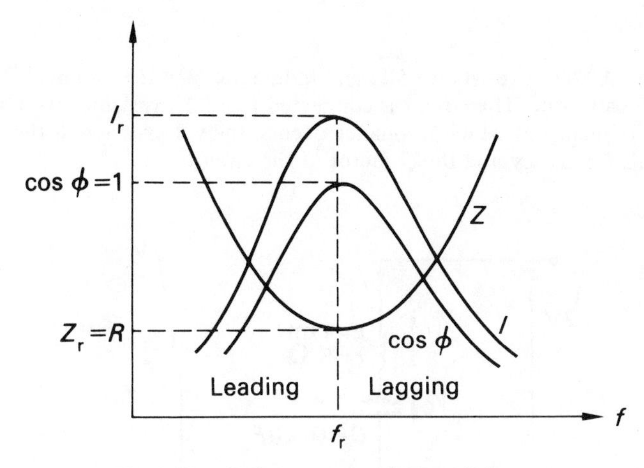

Fig. 7.38. Series resonance characteristics

Due to the peak rates of energy storage in the reactors, it is necessary to enquire into the relation of this statement to the operation of the circuit at unity power factor, which infers that the network has no reactive power. The energy stored by the reactors is a constant and it oscillates between the electric and magnetic modes of storage. It therefore remains with the reactors and it is this oscillation of energy that makes the circuit appear to be purely resistive. A circuit is said to resonate whenever this effect of energy oscillation predominates; this is usually interpreted to be when the peak rate of energy storage in each of the reactors–given by I^2X–is at least ten times the power associated with the circuit resistance. If this condition is not met, then the circuit is a power circuit (usually) that has had its power factor adjusted to unity.

To the electronics and communications engineers, the degree of predomination of the energy oscillation is important since the better the ratio of predominance then the better the circuit is able to accept current (and power) at the resonant frequency to the exclusion of other frequencies. This selection of frequency is used in radio and television receivers and is the process of tuning, e.g. accepting Radio 1 to the exclusion of other stations, each of which has its particular transmitting frequency.

The degree of predomination is given a factor of goodness termed the Q factor. Thus

$$Q \text{ factor} = \frac{I^2 X_L}{I^2 R}$$

$$= \frac{\omega_r L}{R} \qquad (7.30)$$

but $$\omega_r = \frac{1}{(LC)^{\frac{1}{2}}}$$

thus $$Q \text{ factor} = \frac{1}{R}\left(\frac{L}{C}\right)^{\frac{1}{2}} \tag{7.30.1}$$

Again it may be observed that the effect of resonance depends on the resistance of the circuit.

Example 7.14 A coil of resistance 5 Ω and inductance 1 mH is connected in series with an 0·2-μF capacitor. The circuit is connected to a 2-V, variable-frequency supply. Calculate the frequency at which resonance occurs, the voltages across the coil and the capacitor at this frequency and the Q factor of the circuit.

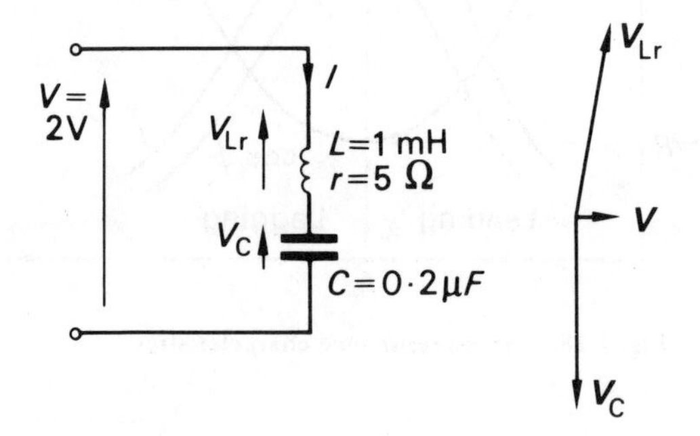

Fig. 7.39.

$$f_r = \frac{1}{2\pi(LC)^{\frac{1}{2}}}$$

$$= \frac{1}{2\pi(1 \times 10^{-3} \times 0{\cdot}2 \times 10^{-6})^{\frac{1}{2}}}$$

$$= 11\ 250 \text{ Hz}$$

$$= \underline{11{\cdot}25 \text{ kHz}}$$

$$I_r = \frac{V}{r} = \frac{2}{5} = 0{\cdot}4 \text{ A}$$

$$X_L = X_C = 2\pi f_r L = 2\pi \times 11\ 250 \times 1 \times 10^{-3} = 70{\cdot}7\ \Omega$$

$$Z_{Lr} = (r^2 + X_L{}^2)^{\frac{1}{2}} = (5^2 + 70{\cdot}7^2)^{\frac{1}{2}} = 71{\cdot}0\ \Omega$$

$$V_{Lr} = I_r Z_{Lr} = 0{\cdot}4 \times 71{\cdot}0 = \underline{28{\cdot}4 \text{ V}}$$

$$V_C = I_r X_C = 0{\cdot}4 \times 70{\cdot}7 = \underline{28{\cdot}3 \text{ V}}$$

$$Q \text{ factor} = \frac{\omega_r L}{r} = \frac{70{\cdot}7}{5} = \underline{14{\cdot}1}$$

Resonance can occur in any circuit containing sufficient inductance and capacitance, thus it is logical to consider parallel circuits as well as series circuits. The most common resonant parallel circuit is that of a coil, with integral resistance, in parallel with a capacitor as shown in Fig. 7.40.

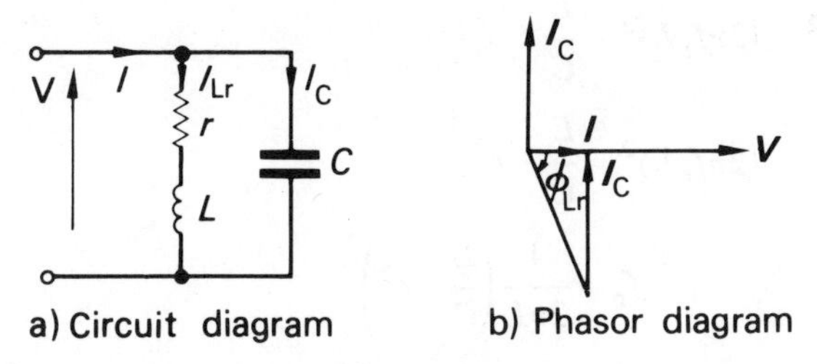

Fig. 7.40. Parallel resonant circuit

The current I_C in the capacitor branch leads the voltage V by 90° whilst the current I_{Lr} in the coil lags by an angle ϕ_{Lr} which is less than 90° depending on the ratio of the resistance to the inductive reactance. From the phasor diagram (Fig. 7.40(*b*)) it can be seen that resonance occurs when

$$I_C = I_{Lr} \sin \phi_{Lr} \tag{7.31}$$

This is termed parallel resonance.

Resonance can again either be arranged by adjustment of the components or by the frequency. In the latter case, when the frequency is zero, i.e. corresponding to a d.c. supply, the coil reactance $X_L = 2\pi fL = 0$ and so only the resistance r limits the current. In the capacitor branch, $X_C = 1/2\pi fC = \infty$ and hence there is no current. Increase of frequency increases the reactance of the coil and consequently the coil impedance increases. The coil current I_{Lr} therefore decreases and lags the voltage V by a progressively greater angle. The capacitive branch current, on the other hand,

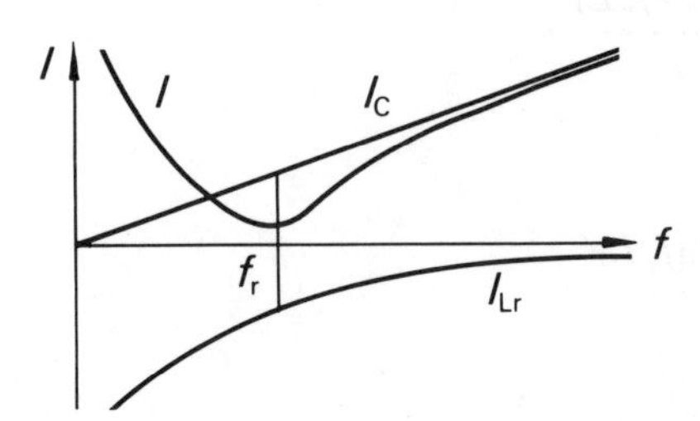

Fig. 7.41. Variation of current and reactance with frequency

increases although it will always lead by 90°. At some frequency, $I_C = I_{Lr} \sin \phi_{Lr}$ and resonance occurs. The variation of current with frequency is shown in Fig. 7.41, which compares with the reactance graph shown in Fig. 7.37.

From the phasor diagram:

$$I_C = I_{Lr} \sin \phi_{Lr}$$

$$\frac{V}{X_C} = \frac{V}{(r^2 + X_L{}^2)^{\frac{1}{2}}} \cdot \frac{X_L}{(r^2 + X_L{}^2)^{\frac{1}{2}}}$$

$$2\pi f_r C = \frac{2\pi f_r L}{r^2 + (2\pi f_r L)^2}$$

$$r^2 + (2\pi f_r L)^2 = \frac{L}{C} \tag{7.32}$$

$$(2\pi f_r L)^2 = \frac{L}{C} - r^2$$

$$f_r = \frac{1}{2\pi L}\left(\frac{L}{C} - r^2\right)^{\frac{1}{2}}$$

$$= \frac{1}{2\pi}\left(\frac{1}{LC} - \frac{r^2}{L^2}\right)^{\frac{1}{2}} \tag{7.32.1}$$

This is the resonant frequency for the general parallel circuit. If r is small then:

$$f_r = \frac{1}{2\pi(LC)^{\frac{1}{2}}} \tag{7.32.2}$$

which is the same solution as in the general series circuit.

Also $\quad I = I_{Lr} \cos \phi_{Lr}$

$$\frac{V}{Z_r} = \frac{V}{(r^2 + X_L{}^2)^{\frac{1}{2}}} \cdot \frac{r}{(r^2 + X_L{}^2)^{\frac{1}{2}}}$$

$$\frac{1}{Z_r} = \frac{r}{r^2 + (2\pi f_r L)^2}$$

$$Z_r = \frac{r^2 + (2\pi f_r L)^2}{r}$$

From relation 7.32:

$$\frac{L}{C} = r^2 + (2\pi f_r L)^2$$

$$Z_r = \frac{L}{Cr} \tag{7.33}$$

It is again of interest to note how the circuit responds to variation of frequency for fixed values of inductance and capacitance. At resonance V and I are in phase and the impedance is a maximum being equal to an effective resistance R_r, hence the minimum current is

$$I_r = \frac{VCr}{L}$$

If r were zero as in the ideal case, the current would be zero. Even so, there would be a current flowing around the circuit made by the coil and the capacitor. This is due to energy being transferred from the inductor to the capacitor and back again. Circuit resistance, however, must be present and this prevents the continual energy transfer taking place without loss. Some current must therefore be drawn from the supply to make good the loss. Even so, the current taken from the supply is small and the branch currents will be much greater than the supply current. At any other frequency, this situation does not arise and the supply current will be greater. Also the impedance and the power factor decrease; this is illustrated in Fig. 7.42. Because the circuit rejects current at resonance, it is called a rejector circuit.

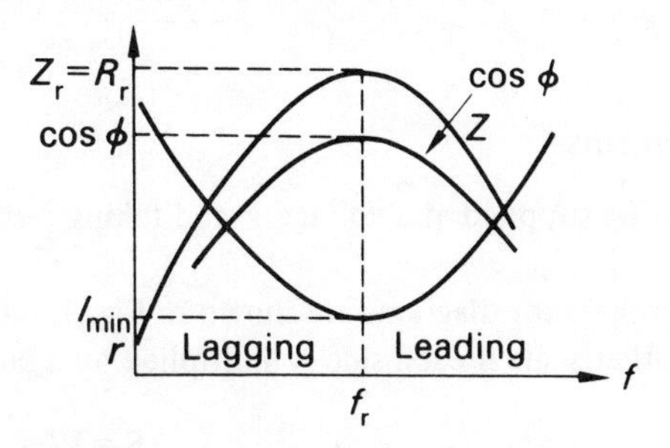

Fig. 7.42. Parallel resonance characteristics

Finally the Q factor of the network may be derived from the peak rate of energy storage, thus

$$Q \text{ factor} = \frac{I_{Lr}^2 X_L}{I_{Lr}^2 r} = \frac{\omega_r L}{r} \text{ as before.}$$

Example 7.15 A coil of 1 kΩ resistance and 0·15 H inductance is connected in parallel with a variable capacitor across a 2-V, 10-kHz a.c. supply. Calculate:

(*a*) The capacitance of the capacitor when the supply current is a minimum.
(*b*) The effective impedance of the network.
(*c*) The supply current.

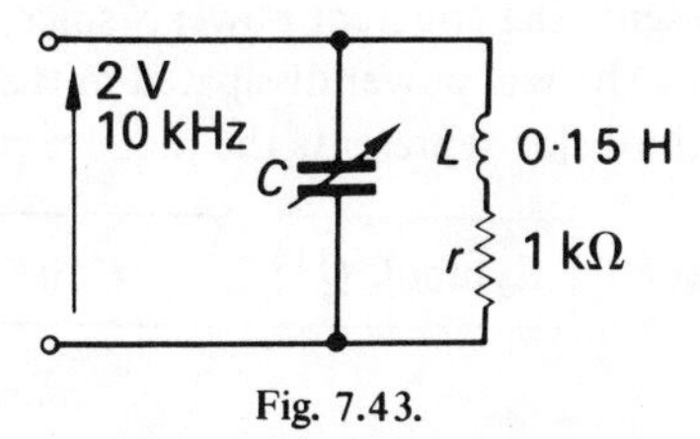

Fig. 7.43.

The network supply current is a minimum when the network is in resonance.

$$f_r = \frac{1}{2\pi}\left(\frac{1}{LC} - \frac{r^2}{L^2}\right)^{\frac{1}{2}}$$

$$\frac{1}{LC} = 4\pi^2 f_r^2 + \frac{r^2}{L^2}$$

$$= 4\pi^2 \times 10^2 \times 10^6 + \frac{1000^2}{0{\cdot}15^2}$$

$$= 3960 \times 10^6$$

$$C = \frac{10^6}{3960L} = 1{\cdot}69 \times 10^{-9}\,\text{F}$$
$$= \underline{1{\cdot}69\,\text{nF}}$$

$$Z_r = \frac{L}{Cr} = \frac{0{\cdot}15}{19 \times 10^{-9} \times 1\,000} = 890 \times 10^3\,\Omega = \underline{890\,\text{k}\Omega}$$

$$I_r = \frac{V}{Z_r} = \frac{2}{890 \times 10^3} = 2{\cdot}25 \times 10^{-6}\,\text{A} = \underline{2{\cdot}25\,\mu\text{A}}$$

7.13 Power diagrams

Consider again a load being supplied at a voltage V and taking a current $\boldsymbol{I}$ at a power factor $\cos \phi$.

The corresponding complexor diagrams are shown in Fig. 7.44(*a*). Similar diagrams can be drawn to some other scale if each side is multiplied by a factor V. It can then

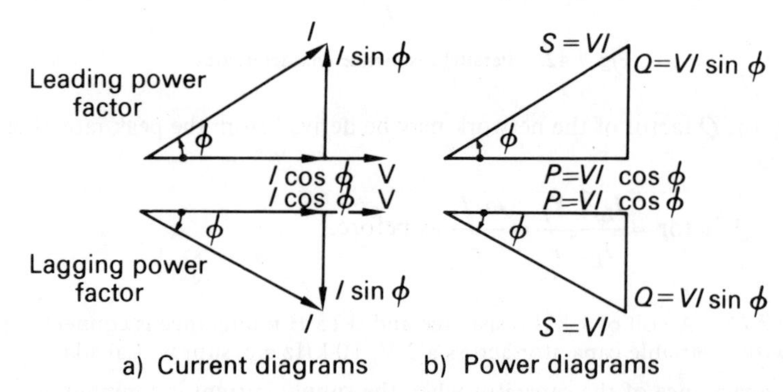

Fig. 7.44. Current and corresponding power diagrams

be seen that one side represents the apparent power S since, by relation (7.16), $S = VI$. Also another side represents the real power dissipated in the circuit since, by relation (7.15), $P = VI \cos \phi$. The third side represents the reactive power in the circuit.

Reactive Power	Symbol: Q	Unit: var (var)

$$Q = VI \sin \phi \tag{7.34}$$

It should be noted that whilst P represents the average real power in the circuit, Q represents the peak reactive power in the circuit, i.e. the peak rate of energy storage. This may be observed from the wave diagram shown in Fig. 7.45 which is drawn for the leading power factor case.

The r.m.s. voltage V corresponds to $0{\cdot}707V_m$ whilst the r.m.s. current I corresponds to $0{\cdot}707I_m$. From the diagram, it can be seen that these values correspond with the peak of the reactive power wave. It should also be noted that the average value of the reactive power wave is zero.

The complexor diagram shown in Fig. 7.44 can be applied to the determination of the combined effective load of several loads connected in parallel, e.g. the effective load of the various equipment installed in a factory. This problem could be determined by adding the various load currents complexorially but it is often more convenient to apply the load power diagram.

The real power and the reactive power components may be added by drawing a complexor diagram to scale. However an analytical method can be applied as follows.

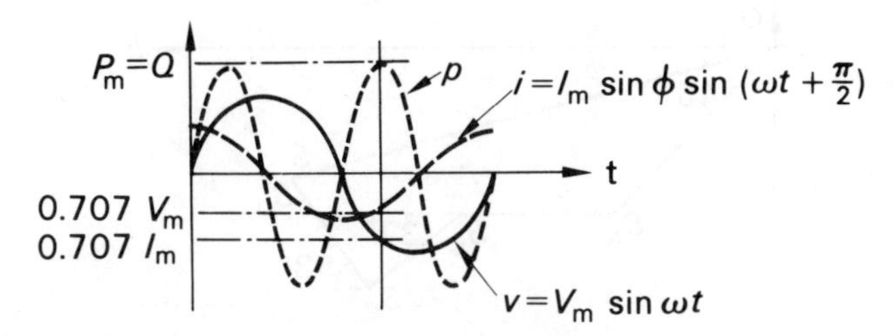

Fig. 7.45. Wave diagram of reactive power

Consider three loads (say) supplied at a voltage V and taking currents I_1, I_2 and I_3 at power factors $\cos\phi_1$, $\cos\phi_2$ and $\cos\phi_3$ respectively.

$$I\cos\phi = I_1 \cos\phi_1 + I_2 \cos\phi_2 + I_3 \cos\phi_3$$
$$I\sin\phi = I_1 \sin\phi_1 + I_2 \sin\phi_2 + I_3 \sin\phi_3$$
$$I = [(I\cos\phi)^2 + (I\sin\phi)^2]^{\frac{1}{2}}$$

If each of the above terms is multiplied by the voltage V then

$$VI\cos\phi = VI_1 \cos\phi_1 + VI_2 \cos\phi_2 + VI_3 \cos\phi_3$$
$$P = P_1 + P_2 + P_3$$
$$VI\sin\phi = VI_1 \sin\phi_1 + VI_2 \sin\phi_2 + VI_3 \sin\phi_3$$
$$Q = Q_1 + Q_2 + Q_3$$
$$S = (P^2 + Q^2)^{\frac{1}{2}} \tag{7.35}$$

Thus the analysis takes the same form as that for parallel-circuit currents. From the geometry of the diagram in Fig. 7.44

$$\phi = \tan^{-1}\frac{Q}{P} = \cos^{-1}\frac{P}{S} \tag{7.35.1}$$

Finally it should be noted that many loads are too large to be measured in watts, vars and volt amperes and hence many power diagrams are in kilowatts, kilovars and kilovolt amperes.

Example 7.16 The following loads are connected to a common supply:

Load A 150 kVA at 0·707 power factor lagging
Load B 100 kW at unity power factor
Load C 75 kVA at 0·8 power factor lagging
Load D 50 kW at 0·6 power factor leading

Determine the total apparent power and the overall power factor.

An estimate of the solution can be made using a complexor-type diagram as indicated below. Conversion of the power factors gives:

Load A is at 45° lag.
Load B is at 0°.
Load C is at 37° lag.
Load D is at 53° lead.

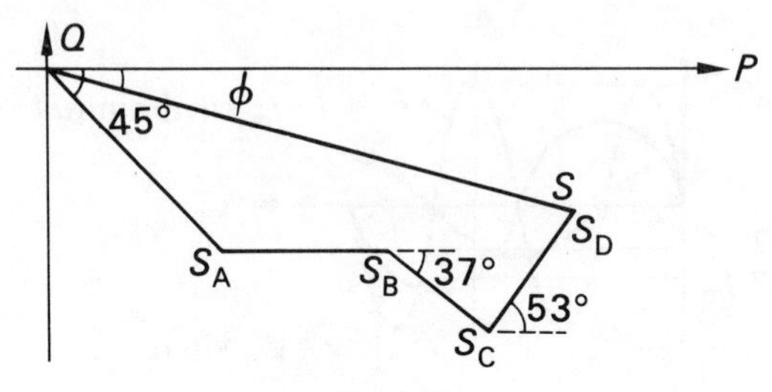

Fig. 7.46.

From the diagram:

$$S = \underline{330 \text{ kVA}}$$

$$\cos\phi = \underline{0{\cdot}96 \text{ lag}}$$

An analytical solution is preferable.

$$\begin{aligned}
P_A &= S_A \cos\phi_A = 150 \times 10^3 \times 0{\cdot}707 = 106{\cdot}1 \times 10^3 \text{ W} = 106{\cdot}1 \text{ kW}\\
P_B &= \qquad\qquad = 100{\cdot}0 \text{ kW}\\
P_C &= S_C \cos\phi_C = 75 \times 10^3 \times 0{\cdot}8 = 60{\cdot}0 \times 10^3 \text{W} = 60{\cdot}0 \text{ kW}\\
P_D &= \qquad\qquad = 50{\cdot}0 \text{ kW}\\
P &= \qquad\qquad = 316{\cdot}1 \text{ kW}
\end{aligned}$$

$$\begin{aligned}
Q_A &= S_A \sin\phi_A = 150 \times 10^3 \times -0{\cdot}707 = -106{\cdot}1 \times 10^3 \text{ var} = -106{\cdot}1 \text{ kvar}\\
Q_B &= P_B \tan\phi_B = 100 \times 10^3 \times 0 = 0 \text{ var} = 0{\cdot}0 \text{ kvar}\\
Q_C &= S_C \sin\phi_C = 75 \times 10^3 \times -0{\cdot}6 = 45{\cdot}0 \times 10^3 \text{ var} = -45{\cdot}0 \text{ kvar}\\
Q_D &= P_D \tan Q_D = 50 \times 10^3 \times 1{\cdot}33 = 66{\cdot}7 \times 10^3 \text{ var} = 66{\cdot}7 \text{ kvar}\\
Q &= \qquad\qquad = -84{\cdot}4 \text{ kvar}
\end{aligned}$$

$$\begin{aligned}
S &= (P^2 + Q^2)^{\frac{1}{2}} = (316{\cdot}1^2 + 84{\cdot}4^2)^{\frac{1}{2}} \times 10^3\\
&= 328 \times 10^3 \text{ VA}\\
&= \underline{328 \text{ kVA}}
\end{aligned}$$

$$\cos\phi = \frac{P}{S} = \frac{316{\cdot}1 \times 10^3}{328 \times 10^3} = \underline{0{\cdot}964 \text{ lag}}$$

7.14 Power factor improvement

It has been noted before that the object of passing a current around a circuit is to transfer energy from one point to another. Consequently much of the emphasis has been laid on the real power content of the apparent power. However, it has also been shown that the lower the power factor, then the higher is the current necessary to deliver the same power. This might be considered to be of little importance until it is remembered that the conductor system for the circuit will have some resistance. A cable is, in fact, rated according to the current it can carry without its temperature exceeding a value appropriate to its insulation. The temperature rise is caused by the I^2R heat loss in the conductor due to its resistance.

On economical grounds, it follows that the power factor is related to the size of cable required. If the power factor is poor, then the current will be greater than the minimum required (i.e. when the power factor is unity) and it may well follow that a larger and more costly cable will be required. In order to keep the cable size to a minimum, it is necessary therefore to improve or correct the power factor.

There is another reason for improving the power factor in the regulation of the voltage. The voltage regulation in a circuit is the change in voltage between the no-load and the loaded condition. The difference between these conditions is due to the volt-drop caused by the current acting on the impedance of the conductor system. This volt-drop has undesirable effects, e.g. lighting becomes appreciably dimmer with a relatively small volt-drop. Generally the less the current, the less the volt-drop so again power factor correction is desirable. In this context, it should be noted that the statutory limits, stated in the Electricity Supply Regulations, are $\pm$ 6·5%.

The degree of power factor improvement is regulated by two factors. The saving in cable cost and also the operating I^2R losses are offset by the cost of the power factor improvement equipment. There is therefore a point of balance at which the best saving can be obtained. Also many industrial tariffs penalise consumers whose power factor is low. This varies throughout the country but English consumers are expected to have power factors of 0·8 to 0·95 if no penalty is to be incurred. Scottish consumers have to have unity power factor.

Power factor improvement may be achieved in two principal ways:

Static capacitors. These are installed to improve the power factor of adjacent equipment. They are almost loss-free. Also they are not variable and on light loads, they tend to overcompensate. As most loads have lagging power factors, this overcompensation is not an important problem.

Use of synchronous motors. These motors can be made to operate at leading power factors and hence they can offset lagging loads within an installation. Whilst the problem of starting these motors can now be easily overcome, they have the disadvantage of being constant-speed machines.

Finally it should be noted that although the power factor correction equipment changes the reactive power and the power factor, it does not change the power taken by the load.

Example 7.17 A 200-V, 50-Hz a.c. motor is loaded to give an output of 11·2 kW operating with an efficiency of 80% and a power factor of 0·75 lagging to the circuit, assuming the motor terminal voltage to remain constant at 200 V. Calculate the reduction in current taken from the supply and the resultant overall power factor obtained when a 250-μF capacitor is connected in parallel with the motor.

If the motor is supplied through a cable of resistance 0·05 Ω, calculate the power loss in the cable before and after the capacitor is connected to the circuit.

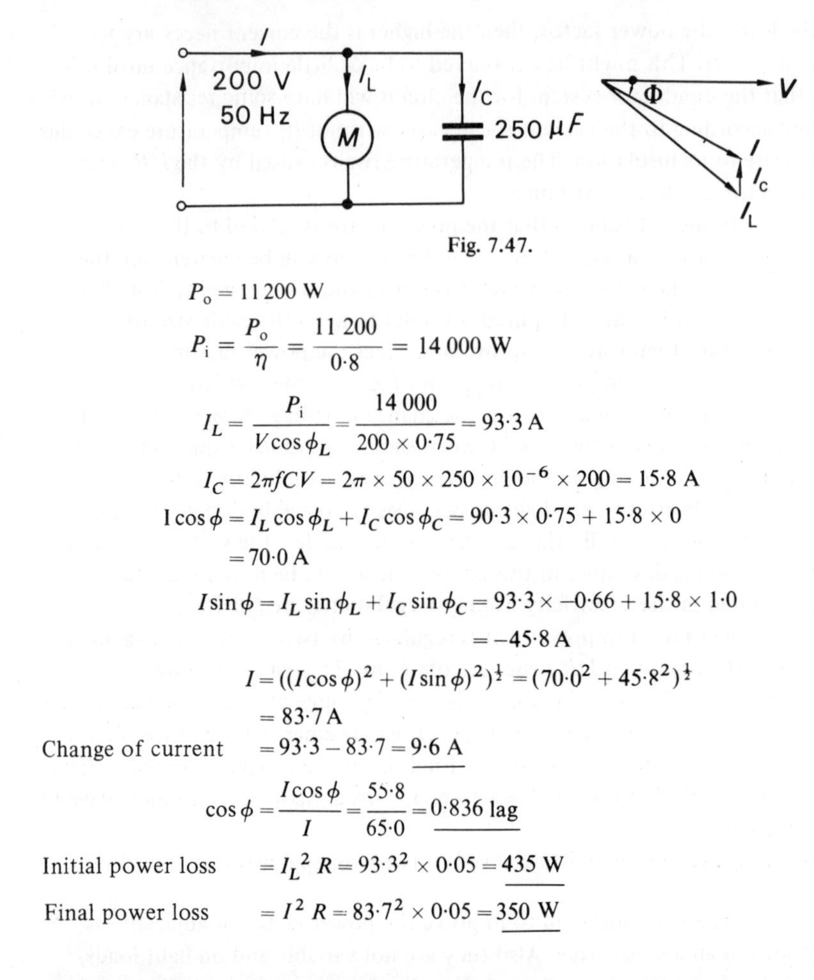

Fig. 7.47.

$$P_o = 11\,200 \text{ W}$$

$$P_i = \frac{P_o}{\eta} = \frac{11\,200}{0{\cdot}8} = 14\,000 \text{ W}$$

$$I_L = \frac{P_i}{V \cos\phi_L} = \frac{14\,000}{200 \times 0{\cdot}75} = 93{\cdot}3 \text{ A}$$

$$I_C = 2\pi f C V = 2\pi \times 50 \times 250 \times 10^{-6} \times 200 = 15{\cdot}8 \text{ A}$$

$$I \cos\phi = I_L \cos\phi_L + I_C \cos\phi_C = 90{\cdot}3 \times 0{\cdot}75 + 15{\cdot}8 \times 0$$
$$= 70{\cdot}0 \text{ A}$$

$$I \sin\phi = I_L \sin\phi_L + I_C \sin\phi_C = 93{\cdot}3 \times -0{\cdot}66 + 15{\cdot}8 \times 1{\cdot}0$$
$$= -45{\cdot}8 \text{ A}$$

$$I = ((I\cos\phi)^2 + (I\sin\phi)^2)^{\frac{1}{2}} = (70{\cdot}0^2 + 45{\cdot}8^2)^{\frac{1}{2}}$$
$$= 83{\cdot}7 \text{ A}$$

Change of current $= 93{\cdot}3 - 83{\cdot}7 = \underline{9{\cdot}6 \text{ A}}$

$$\cos\phi = \frac{I\cos\phi}{I} = \frac{55{\cdot}8}{65{\cdot}0} = \underline{0{\cdot}836 \text{ lag}}$$

Initial power loss $= I_L^2 R = 93{\cdot}3^2 \times 0{\cdot}05 = \underline{435 \text{ W}}$

Final power loss $= I^2 R = 83{\cdot}7^2 \times 0{\cdot}05 = \underline{350 \text{ W}}$

7.15 Power measurement

It has been assumed in this chapter that power can be measured without stating how this may be achieved. Power may be measured by a wattmeter. Whilst there are other methods, the wattmeter is now by far the most important, and it is the only one which will be considered. The construction of the wattmeter is described in para. 17.10 as is its principle of operation. However, in this chapter dealing with a.c. circuits, it is pertinent to consider how the wattmeter is applied.

The power in a circuit depends on the voltage and the current and it is therefore these quantities which have to be supplied to the meter. The circuit connections are shown in Fig. 7.48.

The wattmeter automatically takes care of the phase difference between the voltage and the current and therefore indicates the power appropriate to the applied voltage and the current. The terminals may be lettered in various ways and the diagram shows one of the more common methods. The M stands for mains whilst the L stands for load. If either connection is reversed, the wattmeter will try to read negatively and may be damaged.

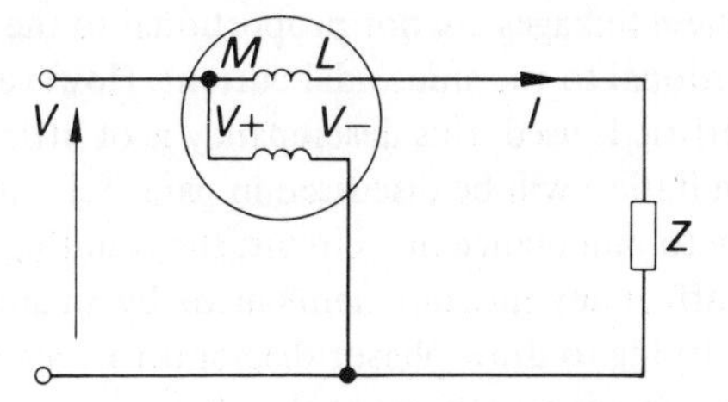

Fig. 7.48. Typical wattmeter connections

If an ammeter and a voltmeter are used in conjunction with a wattmeter, it is then possible to evaluate the active power, the apparent power and the power factor from the readings. A typical arrangement is described in Example 7.18.

Example 7.18 A voltmeter, an ammeter and a wattmeter are connected into a circuit in order to determine the power factor of the circuit. If the ammeter indicates 10 A, the wattmeter indicates 1 800 W and the voltmeter indicates 240 V, calculate the power factor.

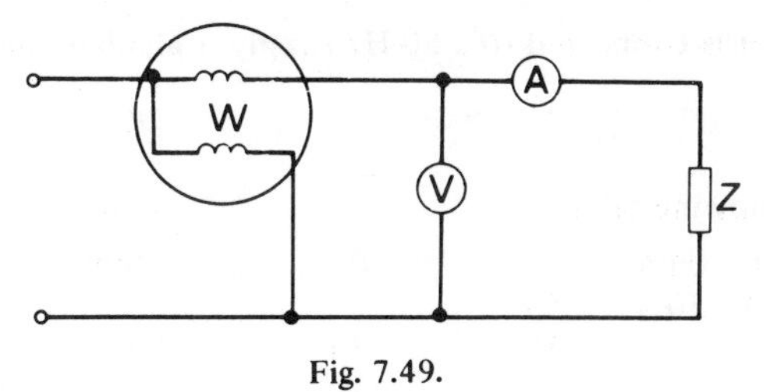

Fig. 7.49.

$$S = VI = 240 \times 10 = 2400 \text{ VA}$$

$$\cos\phi = \frac{P}{S} = \frac{1800}{2400} = \underline{0{\cdot}75}$$

It will be noted that from the information available, it is impossible to say whether the power factor is leading or lagging.

7.16 Sinusoidal waveform assumption

In para. 7.0, it was stated that only sinusoidal waveforms would be considered, and the ensuing theory depends on the validity of this premise. It has been shown that impedance is made up of resistance, inductance and capacitance. Consider each in turn.

In the case of resistance, the basic relation (7.1) is $v = iR$. It was then assumed that the resistance R would remain constant regardless of the applied voltage. This is reasonably true of most materials although there are many exceptions. However, for the normal circuit, which does not undergo radical changes of conditions and does not contain any of these special conductors, the assumption of the sinusoidal relation is acceptable.

The circuit component which is least likely to have sinusoidal relations between the voltage and the current is the inductance. If the inductance is due to flux linkages in air, then both the voltage and the current will be sinusoidal. However, it is more likely to be the case that the flux linkages pass through a ferromagnetic material. Because of the B/H characteristic, these linkages are not proportional to the m.m.f. and hence the back e.m.f. is not proportional to the sinusoidal current. However if only the linear part of the B/H characteristic is used, this descrepancy is of little importance. The effect of ignoring this limitation will be discussed in para. 8.1. Although there is little effect from the error due to inductance in a circuit, the resulting non-sinusoidal nature of the current may well affect any measurements made by an ammeter. This generally becomes apparent when trying to draw phasor diagrams to scale and it is quite possible to find a phase angle apparently more than 90°!

Finally there is the case of capacitance which gives the most reliable relations between voltage and current, i.e. X_c remains constant, being independent of voltage or current.

Thus it may be concluded that the theory of a.c. circuits as presented in this chapter is reasonably valid when applied to most circuits and the most likely source of error in the theory is due to the non-proportionality of the voltage and current in an inductance.

PROBLEMS ON 1-ph A.C. CIRCUITS

1 A 15-mH inductor is connected to a 50-Hz supply. Calculate the inductive reactance of the coil.

4·71 Ω

2 Complete the following table:

Inductance (H)	0·04		0·12	0·008	
Frequency (Hz)	50	50			60
Reactance (Ω)		50	36	4·5	57

3 A reactor allows a current of 15 A to flow from a 240-V, 50-Hz supply. Determine the current that will flow at the same voltage when the frequency changes to:

(*a*) 45 Hz
(*b*) 55 Hz

16·7 A; 13·6 A

4 A coil of wire has a resistance of 8·0 Ω and an inductance of 0·04 H. It is connected to an a.c. supply of 100 V at 50 Hz. Calculate the resulting current.

6·72 A

5 A 200-Ω resistor and a 0·8-H inductor are connected in series to a supply of 240-V, 50-Hz alternating current. Calculate:

(*a*) The circuit current.
(*b*) The power in the resistor.

0·75 A; 112 W

6 A 100-V, 60-W lamp is to be operated on 220-V, 50-Hz supply. Find what values of
(*a*) non-inductive resistance,
(*b*) pure inductance

would be required in order that the lamp is not overrun.

Which would be preferable?

200 Ω; 1·038 H

7 A 200-V, 50-Hz inductive circuit takes a current of 10·0 A lagging the voltage by 30°. Calculate the resistance, reactance and inductance of the circuit.

Sketch in their proper phase relationships the waves of the p.d., current and power.

17·3 Ω; 10·0 Ω; 31·8 mH

8 A 10·0-Ω resistor and a 400-μF capacitor are connected in series to a 60-V sinusoidal a.c. supply. The circuit current is 5·0 A. Calculate the supply frequency and the phase angle between the current and voltage. Draw also the phasor diagram to scale.

60 Hz; 33·6°

9 Find the inductance of a coil of negligible resistance which, when connected in series with a non-reactive resistor of 100 Ω, reduces the current to one half of its original value on a supply of frequency 50 Hz.

552 mH

10 A capacitor of 80 μF takes a current of 1·0 A when the a.c. potential difference across it is 250 V. Find:

(*a*) The supply frequency.
(*b*) The resistance to be connected in series with the capacitor to reduce the current to 0·5 A at this frequency.

7·96 Hz; 433 Ω

11 A coil is connected in series with a non-reactive resistance of 30 Ω across a 240-V, 50-Hz supply. The reading of a voltmeter across the coil is 180 V and across the resistance is 130 V. Calculate the power dissipated in the coil.

138 W

12 A voltage is given by the expression $v = 340 \sin \omega t$ and is applied to a circuit, the current being given by $i = 14{\cdot}14 \sin (\omega t - \pi/6)$. Sketch the waveforms in their correct phase relationship and draw the phasor diagram to scale. Calculate the impedance, resistance and reactance of the circuit and the power supplied.

24·0 Ω; 20·8 Ω; 12·0 Ω; 2 082 W

13 A steel-cored coil connected to a 100-V, 50-Hz supply is found to take current of 5 A and to dissipate a power of 200 W. Find:

(*a*) The impedance.
(*b*) The effective resistance.
(*c*) The inductance.
(*d*) The circuit power factor.

20·0 Ω; 8·0 Ω; 53·7 mH; 0·4 lag

14 A 1-ph circuit takes a power of 4·2 kW at power factor 0·6 lag. Find the value of the apparent input power and the peak reactive power.

7·0 kVA; 5·6 kvar

15 The alternating current flowing through an inductive circuit consists of an active component of 7·2 A and a reactive component of 5·4 A. The supply voltage is 200 V. Find:

(*a*) The value of the supply current.
(*b*) The power factor.
(*c*) The power dissipated.

9·0 A; 0·8 lag; 1·44 kW

16 A capacitor of negligible resistance, when connected to a 220-V, variable-frequency sinusoidal supply takes a current of 10·0 A when the frequency is 50 Hz. A non-inductive resistor when connected to the same supply takes a current of 12·0 A. If the two are connected in series and placed across the supply, calculate:

(*a*) The current taken and its phase angle when the supply frequency is 50 Hz.
(*b*) The supply frequency when the circuit current is 8·0 A.

$7{\cdot}7\angle 39{\cdot}5^\circ$ A; 53·7 Hz

17 A wooden ring having a mean diameter of 250 mm is wound uniformly with 2 000 turns of wire 2·0 mm^2 in cross section. The mean turn length is 60 mm and the effective cross-sectional area is 1000 mm^2. Calculate:

(*a*) The winding inductance.
(*b*) The winding resistance if the conductivity of the wire is 59 MS/m.
(*c*) The coil current if it were connected to a 20-V, 50-Hz alternating current supply.

1·6 mH; 1·02 Ω; 17·6 A

18 A load consisting of a capacitor and a resistor connected in series has an impedance of 50 Ω and a power factor of 0·707 leading. The load is connected in series with a 40-Ω resistor across an a.c. supply and the resulting circuit current is 3·0 A. Determine the supply voltage and the overall phase angle.

250 V; 25·1

19 A steel-cored choking coil has a resistance of 4 Ω when measured by direct current. On a 240-V, 50-Hz mains supply, it dissipates 500 W, the current taken being 10 A. Calculate the core loss, power factor, impedance, reactance and inductance of the coil.

100 W; 0·21 lag; 24 Ω; 23·5 Ω; 75 mH

20 A circuit operating at a power factor of 0·8 lag takes a current of 10 A from a 250-V, 50-Hz sinusoidal supply.

(*a*) Find an expression for the instantaneous values for voltage and current.
(*b*) Calculate the value of current 5 ms after the voltage has reached its positive maximum value.
(*c*) At what time, after being zero and going negative, will the instantaneous current first reach 10 A.

353·6 sin ωt V; 14·14 sin (ωt – 37°) A; 8·48 A; 2·5 ms

21 Define inductance and prove that the sinusoidal voltage across an inductor leads the current through it by 90°.

A series circuit, consisting of a 12-Ω resistor, a 0·15-H inductor and a 100-μF capacitor, is connected across a variable-frequency 100-V sinusoidal supply. Show graphically the effect on the resistance, inductive and capacitive reactances and impedance of variation of frequency. State the frequency at which the circuit will act as a non-inductive resistor and give the values of R, X_L, X_C and Z at this frequency. Hence comment on the current taken from the supply and the voltages across the circuit components.

41·1 Hz; 12 Ω; 38·7 Ω; 38·7 Ω; 12 Ω

22 A resistor of 10 Ω is connected in parallel with a capacitor of 318 μF. A 100-V supply is applied to this circuit. Find the current supplied to this circuit and the power factor if the supply frequency is 50 Hz.

14·14 A; 0·71 lead

23 A resistor of 10 Ω is connected in parallel with a 31·8-mH inductor. A 200-V, 50-Hz supply is applied to the circuit. Find the supply current and the power factor.

28·28 A; 0·71 lag

24 A 10-Ω resistor, a 31·8-mH inductor and a 318-μF capacitor are connected in parallel and supplied from a 200-V, 50-Hz source. Calculate the supply current and the power factor; also calculate the current in each branch.

20 A; 1·0; 20 A; 20 A; 20 A

25 A 10-Ω resistor, a 15·9-mH inductor and a 159-μF capacitor are connected in parallel to a 200-V, 50-Hz source. Calculate the supply current and power factor.

36 A; 0·56 lag

26 The following data was obtained from separate tests on two coils, which were connected separately to a 240-V, 50-Hz supply –

Coil A	5 A	600 W
Coil B	10 A	1 500 W

Find the resistance, reactance and inductance of each coil, the current taken from the supply and the power factor of the circuit if the two coils are connected in parallel across the given supply.

What value of capacitance must be connected across the supply terminals to give unity power factor.

24 Ω; 41·6 Ω; 0·132 H; 15 Ω; 18·7 Ω; 0·060 H; 15·0 A; 0·59 lag; 161 μF

27 The brake on an a.c. hoisting motor is operated from a solenoid contactor with a steel plunger through its coils of resistance 20 Ω. When energised from a 250-V, 50-Hz sinusoidal supply, the initial current taken is 5 A and this decreases to 2 A after the brake has operated. Find:

(*a*) The impedance and the inductance of the coils for each contact in position.
(*b*) The power factor in the operated position.
(*c*) The value of the parallel-connected capacitance to raise the power factor to unity in the operated position.

50 Ω; 0·146 H; 125 Ω; 0·393 H; 0·16 lag; 25·2 μF

28 A coil of effective resistance 12 Ω and inductive reactance 5 Ω is shunted by a resistor and by a capacitor. The supply current is 8 A at unity power factor and 5 A flows through the coil. Calculate the values of the resistance and the capacitive reactance.

19·2 Ω; 33·8 Ω

29 The circuit shown in Fig. 7.50 is connected to a 50-Hz sinusoidal supply. The total current is 12 A leading the supply voltage by a phase angle of 30°. The current in the top branch is 8 A. Calculate:

(*a*) The supply voltage.
(*b*) The unknown branch currents.
(*c*) The values of *R* and *C*.
(*d*) The resistance and reactance of the equivalent series circuit.

104 V; 3·01 A; 9·08 A; 34·6 Ω; 278 μF; 7·5 Ω; 4·3 Ω (SANCAD)

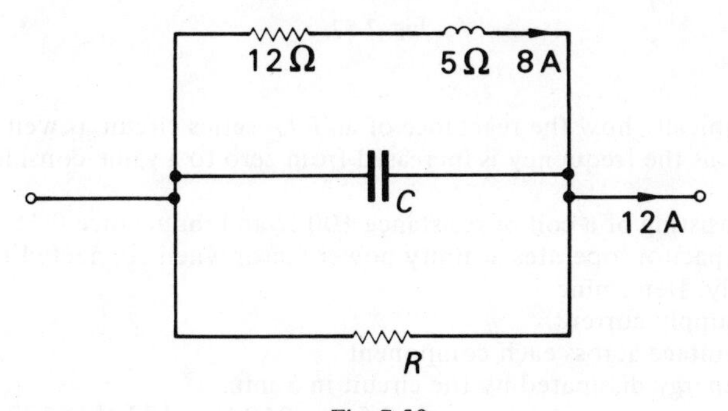

Fig. 7.50.

30 For the circuit shown in Fig. 7.51 calculate for a sinusoidal supply:

(*a*) The current in each branch.
(*b*) The total current.
(*c*) The overall power factor.
(*d*) The equivalent series circuit.
(*e*) The value and nature of reactance to be connected in series with the circuit to bring the overall power factor to unity.

7·2 A; 12·2 A; 12·0 A; 0·92 lead; 17·6 Ω; 417 μF X_L = 7·6 Ω

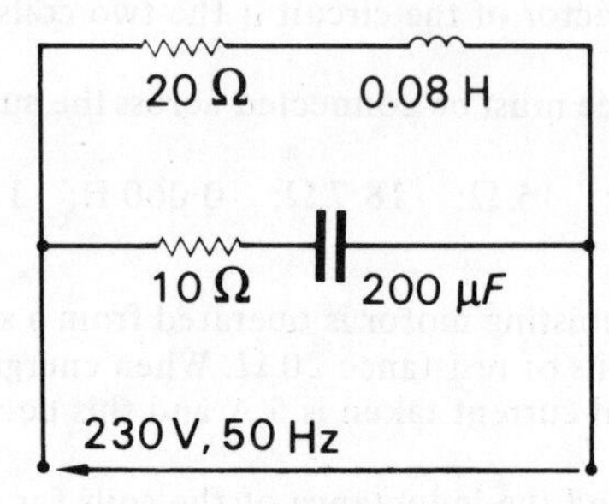

Fig. 7.51.

31 If the voltage across the 100-Ω resistor in the circuit shown in Fig. 7.52 is 50 V, find:

(*a*) The current and the power taken from the supply.
(*b*) The power factor of the circuit.
(*c*) The value of capacitance required to be connected in series with the supply to give unity power factor.

1·40 A; 0·63 lag; 46·2 W; 109 μF

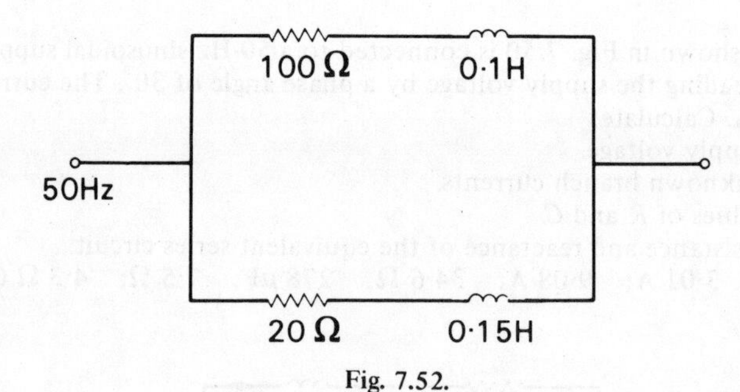

Fig. 7.52.

32 Show graphically how the reactance of an RL_C series circuit as well as the impedance vary as the frequency is increased from zero to a value considerably greater than resonance.

A circuit consisting of a coil of resistance 100 Ω and inductance 0·15 H in series with a 30-μF capacitor, operates at unity power factor when connected to a 250-V sinusoidal supply. Determine:

(*a*) The supply current.
(*b*) The voltage across each component.
(*c*) The energy dissipated by the circuit in 5 min.

2·5A; 250 V; 177 V; 177 V; 187·5 kJ

33 Explain what is meant by electrical resonance in an electrical circuit. A circuit consisting of a coil of resistance 100 Ω and inductance 0·5 H in parallel with a 20-μF capacitor operates at unity power factor when connected to a 250-V variable-frequency sinusoidal supply. Calculate:

(*a*) The supply frequency.
(*b*) The supply current.
(*c*) The coil current.

If the frequency is varied so that resonance occurs when the coil and capacitor are connected in series across the supply, calculate:

(*d*) The change in frequency.
(*e*) The new value of the supply current.

39·0 Hz; 1·0 A; 1·58 A; 2·5 A; 11·3 Hz

34 A choking coil is connected in series with a 20-μF capacitor. With a constant supply voltage of 200 V, it is found that the circuit takes its maximum current of 50 A when the supply frequency is 100 Hz. Calculate the inductance and resistance of the choking coil and the voltage across the capacitor.

127 mH; 4·8 Ω; 3·98 kV

35 A coil whose resistance is not negligible is connected in series with a capacitor across a variable-frequency supply, the frequency being adjusted until the current has a maximum value of 0·5 A. The supply voltage is 20 V and the frequency is 318 Hz. Calculate the resistance of the coil and its inductance if the voltage across the capacitor is 50 V. If the frequency is doubled, calculate the change in capacitance required to keep the current at 0·5 A, the supply voltage remaining at 20 V.

40 Ω; 50 mH; 15 μF

36 Draw a circuit diagram to show a voltmeter, ammeter and wattmeter connected to measure the power input to a steel-cored coil. What information could be obtained from the readings?

If the readings on the instruments are 110 V, 2·5 A and 150 W respectively and the d.c. resistance of the copper windings of the coil is 15 Ω, calculate the inductance of the coil and the power loss in the steel core. The supply is sinusoidal and the frequency is 50 Hz.

0·112 H; 56·25 W

37 With the aid of suitable sketches, explain the action of an electro-dynamic wattmeter. Explain how such an instrument indicates the mean power in an a.c. circuit.

Such a wattmeter, connected in a 1-ph a.c. circuit with its voltage-coil circuit across the load side of the instrument, reads 750 W. The resistance of the voltage coil circuit is 4400 Ω and its inductance may be neglected. The voltage across the load is 440 V. Calculate the correct value of the power in the load and the percentage error, high or low, in the wattmeter reading.

706 W; + 6·23%

38 A cable is required to supply a welding set taking a current of 225 A at 110 V alternating current, the average power factor being 0·5 lagging. An available cable has a rating of 175 A and it is decided to use this cable by installing a capacitor across the terminals of the welding set. Find:

(*a*) The required capacitor current and reactive power to limit the cable current to 175 A.
(*b*) The overall power factor with the capacitor in circuit.

60·8 A; 6·7 kvar; 0·643 lag

39 A substation supplied the following 1-ph loads:

500 kW at unity power factor;
750 kW at 0·8 power factor lagging;
400 kVA at 0·6 power factor lagging;
200 kW at 0·9 power factor leading.

Determine the value of the reactive power to improve the overall power factor of the substation to 0·95 leading.

1341 kvar

40 A 240-V, 50-Hz, 375-W, 1-ph motor has a full-load efficiency of 80% and a power factor of 0·766 lagging. The power factor is improved to 0·94 lagging by connecting a capacitor in parallel with the motor. Calculate:

(*a*) The active and reactive components of the original current.
(*b*) The total current with the capacitor in circuit.
(*c*) The capacitor current and reactance.
(*d*) The capacitance of the capacitor.
(*e*) The peak reactive power taken by the capacitor.

1·95 A; 1·64 A; 2·07 A; 0·93 A; 258 Ω; 12·34 μF; 223 var

8

j notation

Problems involving complexor and phasor diagrams, with particular reference to the representation of currents and voltages that vary sinusoidally, are by now familiar. Using the methods put forward, it is possible to undertake most circuit analyses but the time that would be involved would prove completely impracticable in many cases. Subsequent to a satisfactory understanding of the principles involved, it is now possible to introduce a system of notation which facilitates the manipulation of the calculations although it only adds to the previous work and in no way obviates it. This system is j *notation.*

8.1 Complexor diagrams

It was shown in para. 3.4 that a complexor diagram is one in which a line is defined by a tensor quantity (the magnitude of the line) and a versor quantity (the angle through which the line is rotated with respect to some given unit reference step.) The resulting line, in general termed a complexor, may be used to represent some physical quantity, e.g. voltage or current.

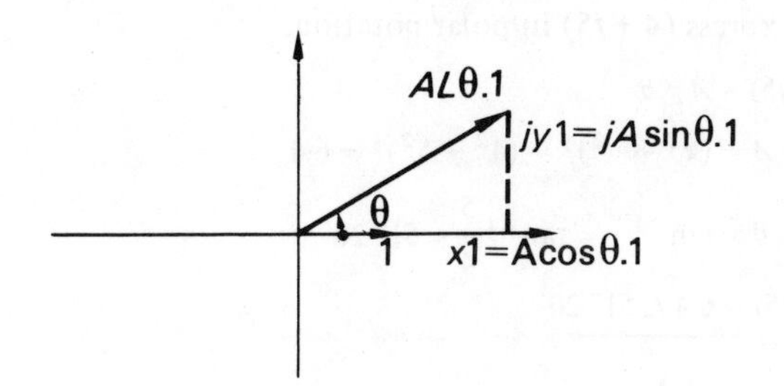

Fig. 8.1. General complexor diagram

Consider again the general diagram shown in Fig. 8.1. A complexor defined as $A\angle\theta\,.\,1$ is shown, the unit reference step again being introduced.

The complexor $A\angle\theta\,.\,\mathbf{1}$ can be resolved into a horizontal component of length $A\,.\cos\theta$ and a vertical component of length $A\,.\sin\theta$. Let the horizontal component be x such that

$$x = A\,.\cos\theta \tag{8.1}$$

The horizontal component in complexor notation is therefore defined as $x\angle 0\,.\,\mathbf{1}$. Similarly let the vertical component by y such that

$$y = A\,.\sin\theta \tag{8.2}$$

The vertical component in complexor notation is therefore defined as $y\angle\frac{1}{2}\pi\,.\,\mathbf{1}$.

It can be seen that the versor of the horizontal component serves no important function and could be omitted without prejudice. The versor of the vertical component denotes a rotation of $+\frac{1}{2}\pi$ radians. Let this operation be denoted by j thus the vertical component can now be defined as $jy\,.\,\mathbf{1}$. From the complexor construction of the diagram, it follows that

$$\boldsymbol{A} = x.\mathbf{1} + jy.\mathbf{1}$$

$$\boldsymbol{A} = (x + jy).\mathbf{1} \tag{8.3}$$

Thus any complexor can be expressed in the form $(x + jy)\,.\,\mathbf{1}$. Since only horizontal and vertical components are involved, the omission of the reference angle with respect to the horizontal component is seen to be justifiable.

An expression such as $(x + jy)$ is called a complex operator since it converts one directed line, such as $\mathbf{1}$, into another of different size and inclination and often of a different physical meaning.

Before leaving polar notation, it will be noted that in general

$$\boldsymbol{A} = A\angle\theta.\mathbf{1}$$

$$= (x^2 + y^2)^{\frac{1}{2}}\angle\tan^{-1}\frac{y}{x}\cdot\mathbf{1} \tag{8.4}$$

Relation 8.4 permits complex operations to be transformed into polar notation. This is illustrated in Example 8.1.

Example 8.1 Express $(4 + j5)$ in polar notation.

Let $\quad (4 + j5) = A\angle\theta$

$$A = (x^2 + y^2)^{\frac{1}{2}} = (4^2 + 5^2)^{\frac{1}{2}} = 6{\cdot}4$$

$$\theta = \tan^{-1} y = \tan^{-1}\frac{5}{4} = 51°\,20'$$

$$(4 + j5) = \underline{6{\cdot}4\angle 51°\,20'}$$

Also $\quad \boldsymbol{A} = (x + jy).\mathbf{1}$

$$= (A\cos\theta + jA\sin\theta).\mathbf{1}$$

$$= A(\cos\theta + j\sin\theta).\mathbf{1} \tag{8.4}$$

The derivation of relation (8.4) gives rise to the method of transforming polar operators into complex notation. This is illustrated by Example 8.2.

Example 8.2 Express $50\angle 60^{\circ}$ in complex notation.

$$\begin{aligned}\text{Let}\quad 50\angle 60^{\circ} &= A\angle\theta \\ &= (A\cos\theta + jA\sin\theta) \\ &= (50\cos 60^{\circ} + j50\sin 60^{\circ}) \\ &= \underline{(25{\cdot}0 + j43{\cdot}3)}\end{aligned}$$

8.2 *j* operator

Since j is defined as an operator which turns a complexor through $+\frac{1}{2}\pi$ radians without changing its size, two operations will turn a complexor through a total of π radians. It follows that the complexor is reversed. Thus

$$\begin{aligned}j(j).\mathbf{1} &= j^2.\mathbf{1} \\ &= -1.\mathbf{1}\end{aligned}$$

This statement can be illustrated by the complexor diagram shown in Fig. 8.2.

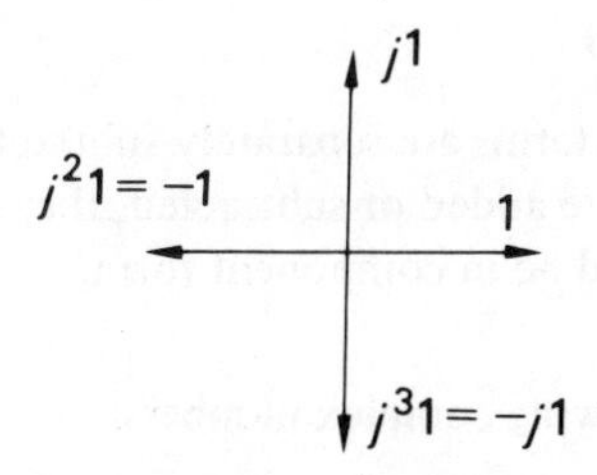

Fig. 8.2. ***j* operator**

It follows that it is convenient to think of

$$j^2 = -1$$

$$j = \sqrt{-1} \tag{8.5}$$

Finally as in chapter 3 it is usual with this type of notation to omit the unit reference step **1** for convenience and to assume that any quantity defined by a complex number has the properties of a complexor, i.e. the complex operator only is used. This omission can be misleading and will be further discussed in para. 8.6. Further it may be assumed that if a voltage or current quantity is defined by a complex number, it has the properties of a phasor.

8.3 Complexor operation

Complex notation has certain important properties, that must be taken into account when the notation is applied. First of all, complexors may be added quite simply. Consider the addition of $A\angle\theta_1 = (a + jb)$ and $B\angle\theta_2 = (c + jd)$ as shown in Fig. 8.3. The horizontal components are termed the real terms and they act in the direction of the reference step. These are a and c. The vertical components are termed the imaginary terms and they act in quadrature with the reference step due to the action of the j operation.

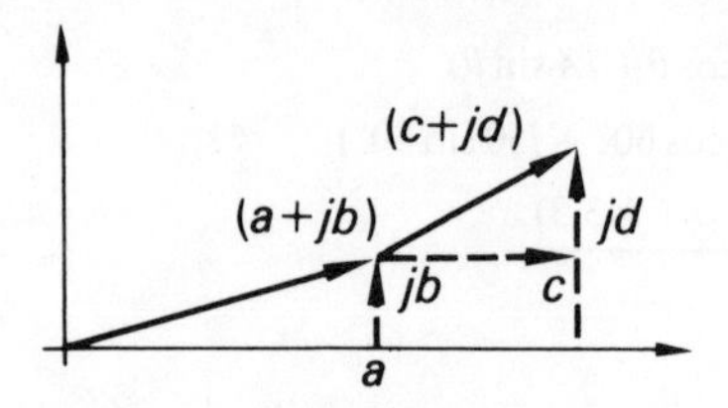

Fig. 8.3. Addition of complexors

From the construction of the diagram, it may be seen that the resulting complexor is given by

$$(a + c) + j(b + d)$$

that is the real and the imaginary terms are added separately.

Similarly it may be shown that had the complexors been subtracted the resulting complexor is given by

$$(a - c) + j(b - d)$$

that is the real and imaginary terms are separately subtracted. It should be noted, however, that when complexors are added or subtracted, they must all represent quantities of the same physical kind, and be in component form.

Example 8.3 Add the following complex numbers:

$$(2 + j5) \quad \text{and} \quad (4 - j3)$$

Also find the value of the first less the second.

$$(2 + j5) + (4 - j3) = (2 + 4) + j(5 + (-3))$$
$$= \underline{(6 + j2)}$$

$$(2 + j5) - (4 - j3) = (2 - 4) + j(5 - (-3))$$
$$= \underline{(-2 + j8)}$$

This last complex number may represent a complexor lying in the second quadrant.

If complex notation is compared with polar notation in the light of the above example, it will be seen that complex notation permits complexors and phasors to be added and subtracted much more easily. This is particularly important when considering the addition and subtraction of voltage and current phasors in accordance with Kirchhoff's Laws.

However, this advantage of manipulation is offset to some extent by the difficulty of successive operations. Consider again the above operators being so applied.

$$(A\angle\theta_1).(B\angle\theta_2) = AB\angle(\theta_1 + \theta_2)$$
$$= (a + jb).(c + jd)$$

It has already been noted that complex notation can be treated algebraically, hence

$$(a+jb).(c+jd) = ac + jad + jbc + j^2 bd$$

but $$j^2 = -1$$

therefore the expression becomes

$$(a+jb).(c+jd) = ac + jad + jbc - bd$$
$$= (ac - bd) + j(ad + bc) \quad (8.6)$$

It should be noted that the combined effect is commutative, i.e. it is independent of the order in which the operators are successively applied.

Also $$\frac{A\angle\theta_1}{B\angle\theta_2} = \frac{A}{B}\cdot\angle(\theta_1 - \theta_2)$$

$$= \frac{a+jb}{c+jd}$$

In order to expand this complex expression, multiply both numerator and denominator by the same factor $(c - jd)$. This does not affect the equivalence of the expression.

$$\frac{a+jb}{c+jd} = \frac{a+jb}{c+jd}\cdot\frac{c-jd}{c-jd}$$

$$= \frac{(ac+bd) + j(bc - ad)}{c^2 + d^2} \quad (8.7)$$

In both cases, it is more simple to deal with the successive operation in polar rather than complex form. It is for this reason that the reader is advised to be familiar with both forms. Each has its own appropriate use. This will become apparent in the worked examples in the remainder of this chapter.

Example 8.4 Expand $(1 + j2).(5 + j6)$.

$$(1 + j2).(5 + j6) = 5 - 12 + j6 + j10$$
$$= \underline{-7 + j16}$$

Example 8.5 Expand $\frac{1 + j2}{3 + j4}$.

$$\frac{1+j2}{3+j4} = \frac{(1+j2)}{(3+j4)}\cdot\frac{(3-j4)}{(3-j4)}$$

$$= \frac{3 + 8 - j4 + j6}{3^2 + 4^2}$$

$$= \frac{11 + j2}{25}$$

$$= \underline{0{\cdot}44 + j0{\cdot}08}$$

Operators such as $A\angle -\theta_1 = a + jb$ and $A\angle\theta_1 = a - jb$ are called conjugate operators. Therefore expression (8.7) was obtained by the application of conjugate operators. In general, the result of the successive application of conjugate operators is $A^2\angle 0$, hence a pure number is obtained.

Operators such as $A\angle\theta_1 = a + jb$ and $\dfrac{1}{A\angle\theta_1} = \dfrac{1}{a + jb}$ are termed inverse operators. They have reciprocal tensors A and $\dfrac{1}{A}$ and they are rotated through equal angles but in opposite directions. The result of their successive operation is unity, i.e. they cancel one another out.

$$(a + jb).\frac{1}{(a + jb)} = 1$$

Finally to obtain the square root of a complex quantity, it is necessary to recall that the square of a complex quantity can be obtained by squaring the tensor and doubling the versor. The square root can only be obtained by reversing this process, i.e. by taking the square root of the tensor and halving the versor. Example 8.6 illustrates this principle. However it will be noted that, although the j operator can be used for most complex operations, there are certain calculations, which can be carried out only by the use of polar notation. This is a further justification for the earlier statement that the systems are complementary.

Example 8.6 Find the square root of $(5{\cdot}0 + j8{\cdot}7)$.

$$5{\cdot}0 + j8{\cdot}7 = A\angle\theta$$

$$A = (5{\cdot}0^2 + 8{\cdot}7^2)^{\frac{1}{2}} = 10{\cdot}0$$

$$\theta = \tan^{-1}\frac{8{\cdot}7}{5{\cdot}0} = 60°$$

$$5{\cdot}0 + j8{\cdot}7 = 10{\cdot}0\angle 60°$$

$$(5{\cdot}0 + j8{\cdot}7)^{\frac{1}{2}} = 10{\cdot}0^{\frac{1}{2}}\angle\frac{60°}{2} = 3{\cdot}16\angle 30°$$

$$= 3{\cdot}16(\cos 30° + j\sin 30°)$$

$$= \underline{2{\cdot}7 + j1{\cdot}6}$$

8.4 Impedance

Phasors representing alternating voltages and currents in a phasor diagram can be expressed as complex operators. All that is necessary is to choose some suitable scale units and a reference direction, usually that due to the supply voltage. The phasors may then be summed with or subtracted from other phasors which represent like quantities, expressed with respect to the same scale unit and reference direction.

Impedance can also be defined in complex notation. However, impedance is not a phasor quantity but it is among those quantities that can be represented by a complex number. This can be illustrated as follows:

$$\text{Impedance} = \frac{\boldsymbol{V}}{\boldsymbol{I}}$$

$$= \frac{V\angle\theta.1}{I\angle\beta.1}$$

$= Z\angle\phi$, where θ and β are the respective versors of the voltage and current and ϕ is their phase angle.

Thus it will be seen that the impedance no longer incorporates the unit reference step and is therefore not a phasor but it can still be expressed as a complex number.

Consider an alternating voltage $\boldsymbol{V}$ applied to a resistor R resulting in a current $\boldsymbol{I}$ through it. If $\boldsymbol{I}$ is taken as reference then

$$\boldsymbol{I} = I\angle 0°$$

and $\boldsymbol{V} = V\angle 0°$, since $\boldsymbol{V}$ is in phase with $\boldsymbol{I}$.

$$Z = \frac{\boldsymbol{V}}{\boldsymbol{I}}$$

$$= \frac{V\angle 0°}{I\angle 0°}$$

$$= \frac{V(1 + j0)}{I(1 + j0)}$$

$$= R \tag{8.8}$$

Thus the impedance of a resistor in complex notation is represented by a pure number R. Now let the voltage be applied to an inductor of inductance L. In this case:

$$\boldsymbol{I} = I\angle 0°$$

and $\boldsymbol{V} = V\angle 90°$, since the voltage leads the current by 90°.

$$Z = \frac{\boldsymbol{V}}{\boldsymbol{I}}$$

$$= \frac{V\angle 90°}{I\angle 0°}$$

$$= \frac{V(0 + j1)}{I(1 + j0)}$$

$$= \frac{jV}{I}$$

$$= j\omega L$$

$$= jX_L \tag{8.9}$$

The effect of *j* in this expression is such that when a current is multiplied by the inductive reactance, the resulting phasor product leads the current phasor by 90°.

Since the product represents the effective volt drop across the reactance, this result compares with that expected.

Finally let the voltage be applied to a capacitor *C*. In this case:

$$\boldsymbol{I} = I\angle 0°$$

and $\boldsymbol{V} = V\angle -90°$, since the voltage lags the current by 90°.

$$Z = \frac{\boldsymbol{V}}{\boldsymbol{I}}$$

$$= \frac{V\angle -90°}{I\angle 0°}$$

$$= \frac{V(0 - j1)}{I(1 + j0)}$$

$$= \frac{-jV}{I}$$

$$= \frac{-j}{w_C}$$

$$= -jX_C \tag{8.10}$$

A series circuit containing resistance and reactance has a phasor diagram of the form shown in Fig. 8.4.

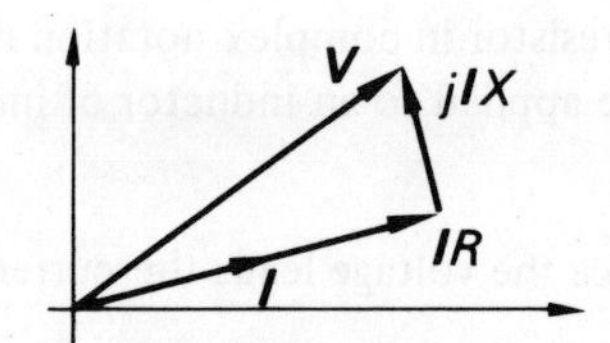

Fig. 8.4. Phasor diagram for a series circuit

By Kirchhoff's 2nd law,

$$\boldsymbol{V} = \boldsymbol{V_R} + \boldsymbol{V_X}$$

$$= \boldsymbol{IR} + j\boldsymbol{IX}$$

$$= \boldsymbol{I}(R + jX)$$

$$Z = \frac{\boldsymbol{V}}{\boldsymbol{I}}$$

$$= R + jX$$

Thus $R + jX$ represents the impedance of the circuit. It can be similarly shown that for a circuit of series resistance R, inductance L and capacitance C,

$$Z = R + jX_L - jX_C$$
$$= R + j(X_L - X_C) \quad (8.11)$$

In more general terms, if more than one impedance are connected in series, then

$$Z = Z_1 + Z_2 + Z_3 \text{ etc} \quad (8.12)$$

Note that in this expression, it must only be made when it is explicit that it forms part of complex notation and that the tensor quantities only are not intended to be applied.

Example 8.7 The current in a circuit is given by $(4{\cdot}5 + j12{\cdot}0)$ A when the applied voltage is $(100 + j150)$ V. Determine, in complex notation, an expression for the impedance and also the phase angle.

$$Z = \frac{V}{I} = \frac{100 + j150}{4{\cdot}5 + j12{\cdot}0} = \frac{(100 + j150)(4{\cdot}5 - j12{\cdot}0)}{4{\cdot}5^2 + 12{\cdot}0^2}$$
$$= \underline{(13{\cdot}7 - j3{\cdot}2)\ \Omega}$$

Since the sign of the reactance component is negative, the reactance must be capacitive. The phase angle must therefore have a value between 0° and 90°.

$$\phi = \tan^{-1}\frac{3{\cdot}2}{13{\cdot}7} = \underline{13{\cdot}1°}$$

Example 8.8 A resistor of 5 Ω and an inductive reactance of 10 Ω are connected in series. Find the current and the power dissipated in the 5-Ω resistor if an alternating voltage of 200 V is applied across the circuit.

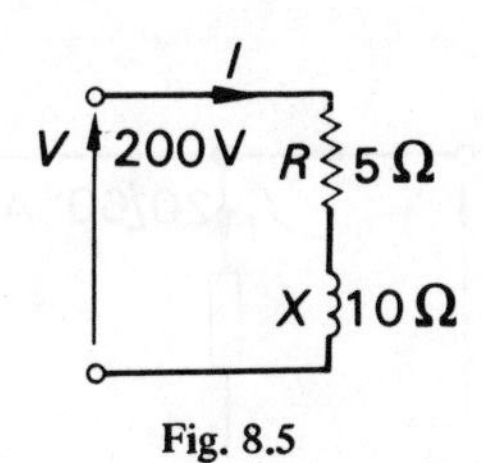

Fig. 8.5

$$Z = R + jX = (5 + j10)\ \Omega$$

Take the voltage as reference, i.e. $V = 200(1 + j0)$ V

$$I = \frac{V}{Z} = \frac{200(1 + j0)}{5 + j10} = \frac{200(5 - j10)}{5^2 + 10^2} = (8 - j16)\ \text{A}$$

It is not particularly good practice to express a current in complex notation and in this case only the magnitude of the current is required, no reference having been given.

$$I = (8^2 + 16^2)^{\frac{1}{2}} = \underline{17{\cdot}9\ \text{A}}$$

$$P = I^2 R = 17{\cdot}9^2 \times 5 = \underline{1600\ \text{W}}$$

Consider again expression (8.12). Provided the relation is expressed in complex notation, the impedances can now be manipulated in the same manner that series resistors are analysed. The only difference is that instead of adding pure numbers, it is now complex numbers which are being added. By replacing the impedances in polar notation in any of the other relative relations by impedances in complex notation, it can be seen that this similarity between resistance circuit analysis and complex impedance circuit analysis is maintained. This applies to all the circuit theorems of chapter 2.

In the case of parallel circuits, provided the impedances are in complex notation,

$$\frac{1}{Z}=\frac{1}{Z_1}+\frac{1}{Z_2}+\frac{1}{Z_3} \text{ etc} \qquad (8.13)$$

It follows that in the case of two parallel impedances

$$\frac{1}{Z}=\frac{1}{Z_1}+\frac{1}{Z_2}$$

$$Z=\frac{Z_1 Z_2}{Z_1+Z_2} \qquad (8.14)$$

Again this compares with the case of two parallel resistors. It should be noted that parallel circuits need not be analysed through calculating the effective impedance of the circuit, and it is still valid to think of the circuits as series circuits in parallel, a solution being obtained by applying Kirchhoff's 1st law and adding the currents in the separate branches. Both approaches are illustrated in the following examples.

Example 8.9 A voltage of $200\angle 30°$ V is applied to two circuits connected in parallel. The currents in the branches are $20\angle 60°$ A and $40\angle -30°$ A. Find the total effective impedance of the circuit.

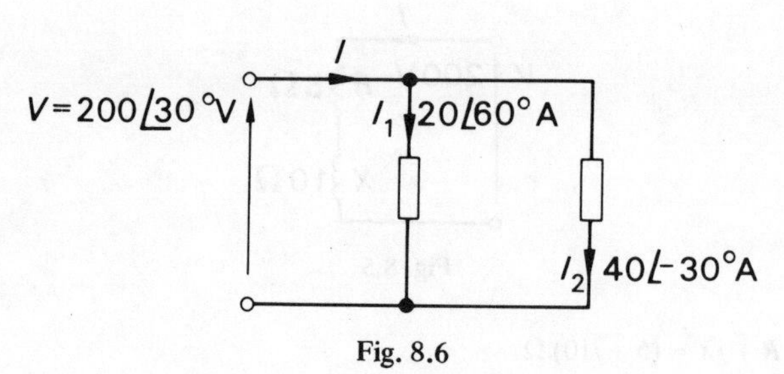

Fig. 8.6

$$V=200\angle 30° = 200(\cos 30° + j\sin 30°) = (173 + j100)\text{ V}$$

$$I_1 = 20\angle 60° = 20(\cos 60° + j\sin 60°) = (10{\cdot}0 + j17{\cdot}3)\text{ A}$$

$$I_2 = 40\angle -30° = 40(\cos -30° + j\sin -30°) = (34{\cdot}6 - j20{\cdot}0)\text{ A}$$

$$I = I_1 + I_2 = 10{\cdot}0 + j17{\cdot}3 + 34{\cdot}6 - j20{\cdot}0 = (44{\cdot}6 - j2{\cdot}7)\text{ A}$$

$$Z=\frac{V}{I}=\frac{173+j100}{44{\cdot}6-j2{\cdot}7}=\frac{(173+j100)(44{\cdot}6+j2{\cdot}7)}{44{\cdot}6^2+2{\cdot}7^2}$$

$$= (3{\cdot}74 + j2{\cdot}47)\ \Omega$$

Since the question was given in polar notation, the answer should be similarly expressed.

$$Z = (3{\cdot}74^2 + 2{\cdot}47^2)^{\frac{1}{2}} \angle \tan^{-1}\frac{2{\cdot}47}{3{\cdot}74}$$

$$= \underline{4{\cdot}48 \angle 33{\cdot}3^\circ\,\Omega}$$

Example 8.10 For the following network, find the input supply voltage and current.

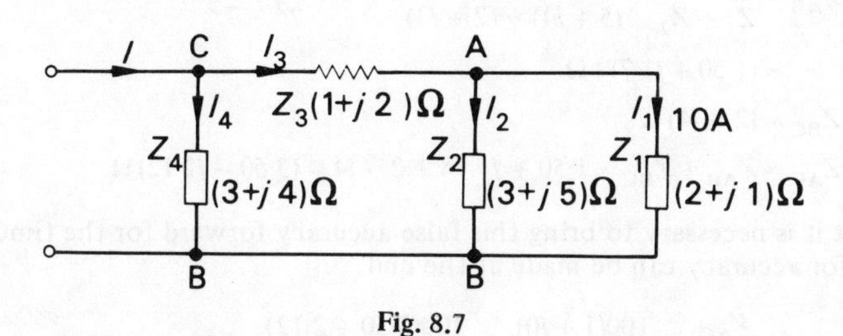

Fig. 8.7

Take $I_1 = (10 + j0)$ A as reference.

$$V_{AB} = I_1 Z_1 = 10(2 - j1) = (20 - j10)\text{ V}$$

$$I_2 = \frac{V_{AB}}{Z_2} = \frac{20 - j10}{3 + j5} = \frac{(20 - j10)(3 - j5)}{3^2 + 5^2} = (0{\cdot}29 - j3{\cdot}82)\text{ A}$$

$$I_3 = I_1 + I_2 = 10 + 0{\cdot}29 - j3{\cdot}82 = (10{\cdot}29 - j3{\cdot}82)\text{ A}$$

$$V_{CA} = I_3 Z_3 = (10{\cdot}29 - j3{\cdot}82)(1 + j2) = (17{\cdot}93 + j16{\cdot}77)\text{ V}$$

$$V_{CB} = V_{CA} + V_{AB} = 20 - j10 + 17{\cdot}93 + j16{\cdot}77 = (37{\cdot}93 + j6{\cdot}77)\text{ V}$$

$$V_{CB} = (37{\cdot}93^2 + 6{\cdot}77^2)^{\frac{1}{2}} = \underline{38{\cdot}5\text{ V}}$$

$$I_4 = \frac{V_{CB}}{Z_4} = \frac{37{\cdot}93 + j6{\cdot}77}{3 + j4} = \frac{(37{\cdot}93 + j6{\cdot}77)(3 - j4)}{3^2 + 4^2}$$

$$= (5{\cdot}63 - j5{\cdot}28)\text{ A}$$

$$I = I_3 + I_4 = 10{\cdot}29 - j3{\cdot}82 + 5{\cdot}63 - j5{\cdot}28 = (15{\cdot}92 - j9{\cdot}10)\text{ A}$$

$$I = (15{\cdot}92^2 + 9{\cdot}10^2)^{\frac{1}{2}} = \underline{18{\cdot}3\text{ A}}$$

Example 8.11 An impedance of $(5 + j4)\ \Omega$ is connected in parallel with another of $(2 + j3)\ \Omega$. This circuit is connected in series with another comprising a resistance of

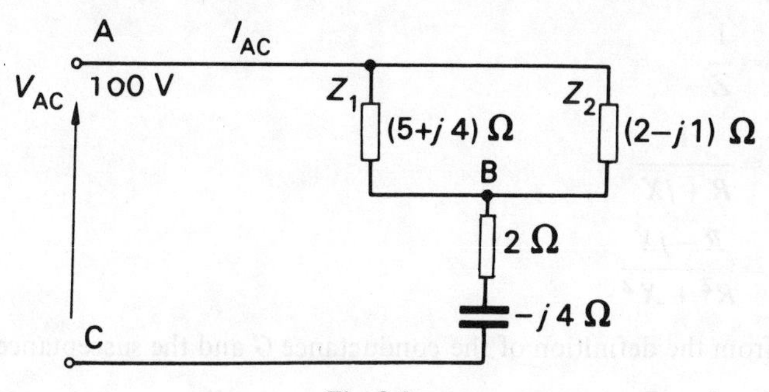

Fig. 8.8

2 Ω in series with a capacitive reactance of 4 Ω. The complete circuit is connected to a 100-V a.c. supply. Find:

(*a*) The supply current.
(*b*) The circuit power factor.
(*c*) The p.d. across each circuit.

Take the supply voltage $V_{AC} = 100(1+j0)$ *V* as reference.

$$Z_{AB} = \frac{Z_1 Z_2}{Z_1+Z_2} = \frac{(5+j4)(2+j3)}{(5+j4)+(2+j3)} = \frac{(-2+j23)(7-j7)}{7^2+7^2}$$

$$= (1{\cdot}50 + j1{\cdot}78)\ \Omega$$

$$Z_{BC} = (2-j4)\ \Omega$$

$$Z_{AC} = Z_{AB} + Z_{BC} = 1{\cdot}50 + j1{\cdot}78 + 2 - j4 = (3{\cdot}50 - j2{\cdot}12)\,\Omega$$

Note that it is necessary to bring this false accuracy forward for the time being. Correction for accuracy can be made at the end.

$$I_{AC} = \frac{V_{AC}}{Z_{AC}} = \frac{100(1+j0)}{(3{\cdot}50-j2{\cdot}12)} = \frac{100(3{\cdot}50+2{\cdot}12)}{(3{\cdot}50^2+2{\cdot}12^2)}$$

$$= (20{\cdot}9 + j12{\cdot}7)\ \text{A}$$

$$I_{AC} = (20{\cdot}9^2 + 12{\cdot}7^2)^{\frac{1}{2}} = \underline{24{\cdot}4\ \text{A}}$$

$$\phi = \tan^{-1}\frac{X}{R} = \tan^{-1}\frac{2{\cdot}12}{3{\cdot}50} = 31^\circ$$

$$\cos\phi = \cos 31^\circ = \underline{0{\cdot}86\ \text{lead}}$$

$$V_{AB} = I_{AB} Z_{AB} = (20{\cdot}9 + j12{\cdot}7)(1{\cdot}50 + j1{\cdot}78)$$

$$= (8{\cdot}75 + j56{\cdot}2)\ \text{V}$$

$$V_{AB} = (8{\cdot}75^2 + 56{\cdot}2^2)^{\frac{1}{2}} = \underline{56{\cdot}9\ \text{V}}$$

$$V_{BC} = V_{AC} - V_{AB} = 100 - (8{\cdot}75 + j56{\cdot}2) = (91{\cdot}3 - j56{\cdot}2)\ \text{V}$$

$$V_{BC} = (91{\cdot}3^2 + 56{\cdot}2^2)^{\frac{1}{2}} = \underline{107{\cdot}3\ \text{V}}$$

8.5 Admittance

In the analysis of parallel circuits, it is quite often useful to apply admittance values instead of impedance values. This is more easily applied in complex notation as follows:

$$Y = \frac{1}{Z}$$

$$= \frac{1}{R+jX}$$

$$= \frac{R-jX}{R^2+X^2}$$

However, from the definition of the conductance *G* and the susceptance *B*,

$$Y = G + jB$$

Equating reference and quadrate terms:

$$G = \frac{R}{R^2 + X^2} \tag{8.15}$$

and
$$B = \frac{-X}{R^2 + X^2} \tag{8.16}$$

By substituting $Y = 1/Z$ in relation (8.13),

$$Y = Y_1 + Y_2 + Y_3 \text{ etc} \tag{8.17}$$

Finally it will be noted that the sign of the susceptance is the opposite of that of the reactance. Thus:
If X is + *ve* for inductance then B is −*ve*.
If X is − *ve* for capacitance then B is + *ve*.

Example 8.12 Find the resistance and reactance, which, when connected in parallel with each other, will be equivalent to a circuit consisting of a series resistance of 10 Ω and inductive reactance 5 Ω.

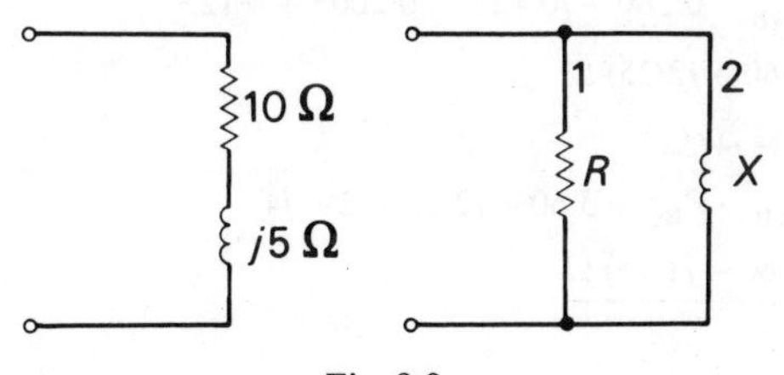

Fig. 8.9

For the series circuit,

$$Z = (10 + j5)\ \Omega$$

$$Y = \frac{1}{Z} = \frac{1}{10 + j5} = \frac{10 - j5}{10^2 + 5^2} = (0{\cdot}08 - j0{\cdot}04)\ \text{S}$$

For the equivalent parallel circuit:

$$Y = Y_1 + Y_2 = \frac{1}{Z_1} + \frac{1}{Z_2} = \frac{1}{R} + \frac{1}{jX}$$

Equating reference and quadrate terms:

$$\frac{1}{R} = 0{\cdot}08\ \text{S}$$

$$R = 12{\cdot}5\ \Omega$$

$$\frac{1}{jX} = -j0{\cdot}04\ \text{S}$$

$$X = \frac{-j}{-j0{\cdot}04} = \underline{25{\cdot}0\ \Omega}$$

Example 8.13 A resistance of 5 Ω is connected in parallel with an inductance of reactance 8 Ω. This circuit is connected in series with another consisting of a resistance of 2 Ω in series with a capacitive reactance of 4 Ω. Calculate the effective impedance of the complete circuit.

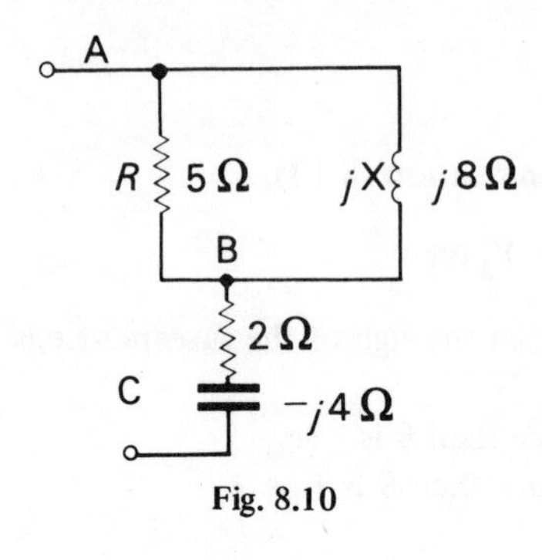

Fig. 8.10

$$Y_{AB} = \frac{1}{R} + \frac{1}{jX} = \frac{1}{5} + \frac{1}{j8} = (0{\cdot}200 - j0{\cdot}125)\ \text{S}$$

$$Z_{AB} = \frac{1}{Y_{AB}} = \frac{1}{0{\cdot}200 - j0{\cdot}125} = \frac{0{\cdot}200 + j0{\cdot}125}{0{\cdot}200^2 + 0{\cdot}125^2}$$

$$= (3{\cdot}60 + j2{\cdot}25)\ \Omega$$

$$Z_{BC} = (2 - j4)\ \Omega$$

$$Z_{AC} = Z_{AB} + Z_{BC} = 3{\cdot}60 + j2{\cdot}25 + 2 - j4$$

$$= \underline{(5{\cdot}60 - j1{\cdot}75)\ \Omega}$$

Example 8.14 Three impedances of $(3{\cdot}0 + j2{\cdot}5)$, $(4{\cdot}0 - j3{\cdot}0)$ and $(4{\cdot}0 + j5{\cdot}0)\Omega$ are connected in parallel. Calculate the current in each branch when the total supply current is 20·0 A.

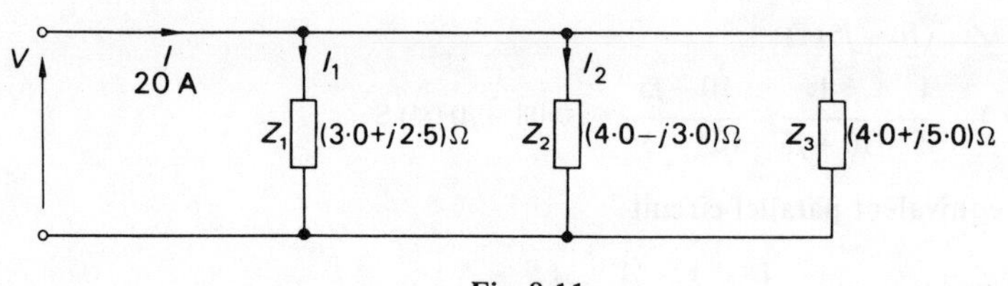

Fig. 8.11

$$Y_1 = \frac{1}{Z_1} = \frac{1}{3{\cdot}0 + j2{\cdot}5} = \frac{3{\cdot}0 - j2{\cdot}5}{3{\cdot}0^2 + 2{\cdot}5^2} = (0{\cdot}1966 - j0{\cdot}1639)\ \text{S}$$

Similarly:

$$Y_2 = (0{\cdot}1600 + j0{\cdot}1200)\ \text{S}$$

$$Y_3 = (0{\cdot}0976 - j0{\cdot}1220)\ \text{S}$$

$$Y = Y_1 + Y_2 + Y_3 = 0{\cdot}1966 - j0{\cdot}1639 + 0{\cdot}1600 + j1200$$

$$(+0{\cdot}0976 - j0{\cdot}1220)$$

$$= (0{\cdot}4542 - j0{\cdot}1659)\ \text{S}$$

Take $\boldsymbol{I} = (20 + j0)$ A as reference.

$$\boldsymbol{I} = \boldsymbol{V}\boldsymbol{Y}$$

$$\boldsymbol{I}_1 = \boldsymbol{V}\boldsymbol{Y}_1$$

$$\frac{\boldsymbol{I}}{\boldsymbol{I}_1} = \frac{\boldsymbol{V}\boldsymbol{Y}}{\boldsymbol{V}\boldsymbol{Y}_1}$$

$$\boldsymbol{I}_1 = \frac{\boldsymbol{Y}_1\boldsymbol{I}}{\boldsymbol{Y}} = \frac{(0{\cdot}1966 - j0{\cdot}1639)(20 + j0)}{0{\cdot}4542 - j0{\cdot}1659}$$

$$= (9{\cdot}968 - j3{\cdot}575) \text{ A}$$

$$I_1 = (9{\cdot}968^2 + 3{\cdot}575^2)^{\frac{1}{2}} = \underline{10{\cdot}6 \text{ A}}$$

Similarly:

$$I_2 = \underline{8{\cdot}3 \text{ A}}$$

$$I_3 = \underline{6{\cdot}5 \text{ A}}$$

8.6 Power

Given a voltage $\boldsymbol{V} = (a + jb)$ and a current $\boldsymbol{I} = (c + jd)$, one important quantity which may very well be required is the resulting power dissipated due to these electrical quantities. One method of obtaining the power is as follows:

$$\boldsymbol{Z} = \frac{\boldsymbol{V}}{\boldsymbol{I}}$$

$$= \frac{a + jb}{c + jd}$$

$$= \frac{(ac + bd) + j(ad - bc)}{c^2 + d^2}$$

$$= R + jX$$

Equating reference and quadrate terms:

$$R = \frac{ac + bd}{c^2 + d^2}$$

and $$X = \frac{ad - bc}{c^2 + d^2}$$

Also $$I = (c^2 + d^2)^{\frac{1}{2}}$$

hence $$I^2 \boldsymbol{Z} = I^2 R + jI^2 X$$

$$\boldsymbol{S} = P + jQ \qquad (8.18)$$

This result is in accordance with the power diagrams drawn in the previous chapter. Consider now the power term:

$$P = I^2 R$$
$$= (c^2 + d^2) \cdot \frac{ac + bd}{c^2 + d^2}$$
$$= ac + bd \qquad (8.19)$$

Similarly

$$Q = ad - bc$$

This type of solution is devious. It is therefore necessary to consider a more direct method of solution. Consider the effect of multiplying the voltage and current.

$$VI = (a + jb)(c + jd)$$
$$= (ac - bd) + j(ad + bc)$$

There is obviously a similarity between this expression and relation (8.19) but clearly it does not evaluate the power correctly. Since this approach would be expected to yield a satisfactory relation, this form of solution therefore merits further consideration. In particular, the question must be asked—why is this solution not compatible with that expected? To answer this question, it is necessary to introduce one final method of expressing a complexor or phasor. This depends on the mathematical identity

$$e^{j\omega t} = \cos \omega t + j . \sin \omega t$$

also $$e^{-j\omega t} = \cos \omega t - j . \sin \omega t$$

These are conjugate expressions.

It follows that a phasor may be expressed by any of the methods indicated below:

$$V_m = V_m \angle \omega t$$
$$= V_m(\cos \omega t + j . \sin \omega t)$$
$$= V_m . e^{j\omega t}$$

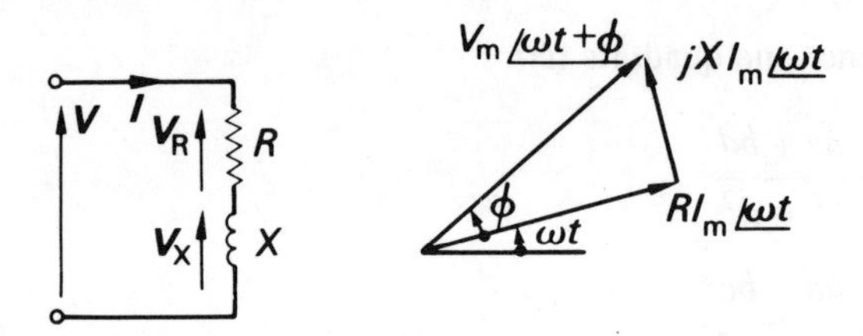

Fig. 8.12. General impedance circuit and corresponding instantaneous phasor diagram

For a general impedance circuit, as shown in Fig. 8.12,

$$V_m e^{j(\omega t + \phi)} = (R + jX) I_m e^{j\omega t}$$

For sinusoidal waveforms, $V = \dfrac{V_{\mathrm{m}}}{\sqrt{2}}$ and $I = \dfrac{I_{\mathrm{m}}}{\sqrt{2}}$, hence

$$Ve^{j(\omega t+\phi)} = (R+jX)I_{\mathrm{m}}e^{j\omega t} \tag{8.20}$$

Note that at this point both sides of the equation can be divided by $e^{j\omega t}$ to give

$$Ve^{j\phi} = (R+jX)Ie^{j0}$$

This gives the phasor diagram shown in Fig. 8.13 in which the current is taken as reference.

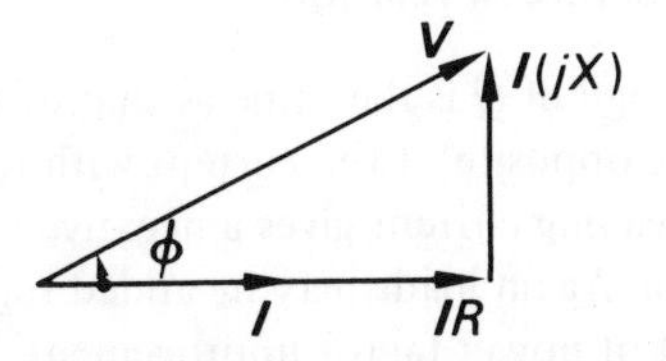

Fig. 8.13. Phasor diagram for general impedance circuit

Consider again relation (8.20). As before, multiply both sides by $\boldsymbol{I} = Ie^{j\omega t}$.

$$\boldsymbol{VI} = Ve^{j(\omega t+\phi)}Ie^{j\omega t} = (R+jX)Ie^{j\omega t}Ie^{j\omega t}$$

$$VIe^{j(2\omega t+\phi)} = (R+jX)I^2e^{j2\omega t}$$

This gives an expression which retains an element of time. However the voltage and current are expressed in r.m.s. values and the power relation is an average value divorced from instantaneous time values. Therefore the desired product should have removed the angular velocity factor; this would compare with the manner in which it was shown that the phasor diagram could be obtained without there being a factor of angular velocity. To achieve this aim, an artifice is introduced, viz. that the polarity of the power of one of the exponential expressions should be reversed. This will cause the angular velocity factor to disappear.

Let $\boldsymbol{I}^*$ be the conjugate of $\boldsymbol{I}$. Hence:

$$\boldsymbol{I} = Ie^{j\omega t}$$

$$\boldsymbol{I}^* = Ie^{-j\omega t}$$

This satisfies the suggested artifice. Now multiply both sides of this relation by $V = Ve^{j(\omega t+\phi)} = (R + jX)Ie^{j\omega t}$

$$\begin{aligned}\boldsymbol{VI}^* = Ve^{j(\omega t+\phi)}Ie^{-j\omega t} &= (R+jX)Ie^{j\omega t}Ie^{-j\omega t}\\ &= I^2R + jI^2X\\ &= P + jQ\\ &= S\end{aligned}$$

$$\boldsymbol{VI}^* = S = P + jQ \tag{8.21}$$

It may therefore be observed that $\boldsymbol{VI}$ did not give the desired results because it did not remove the time factor but instead gave a function based on instantaneous values.

It will also be observed that relation (8.21) gives rise to a complexor diagram the reference function having been removed.

The effect of relation (8.21) may be observed by applying it to the initial problem, i.e. when $\boldsymbol{V} = (a + jb)$ and $\boldsymbol{I} = (c + jd)$. Hence

$$
\begin{aligned}
S &= \boldsymbol{VI}^* \\
&= (a + jb)(c - jd) \\
&= (ac + bd) - j(ad - bc) \\
&= P + jQ \text{ as before in relation} \qquad (8.19)
\end{aligned}
$$

It will be noted that the sign of Q is the same as that of the reactive element of the load impedance and also the opposite of the current with respect to the voltage when taken as reference. Thus a leading current gives a negative value for Q and a lagging current gives a positive value. Again loads may be added together for the purposes, say, of estimating the effect of power factor improvement. This interpretation of the sign of Q is at variance with that in chapter 7. However, it would have been difficult to justify the reversal at that stage and the significance of the polarity only affects j-notation calculations.

Finally the reader may find the above derivation of relation (8.21) relatively long; shorter statements of this relation only observe the result without justification. The result may be observed from relations (8.19) et seq. The expression for the power can also be obtained by taking the conjugate of the voltage instead of the current, but this results in a reversal of the sign of Q and this would be contrary to the recommendations of the I.E.C.

Example 8.15 An impedance of $(3 + j4)\ \Omega$ is connected in series with another impedance of $(5 + j8)\ \Omega$. If a voltage, which may be represented by $(80 + j60)$ V, is applied to the resulting circuit, find the power dissipated in the circuit.

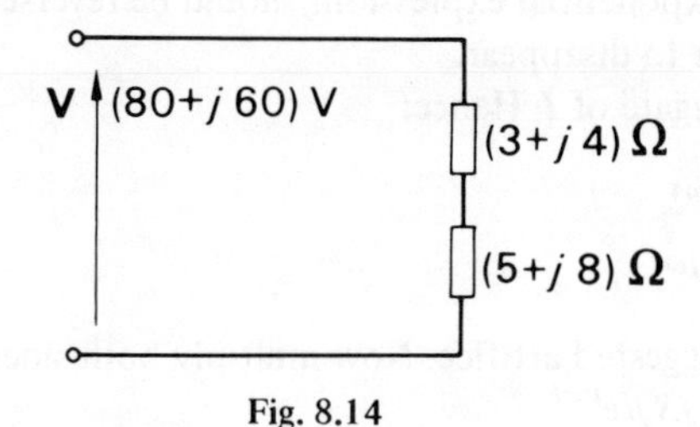

Fig. 8.14

$$
\begin{aligned}
Z &= Z_1 + Z_2 = 3 + j4 + 5 + j8 = (8 + j12)\ \Omega \\
\boldsymbol{I} &= \frac{V}{Z} = \frac{80 + j60}{8 + j12} = \frac{20(4 + j3)}{4(2 + j3)} = \frac{5(4 + j3)(2 - j3)}{2^2 + 3^2} \\
&= (6{\cdot}53 - j2{\cdot}31)\ \text{A} \\
S &= \boldsymbol{VI}^* = (80 + j60)(6{\cdot}53 + j2{\cdot}31) \\
&= (383 + j477)\ \text{VA} \\
&= P + jQ \\
P &= \underline{383\ \text{W}}
\end{aligned}
$$

Since $Q = 477$ kVar is positive, this indicates that the circuit power factor is a lagging one, which is to be expected from the expression for the circuit impedance.

Example 8.16 Determine the total load when the following loads are connected in parallel:
(*a*) 200 kVA at 0·8 power factor lagging.
(*b*) 100 kW at unity power factor.
(*c*) 200 kW at 0·8 power factor leading.
(*d*) 500 kVA at 0·6 power factor lagging.

Load (*a*): $\cos\phi_a = 0{\cdot}8$

$$\phi_a = 53{\cdot}2^\circ$$

$$\sin\phi_a = 0{\cdot}6$$

$$S_a = S(\cos\phi_a + j\sin\phi_a)$$
$$= 200 \times 10^3(0{\cdot}8 + j0{\cdot}6)\ \text{VA}$$
$$= (160 + j120)\ \text{kVA}$$

Similarly the other loads may be expressed as:

$$S_b = (100 + j0)\ \text{kVA}$$
$$S_c = (200 - j150)\ \text{kVA}$$
$$S_d = (300 + j400)\ \text{kVA}$$

In calculating S_c, it is necessary to find the magnitude of the apparent power since it is the true power that is given. This is obtained by dividing the true power P_c by the power factor $\cos\phi_c$, hence $S_c = 250$ kVA.

$$S = S_a + S_b + S_c + S_d = (760 + j370)$$

$$\cos\phi = \frac{P}{(P^2 + Q^2)^{\frac{1}{2}}} = \frac{760}{(760^2 + 370^2)^{\frac{1}{2}}} = 0{\cdot}9$$

Total load is 760 kW at 0·9 power factor lagging.

PROBLEMS ON *j*-NOTATION

1 Two impedances $(3 + j5)\ \Omega$ and $(4 - j7)\ \Omega$ are connected in parallel to an alternating sinusoidal voltage supply of 200 V. Find the current in each branch and the total current taken from the supply.

34·3 A; 24·8 A; 31·0 A

2 A coil of resistance 5 Ω and inductance 31·8 mH is connected in series with an impedance of value $(10 - j15)\ \Omega$ at 50 Hz. If the circuit is connected to a 150-V a.c. supply, calculate:
(*a*) The current through the circuit when the supply frequency is 50 Hz.
(*b*) The frequency at which the impedance is a minimum.
(*c*) The p.d. across the coil for this latter frequency.

9·49 A; 61·2 Hz; 132 V

3 A circuit has two parts AB and BC connected in series. AB consists of a 5-Ω resistor connected in parallel with an impedance of $(5 - j5)\ \Omega$ and BC is an impedance of $(2 + j4)\ \Omega$. If the current in the $(5 - j5)$-Ω branch is $20\angle 0^\circ$ A, determine the corresponding value of V_{AC}.

$261\angle 4{\cdot}4^\circ$ V

4 An impedance of $(6 + j12)\ \Omega$ is connected in parallel with an impedance of $(24 + j12)\ \Omega$. The resulting impedance is connected in series with an impedance of $(24 - j5)\ \Omega$, all values being taken with respect to a frequency of 50 Hz. A supply voltage of 100 V at 50 Hz, being sinusoidal, is connected to the above load. Calculate the supply current and its phase angle with respect to the voltage.

$3{\cdot}34\angle{-4{\cdot}5°}$ A

5 Two coils of resistance 5 Ω and 10 Ω and inductance 10 mH and 15 mH respectively are connected in parallel across a 100-V, 50-Hz a.c. supply. Calculate the resistance and inductance of a single coil which will take the same current at the same power factor as the coils in parallel.

3·34 Ω; 6·18 mH

6 Three impedances are connected in series, being:

$Z_1 = 10 + jX_1, Z_2 = 15 + jX_2, Z_3 = 30 + jX_3$

This circuit is supplied from a 200-V, 50-Hz supply, the supply current being 2·95 A. If the voltage across Z_1 is 55 V whilst that across Z_2 is 70 V, find the numerical complex expressions for each of the impedances. Also find the total inductance of the circuit and the power factor.

$(10{\cdot}0 + j15{\cdot}8)\ \Omega$; $(15{\cdot}7 + j18{\cdot}4)\ \Omega$; $(30{\cdot}0 + j5{\cdot}5)\ \Omega$; 126 mH; 0·81 lag

7 Three impedances $Z_A = (5 + j8)\ \Omega$, $Z_B = (4 - j6)\ \Omega$ and $Z_C = (7 + j10)\ \Omega$ are connected in parallel to an a.c. supply. If the power dissipated in Z_A is 100 W, find the supply voltage, the total power dissipated in the combination and the power factor.

3·20 W; 3·4 lag; 42·2 V

8 Two impedances $R + jX$ and $5 + j8$, measured in ohms, take a lagging current of 5·23 A and a power of 356 W when connected in series to a 100-V a.c. supply. Calculate the current and power taken when they are connected in parallel across the same supply.

20·2 A; 1·36 kW

9 Three impedances $(4 - j5)\ \Omega$, $(6 + j4)\ \Omega$ and $(4 + j3)\ \Omega$ are connected in parallel across a 1-ph a.c. supply. If the total load is 2 kW, calculate the power taken by each branch and also the power factor of the combination.

524 W; 618 W; 858 W; 0·978 lag

10 Two impedances $(4 + j10)\ \Omega$ and $(6 + j4)\ \Omega$ are connected in parallel across an a.c. supply and dissipate 600 W. Calculate the power taken when the impedances are joined in series across the same supply.

134 W

11 An alternating voltage is applied between points A and C of a circuit which consists of an impedance $(10 + j10)\ \Omega$ connected between A and B and $(5 + j10)\ \Omega$ and $(10 - j5)\ \Omega$ connected in parallel between B and C. The power dissipated in Z_{AB} is 360 W. Calculate the supply voltage and the power dissipated in each of the other impedances. Determine the difference in phase between the currents in the parallel branches.

129 V; 180 W; 90 W; 90°

12 A circuit consists of parallel impedances $(2 + j4)\ \Omega$ and $(4 - j2)\ \Omega$ and is connected in series with an impedance of $(1 + j2)\ \Omega$. If a 100-V a.c. supply is connected across the combination, calculate the supply current and the power.

20·0 A; 1600 W

13 Two coils of impedance $Z_1 = (6 + j8)\ \Omega$ and $Z_2 = (R_2 + j20)\ \Omega$ are connected in series to a 50-Hz supply. Impedance Z_1 is shunted by a capacitor of reactance 12·5 Ω. The current drawn from the supply is 8·5 A and the power dissipated in Z_2 is 1·5 times that dissipated in Z_1. Calculate:

(*a*) The applied voltage.
(*b*) Its phase angle with respect to the current.
(*c*) The total power dissipated in the circuit.

394 V; 25·6°; 3 020 W

14 Using the *j*-operator, determine the minimum current rating for an 11-kV, 1-pH feeder to supply the following combined load.

(*a*) 1 000-kW lighting load.
(*b*) 2 000-kVA synchronous motor load at 0·707 power factor leading.
(*c*) 3 000-kVA induction motor load at 0·8 power factor lagging.
(*d*) 1 500-kVA induction motor load at 0·2 power factor lagging.

495 A

15 A load of 2 000 kW is supplied at 11 kV, 50 Hz through a 1-ph line with a loop resistance of 4 Ω and loop inductive reactance of 6 Ω. Determine the sending-end voltage and power factor if the load is at:

(*a*) 0·8 power factor lagging.
(*b*) Unity power factor.
(*c*) 0·8 power factor leading.

12·2 kV 0·778 lag; 11·8 kV 0·995 lag; 11·0 kV 0·865 lag

16 A circuit comprising of a 30-Ω resistor, a 60-Ω inductor and a 20-Ω capacitor in series is shunted by an 80-Ω resistor in series with a 60-Ω capacitor. The combination is connected in series with a 30-Ω inductor and the complete circuit is supplied from a 100-V a.c. source. A wattmeter is connected with its current coil in series with the 80-Ω resistor and the voltage coil across the 30-Ω inductor. Calculate the wattmeter reading.

29·4 W

9

A. C. polyphase circuits

The 1-phase a.c. circuit is suitable for most applications but there are two fields of electrical engineering for which it is not so well suited–power transmission and electromechanical energy conversion by machines. In the case of power transmission, the 1-ph circuit does not make the best use of the conduction system. In the case of energy conversion, the 1-ph machine provides a pulsating torque, operates at a very poor power factor and often requires additional apparatus for starting. This can be overcome by using a polyphase (i.e. two or more phases) system.

9.1 Polyphase systems

There are two polyphase systems of importance, i.e. 2-ph and 3-ph systems. The 2-ph system was quite common at one time but it is less flexible and less economical than the 3-ph system which has consequently been implemented as the standard national system. The 2-ph system has therefore disappeared almost completely–there are still a number of rural consumers supplied by this system–as a system for power transmission

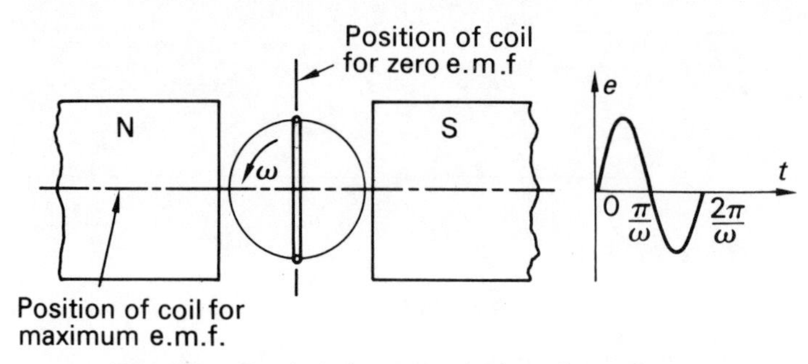

Fig. 9.1. Simple 1-ph generator and e.m.f. waveform

although it is now an important system for servomechanisms. For the majority of power applications, the 3-ph system is that used and therefore most of the following discussion will concern this system.

It has been shown in para. 5·9 that when a coil is rotated with a uniform angular velocity in a uniform magnetic field, an e.m.f. is induced in it, which varies sinusoidally. This simple form of generator is shown in Fig. 9.1 along with the resulting e.m.f. wave.

If two coils had been mounted at right-angles to one another, as shown in Fig. 9.2, then each would have a sinusoidally varying e.m.f. induced in it, the waves being displaced by $\pi/2\omega$ seconds.

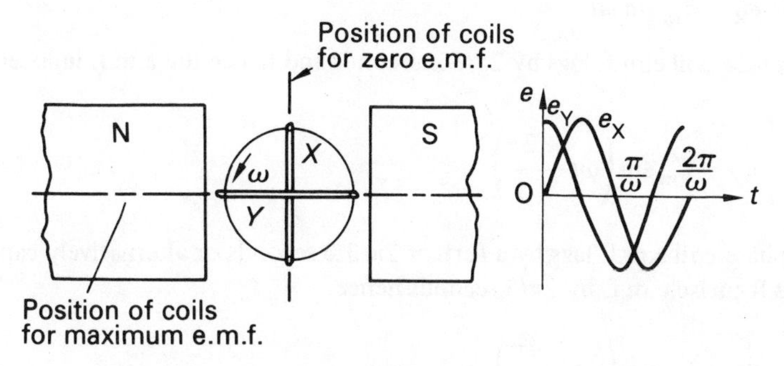

Fig. 9.2. Simple 2-ph generator and e.m.f. waveforms

If three coils had been mounted each displaced by $2\pi/3$ radians from one another, as shown in Fig. 9.3, each would again have a sinusoidally varying e.m.f. induced in it, the waves being displaced by $2\pi/3\omega$ seconds.

The 3-ph system coils are designated red, yellow and blue (R, Y, B). For the arrangement shown, first the red coil, then the yellow coil and finally the blue coil reach their maximum positive values in that order. This is termed the RYB sequence and is now the national standard sequence.

In each of the above cases, the importance of the polarity of the induced e.m.f. should be noted. So that the diagrams conform with one another, the start of each coil has been marked with a 1 and the finish marked with a 2. In Fig. 9.1, if the ends had been interchanged then the sine wave of the e.m.f. would effectively have been

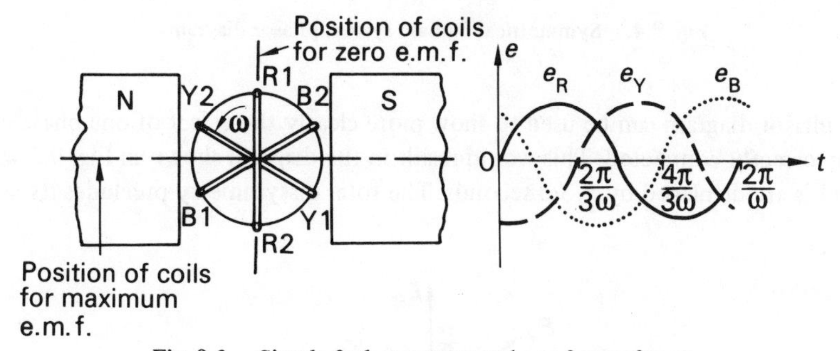

Fig. 9.3. Simple 3-ph generator and e.m.f. waveforms

retarded in time by π/ω seconds. In Fig. 9.2, a more important effect could have been brought about by one coil having its connections reversed whereby e_X would lead e_Y instead of the standard sequence shown. In Fig. 9.3, the reversal of any one phase coil would destroy the symmetry of the resulting waveforms; the symmetry is one of the outstanding features of the 3-ph system.

In a polyphase system, it is the normal case to generate a completely symmetrical system in which each of the e.m.f.'s has the same effective value. This is achieved by using identical coils in each phase and hence the maximum e.m.f. in each will be the same $-e_m$. Taking the R-phase coil e.m.f. as reference:

$$e_R = E_m \sin \omega t \tag{9.1}$$

The Y-phase coil e.m.f. lags by $2\pi/3\omega$ seconds and hence the e.m.f. induced in it is given by:

$$e_Y = E_m \sin\left(\omega t - \frac{2\pi}{3}\right) \tag{9.2}$$

The B-phase coil e.m.f. lags by a further $2\pi/3\omega$ seconds or alternatively can be said to lead the R-phase e.m.f. by $2\pi/3$ seconds hence:

$$e_B = E_m \sin\left(\omega t - \frac{4\pi}{3}\right) \tag{9.3}$$

$$= E_m \sin\left(\omega t + \frac{2\pi}{3}\right) \tag{9.3.1}$$

These e.m.f's are termed the phase e.m.f.'s. The corresponding phasor diagram is shown in Fig. 9.4.

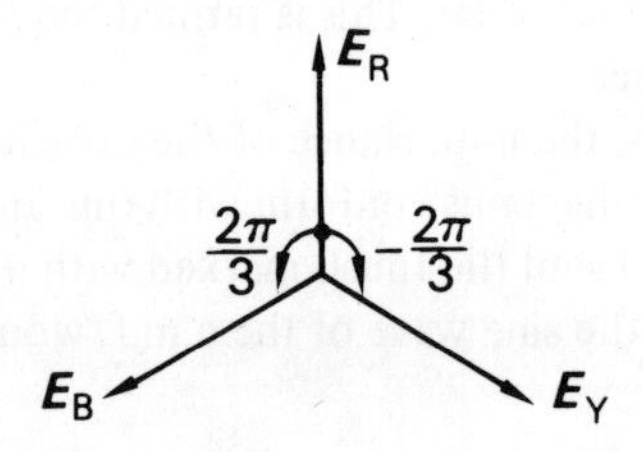

Fig. 9.4. Symmetrical 3-phase system phasor diagram

The phasor diagram can be used to show more clearly the effect of one phase coil being incorrectly connected. This would result in the diagram shown in Fig. 9.5 whereby the e.m.f.'s are displaced by $\pi/3\omega$ seconds. The total dissymmetry precludes its use.

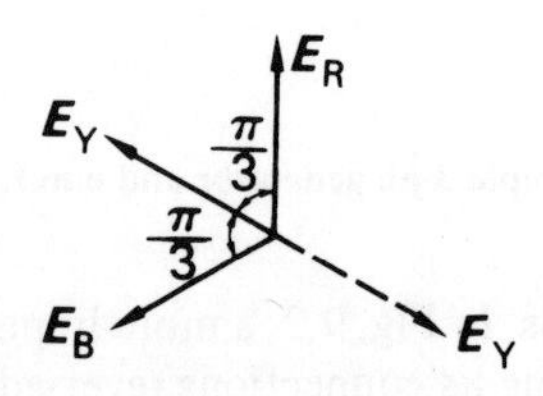

Fig. 9.5. Effect of incorrect connection of one phase

9.2 Symmetrical star-connected systems

The construction of the simple 3-ph generator is so arranged that the e.m.f.'s can be tapped through slip-rings. Consider now if a load is connected by this means across each coil. The arrangement is shown in Fig. 9.6.

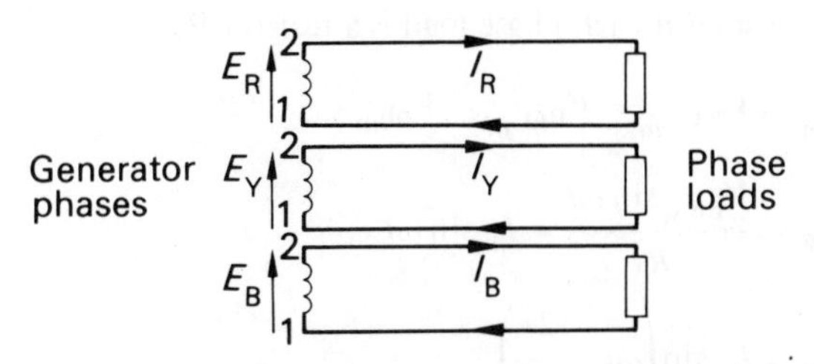

Fig. 9.6. Three loads supplied separately from three phases

The resulting transmission system requires six wires and is equivalent to three 1-ph systems. Consider if each of the coil ends marked 1 is connected together and brought out from the generator through one common slip-ring. The resulting circuit is shown in Fig. 9.7 which is termed a 4-wire, star-connected system. The diagram is drawn to simulate the phasor diagram hence the term 'star'.

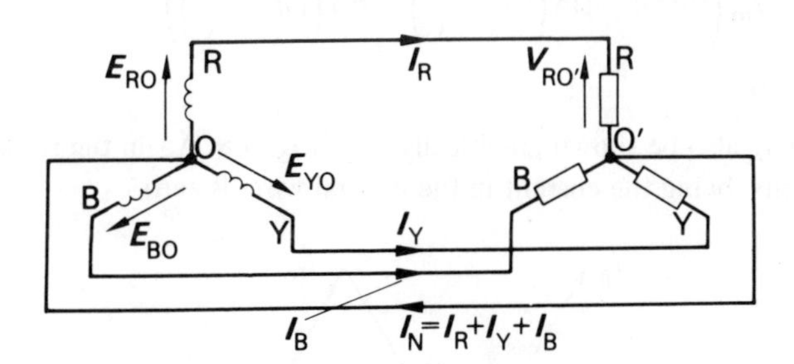

Fig. 9.7. 3-phase, 4-wire star-connected load supplied from a star-connected generator

The fourth wire acts as a common 'return' wire for the currents in the three phases and is called the neutral wire. This part of the circuit is also referred to as the star or neutral point when discussing the e.m.f.'s and potential differences within the circuit.

The e.m.f. between any line and the star point is termed the phase e.m.f. E_{ph}. The R-phase e.m.f. is designated E_{RO} but reference to the star point may be omitted hence it is then designated E_R. The other phase e.m.f.'s are E_{YO} and E_{BO}. These e.m.f.'s are equal in magnitude and differ only in time phase. Similarly the volt drops across the phase loads are $V_{RO'}$, $V_{YO'}$ and $V_{BO'}$. Also the phase voltages are equal to the phase e.m.f.'s, i.e. $V_{ph} = E_{ph}$. The current in the R-phase is given by:

$$I_R = \frac{V_{RO'}}{Z} \tag{9.4}$$

The other phase currents can be found in a similar manner. The phase current I_{ph}, i.e. the current in each of the phase loads, is therefore given by:

$$I_{ph} = \frac{V_{ph}}{Z_{ph}} \tag{9.4.1}$$

An important condition is that whereby each phase impedance is identical, i.e. the load is balanced. Consider if each phase load is a resistor R.

$$V_{RO'_m} = V_{YO'_m} = V_{BO'_m} = V_{ph_m}$$

$$i_R = \frac{V_{ph_m} \sin \omega t}{R} = I_m \sin \omega t$$

$$i_Y = I_m \sin\left(\omega t - \frac{2\pi}{3}\right)$$

$$i_B = I_m \sin\left(\omega t - \frac{4\pi}{3}\right)$$

By Kirchhoff's 1st Law:

$$i_N = i_R + i_Y + i_B$$

$$= I_m \left(\sin \omega t + \sin\left(\omega t - \frac{2\pi}{3}\right) + \sin\left(\omega t - \frac{4\pi}{3}\right)\right)$$

$$= 0 \tag{9.5}$$

This result may also be shown graphically as in Fig. 9.8. Again the summation of the phase currents, being the current in the neutral wire, is zero.

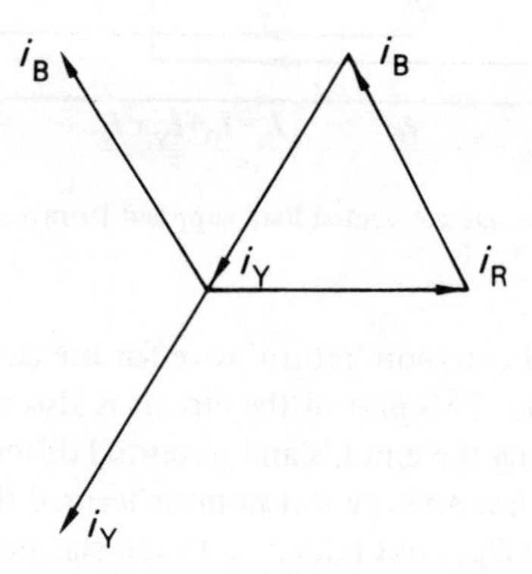

Fig. 9.8. Summation of phase currents in neutral wire

Thus the current in the neutral wire of a balanced 3-ph, 4-wire star-connected system is zero and the conductor may be omitted, with a further economic saving. Although the case taken was that for equal load resistors, the same result would have been obtained for any system of star-connected similar impedances. The 3-ph, 3-wire system is shown in Fig. 9.9.

Although the fourth wire has been omitted, O and O′ are at the same potential and hence for the balanced system shown, E_{ph} is still equal to V_{ph}. However, it will be seen that it is no longer possible to measure the phase e.m.f.'s in this system and only

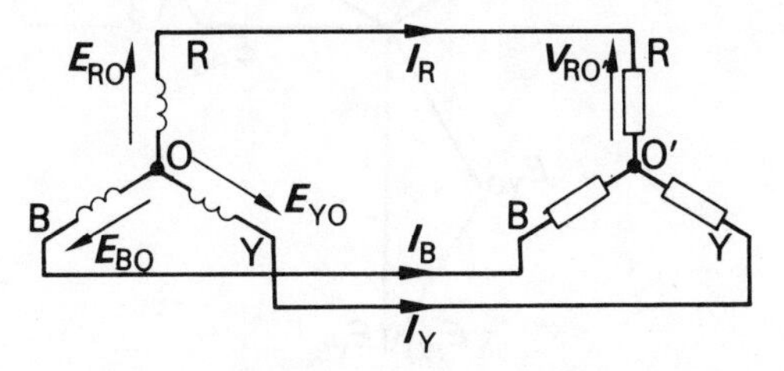

Fig. 9.9. 3-phase, 3-wire star-connected circuit

the e.m.f.'s between lines E_l can be measured. These are termed the line e.m.f.'s and are E_{RY}, E_{YB} and E_{BR}.

The line e.m.f.'s and the corresponding line voltages are important since they are the voltages given when describing a 3-ph system, e.g. a 415-V, 3-ph supply is one in which the line voltage is 415 V.

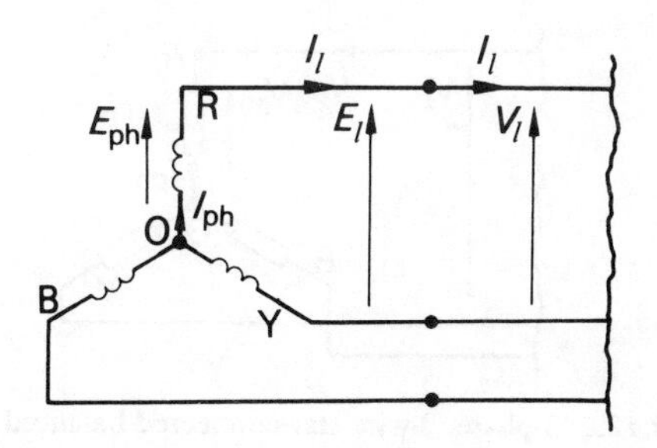

Fig. 9.10. 3-phase, 3-wire star-connected generator supplying a balanced load

The directions of current and voltage shown in Fig. 9.10 are conventional. Also it will be seen from the diagram that the phase current is equal to the current in each line, i.e. the line current I_l.

$$I_l = I_{ph} \tag{9.6}$$

The phasor diagram is constructed by the application of Kirchhoff's 2nd Law and is shown in Fig. 9.11. Because of the convention used:

$$\left.\begin{aligned} & V_{RY} = E_{RY} = E_{RO} + E_{OY} = E_{RO} - E_{YO} \\ \text{Also}\quad & V_{YB} = E_{YB} = E_{YO} - E_{BO} \\ \text{and}\quad & V_{BR} = E_{BR} = E_{BO} - E_{RO} \end{aligned}\right\} \tag{9.7}$$

The required subtraction is carried out in the normal phasor diagram manner by reversing the required phasor and then adding this to the other phasor. The phase angle between E_{RO} and $-E_{YO}$ is $\pi/3$ rad and hence by the symmetry of the phasor diagram,

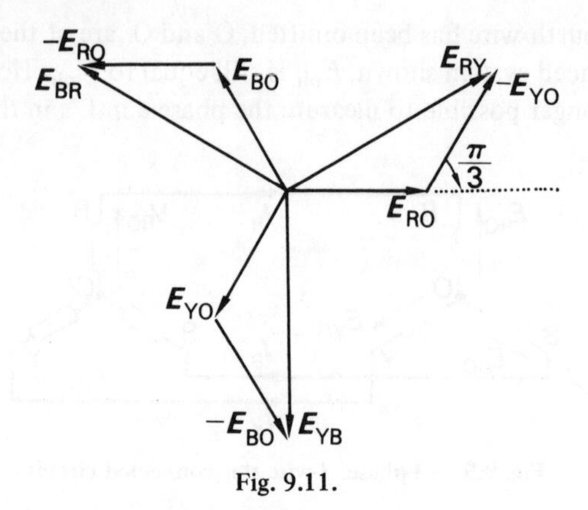

Fig. 9.11.

i.e. the phasor triangle is isosceles, E_{RY} leads E_{RO} by $\pi/6$ rad. Also by the geometry of the diagram.

$$V_{RY} = E_{RY} = E_{RO}\cos\frac{\pi}{6} + E_{YO}\cos\frac{\pi}{6} = 2.E_{RO}\cos\frac{\pi}{6}$$

$$V_l = E_l = \sqrt{3}E_{ph} \tag{9.8}$$

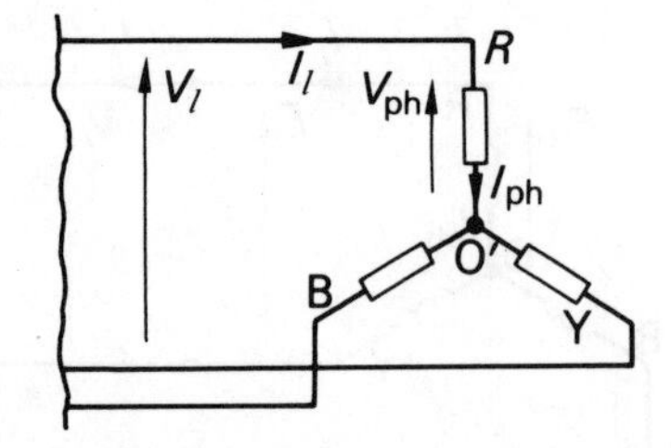

Fig. 9.12. 3-phase, 3-wire star-connected balanced load

In the case of the balanced load shown in Fig. 9.12, since O and O′ are at the same potential, $E_{ph} = V_{ph}$.

$$V_l = \sqrt{3}V_{ph} \tag{9.9}$$

In each phase, the line voltage leads the phase voltage by $\pi/6$ rad but as always there is a mutual displacement of $2\pi/3$ rad between the line voltages and the phase voltages.

The power developed in the generator is three times that developed in each phase.

$$P = 3E_{ph}I_{ph}\cos\phi$$

The power dissipated in the load is three times that dissipated in each phase.

$$P = 3V_{ph}I_{ph}\cos\phi$$

In either case:

$$P = 3\cdot\frac{V_l}{\sqrt{3}}\cdot I_l\cos\phi$$

$$= \sqrt{3}V_l I_l\cos\phi \tag{9.10}$$

The phase angle ϕ is the angle between the phase voltage and the phase current. It is also the phase angle of the load. The expression (9.10) can be applied to both the load and the generator.

Example 9.1 Three similar coils, each of resistance 7·0 Ω and inductance 0·03 H, are connected in star to a 415-V, 3-ph, 50-Hz supply. Calculate the line current and the total power absorbed.

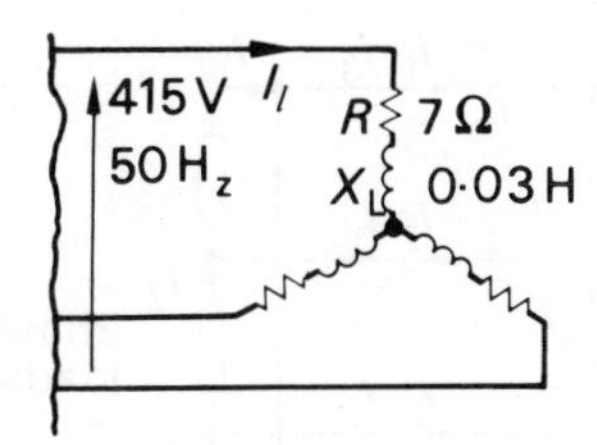

Fig. 9.13

$$X_L = 2\pi fL = 2\pi 50 \times 0{\cdot}03 = 9{\cdot}4\ \Omega$$

$$Z_{ph} = (R^2 + X_L{}^2)^{\frac{1}{2}} = (7{\cdot}0^2 + 9{\cdot}4^2)^{\frac{1}{2}} = 11{\cdot}7\ \Omega$$

$$V_{ph} = \frac{V_l}{\sqrt{3}} = \frac{415}{\sqrt{3}} = 240\ \text{V}$$

$$I_l = I_{ph} = \frac{V_{ph}}{Z_{ph}} = \frac{240}{11{\cdot}7} = \underline{20{\cdot}5\ \text{A}}$$

$$\cos\phi = \frac{R}{Z} = \frac{7{\cdot}0}{9{\cdot}4} = 0{\cdot}743\ \text{lag}$$

$$P = \sqrt{3}V_l I_l \cos\phi = \sqrt{3} \times 415 \times 20{\cdot}5 \times 0{\cdot}743$$

$$= \underline{10950\ \text{W}}$$

9.3 Symmetrical delta-connected systems

An alternative modification to the 3-ph, 6-wire system shown in Fig. 9.14(*a*) is to connect the end of each coil to the beginning of the next coil. Effectively this is combining adjacent pairs of wires and the resulting circuit is shown in Fig. 9.14(*b*).

The current in each line is the phasor difference between the phase currents. It will be noted that the resulting circuit has again a considerable economic advantage over the 6-wire system.

The conventional method of drawing the circuit for this arrangement is shown in Fig. 9.15. Because of the layout, both the generator and the load are said to be delta-connected. The numerical notation used in Fig. 9.14 has been replaced by the conventional notation.

By Kirchhoff's 2nd Law:

$$\left.\begin{aligned} \boldsymbol{I}_{\mathbf{R}} &= \boldsymbol{I}_{\mathbf{RY}} - \boldsymbol{I}_{\mathbf{BR}} \\ \boldsymbol{I}_{\mathbf{Y}} &= \boldsymbol{I}_{\mathbf{YB}} - \boldsymbol{I}_{\mathbf{RY}} \\ \boldsymbol{I}_{\mathbf{B}} &= \boldsymbol{I}_{\mathbf{BR}} - \boldsymbol{I}_{\mathbf{YB}} \end{aligned}\right\} \qquad (9.9)$$

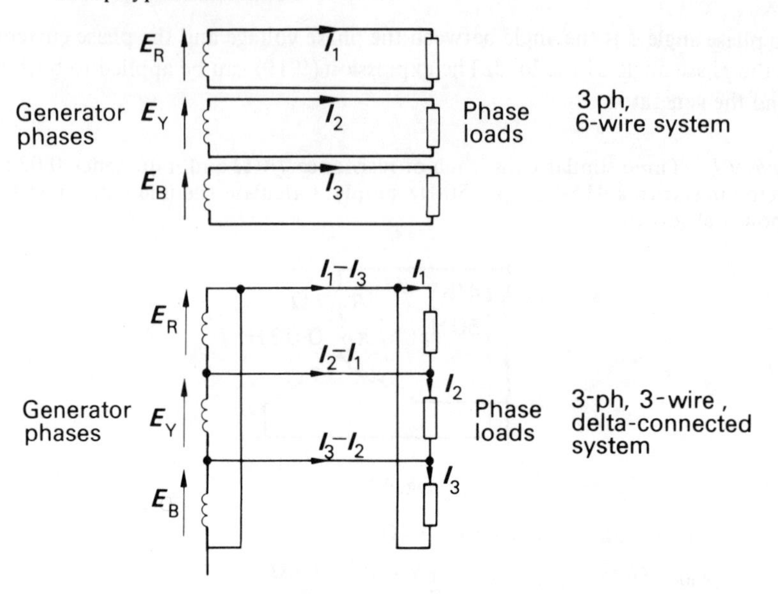

Fig. 9.14. Derivation of the delta-connected system

It can also be seen that the phase e.m.f.'s act in the same direction around the delta-connected coils of the generator. However there is no resulting circulating current around the loop since at any instant:

$$\begin{aligned} e_R + e_Y + e_B &= E_m\left(\sin \omega t + \sin\left(\omega t - \frac{2\pi}{3}\right) + \sin\left(\omega t - \frac{4\pi}{3}\right)\right) \\ &= E_m \times 0 \\ &= 0 \end{aligned} \tag{9.10}$$

Hence there is no circulating current in the loop. This phasor addition of the e.m.f.'s to zero can also be shown by adding them on a phasor diagram. This is shown in Fig. 9.16.

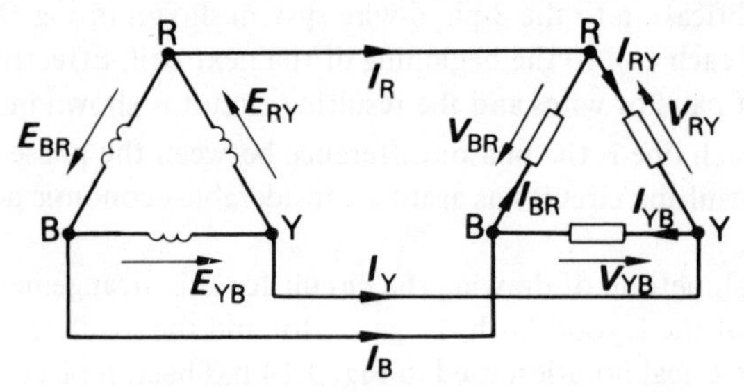

Fig. 9.15. Delta-connected generator supplying a delta-connected load

In the delta-connected system, the term 'phase' takes on a different effect from that used in the star-connected system. In the generator shown in Fig. 9.17, the phase coils are now connected between lines therefore

$$E_{ph} = E_l = V_l \tag{9.11}$$

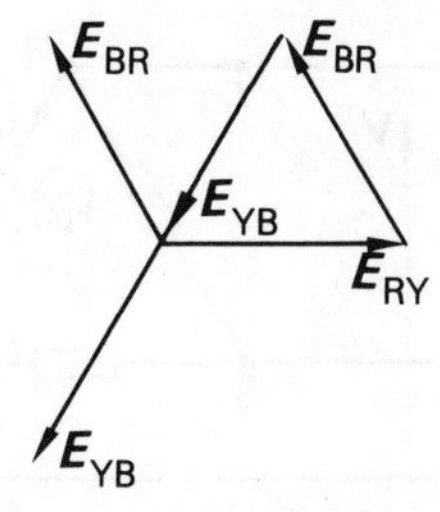

Fig. 9.16. Phasor diagram showing the addition of the e.m.f.'s in a delta-connected generator

The relation between the phase and the line currents has already been stated in relation (9.9). The corresponding phasor diagram in Fig. 9.17(b) shows the appropriate subtraction of the phasors. The phase angle between I_{RY} and $-I_{BR}$ is $\pi/3$ rad and hence, by the symmetry of the phasor diagram, i.e. the phasor diagram is isoscles, I_{RY} leads I_R by $\pi/6$ rad. Also by the geometry of the diagram:

$$I_R = I_{RY} \cos\frac{\pi}{6} + I_{BR} \cos\frac{\pi}{6}$$

$$I_l = \sqrt{3} I_{ph} \qquad (9.12)$$

In the case of the balanced load shown is Fig. 9.17, again:

$$I_l = \sqrt{3} I_{ph} \qquad (9.12.1)$$

In each phase, the line current lags the phase current by $\pi/6$ rad but, as always, there is a mutual displacement of $2\pi/3$ rad between the line currents and the phase currents.

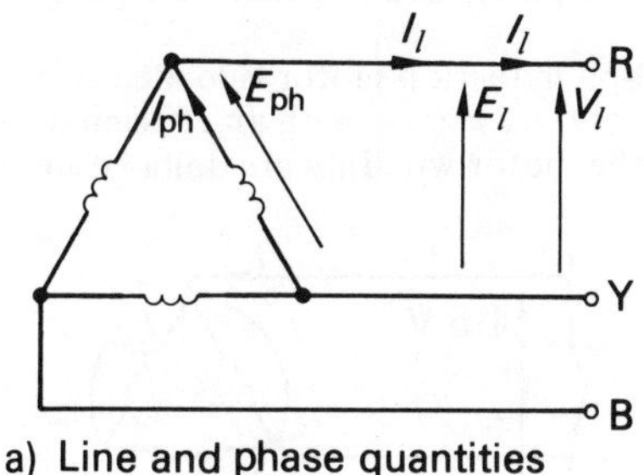

a) Line and phase quantities

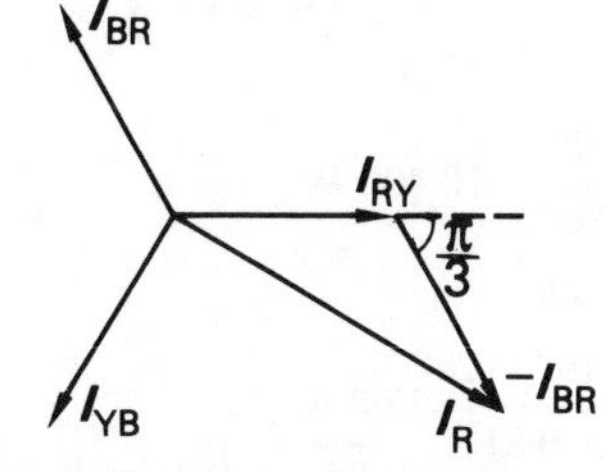

b) Phasor diagram of currents

Fig. 9.17. Delta-connected generator supplying a balanced load

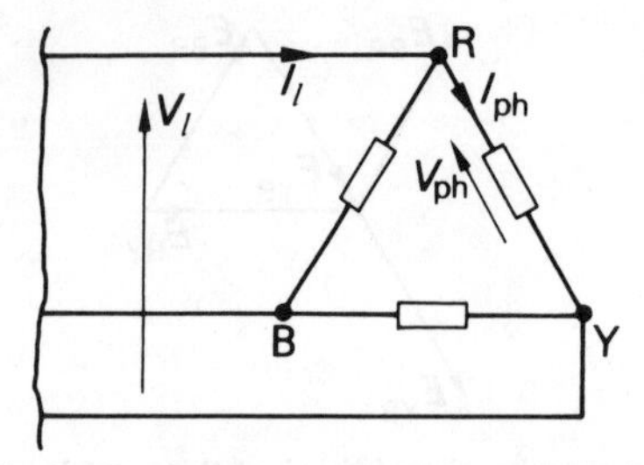

Fig. 9.18. Delta-connected balanced load

The power developed in the generator, when supplying a balanced load, is three times that developed in each phase, where again ϕ is the angle between the phase voltage and the phase current.

$$P = 3E_{ph}\,I_{ph}\cos\phi$$

The power dissipated in the balanced load is three times that dissipated in each phase. ϕ in this case is also the phase angle of the load.

$$P = 3V_{ph}\,I_{ph}\cos\phi$$

In either case:

$$P = 3V_l\,.\frac{I_l}{\sqrt{3}}.\cos\phi$$

$$= \sqrt{3}V_l\,I_l\cos\phi \qquad (9.13)$$

This relation is the same as that for the power in a star-connected system and therefore it holds true for the power in any balanced 3-ph system.

Example 9.2 A 415-V, 3-ph induction motor is loaded to 8·94 kW output and operates at a power factor of 0·81 lagging with an efficiency of 86%. Calculate the line and phase currents if the motor windings are delta-connected.

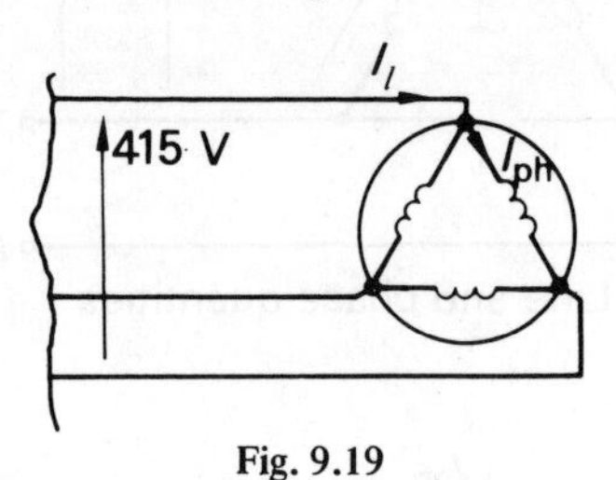

Fig. 9.19

$$P_o = 8{\cdot}94\text{ kW}$$

$$P_i = \frac{P_o}{\eta} = \frac{8\,940}{0{\cdot}86} = 10\,400\text{ W}$$

$$= \sqrt{3}V_l\,I_l\cos\phi$$

$$I_l = \frac{10{\cdot}4\times10^3}{\sqrt{3}\times415\times0{\cdot}81} = \underline{17{\cdot}8\text{ A}}$$

$$I_{ph} = \frac{I_l}{\sqrt{3}} = \frac{17{\cdot}8}{\sqrt{3}} = \underline{10{\cdot}3\text{ A}}$$

9.4 Interconnection of star- and delta-connected systems

If the diagrams for the star- and delta-connected generators and loads (Figs. 9.10, 9.11, 9.16 and 9.17) are studied, it will be seen that the output and input information in each case is the same, i.e. the line voltages and currents are given. It follows that a delta-connected generator can supply a star-connected load and vice versa. Also a star-connected load can operate in parallel with a delta-connected load.

The case of the delta-connected generator is of little importance in practice for the following reasons:

1. For the same line voltage, the e.m.f. generated in each phase of the delta-connected generator is the full line voltage, whilst the e.m.f. generated per phase of the star-connected generator is only $1/\sqrt{3}$ of the line voltage. Thus for a given phase e.m.f., the star-connected generator gives a greater line voltage. This is economically advantageous.

2. The star-connected generator gives a star point which is advantageous when supplying an unbalanced star-connected load. The neutral wire ensures that each phase of the load receives the same applied voltage.

3. If the generator is not ideal and therefore does not produce a pure sinusoidal e.m.f., it will generate excess heat losses.

The loads may either be star-connected or delta-connected. In the former case, either the 3-wire or the 4-wire connection may be used. As previously stated, the 4-wire connection is useful when the load is not balanced. This most commonly occurs when 1-ph loads are connected to a 3-ph system, e.g. 1-ph domestic consumers being supplied from the 3-ph mains of the electricity authority. A typical arrangement is shown in Fig. 9·20.

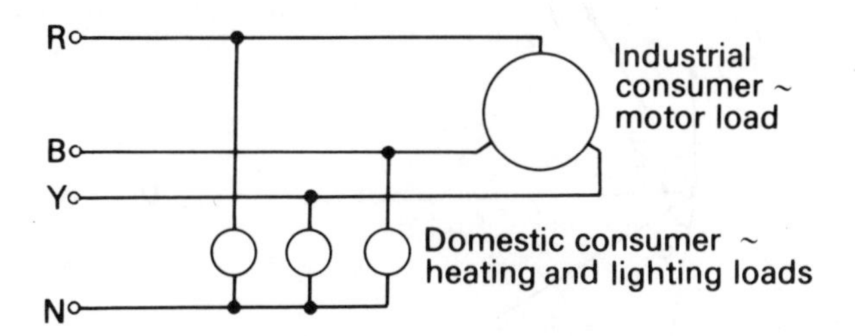

Fig. 9.20. 1-ph loads taken from a 3-ph system

The 3-ph motor of the industrial consumer forms a balanced load supplied at 415 V between lines. The lighting and heating loads of the domestic consumer may well be unbalanced but nevertheless each still receives the required 240-V supply. Only if each of the 1-ph loads were equal would the system be balanced again.

It is possible to analyse the operation of unbalanced loads using polar notation but this is much more easily done using complex notation.

Finally it should be noted that in many cases the impedance of a 3-ph circuit is given in per phase values. This may be interpreted as being the impedance of each phase of a star-connected load unless specific reference is made to the impedance values being those of a delta-connected load.

Example 9.3 Each phase of a star-connected load consists of a coil of resistance 20 Ω and inductance 0·07 H. The load is connected to a 415-V, 3-ph, 50-Hz supply. Calculate:

(*a*) The line current, the power and the power factor.
(*b*) The capacitance per phase of a delta-connected capacitor bank which would improve the overall power factor to unity.

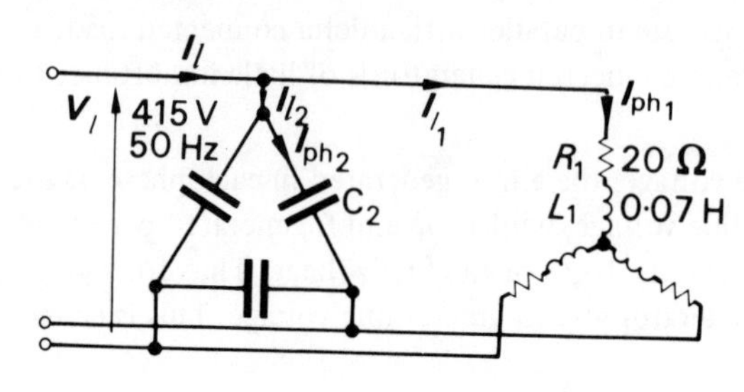

Fig. 9.21

$$X_{L_1} = 2\pi fL = 2\pi 50 \times 0{\cdot}07 = 22{\cdot}0\ \Omega$$

$$Z_{ph_1} = (R_1{}^2 + X_{L_1}{}^2)^{\frac{1}{2}} = (20{\cdot}0^2 + 22{\cdot}0^2)^{\frac{1}{2}} = 29{\cdot}8\ \Omega$$

$$I_{l_1} = I_{ph_1} = \frac{V_{ph}}{Z_{ph_1}} = \frac{V_l}{\sqrt{3}Z_{ph_1}} = \frac{415}{\sqrt{3} \times 29{\cdot}8} = \underline{8{\cdot}05\text{ A}}$$

$$P_1 = 3I_{ph_1}{}^2R_1 = 3 \times 8{\cdot}05^2 \times 20 = \underline{3890\text{ W}}$$

$$\cos\phi_1 = \frac{R_1}{Z_{ph_1}} = \frac{20{\cdot}0}{29{\cdot}8} = \underline{0{\cdot}672\text{ lag}}$$

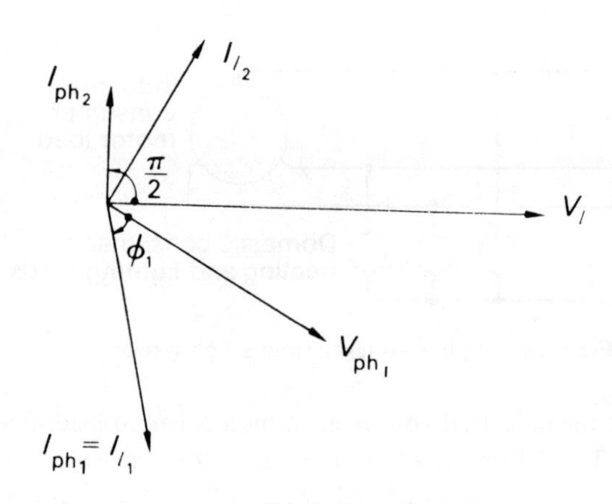

Fig. 9.22 $V_1 = V_{ph_2}$

For unity power factor

$$I_l \sin\phi = 0$$

$$I_{l_2} = I_{l_1}\sin\phi_1 = I_{l_1}\frac{X_{L1}}{Z_{ph_1}} = 8{\cdot}05 \times \frac{22{\cdot}0}{29{\cdot}8} = 5{\cdot}95\text{ A}$$

$$I_{ph_2} = \frac{I_l}{\sqrt{3}}2\ \frac{5{\cdot}95}{\sqrt{3}} = 3{\cdot}44\text{ A}$$

$$X_{C_2} = \frac{V_{ph2}}{I_{ph2}} = \frac{415}{3{\cdot}44} = 121\ \Omega$$

$$C_2 = \frac{1}{2\pi f X_{C_2}} = \frac{1}{2\pi 50 \times 121} = 26{\cdot}3 \times 10^{-6}\ \text{F}$$
$$= \underline{26{\cdot}3\ \mu\text{F}}$$

It is of interest to compare the power dissipated by three identical phase loads when first connected in star and then in delta.

In star: $$P = 3V_{ph} I_{ph} \cos\phi$$
$$= 3 \cdot \frac{V_l}{\sqrt{3}} \cdot \frac{V_{ph}}{Z} \cdot \frac{R}{Z}$$
$$= 3 \cdot \frac{V_l}{\sqrt{3}} \cdot \frac{V_l}{\sqrt{3}} \cdot \frac{R}{Z^2}$$
$$= \frac{V_l^2 R}{Z^2} \tag{9.14}$$

In delta: $$P = 3V_{ph} I_{ph} \cos\phi$$
$$= 3V_l \cdot \frac{V_l}{Z} \cdot \frac{R}{Z}$$
$$= \frac{3V_l^2 R}{Z^2} \tag{9.15}$$

Hence it can be seen that the load connected in delta dissipates three times more power than it would do when connected in star.

Example 9.4 Three identical resistors of 30 Ω are connected in star to a 415-V, 3-ph, 50-Hz sinusoidal supply. Calculated in the load. Also calculate the power dissipated in the resistors if they are reconnected in delta to the same supply.

If one resistor were to be open-circuited in each case, calculate the power dissipated.

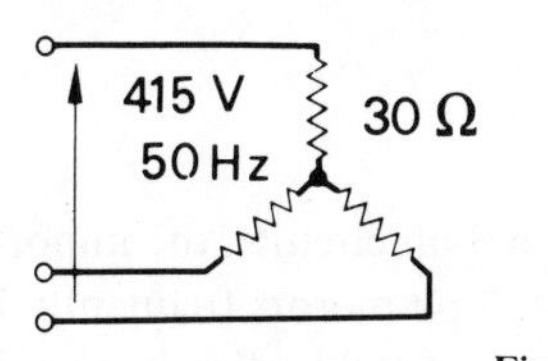

Fig. 9.23

In star: $$V_{ph} = \frac{V_l}{\sqrt{3}} = \frac{415}{\sqrt{3}} = 240\ \text{V}$$

$$I_{ph} = \frac{V_{ph}}{R} = \frac{240}{30} = 8{\cdot}0\ \text{A}$$

$$P = 3I_{ph}^2 R = 3 \times 8^2 \times 30 = \underline{5760\ \text{W}}$$

In delta: $$P = 3 \times 5760 = \underline{17280\ \text{W}}$$

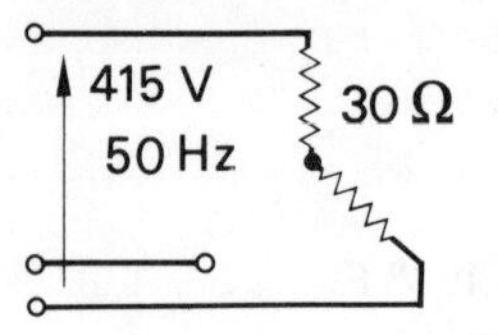

Fig. 9.24

In star: $I = \frac{V_l}{2R} = \frac{415}{2 \times 30} = 6{\cdot}9 \text{ A}$

$$P = I^2(2R) = 6{\cdot}9^2 \times 2 \times 30 = \underline{2870 \text{ W}}$$

In delta: $P = \frac{2}{3} \times 17280 = \underline{11\,520 \text{ W}}$

9.5 Power in a balanced 3-ph system

It has already been shown that the power in a balanced 3-ph load is given by:

$$P = \sqrt{3} V_l I_l \cos\phi$$

By virtue of its definition, the power factor $\cos\phi$ is given by:

$$\cos\phi = \frac{P}{S}$$

Hence

$$S = \sqrt{3} V_l I_l \qquad (9.16)$$

It must again be stated that this only applied to a balanced load. It can also be shown that:

$$Q = \sqrt{3} V_l I_l \sin\phi \qquad (9.17)$$

9.6 Power factor improvement

Because of the much greater power involved in 3-ph circuits, the importance of power factor correction is correspondingly increased. 3-ph motors frequently have power factors as poor as 0·5 lagging and a power factor of 0·8 lagging is considered to be good for a motor. Other power plant with highly inductive components may have even poorer power factors, e.g. welding plant. As in 1-ph circuits, the effects are:

(*a*) large apparent power relative to true power,

(*b*) poor voltage regulation.

The design therefore of an installation must take into account consideration of the economic saving in plant cost as a consequence of poor power factor and also the saving by virtue of the penalty tariffs which may be applied by the supply authority.

This may be achieved by any one of the following methods:

Static capacitors. These were described in para. 7.14.

Synchronous motors. These motors can be made to operate at leading power factors and hence they can offset lagging loads within the installation. Whilst the problem of starting these motors can be easily overcome, they have the disadvantage of being constant-speed machines.

Again it should be noted that although the power factor correction equipment changes the reactive power and the power factor, it does not change the power taken by the load.

Example 9.5 A factory is supplied at 11-kV, 3-ph, 50-Hz system and has the following balanced loads:

Load A 1·5 MW at 0·9 power factor lagging
Load B 600 kW at unity power factor
Load C 2·0 MVA at 0·98 power factor lagging
Load D 3·0 MVA at 0·8 power factor lagging

A 3-ph bank of star-connected capacitors is connected at the supply terminals to give power factor correction. Find the required capacitance per phase to give an overall power factor of 0·98 lagging when the factory is operating at maximum load.

$$P_A = \qquad = 1{\cdot}50 \text{ MW}$$
$$P_B = \qquad = 0{\cdot}60 \text{ MW}$$
$$P_C = S_C \cos\phi_C = 2{\cdot}0 \times 10^6 \times 0{\cdot}98 = 1{\cdot}96 \times 10^6 \text{ W} = 1{\cdot}96 \text{ MW}$$
$$P_D = S_D \cos\phi_D = 3{\cdot}0 \times 10^6 \times 0{\cdot}8 = 2{\cdot}4 \times 10^6 \text{ W} = 2{\cdot}40 \text{ MW}$$
$$6{\cdot}46 \text{ MW}$$

$$Q_A = P_A \tan\phi_A = 1{\cdot}5 \times 10^6 \times 0{\cdot}47 = 0{\cdot}72 \times 10^6 \text{ var} = 0{\cdot}72 \text{ Mvar}$$
$$Q_B = P_B \tan\phi_B = 0{\cdot}6 \times 10^6 \times 0 = 0 = 0{\cdot}00 \text{ Mvar}$$
$$Q_C = S_C \sin\phi_C = 2{\cdot}0 \times 10^6 \times 0{\cdot}2 = 0{\cdot}4 \times 10^6 \text{ var} = 0{\cdot}40 \text{ Mvar}$$
$$Q_D = S_D \sin\phi_D = 3{\cdot}0 \times 10^6 \times 0{\cdot}6 = 1{\cdot}8 \times 10^6 \text{ var} = 1{\cdot}80 \text{ Mvar}$$
$$2{\cdot}92 \text{ Mvar}$$

But $Q = P \tan\phi = 6{\cdot}46 \times 10^6 \times 0{\cdot}2 = 1{\cdot}29 \times 10^6$ var

$= 1{\cdot}29$ Mvar

For capacitor bank:

$$Q_E = 2{\cdot}92 - 1{\cdot}29 = 1{\cdot}63 \text{ Mvar}$$
$$= \sqrt{3} V_l I_{l\,E} = \sqrt{3} V_l I_{ph\,E}$$

$$I_{ph\,E} = \frac{1{\cdot}63 \times 10^6}{\sqrt{3} \times 11 \times 10^3} = 85{\cdot}5 \text{ A}$$

$$X_E = \frac{V_{ph}}{I_{ph\,E}} = \frac{11 \times 10^3}{\sqrt{3} \times 85{\cdot}5} = 74{\cdot}2\ \Omega$$

$$C_E = \frac{1}{2\pi f X_E} = \frac{1}{2\pi 50 \times 74{\cdot}2} = 42{\cdot}9 \times 10^{-6} \text{ F}$$
$$= 42{\cdot}9\ \mu\text{F}$$

9.7 3-ph power measurement

If the load is balanced, the power in any one phase may be measured. The total circuit power is therefore given by multiplying the wattmeter reading by three. The circuits for star- and delta-connected loads are shown in Fig. 9.25. In each case, the wattmeter current coil is connected to any phase current and the potential coil is connected across any phase voltage. This method is termed the 'One-Wattmeter Method'. Its principle disadvantage is that it is not always possible to make the required connections.

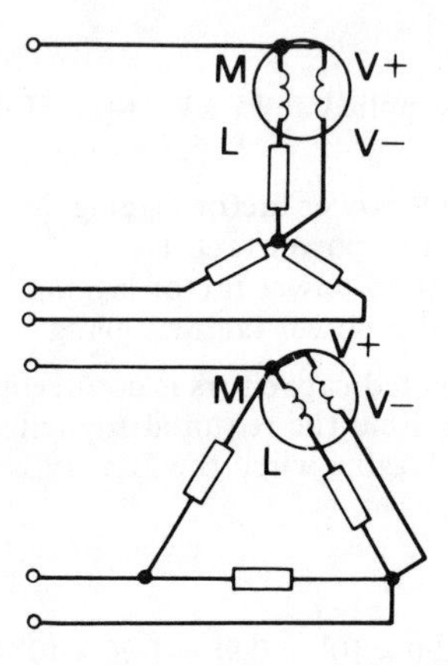

Fig. 9.25. Connections for one wattmeter method of power measurement

When the required connections for the one-wattmeter method are not available, an alternative method is the 'Two-Wattmeter Method'. The current coils of the wattmeters are connected into any two lines while their potential coils are connected between the corresponding lines and the third line. The circuit diagram is shown in Fig. 9.26.

Although the load shown in Fig. 9.26 is star-connected, the method of connecting the wattmeters would be the same for a delta-connected load. In either case the total power dissipated by the load is given by the sum of the two wattmeter readings.

$$\begin{aligned} p_1 + p_2 &= v_{RB}\, i_R + v_{YB}\, i_Y \\ &= (v_{RO'} - v_{BO'})\, i_R + (v_{YO'} - v_{BO'})\, i_Y \\ &= v_{RO'}\, i_R + v_{YO'}\, i_Y - v_{BO'}(i_R + i_Y) \\ &= v_{RO'}\, i_R + v_{YO'}\, i_Y + v_{BO'}\, i_B \\ &= p_R + p_Y + p_B \end{aligned}$$

Therefore at any instant, the power indicated by the wattmeters is equal to the power dissipated in the load. Hence the average power indicated by the wattmeters is equal to the average power dissipated in the load. The wattmeters indicate average power and therefore indicate the active power supplied to the load. This relation holds true whether the load is balanced or not.

An alternative proof can be shown for the case of the balanced load. In this case

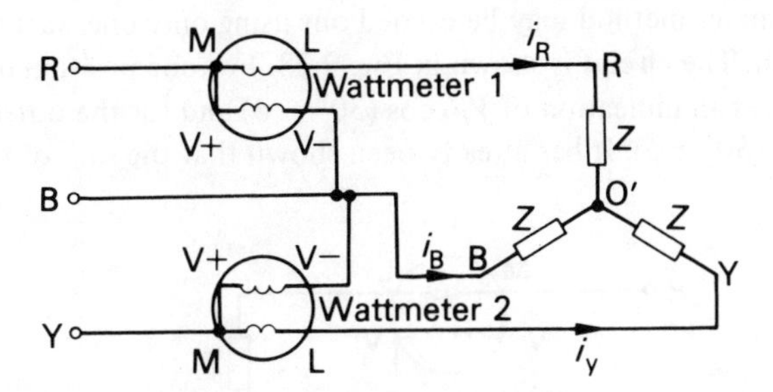

Fig. 9.26. Connections for two-wattmeter method of power measurement

the currents are equal and in the typical phasor diagram shown in Fig. 9.27, they lag behind the respective phase voltages by the same angle ϕ.

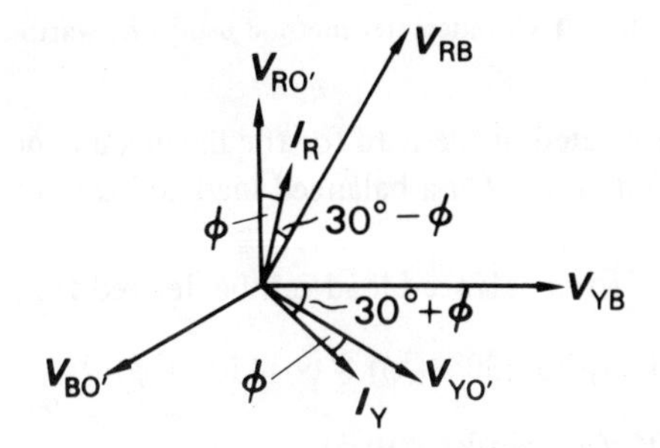

Fig. 9.27. Phasor diagram for two-wattmeter method of measurement applied to a balanced load

$$P_1 = V_{RB}\, I_R \cos(30° - \phi) = V_l\, I_l \cos(30° - \phi)$$
$$P_2 = V_{YB}\, I_Y \cos(30° + \phi) = V_l\, I_l \cos(30° + \phi)$$
$$P_1 + P_2 = V_l\, I_l(\cos(30° - \phi) + \cos(30° + \phi))$$
$$= V_l\, I_l(2 . \cos 30° . \cos\phi)$$
$$= \sqrt{3} V_l\, I_l \cos\phi$$

Thus again the sum of the wattmeter readings is equal to the power dissipated in the load. Consideration of the analysis shows that one wattmeter reads $V_l I_l \cos(30° + \phi)$ and the other reads $V_l I_l \cos(30° - \phi)$. The following points should therefore be noted:

1. If the load power factor is greater than 0·5, i.e. the phase angle is less than 60°, then both wattmeters will read positively.

2. If the load power factor is 0·5, i.e. the phase angle is 60°, then one wattmeter will read zero and the other will read the total power.

3. If the load power factor is less than 0·5, i.e. the phase angle is greater than 60°, one wattmeter will give a negative indication.

In the last case, this deflection must be recorded as a negative power reading. To do so, the direction of the deflection must be changed–most wattmeters are adapted with a changeover switch for this purpose, or else the connections to the current coil have to be interchanged.

The two-wattmeter method may be carried out using only one wattmeter and a changeover switch. The circuit is shown in Fig. 9.28. For one position of the switch, the wattmeter gives an indication of $V_l I_l \cos(30^\circ - \phi)$ and for the other position it indicates $V_l I_l \cos(30^\circ + \phi)$. It has already been shown that the sum of these readings

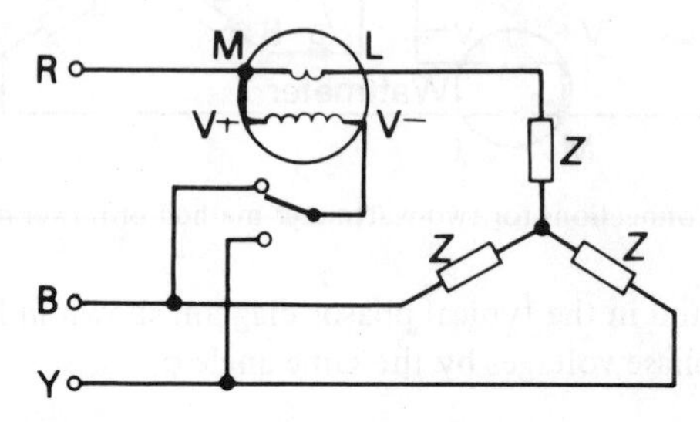

Fig. 9.28. Two-wattmeter method using one wattmeter

is equal to the power dissipated in the load for the balanced condition. However, this method of connection only holds for a balanced load and cannot be applied to an unbalanced load.

The load power factor for a balanced load can be derived from the wattmeter readings.

$$P_1 - P_2 = V_l I_l(\cos(30^\circ - \phi) - \cos(30^\circ + \phi))$$

$$= V_l I_l(2.\sin 30^\circ.\sin\phi)$$

$$= V_l I_l \sin\phi$$

$$\frac{P_1 - P_2}{P_1 + P_2} = \frac{V_l I_l \sin\phi}{\sqrt{3} V_l I_l \cos\phi} = \frac{\tan\phi}{\sqrt{3}}$$

$$\tan\phi = \frac{\sqrt{3}(P_1 - P_2)}{P_1 + P_2} \tag{9.16}$$

From this relation, the phase angle ϕ and hence the power factor may be calculated.

Finally a wattmeter may also be used to read the reactive power in a balanced 3-ph load. The method of connection is shown in Fig. 9.29.

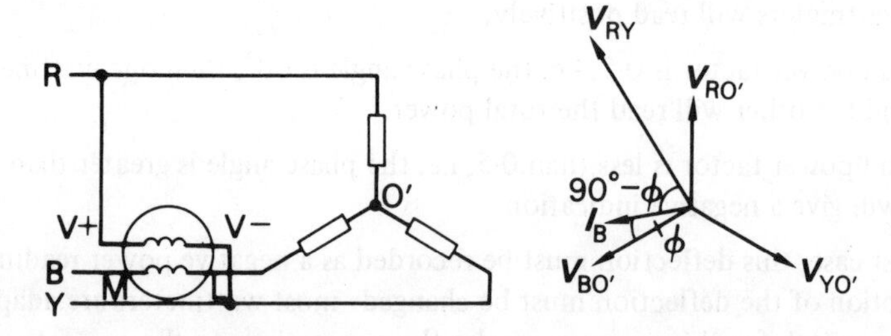

Fig. 9.29. Reactive power measurement in balanced load using one wattmeter

Wattmeter reading $= V_{RY} I_B \cos(90° - \phi) = V_l I_l \sin\phi$

But $Q = \sqrt{3} V_l I_l \sin\phi$

$= \sqrt{3} \times$ wattmeter reading.

Example 9.6 Two wattmeters are connected to measure the input power to a balanced 3-ph load which has a unity power factor. Each meter indicates 10 kW. If the power factor falls to 0·866 lagging but the power remains unchanged, calculate the readings on the wattmeters.

$$P = P_1 + P_2 = 10 + 10 = 20 \text{ kW regardless of power factor}$$

$$\cos\phi = 0{\cdot}866 = \frac{\sqrt{3}}{2}$$

$$\tan\phi = \frac{1}{\sqrt{3}}$$

$$= \frac{\sqrt{3}(P_1 - P_2)}{P_1 + P_2} = \frac{\sqrt{3}(P_1 - P_2)}{20}$$

$$3P_1 - 3P_3 = 20$$

but $3P_1 + 3P_2 = 60$

$$6P_1 = 80$$

$$P_1 = \underline{13{\cdot}33 \text{ kW}}$$

$$P_2 = 20 - P_1 = 20 - 13{\cdot}33 = \underline{6{\cdot}67 \text{ kW}}$$

PROBLEMS ON A.C. POLYPHASE CIRCUITS

1 A symmetrical 3-ph system supplies a balanced load of 24·9 kVA at a line voltage of 415 V. Calculate the line current.

34·6 A

2 A 3-ph generator supplies a balanced 3-ph load of 50·4 kVA. If the line current is 69·28 A, calculate the line voltage.

420 V

3 A balanced 3-ph load is connected to a 415-V, 3-wire system. The input line current is 35 A and the total input power 12 kW. Calculate the power factor of the load.

0·477 lead or lag

4 A 3-ph, star-connected generator delivers a power of 4·6 MW at a power factor of 0·92 lagging and at this load the terminal line voltage is 11 kV. Calculate:

(*a*) The apparent output power.
(*b*) The line current.

5·0 MVA; 262 A

5 A 3-ph star-connected alternator has a line voltage of 11 kV. The output of the generator is 12 MVA at a power factor of 0·85 lagging. Calculate:

(*a*) The phase voltage.
(*b*) The power output.
(*c*) The line current.

6·36 kV; 10·2 MW; 631 A

6 Three 100-Ω resistors are connected first in star and then in delta across 415-V, 3-ph lines. Calculate the line currents in each case and also the power taken from the source.

Find also what these values would be in each case if one of the resistors were disconnected.

2·40 A; 170 W; 4·15 A; 7·18 A; 5 170 W;
2·08 A; 2·08 A; OA; 860 W; 4·15 A; 4·15 A; 7·18 A; 3 450 W

7 Three similar resistors are connected in star across 400-V, 3-ph lines. The line current is 10 A. Calculate:

(*a*) The value of each resistor.
(*b*) The line voltage required to give the same line current, if the resistors were connected in delta.

23·1 Ω; 133·3 V

8 Deduce an expression for the power in a 3-ph, 3-wire balanced load. Each phase of a 3-ph star-connected load is a coil of resistance 20 Ω and inductance 0·05 H. Calculate the power drawn from a 400-V, 3-ph, 50-Hz sinusoidal supply.

What must be the resistance and inductance of each phase of a delta-connected load which will take the same power at the same power factor when connected to the above supply? Find also the line current in the delta connection.

4.95 kW; 60 Ω; 48 Ω; 9·1 A

9 A 3-ph, star-connected generator with a line voltage of 1·5 kV supplies a delta-connected load, each phase of which has a resistance of 40 Ω and an inductive reactance of 15 Ω. Calculate:

(*a*) The alternator current, the line current and the load current.
(*b*) The power, the apparent power and the peak vars supplied by the generator, assuming the line losses are negligible.

61 A; 149 kW; 160 kVA; 55 kvar

10 Three identical resistors each of value 52 Ω are connected first in star and then in delta across a 415-V, 3-ph, 50-Hz supply. Calculate the line and phase currents and the power taken from the supply in each case.

If one resistor is taken out of the circuit in each case, what would be the new values of line and phase currents.

4·62 A; 3·32 kW; 7·96 A; 13·8 A; 9·96 kW; 3·98 A; 7·96 A; 13·8 A

11 Each phase of a 3-ph balanced load consists of a coil of resistance 10 Ω and inductance 0·02 H in parallel with a capacitor of 31·8 μF. Calculate the line current, the power absorbed, the apparent power, and the power factor when connected in star and in delta to a 415-V, 3-ph, 50-Hz sinusoidal supply.

19·1 A; 12·4 kW; 13·8 kVA; 0·9 lag; 53·7 A; 37·2 kW; 41·3 kVA; 0·9 lag

12 Show that the power in a 3-ph balanced circuit can be measured by two wattmeters and deduce an expression giving the total power in terms of the wattmeter readings. A 3-ph, 500-V motor operating at a power factor of 0·4 lag takes 30 kW. Find the reading on each of two wattmeters correctly connected to measure the input power.

– 4·9 kW; + 34·9 kW

13 A factory has the following simultaneous demands balanced across the three phases of a 3-ph, 50-Hz supply:

200 kW of lighting and heating.
150 kW synchronous motor of 85% efficiency.
750 kW induction motor also of 85% efficiency at 0·8 power factor lagging.

If the total demand is 1 380 kVA, find the power factor at which the synchronous motor must be operated. Also determine the power factor at which the synchronous motor must operate to bring the overall power factor to unity, the load remaining constant.

0·916 lead; 0·259 lead

14 An 11-kV, 3-ph, 50-Hz system supplies, at a power factor of 0·95 lagging, a simultaneous demand of 2 000 kVA made up of five distinct loads. Four of the five loads are:

600 kW at 0·8 power factor lagging.
400 kW at unity power factor.
500 kW at 0·9 power factor leading.
200 kvar power factor correction plant.

Find the value of the fifth load and its power factor. Also calculate the capacitance per phase of a delta-connected capacitor bank to improve the overall power factor to unity.

400 kW; 0·54 lag; 8·12 μF

15 The following balanced loads are connected to a 500-V, 3-ph, 50-Hz sinusoidal supply:

500 kW at unity power factor.
1 500 kVA at 0·9 power factor lagging.
1 000 kVA at 0·8 power factor lagging.
A star-connected bank of capacitors.

Determine the rating and capacitance per phase of the last load if the overall power factor is 0·95 lagging.

385 kVA; 4 900 μF

16 Each phase of a 3-ph, star-connected load consists of a coil of resistance 25 Ω and reactance of 0·1 H inductance. Calculate:

(*a*) The line current, the power absorbed, and the power factor when this load is connected to a 415-V, 3-ph, 50-Hz supply.
(*b*) The capacitance per phase of a delta-connected capacitor bank which would improve the overall power factor to unity.

5·98 A; 2·68 kW; 0·623 lag; 20·7 μF

17 Two 3-ph, 450-V, 50-Hz generators operating in parallel supply the following balanced loads:

40 kW at unity power factor.
100 kVA at 0·95 power factor lagging.
80 kVA at 0·8 power factor lagging.

The first generator supplies 100 kW at 0·85 power factor lagging.

Determine:

(*a*) The power, the apparent power and the peak vars supplied by the second generator.
(*b*) The capacitance per phase required of a delta-connected capacitor bank to improve the overall power factor of the load to 0·95 lagging.

99 kW; 105 kVA; 17·8 kvar; 76 μF

18 State the disadvantages of a load with a low lagging power factor from the point of view of:

(*a*) The supply authority.
(*b*) The consumer.

Why are capacitors for improving the power factor usually connected in parallel and not in series with the load?

A capacitor bank taking a total of 1 200 kvar is necessary to improve the power factor of a 3-ph balanced load in a factory from 0·75 to 0·95 lagging. Find the factory load in kilowatts and kilovoltamperes before and after the connection of the capacitors.

2 150 kW; 2 260 kVA; 2 870 kVA

19 A factory is supplied from an 11-kV, 3-ph, 50-Hz system and has the following balanced loads:

1·5 MW at 0·9 power factor lagging.
600 kW at unity power factor.
2 MVA at 0·98 power factor lagging.
3 MVA at 0·8 power factor lagging.

A 3-ph bank of star-connected capacitors is connected at the supply point terminals to give automatic power factor correction. Find the capacitance required per phase to give an overall system power factor of 0·98 lagging when the factory is operating at maximum load.

42·5 μF

10

Transformers

A transformer is a machine that has no moving parts but is able to transform alternating voltages and currents from high to low values and vice versa. Transformers are used extensively in all branches of electrical engineering from the large power transformer employed in the national grid to the small signal transformer of an electronic amplifier.

10.1 Principle of operation

When an alternating voltage is applied to a concentrated coil wound on a ferromagnetic core, a back e.m.f. is induced in the coil due to the continual alternation of the self flux linkage. This is the principle of the simple-inductor. The e.m.f. is due to the rate of change of flux linkage within the coil. If another coil were placed around the core, as shown in Fig. 10.1, this coil also would have an e.m.f. induced in it. This e.m.f. is caused by the effect of mutual inductance. It should be noted that the two coils are insulated from one another and are therefore electrically separate.

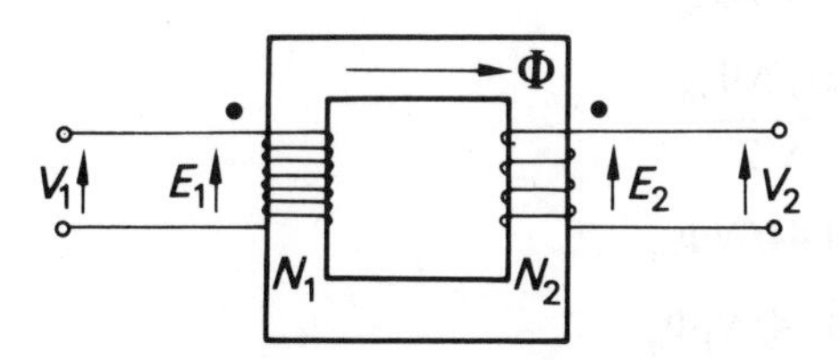

Fig. 10.1. Simple transformer

Neglecting any losses which may occur, the induced e.m.f. of the first or primary winding must equal the applied voltage. This is termed the primary e.m.f. E_1. Each turn provides its own proportion of the total voltage and, if there are N_1 turns in the primary winding, the induced e.m.f. per turn is E_1/N_1.

Each turn on the other or secondary winding will also have the same e.m.f. induced in it, i.e. E_1/N_1. It will be remembered that it has been assumed that there are no losses and this includes that there will be no flux leakage. If there are N_2 turns in the secondary winding, the total induced e.m.f. E_2 is given by

$$E_2 = N_2 . \frac{E_1}{N_1}$$

$$\therefore \quad \frac{E_1}{E_2} = \frac{N_1}{N_2} \tag{10.1}$$

E_2 is termed the secondary e.m.f. The ratio of the e.m.f.'s and thus the voltages in the two windings is therefore seen to be the same as the ratio of the turns. A simple means is thus provided of transforming from one voltage to another.

In the general case, the applied voltage is sinusoidal. For the ideal transformer, the flux produced in the core will also be sinusoidal. Let the flux be represented by:

$$\phi = \Phi_m \sin \omega t$$

The instantaneous e.m.f. induced in a coil of N turns linked by this flux ϕ is given by:

$$e = \frac{d\psi}{dt}$$

$$= N\frac{d\phi}{dt}$$

$$= N\frac{d}{dt}(\Phi_m \sin \omega t)$$

$$= \omega N \Phi_m \cos \omega t$$

$$= 2\pi f N \Phi_m \sin (\omega t + 90°)$$

This represents an induced e.m.f. of maximum value $2\pi f N \Phi_m$, which leads the flux by 90°.

$$E_m = 2\pi f N \Phi_m$$

$$\therefore \quad E = \frac{2\pi f N \Phi_m}{\sqrt{2}}$$

$$E = 4{\cdot}44 f N \Phi_m \tag{10.2}$$

Thus $\quad E_1 = 4{\cdot}44 f N_1 \Phi_m$

and $\quad E_2 = 4{\cdot}44 f N_2 \Phi_m$

If these expressions are divided, the one into the other, relation (10.1) is again obtained.

If a load is connected across the secondary winding, the secondary e.m.f. E_2 will cause a current I_2 to flow in the resulting circuit. Such a current flow gives rise to an m.m.f. in the ferromagnetic core, the m.m.f. being in phase with the current and given by

$$F_2 = I_2 N_2$$

The loaded-transformer circuit is shown in Fig. 10.2.

Such an m.m.f. will alter the flux in the core from its original value. However the flux in the core must remain unchanged from the previous no-load condition because this is the flux, which, when linking the primary winding and varying at the supply

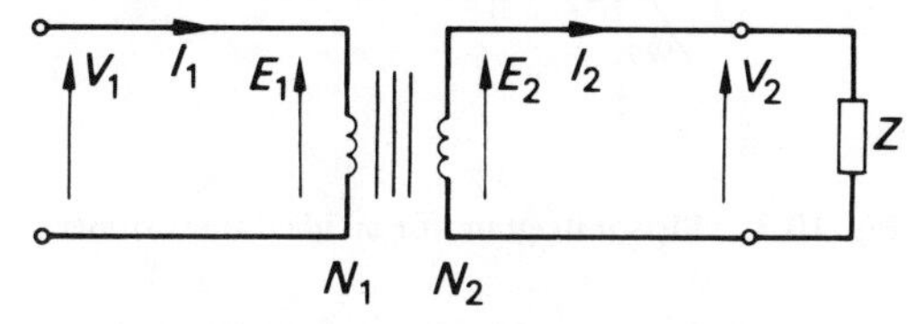

Fig. 10.2. Transformer on load

frequency, gives rise to the primary back e.m.f. equal to the supply voltage. This equality must be maintained at all times in order to comply with Kirchhoff's 2nd Law. In order to conserve this equality, an additional current must flow in the primary winding to give rise to an additional primary m.m.f. equal in magnitude but opposite in direction of action to the secondary m.m.f. Let the primary current be I_1.

$$F_1 = N_1 I_1$$

but

$$F_1 = F_2$$

$$I_1 N_1 = I_2 N_2$$

$$\frac{N_1}{N_2} = \frac{I_2}{I_1} \qquad (10.3)$$

This relation may be combined with relation (10.1):

$$\frac{E_1}{E_2} = \frac{N_1}{N_2} = \frac{I_2}{I_1} \qquad (10.3.1)$$

10.2 Ideal transformer

The ideal transformer operates on the principles outlined in para. 10.1 and has negligible losses either in the electric circuits or in the magnetic circuit. Moreover it has the important property that it requires no current to magnetise the core. It has been noted that when a transformer is loaded, an m.m.f. balance is set up between the primary and secondary circuits in order that the core flux remains unchanged. No mention was made of how this initial core flux was created. If the inductance of the primary winding is infinite, then zero current is required to create the required core flux. It follows that when the transformer is loaded, all of the primary current is used to balance the secondary current in accordance with relation (10.3). Relation (10.3.1) therefore applies to an ideal transformer; the appropriate phasor diagram is shown in Fig. 10.3.

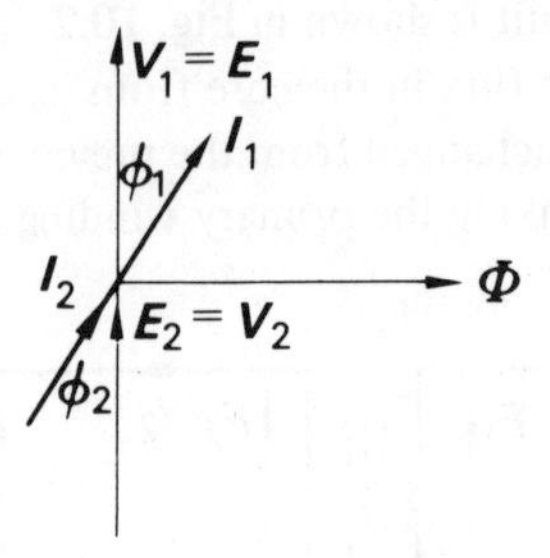

Fig. 10.3. Phasor diagram for an ideal transformer

In the ideal case, the applied voltage V_1 is equal to the primary back e.m.f. E_1 and similarly $E_2 = V_2$, where V_2 is the voltage applied to the load by the secondary. By relation (10.3.1):

$$\frac{V_1}{V_2} = \frac{E_1}{E_2} = \frac{I_2}{I_2}$$

$$\therefore \quad V_1 I_1 = V_2 I_2$$

$$\therefore \quad S_1 = S_2$$

But the primary and secondary voltages are in phase with one another and the primary and secondary currents are in phase with one another hence

$$\cos\phi_1 = \cos\phi_2 = \cos\phi$$

$$S_1 \cos\phi = S_2 \cos\phi$$

$$P_1 = P_2 \tag{10.4}$$

Thus the input power is equal to the output power. This is in keeping with the result to be expected from an ideal transformer.

Example 10.1 An ideal 25-kVA transformer has 500 turns on the primary winding and 40 turns on the secondary winding. The primary winding is connected to a 3-kV, 50-Hz supply. Calculate:

(*a*) the primary and secondary currents on full load.
(*b*) the secondary e.m.f.
(*c*) the maximum core flux.

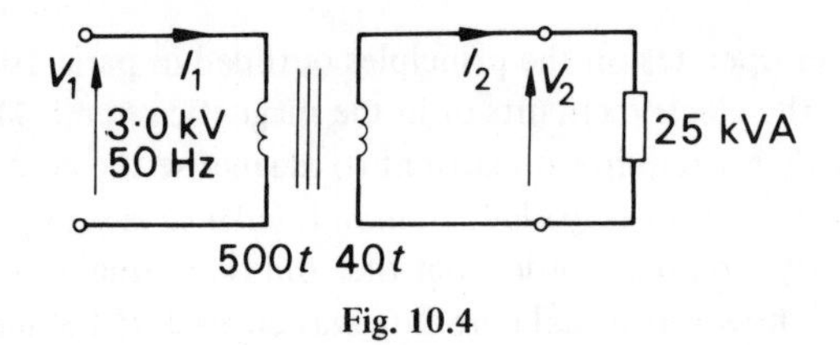

Fig. 10.4

$$I_1 = \frac{S}{V_1} = \frac{25 \times 10^3}{3 \times 10^3} = \underline{8{\cdot}33 \text{ A}}$$

$$I_2 = \frac{I_1 N_1}{N_2} = \frac{8{\cdot}33 \times 500}{40} = \underline{104{\cdot}2 \text{ A}}$$

$$E_2 = E_1 \cdot \frac{N_2}{N_1} = V_1 \cdot \frac{N_2}{N_1} = 3000 \times \frac{40}{500} = \underline{240 \text{ V}}$$

$$E_1 = 4{\cdot}44 f N \Phi_m$$

$$3000 = 4{\cdot}44 \times 50 \times 500 \times \Phi_m$$

$$\Phi_m = 27 \times 10^{-3} \text{ Wb}$$

$$= \underline{27 \text{ mWb}}$$

Example 10.2 A 1-ph, 415/22 000-V, 50-Hz transformer has a nett core area of 10 000 mm² and a maximum flux density of 1·3 T. Estimate the number of turns in each winding.

$$N_1 = \frac{E_1}{4{\cdot}44 f \Phi_m} = \frac{415}{4{\cdot}44 \times 50 \times 1{\cdot}3 \times 10000 \times 10^{-6}}$$

$$= 143{\cdot}6 \text{ turns}$$

The number of turns must be an integer hence

$$N_1 = \underline{144 \text{ turns}}$$

$$N_2 = \frac{E_2}{E_1} \cdot N_1 = \frac{22000 \times 144}{415} = \underline{7620 \text{ turns}}$$

It should be noted that in a calculation of this type, the number of turns in the low-voltage winding is always estimated first. This is because any correction to the number of turns to give an integer is relatively more important to the smaller number. When adjusting the number of turns to give an integer, this should be made to the next largest integer, otherwise the maximum permitted flux density will be exceeded.

10.3 Practical transformer

The practical transformer differs from the ideal transformer on two points–it requires a current to magnetise it and it does have losses.

Arising from the first point, there must always be a current in the primary winding to magnetise the core. When a load is connected to the secondary winding of the transformer, an additional current flows in the primary to maintain the m.m.f. balance in the core. Thus the total primary current now consists of two components–the current to magnetise the core and the load current. Since the primary current is designated I_1, some other symbol must be given to the current required to maintain the m.m.f. balance. Let it be I_2' which is described as I_2 referred. In relation (10.3), it follows that

$$\frac{N_1}{N_2} = \frac{I_2}{I_2'} \qquad (10.5)$$

The current that magnetises the core exists at all times and remains virtually constant regardless of the secondary load. Because it can be observed on its own when there is no load connected to the secondary winding, it is termed the no-load current I_0.

$$I_1 = I_0 + I_2' \qquad (10{\cdot}6)$$

The no-load current has, in its turn, two components. The first of these is the magnetising current I_{0m}, which gives rise to the core flux. This is a purely reactive current and is in phase with the flux that it creates. The second component is an active one due to the hysteresis and eddy-current losses in the core. It is designated I_{0l}. The total no-load current is given by the phasor sum of the components.

$$I_0 = I_{0m} + I_{0l} \tag{10·7}$$

It should be noted that, whilst the existence of the no-load current should be acknowledged, it does not always play a part in the solution of the accompanying problems. In many transformers, particularly power transformers, the omission of the no-load current will not materially affect the validity of any on-load solution, due to the relative smallness of the no-load current.

Provided that there are no losses in the windings, $E_1 = V_1$ and $E_2 = V_2$. If there were a volt drop within the winding due either to the resistance of the winding conductor or due to leakage reactance created by imperfect coupling magnetically of the windings, these relations would not hold. The phasor diagram shown in Fig. 10.5

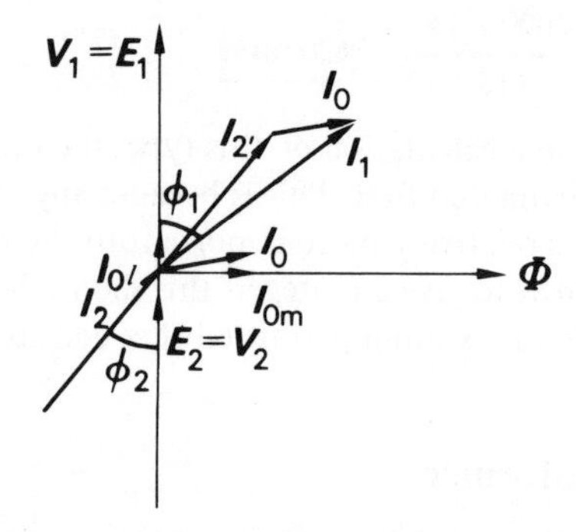

Fig. 10.5. Phasor diagram for a practical transformer without winding volt drops

is drawn on the assumption that there are no volt drops in the windings. It should be noted that E_1 and E_2 are shown to be in phase since this is the normal method of winding connection.

10.4 Magnetisation of core

It has been shown that if the applied voltage across a winding wound on an ferromagnetic core is sinusoidal, then the rate of change of the core flux must also be sinusoidal. It follows that the flux must vary sinusoidally. However, it does not follow that the current which creates this flux varies sinusoidally. If the B/H curve is redrawn as a ϕ/i curve, the relation between flux and current can be derived as shown in Fig. 10.6. The change of scale for the axes is permissible since $\phi = BA$ where the c.s.a. of the core A is constant and similarly $i = Hl/N$, where the core length l and the number of turns on the coil N are constant. Using this ϕ/i curve, by plotting instantaneous values of flux against current, the current waveform can be derived.

Whilst it is not the object of this text to explain the analysis of this waveform, its significance should be noted. Since it is not sinusoidal, it cannot be measured using standard instruments such as moving-iron instruments unless the readings obtained are correctly interpreted. For instance, if a ferromagnetic-cored coil is connected in parallel

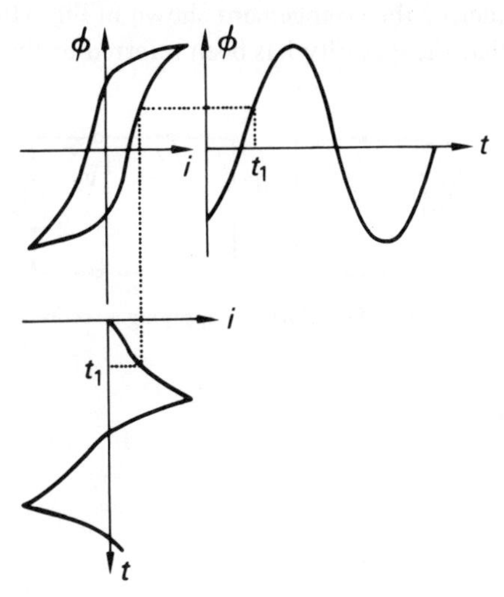

Fig. 10.6. Waveform of magnetising current

with a capacitor and the supply and the branch currents are measured using an ammeter, the supply current will not be obtained from the phasor summation of the currents because neither the supply current nor the branch current are sinusoidal. This is a common source of error in measurements connected with ferromagnetic-cored coils.

Finally the effect of saturation of the core on the back e.m.f. should be noted. Consider a ferromagnetic-cored coil that also has some winding resistance. If the core has an ideal B/H characteristic as shown in Fig. 10.7(*a*), then provided the flux density of the knee point is not exceeded, a back e.m.f. will oppose the applied voltage at all times. However, if this flux density value is exceeded, the flux in the core can only increase if the coil current increases by a very large amount. This increase is limited by the coil resistance and so the core flux is unable to change at the required rate. This results in the collapse of the back e.m.f. The appropriate waveforms are shown in Fig. 10.7(*b*). In order to avoid this effect, transformers and similar ferromagnetic-cored machines are designed so that the core flux density does not exceed the value appropriate to the knee point.

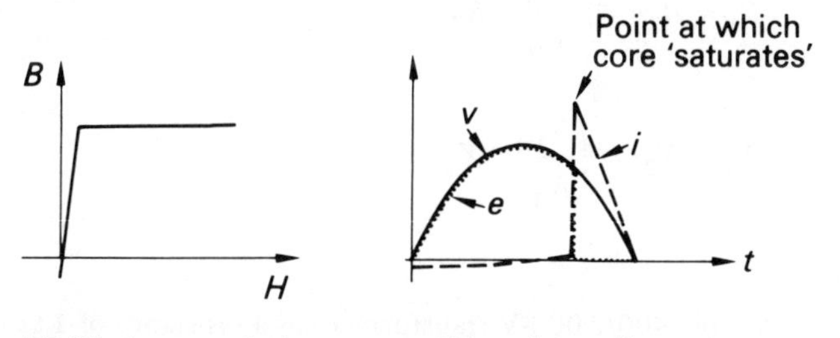

Fig. 10.7. Effect of excessive flux in a ferromagnetic-core used for an a.c. machine

10.5 Impedance transformation

Consider the effect of the secondary load as seen from the primary winding. This may be analysed by considering the arrangement shown in Fig. 10.8. As before, the subscript prime means that the quantity has been referred to the other winding of the transformer.

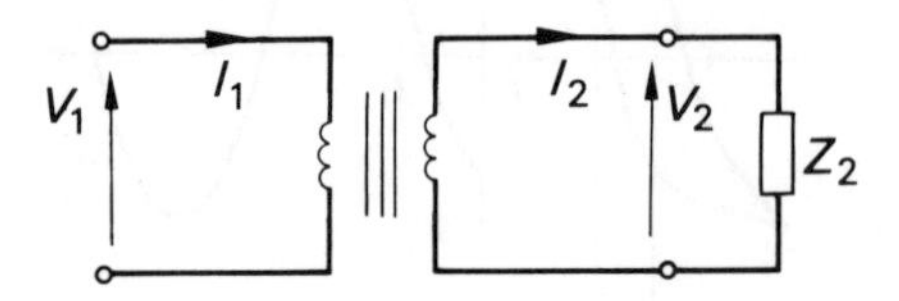

Fig. 10.8. Transformer supplying a secondary load

$$Z_2 = \frac{V_2}{I_2}$$

$$Z_1 = \frac{V_1}{I_1}$$

$$= V_2 \cdot \frac{N_1}{N_2} \cdot \frac{1}{I_2} \cdot \frac{N_1}{N_2}$$

$$= \left(\frac{N_1}{N_2}\right)^2 \cdot Z_2$$

$$= Z_2'$$

$$Z_1 = Z_2' = \left(\frac{N_1}{N_2}\right)^2 \cdot Z_2 \tag{10.8}$$

Since the transformer is ideal:

$$Z_1 \cos\phi = Z_2' \cos\phi = \left(\frac{N_1}{N_2}\right)^2 \cdot Z_2 \cos\phi$$

$$R_1 = R_2' = \left(\frac{N_1}{N_2}\right)^2 \cdot R_2 \tag{10.8.1}$$

Also

$$X_1 = X_2' = \left(\frac{N_1}{N_2}\right)^2 . X_2 \tag{10.8.2}$$

Example 10.3 A 1-ph, 400/2 000-V transformer has a resistance of 1 Ω connected in series with its primary winding and a 225-Ω resistor connected across its secondary winding. Calculate the primary current when the circuit is supplied at 400 V.

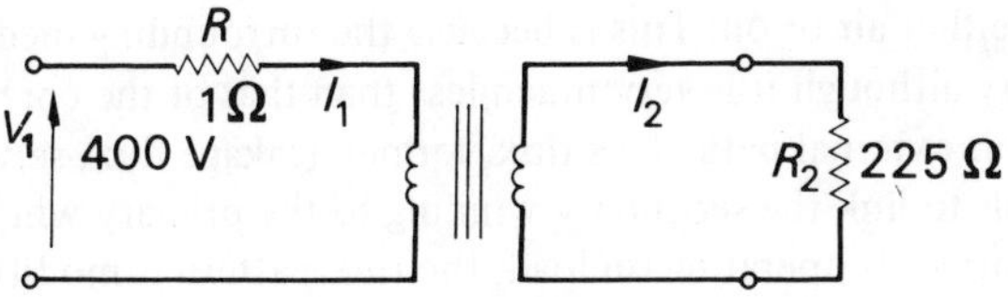

Fig. 10.9

$$R_2 = 225\ \Omega$$

$$R_2{}' = \left(\frac{N_1}{N_2}\right)^2 \cdot R_2 = \left(\frac{E_1}{E_2}\right)^2 \cdot R_2 = \frac{400^2}{2000^2} \times 225 = 9\ \Omega$$

$$R_i = R + R_2{}' = 1 + 9 = 10\ \Omega$$

$$I_1 = \frac{V_1}{R_1} = \frac{400}{10} = \underline{40\ \text{A}}$$

10.6 Transformer losses

There are three sources of power loss in a transformer, all of which have already been mentioned. They are:

1. Hysteresis loss.
2. Eddy-current loss.
3. Winding i^2R losses.

The first of these was discussed in para. 4.8; its effect is minimised by using the best possible steel, bearing in mind that the better the steel, the higher will be the cost. The eddy-current loss is minimised again be the choice of steel, as described in para. 5.13, and also by lamination. The winding i^2R loss is due to the resistance of the winding conductors, which, although generally small, is not always negligible. In modern transformers, the i^2R loss may account for 90% of the total losses.

10.7 Flux leakage in a transformer

Consider again the magnetic circuit of the simple transformer. This is shown in Fig. 10.10.

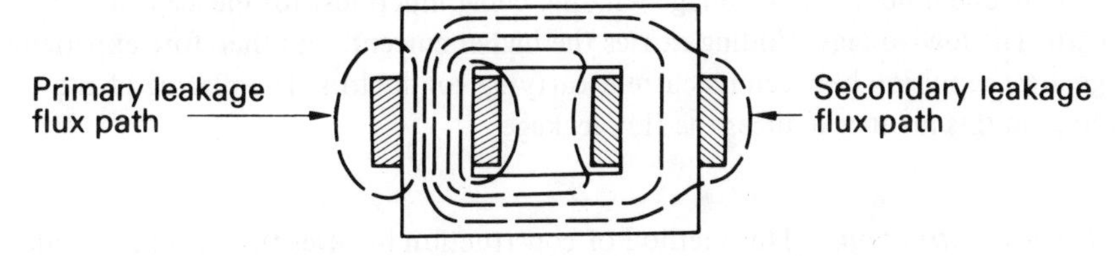

Fig. 10.10. Magnetic circuit of a simple transformer

The major portion of the magnetising flux set up by the no-load current passes through both primary and secondary windings linking them. However, not all of the flux remains within the ferromagnetic path: some is diverted into the surrounding medium, generally either air or oil. This is because the surrounding medium also has a definite permeability although it is very much less than that of the core. Thus a little of the flux traverses this external path. This flux, termed leakage flux, serves no useful purpose, since it fails to link the secondary winding to the primary winding.

When the transformer is operating on load, the flux pattern is modified to that shown in Fig. 10.10. Again there is leakage flux emanating from the primary winding. This is termed the primary leakage flux. However, the secondary current sets up an m.m.f. which opposes the main flux and causes a portion of the flux to be diverted in secondary leakage paths. Thus there is formed a flux which links only the secondary winding. This is termed the secondary leakage flux.

The effect of these leakage fluxes is purely self inductive, hence both windings appear to have self-inductive reactance. This serves no useful purpose but acts as an effective volt drop within the windings. The main useful flux decreases very slightly as the load increases, but the leakage fluxes are practically proportional to the currents in the respective windings. The effect of flux leakage upon the ratio of transformation is thus to reduce the secondary terminal voltage for a given primary applied voltage.

To minimise the leakage, transformers are constructed in one of the two following designs:

Core-type construction. This is a modification of the simple transformer and has a single magnetic circuit as shown in Fig. 10.11. It is usual to wind one half of each

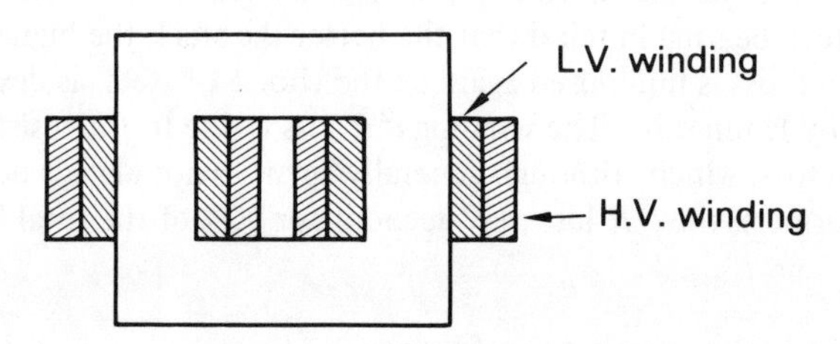

Fig. 10.11. Core-type transformer

winding on one limb, the low-voltage winding being innermost for mechanical strength. The low-voltage winding carries the higher currents and therefore experiences the greatest repulsion between its current-carrying conductors. The placing of the windings in this manner reduces the flux leakage.

Shell-type construction. This method of construction involves the use of a double magnetic circuit. It is illustrated in Fig. 10.12 showing how the windings are again wound concentrically but are complete on the central limb.

Finally it should be noted that the leakage flux forms only a very small part of the total flux. Modern transformer generally have coefficients of coupling in excess of 0·99.

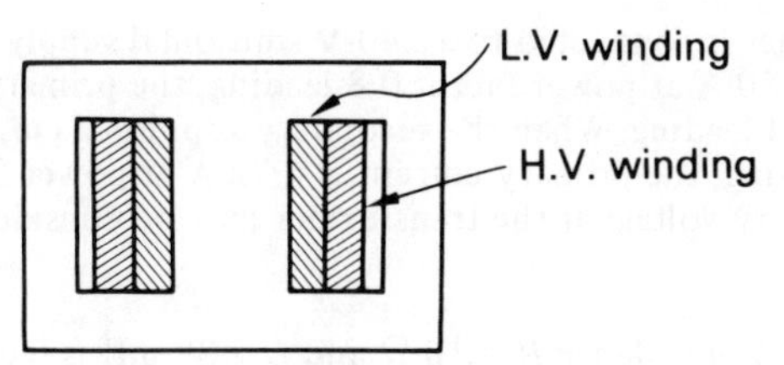

Fig. 10.12. Shell-type transformer

PROBLEMS ON TRANSFORMERS

1 A 3 300/250-V, 50-Hz, 1-ph transformer has a core of effective cross-sectional area 13 000 mm^2 and a low-voltage winding of 80 turns. Determine:

(*a*) The number of turns of the high-voltage winding.

(*b*) The maximum flux density in the core.

If the transformer supplied a load of 25 kW at 0·8 power factor lagging when connected to a 3·3-kV supply, calculate the approximate values of secondary and primary currents.

1 056; 1·08 T; 125 A; 9·5 A

2 Describe the operation of an ideal 1-ph transformer on no-load and on load. State for each condition where the effective losses occur in a practical transformer and explain how these losses occur.

A 20-kVA, 1-ph transformer has a turns ratio of 44:1. If the primary winding has 4 000 turns and is connected to an 11-kV, 50-Hz sinusoidal supply, calculate for full-load.

(*a*) The approximate values of primary and secondary currents.

(*b*) The maximum value of core flux.

1·8 A; 80·0 A; 12·4 mWb

3 A 1-ph transformer has a ratio of 1:10 and a secondary winding of 1 000 turns. The primary winding is connected to a 25-V sinusoidal supply. If the maximum core flux is 2·25 mWb, determine:

(*a*) The secondary voltage on open-circuit.

(*b*) The number of primary turns.

(*c*) The frequency of the supply.

250 V; 100 turns; 25 Hz

4 A 2·4-kV, 1-ph, 50-Hz induction furnace for melting aluminium consists of a 3 560-turn primary winding wound on a transformer core. The secondary winding comprises the molten aluminium contained in an annular channel of firebrick which encircles the core. This ring of aluminium has a resistivity of 0·028 7 $\mu\Omega$ m, a mean length of 2·5 m and a cross-sectional area of 17 000 mm^2. Estimate the primary current and the power dissipation in the aluminium.

If the cross-sectional area of the core is 2 500 mm^2, estimate the peak flux density in the core.

45 A; 108 kW; 1·22 T (SANCAD)

5 A transformer on no-load requires a magnetising current but is otherwise ideal. When the secondary current is 100 A at a power factor of 0·866 lagging, the primary current is 11 A at a power factor of 0·788 lagging. The primary voltage is 200 V. What is the secondary voltage?

20 V (SANCAD)

6 A 1-ph transformer is connected to a 240-V sinusoidal supply. When the secondary supplies a current of 50 A at power factor 0·8 leading, the primary current is 4·62 A at power factor 0·974 leading. When the secondary supplies a current of 50 A at power factor 0·8 lagging, the primary current is 6·72 A at power factor 0·67 lagging. Calculate the secondary voltage if the transformer may be considered ideal.

24·2 V

7 A load of constant impedance $R = 16\ \Omega$ and $L = 38$ mH is fed from a constant-voltage source through the primary winding of an ideal transformer of turns ratio 1:10. What power will be dissipated in the load with terminals P and Q:

(*a*) Short-circuited.
(*b*) Open-circuited.
(*c*) Connected by a 1 150-Ω resistor.
(*d*) Connected by a 2·65-μF capacitor?

12 A; 0; 8 A; 15 A (SANCAD)

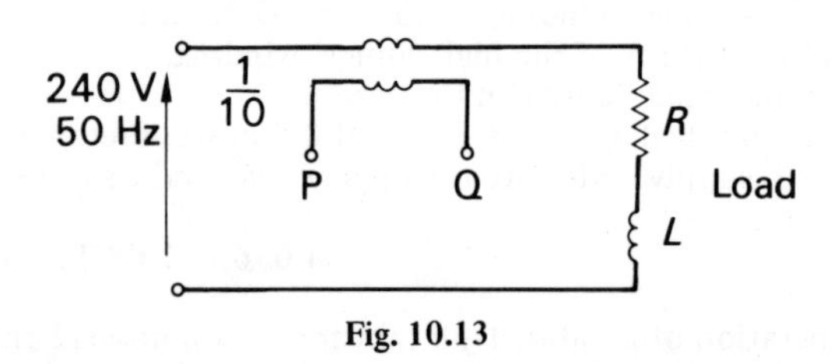

Fig. 10.13

8 From the first principles of an ideal, 1-ph transformer, derive an expression for the input impedance of a transformer with an impedance Z_2 connected across its secondary terminals. (The input impedance should be expressed in terms of Z_2.)

An ideal transformer has 100 primary turns and 200 secondary turns, the secondary winding being centre-tapped. The transformer is used in the circuit shown in Fig. 10.14. Explain why $V_{PQ} = 0$.

Calculate the total power dissipated in the two 10-Ω resistors.

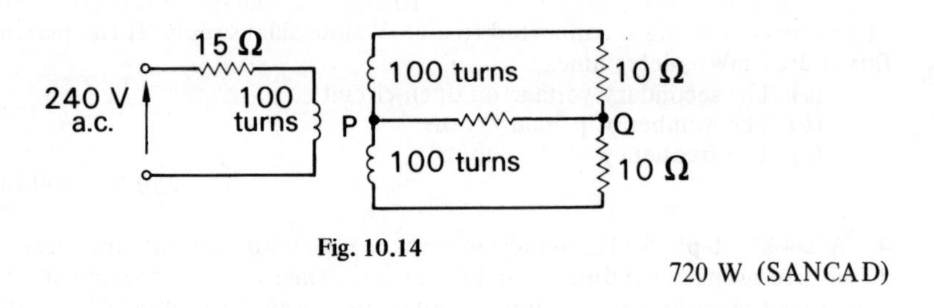

Fig. 10.14

720 W (SANCAD)

11

Unified machine theory

All electromagnetic machines operate on a common set of fundamental principles. These principles may be applied whether the machine is operated by alternating current or direct current, whether the machine acts as a motor, a generator or a converter, and whether the machine operates under steady-state conditions or transient conditions. This chapter deals mainly with the principles and uses the simple linear machines to illustrate this. The subsequent chapter deals with the rotational machines which are more complex forms of electromagnetic machines yet which operate on the same fundamental principles.

11.1 Conversion process in a machine

An electromagnetic machine is one that links an electrical energy system to another energy system by providing a reversible means of energy flow in its magnetic field. The magnetic field is therefore the coupling between the two systems and is the mutual link. The energy transferred from the one system to the other is temporarily stored in the field and then released to the other system.

Usually the energy system coupled to the electrical energy system is a mechanical one; the function of a motor is to transfer electrical energy into mechanical energy whilst a generator converts mechanical energy into electrical energy. Converters transfer electrical energy from one system to another as in the transformer. Not all machines deal with large amounts of energy–those operating at very low power levels are often termed transducers, particularly when providing 'signals' with which to activate electronic control devices.

Discussion of a mechanical energy system implies that a mechanical force is associated with displacement of its point of action. An electromagnetic system can develop a mechanical force in two ways:

1. By alignment.
2. By interaction.

The force of alignment can be illustrated by the arrangements shown in Fig. 11.1. In the first, two poles are situated opposite one another; each is made of a ferromagnetic material and a flux passes from the one to the other. The surfaces through which the flux passes are said to be magnetised surfaces and they are attracted towards one another as indicated in the diagram.

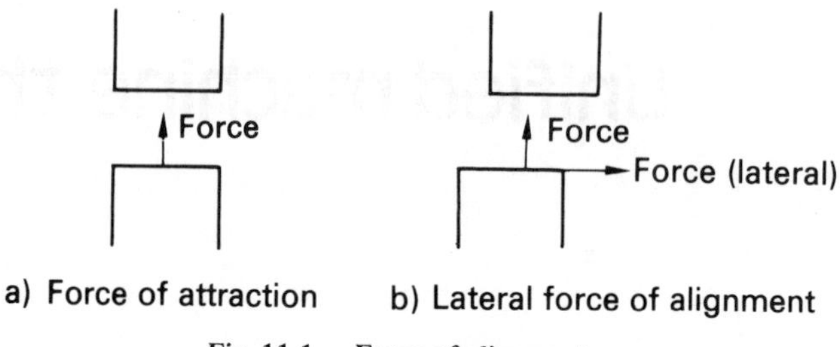

Fig. 11.1. Force of alignment

It will be shown that the force of alignment acts in any direction that will increase the magnetic energy stored in the arrangement. In the first case, it will try to bring the poles together since this decreases the reluctance of the air gap in the magnetic circuit and hence will increase the flux and consequently the stored energy. This principle of increasing the stored magnetic energy is a most important one and it is the key to the unified machine theory that follows.

In the second case, shown in Fig. 11.1(*b*), the poles are not situated opposite one another. The resultant force tries to achieve greater stored magnetic energy by two component actions:

(*a*) By attraction of the poles towards one another as before.
(*b*) By aligning the poles laterally.

If the poles move laterally, the cross-sectional area of the air gap is increased and the reluctance is reduced with consequent increase in the stored magnetic energy as before.

Both actions attempt to align the poles to the point of maximum stored energy, i.e. when the poles are in contact with maximum area of contact. It should be noted that the force of alignment does not necessarily act in the direction of the lines of flux.

Many devices demonstrate the principle of the force of alignment. Probably the one with which the reader is most familiar is that of the permanent magnet and its attraction of ferromagnetic materials. Electromagnetic devices such as the relay shown in Fig. 11.2 demonstrate the force of alignment giving rise to linear motion.

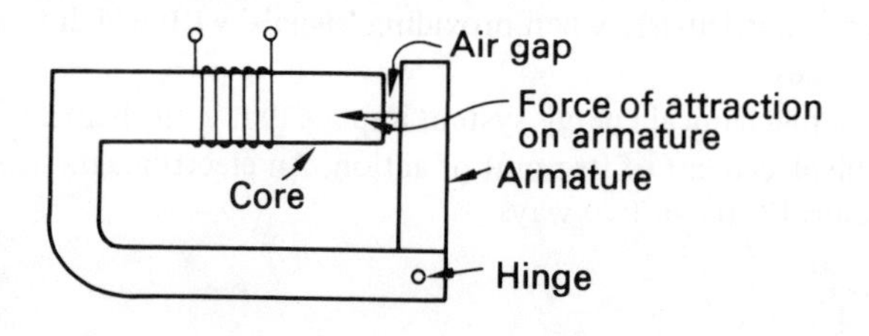

Fig. 11.2. Electromagnetic relay

When the coil is energised, a flux is set up in the relay core and the air gap. The surfaces adjacent to the air gap become magnetised and are attracted hence pulling the armature plate in the direction indicated. The relay will be analysed later in para. 11.8. Its function is to operate switches and it is used extensively in telephone exchanges.

The force of alignment can also be used to produce rotary motion, as in the reluctance motor shown in Fig. 11.3. In this case, the rotating piece, termed the rotor,

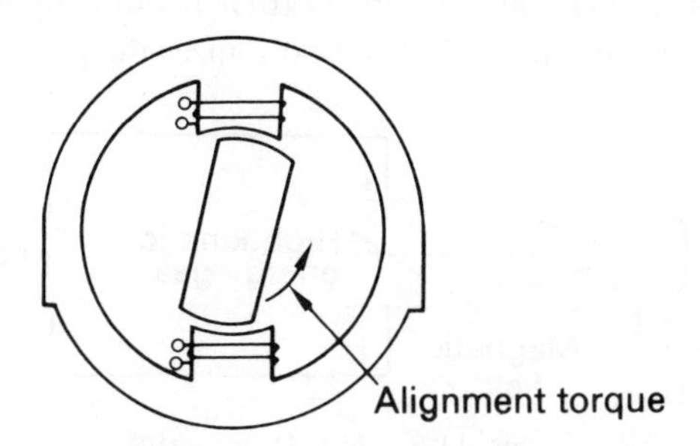

Fig. 11.3. Reluctance motor

experiences radial forces in opposite directions thereby cancelling each other out. The rotor also experiences a torque due to the magnetised rotor and pole surfaces attempting to align themselves. This alignment torque occurs in any rotating machine which does not have a cylindrical rotor, i.e. in a rotor which is salient as in the case shown in Fig. 11.3.

The force of interaction has already been discussed in chapters 4 and 5, being the principle of the force on a current-carrying conductor. This is another form of the principle of maximum stored magnetic energy—the relationship between these principles will be given extensive discussion later in para. 12.3, when the case of the doubly-excited rotating machine is analysed.

The force of interaction has the advantage of simplicity in its application. To calculate, or even to estimate, the energy stored in the magnetic fields of many arrangements is difficult if not impossible. Many of these cases, however, can be dealt with by the relationship $F = Bli$. This includes the case of a beam of electrons being deflected by a magnetic field in, say, a cathode-ray tube.

There are not many devices that involve the force of interaction to give rise to linear motion but this is a section of machine technology which is rapidly increasing due to the linear induction motor. A simple application is illustrated by the electromagnetic pump shown in Fig. 11.4.

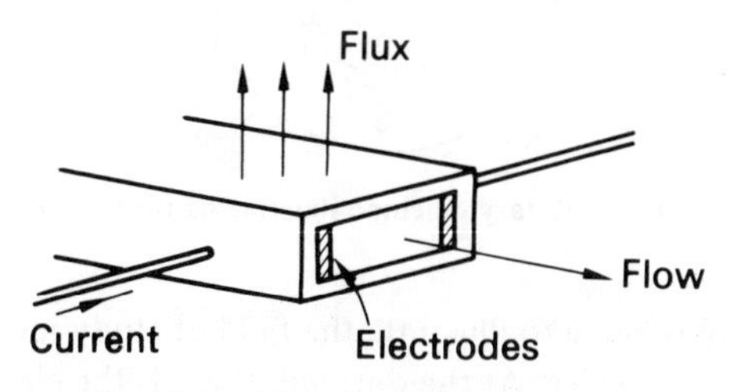

Fig. 11.4. Electromagnetic pump

This pump is used to move liquid metals. A current is passed through the metal at right angles to the intended direction of flow. A magnetic field is set up mutually at right angles to the other directions as shown in the diagram. The current-carrying conductor—the liquid metal—therefore experiences a force hence there is the required pumping action.

To show that a conductor need not be involved, consider the magneto-hydrodynamic generator (m.h.d. generator). This converts heat energy directly into electrical energy and is shown in principle in Fig. 11.5. The heat ionizes the gas ejected through the

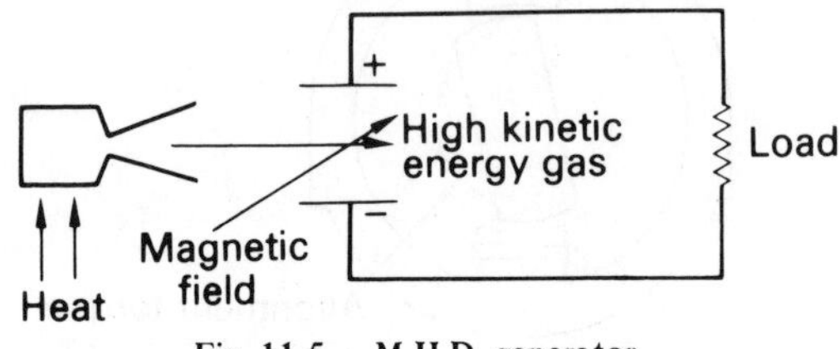

Fig. 11.5. M.H.D. generator

nozzle. This gas has high kinetic energy and, when it comes under the influence of the magnetic field, the high-energy electrons are deflected onto one of the plates making it negatively charged. The electrons flow round the circuit to the other plate where they combine with gas ions, which have been deflected to that plate by the magnetic field.

There are many applications involving the force of interaction to give rise to rotary motion. These include the synchronous and induction machines as well as the commutator machines all of which are given further consideration in chapter 12. All are variations on the same theme, each giving a different characteristic suitable for the various industrial drives required.

A simple machine illustrating the principle involved is shown in Fig. 11.6. By passing a current through the coil, it experiences a force on each of the coil sides resulting in a torque about the axis of rotation. A practical machine requires many conductors in order to develop a sufficient torque; depending on how the conductors are arranged, the various machines mentioned above and many others are created.

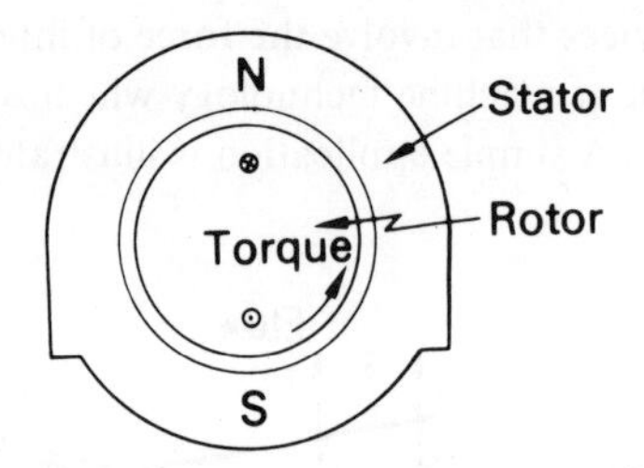

Fig. 11.6. Rotary machine illustrating force of interaction

These examples only serve to illustrate the field of study involved in machine theory. Each is a system-linking device. At the one end, there is the electrical system, at the other end is the mechanical (or other) system. In between there is a magnetic field forming a two-way link between them. If there is to be a flow of energy, all three will be involved simultaneously. Note that the reaction in the electrical system, apart from the flow of current, is the introduction of an e.m.f. into that system; the product of e.m.f. and current gives the rate of electrical energy conversion.

Before progressing to analyse the energy conversion process, consideration must be made as to how to approach the analysis. There are three methods of approach, each of which has to allow for the imperfections of a machine. No machine gives out as much energy as it takes in. The difference is termed the losses—the losses in the electrical system, the mechanical system and the magnetic system.

11.2 Methods of analysis of machine performance

The basic energy conversion process is one involving the magnetic coupling field and its action and reaction on the electrical and mechanical systems. There are three possible approaches to analysing the energy conversion:

The so-called classical approach. This dates back to the end of the nineteenth century when it was found that the operation of a machine could be predicted from a study of the machine losses. Such an approach is generally simple but it has two major disadvantages. First it deals almost exclusively with the machine operating under steady-state conditions, thus transient response conditions are virtually ignored, i.e. when it is accelerating and decelerating. Second the losses of each machine are different. It follows that each type of machine required to be separately analysed. As a consequence, much attention has been given in the past to the theory of d.c. machines, 3-ph induction machines and 3-ph synchronous machines to the exclusion of most others. Further, if any other machine were to be considered, its mode of operation would then require to be explained and the losses consequently analysed. Such a process is both tedious and unnecessarily repetitive, and hence is wasteful.

The generalised-machine approach. This approach depends on a full analysis of the coupling field as observed from the terminals of the machine windings. The losses are recognised as necessary digressions from the main line of the analysis. The coupling field is described in terms of mutual inductance. Such an approach therefore considers any machine to be merely an arrangement of coils which are magnetically linked. No attention is paid as to the form of machine construction initially and it exists as a box with terminals from which measurements may be made to determine the performance of the machine. The measured quantities are voltage, current, power, frequency, torque and rotational speed from which may be derived the resistance and inductance values for the coils. In the light of the derived values, it is possible to analyse the performance both under steady-state and transient conditions.

It should not be taken from this description of the generalised theory that the design arrangement of a machine is considered to be unimportant. However, if too much emphasis is given to the constructional detail, the principle of the energy transfer may well become ignored.

The generalised approach has been universally appreciated only since the late 1950's. Since the approach has so much to offer, one may well enquire as to why it has taken so long to attain recognition. The difficulty, that precluded its earlier use, is the complexity of the mathematics involved. The initial complexity of the approach remains but once the appropriate mathematical manipulations have been carried out, they need not be repeated. The results of this preliminary step can then be separately modified to analyse not only most types of machines but also the different modes of operation. Thus an analysis is being made of the electromechanical machine in general. The individual machine is considered by making simplifying assumptions at the end of the analysis but the effect of the theory is to concentrate attention upon the properties common to all machines.

The unified approach. This approach recognises the advantages of the generalised machine approach in that the fundamental principle of machine operation is common to all machines. However, it also recognises that the mathematics may well prevent this principle from becoming readily apparent to an engineer approaching the subject for the first time. The unified approach therefore seeks to explain the principle of operation more simply but it does so at the expense of permitting a full analysis of individual machines becoming directly available. Nevertheless, once the principle has been grasped, the engineer may then proceed to either of the previous approaches with a clear understanding of the objectives in machine performance analysis.

Since this book serves as an introductory text, it is therefore to this last approach that it must look. The unified approach concerns itself not with the inductive element of the coupling field but with the field energy, i.e. it observes the machine from the air gap and not from the terminals. The procedure involves two steps:

1. Obtaining an expression for the energy stored in the coupling field in terms of the electrical variables and the configuration of the mechanical parts.

2. Investigating the manner in which the electrical and field energy terms are affected by changes in the configuration of the mechanical parts.

Before investigating this procedure, attention requires to be drawn to some relevent magnetic circuit relations.

11.3 Magnetic field energy

It was shown in para. 5.7 (p. 107) that the energy in a magnetic field is given by

$$W_f = \tfrac{1}{2}Li^2$$

However, there are a number of ways in which the inductance can be expressed.

$$L = \frac{N\phi}{i} = \frac{\psi}{i} = \frac{N^2}{S} = \frac{N^2 \mu_0 \mu_r A}{l}$$

These expressions can be substituted in the energy relation to give

$$W_f = \tfrac{1}{2}i\psi = \tfrac{1}{2}F\phi = \tfrac{1}{2}S\phi^2$$

It should be noted that all of these expressions for the energy depend on the flux and the m.m.f. being directly proportional, i.e. the inductance is constant. This is generally limited to the case of air, which is the most important one in electrical machines.

Sometimes the energy density can be of greater importance. In the proof of the relationship of hysteresis loss in a magnetic core, it was shown that the energy stored was proportional to the shaded area due to the B/H curve. This is shown in Fig. 11.7.

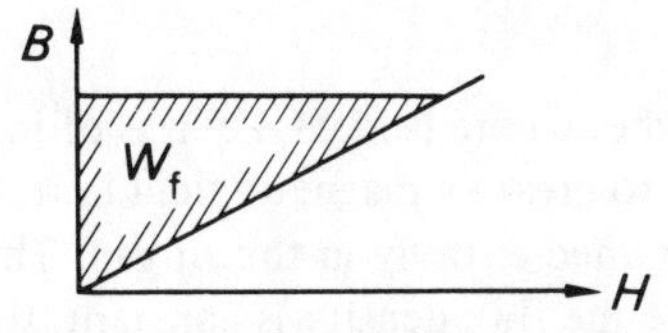

Fig. 11.7. Stored-energy diagram

In the case of an air gap, the B/H characteristic is straight and the energy stored (see p. 116) is given by

$$W_f = \tfrac{1}{2}BH \times \text{Volume of air gap}$$

If the air gap has a cross-sectional area A and length l,

$$W_f = \tfrac{1}{2}BH \times Al$$
$$= \tfrac{1}{2}F\phi$$

The stored energy density is thus given by

$$w_f = \tfrac{1}{2}BH$$
$$= \frac{\mu_0 H^2}{2}$$
$$= \frac{B^2}{2\mu_0} \qquad (11.1)$$

It will be noted that all the expressions have been shown for instantaneous values and not for steady-state conditions. However, this does not always draw attention to some quantities which can be variable. In particular, the flux density B, the magnetic field strength H and the inductance L are all variable quantities and so instantaneous quantities are implied in these cases.

11.4 Simple analysis of force of alignment

Consider the force of alignment between two poles of a magnetic circuit as shown in Fig. 11.8.

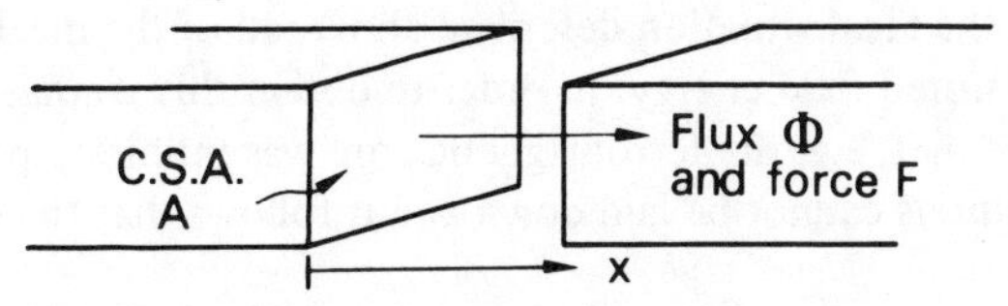

Fig. 11.8. Force of alignment between two poles

Let there be a flux Φ in the air gap and let there be no fringing of the flux. The uniform flux density in the air gap is then given by

$$B = \frac{\Phi}{A}$$

Suppose that the poles can be separated by a small distance *dx* without there being a change in the flux and the flux density. This can be effectively realised in practice. Because there is a mechanical force experienced by the poles, the mechanical work done is

$$\mathrm{d}W_{\mathrm{M}} = F.\mathrm{d}x$$

Assume now that the magnetic core is ideal, i.e. it is of infinite permeability and therefore requires no m.m.f. to create a magnetic field in it. The stored magnetic energy will therefore be contained entirely in the air gap. The air gap has been increased by a volume $A . dx$, yet, since the flux density is constant, the energy density must remain unchanged. There is therefore an increase in the stored energy,

$$\mathrm{d}W_{\mathrm{f}} = \frac{B^2}{2\mu_0} \times A.\mathrm{d}x$$

Since the system is ideal and the motion has taken place slowly from one point of rest to another, this energy must be due to the input of mechanical energy, i.e.

$$\mathrm{d}W_{\mathrm{M}} = \mathrm{d}W_{\mathrm{f}}$$

$$F.\mathrm{d}x = \frac{B^2 A}{2\mu_0} \cdot \mathrm{d}x$$

$$F = \frac{B^2 A}{2\mu_0} \tag{11.2}$$

A more important relationship, however, may be derived from the above argument, i.e.

$$\mathrm{d}W_{\mathrm{M}} = F.\mathrm{d}x$$

$$= \mathrm{d}W_{\mathrm{f}}$$

$$F = \frac{\mathrm{d}W_{\mathrm{f}}}{\mathrm{d}x} \tag{11.3}$$

It will be shown later that this relation only holds true provided the flux (or flux linkage) remains constant. However the principle demonstrated is more important, i.e. the force is given by the rate of change of stored field energy with distortion of the arrangement of the ferromagnetic poles.

In order to develop this principle, more attention must now be given to the energy transfer process. In the ideal situation described above, all of the mechanical energy was converted into stored field energy. In order to obtain this arrangement, several conditions had to be met, e.g. the ferromagnetic core was infinitely permeable. In practice, such conditions cannot be laid down and it follows that the energy balance must be modified.

11.5 Energy balance

Consider the operation of a simple attracted-armature relay such as that shown in Fig. 11.9. Assume that initially the switch is open and that there is no stored field energy. These conditions are quite normal and in no way unreasonable.

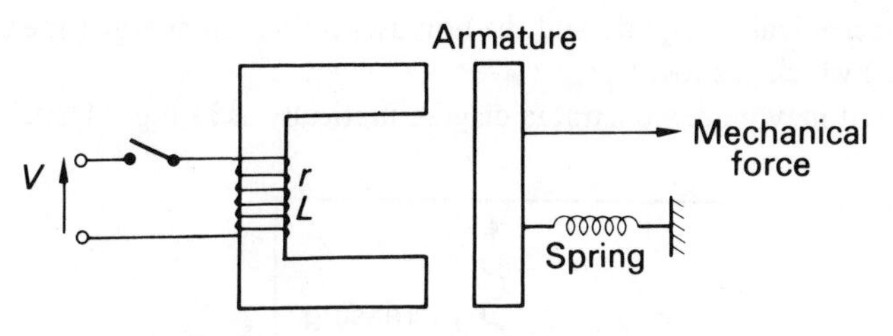

Fig. 11.9. Attracted-armature relay

After the switch is closed, the sequence of events falls into four distinct groups.

1. After the switch is closed, the current rises exponentially in the manner described in chapter 5. If L_1 is the inductance of the coil for the initial position of the armature, the initial rate of rise of current is given by V/L_1. The electrical energy from the source is partly dissipated in i^2R loss in the magnetising coil whilst the remainder is converted into stored energy in the magnetic field. During this period, the armature experiences an attractive force but the various mechanical restraints prevent it from moving. It is not unusual to find that the steady-state value of coil current has almost been reached before the armature starts to move.

2. At some appropriate value of current, the armature begins to move. This occurs when the force of attraction f_E balances the mechanical force f_M. During the motion of the armature, there are many changes of energy in the system. On the mechanical side, energy is required to stretch the spring, to drive the external load and to supply the kinetic energy required by the moving parts. At the same time, the air gaps are being reduced with consequent increase in the inductance of the arrangement. This causes a reaction in the electrical system in the form of an induced e.m.f.; this e.m.f. tends to reduce the coil current and also permits the conversion of electrical energy, i.e. it is the reaction to the action.

3. The armature cannot continue to move indefinitely, but, instead, it hits an end stop. This causes the kinetic energy of the system to be dissipated as noise, deformation of the poles and vibration.

4. Now that there is no further motion of the system, the inductance now becomes constant at a new higher value L_2. The current increases exponentially to a value V/r. It should be noted that the rate of rise is less than the initial rate of rise since the inductance is now much greater.

The energy flow processes are therefore many and yet they are typical of many machines–in rotating machines, there is no sudden stop but otherwise the processes are similar. To be able to handle so many at one time, it is necessary to set up an energy balance involving the convention which will remain during all the subsequent machine analysis.

Consider first the external systems, i.e. the electrical and mechanical systems. Because the conversion process can take place in either direction, let both electrical and mechanical energies be input energies to the system. It follows that an output energy is mathematically a negative input energy. The electrical and mechanical energies are W_E and W_M.

Second there is the internal system comprising the stored magnetic field energy

W_f, the stored mechanical energy W_s and the non-useful thermal energy (due to i^2R loss, friction, etc.) which is a loss W_l.

The arrangement may be demonstrated diagrammatically as in Fig. 11.10.

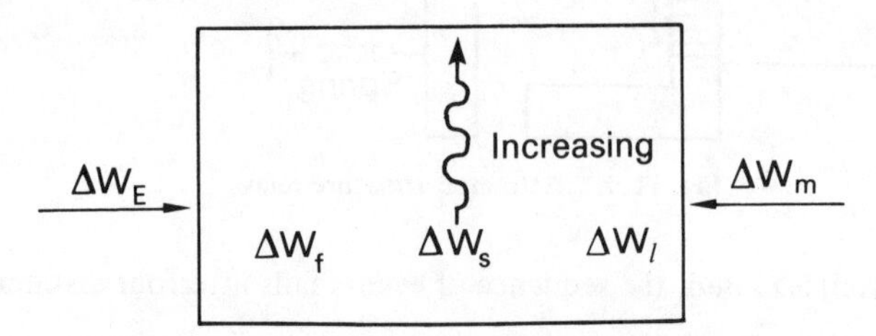

Fig. 11.10. Energy balance diagram

Between any two states of the system, the energy balance may be expressed as

$$\Delta W_{\mathrm{E}} + \Delta W_{\mathrm{M}} = \Delta W_{\mathrm{f}} + \Delta W_{\mathrm{s}} + \Delta W_l \tag{11.4}$$

Alternatively the energy rates of flow may be expressed as

$$p_{\mathrm{E}} + p_{\mathrm{M}} = \frac{\mathrm{d}W_{\mathrm{f}}}{\mathrm{d}t} + \frac{\mathrm{d}W_{\mathrm{s}}}{\mathrm{d}t} + \frac{\mathrm{d}W_l}{\mathrm{d}t} \tag{11.5}$$

In this chapter, the basic machine shall be idealised to a limited extent by separating some of the losses as indicated in Fig. 11.11.

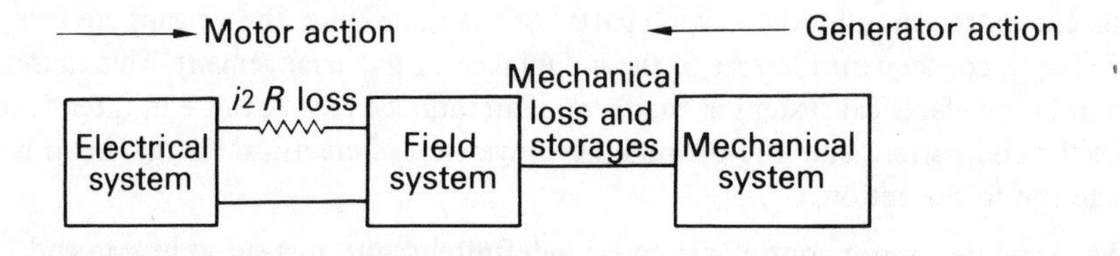

Fig. 11.11. Idealised energy balance diagram

It follows that the ideal and essential energy balance may be expressed as

$$\mathrm{d}W_{\mathrm{E}} + \mathrm{d}W_{\mathrm{M}} = \mathrm{d}W_{\mathrm{f}} \tag{11.4a}$$

and hence the power balance may be expressed as

$$p_{\mathrm{E}} + p_{\mathrm{M}} = \frac{\mathrm{d}W_{\mathrm{f}}}{\mathrm{d}t} \tag{11.5a}$$

Finally there are the actions and reactions to consider. These are indicated in Fig. 11.12.

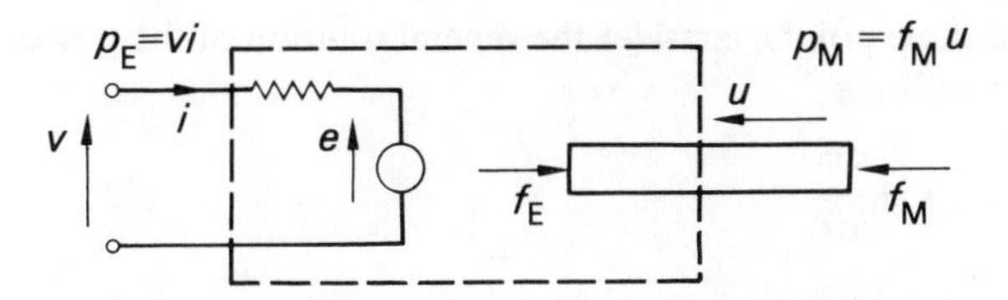

Fig. 11.12. Actions and reactions in practical conversion system

On the electrical side, the applied voltage is v and this is opposed by the reaction in the form of the back e.m.f. e. The electrical power is

$$p_E = vi$$

whilst the rate of electrical energy conversion is ei. These two terms are only equal when the i^2R loss is either neglected or considered external to the conversion process as in the idealised system of Fig. 11.11.

On the mechanical side, the mechanical input force f_M acts towards the conversion system and moves in a similar direction, say with a velocity u. The reaction to this is the magnetically developed force f_E. These two forces are equal and opposite only when the mechanical system is at rest or moves with uniform velocity. The difference would otherwise give rise to acceleration and hence there could not be steady-state conditions. There is also a slight difference between the forces when the mechanical system is moving due to friction. However, it has been noted that the effect of friction is to be neglected for the time being.

11.6 Division of converted energy and power

There are several methods of demonstrating the division of energy and power of which the following three cases are sufficient. Consider first of all the attracted-armature-relay shown in Fig. 11.13. As before, it is assumed that the relay is to be energised with the result that the relay closes.

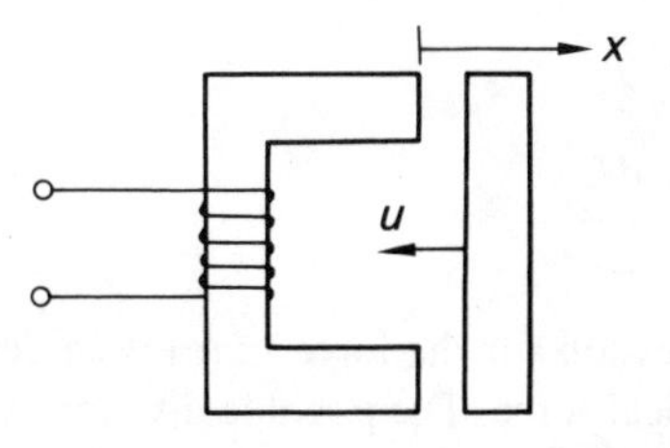

Fig. 11.13. Attracted-armature relay operation

The movement of the armature is in the direction $-x$. Because the lengths of the air gaps are decreasing, the inductance of the system increases; it can also be noted that the velocity is

$$u = -\frac{dx}{dt}$$

Having noted these points, consider the general solution of the power balance. At any instant,

$$e = \frac{d\psi}{dt}$$

$$= \frac{d}{dt}(Li)$$

$$= L \cdot \frac{di}{dt} + i \cdot \frac{dL}{dt}$$

$$p_E = ei$$

$$= Li \cdot \frac{di}{dt} + i^2 \cdot \frac{dL}{dt}$$

also

$$W_f = \tfrac{1}{2} Li^2$$

$$\frac{dW_f}{dt} = Li \cdot \frac{di}{dt} + \tfrac{1}{2} i^2 \cdot \frac{dL}{dt}$$

and

$$p_E + p_M = \frac{dW_f}{dt}$$

hence

$$p_M = -\tfrac{1}{2} i^2 \cdot \frac{dL}{dt}$$

$$= -\tfrac{1}{2} i^2 \cdot \frac{dL}{dx} \cdot \frac{dx}{dt}$$

$$= -\tfrac{1}{2} i^2 \cdot \frac{dL}{dx} \cdot (-u)$$

$$= \tfrac{1}{2} i^2 \cdot \frac{dL}{dx} \cdot u$$

$$= -f_E u$$

$$f_E = -\tfrac{1}{2} i^2 \cdot \frac{dL}{dx} \qquad (11.6)$$

Relation (11.6) is an expression for the force of reaction developed by the magnetic field and not for the mechanical force. The possible difference between these forces has been noted above but is again emphasised in view of the importance in this and other subsequent relations.

It will be noted that the expression for the rate of mechanical energy conversion p_M is negative, i.e. the machine acts as a motor and p_M is an output. This requires that the mechanical force acts away from the machine and is negative. Relation (11.6) does not readily indicate that the force has a negative value but it should be remembered that the inductance decreases with increase of the air gaps.

The output mechanical power accounts for both any mechanical storages as well as the useful power.

It will also be noted that the rate of mechanical energy conversion is equal to one part of the expression for the rate of field energy storage—that partial differential due to constant current. It follows that the expression for the force of reaction can also be expressed as

$$f_E = \left.\frac{dW_f}{dx}\right|_{i \text{ constant}} \tag{11.7}$$

Thus, again the argument has returned to relating the force to the rate of change of stored energy with distortion of the system. Since most of the stored energy is retained in the air gap and all of the energy is stored within the magnetic circuit, it can be seen that the force is also due to change in the reluctance of the magnetic circuit hence the force is alternatively known as the reluctance force.

There are two particular cases of the division of energy that merit special consideration. In terms of flux linkage, the power balance becomes

$$p_E + p_M = \frac{dW_f}{dt}$$

$$i \cdot \frac{d\psi}{dt} - f_E u = \frac{dW_f}{dt}$$

but

$$\frac{dW_f}{dt} = \frac{d}{dt}(\tfrac{1}{2}\psi i)$$

$$= \tfrac{1}{2}\psi \cdot \frac{di}{dt} + \tfrac{1}{2}i \cdot \frac{d\psi}{dt}$$

hence

$$i \cdot \frac{d\psi}{dt} - f_E u = \tfrac{1}{2}\psi \cdot \frac{di}{dt} + \tfrac{1}{2}i \cdot \frac{d\psi}{dt}$$

The conditions of interest are:

Constant current:

$$I \cdot \frac{d\psi}{dt} - f_E u = \tfrac{1}{2}I \cdot \frac{d\psi}{dt}$$

$$f_E u = \tfrac{1}{2}I \cdot \frac{d\psi}{dt}$$

$$f_E = -\tfrac{1}{2}I \cdot \frac{d\psi}{dx} \tag{11.8}$$

$$= -\tfrac{1}{2}I \cdot \frac{d}{dx}(LI)$$

$$= -\tfrac{1}{2}I^2 \cdot \frac{dL}{dx}$$

which is as before in relation (11.6). However, on this occasion

$$p_M = -\left.\frac{dW_f}{dt}\right|_{i \text{ constant}} \tag{11.9}$$

The input of electrical energy is equally divided, half being converted to mechanical energy and half to the energy stored in the magnetic field. This division of the energy is sometimes termed the 50–50 rule. The converse of the foregoing relates to the case when work is done on the circuits by changing their form against the action of electromagnetic forces or by pulling them apart. In the latter case, the energy returned to the source is twice the amount of mechanical work that is done, the second half being restored to the source by a reduction in the stored magnetic energy in the field.

This 50-50 rule is not applicable when ferromagnetic materials exist in the neighbourhood of the currents. This is particularly the case if magnetic saturation in the core is liable to take place. For the purposes of the unified approach, the core has been assumed ideal thus only the energy in the air gaps need be considered. For the magnetic circuit to be ideal, the ferromagnetic core of it should have an infinitely high permeability and thus have zero reluctance.

Constant flux linkage:

$$-f_E u = \tfrac{1}{2}\Psi \cdot \frac{\mathrm{d}i}{\mathrm{d}t}$$

$$f_E = \tfrac{1}{2}\Psi \cdot \frac{\mathrm{d}i}{\mathrm{d}x} \qquad (11.10)$$

In this case, there is no conversion of electrical energy because the constant flux linkage cannot induce an e.m.f. reaction in the electrical system. Instead the field energy is converted into mechanical energy. Thus

$$p_M = \frac{\mathrm{d}W_f}{\mathrm{d}t}$$

$$\mathrm{d}W_M = \mathrm{d}W_f$$

but

$$\mathrm{d}W_M = f_E \cdot \mathrm{d}x$$

$$f_E = \left.\frac{\mathrm{d}W_f}{\mathrm{d}x}\right|_{\Psi\,\text{constant}} \qquad (11.11)$$

At this point the polarity of the mechanical force requires some further discussion. The length of the air gap(s) is taken as x and an increase in the gap length(s) is taken to be positive. It follows that when the armature moves to close the gap(s), the velocity has a negative value.

In the constant-current case, the electrical energy is converted to supply the increase in the stored field energy and to supply the mechanical energy input, which consequently has a negative value. The force acts in a negative direction, i.e. to close the gap. Alternatively the mechanical force attains a negative value because the inductance increases with decrease in gap length(s) and hence $\mathrm{d}L/\mathrm{d}x$ is negative.

The mechanical force in the constant-flux linkage case has a negative value since, in relation (11.10), the current decreases with decrease in gap length(s) and hence $\mathrm{d}i/\mathrm{d}x$ is positive. This gives a negative value to the mechanical energy and this is equal to the loss in stored energy in the magnetic field, which, being a loss, has a negative value. There is therefore an exchange of energy as noted above.

Returning to the general power balance relation, it can be seen that the 50-50 rule requires to be defined in that the energy transferred from the mechanical system during any period of time is equal to that part of the change of field energy stored due to change in the magnetic circuit, the current being considered constant. This may be expressed as

$$\mathrm{d}W_{\mathrm{M}} = \mathrm{d}W_{\mathrm{f}}\Big|_{\mathrm{I}} \tag{11.12}$$

The constant-flux linkage is a special case of the same principle.

The expressions involving the rate of change of inductance or part of the field energy are not necessarily convenient. Sometimes an expression involving the reluctance or else the permeance is useful. Consider again the general expression for the force in terms of the rate of change of inductance with gap length.

$$f_{\mathrm{E}} = -\tfrac{1}{2}i^2 \cdot \frac{\mathrm{d}L}{\mathrm{d}x}$$

$$= -\tfrac{1}{2}i^2 \cdot \frac{\mathrm{d}L}{\mathrm{d}S} \cdot \frac{\mathrm{d}S}{\mathrm{d}x}$$

but

$$L = \frac{N^2}{S}$$

$$\frac{\mathrm{d}L}{\mathrm{d}S} = -\frac{N^2}{S^2}$$

$$f_{\mathrm{E}} = \frac{i^2 N^2}{2S^2} \cdot \frac{\mathrm{d}S}{\mathrm{d}x}$$

$$= \tfrac{1}{2}\phi^2 \cdot \frac{\mathrm{d}S}{\mathrm{d}x} \tag{11.13}$$

It can similarly be shown that

$$f_{\mathrm{E}} = -\tfrac{1}{2}F^2 \cdot \frac{\mathrm{d}\Lambda}{\mathrm{d}x} \tag{11.14}$$

Again it will be noted that the reluctance decreases with decrease in the gap length(s) and thus the rate of change is positive and the force is negative, i.e. the device again acts as a motor.

This prolonged derivation of the general expressions for the forces due to change in the reluctance of the magnetic circuit and the consequent force of alignment is necessary to give understanding of the applications which follow. Its significance becomes fully apparent if it is again read after the rest of the chapter has been considered. However, it is now possible to proceed to some simple arrangements which illustrate the principles involved.

11.7 Force of alignment between parallel magnetised surfaces

To find the force of alignment between the core members of the arrangement shown in Fig. 11.14, consider again relation (11.13).

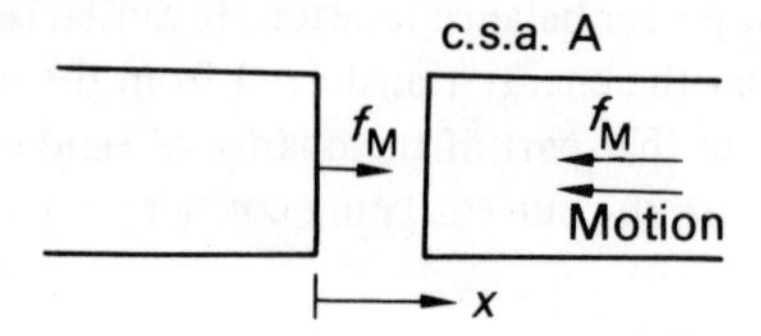

Fig. 11.14. Force of alignment between parallel magnetised surfaces

$$f_E = \tfrac{1}{2}\phi^2 . \frac{dS}{dx}$$

Since the remainder of the core is assumed ideal, the total reluctance appears at the air gap, hence

$$S = \frac{l_g}{\mu_0 A}$$

It will be noted that $\mu_r = 1$ for air and that l_g is the particular value of x for the arrangement considered.

$$\frac{dS}{dx} = \frac{1}{\mu_0 A}$$

$$f_E = \tfrac{1}{2}\phi^2 . \frac{1}{\mu_0 A}$$

$$= \tfrac{1}{2}B^2 A^2 . \frac{1}{\mu_0 A}$$

$$= \frac{B^2 A}{2\mu_0} \tag{11.15}$$

This is almost the same as relation (11.2) but the force has now been interpreted in the effect it will have, i.e. the positive polarity shows that the poles are attracted towards one another.

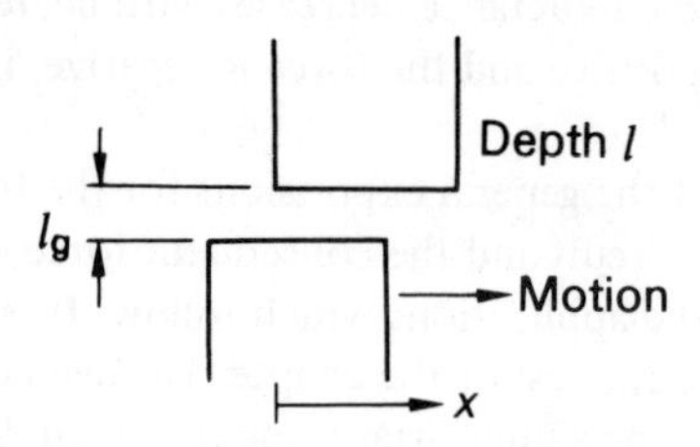

Fig. 11.15. Lateral force between magnetised surfaces

The development of the theory has now completed a full cycle. Now consider a different arrangement and analyse it using the same principles. If the poles had been laterally displaced, a force of alignment would be experienced trying to align the poles. This arrangement is shown in Fig. 11.15.

Ignoring the effect of leakage flux, let the area covered by the gap be xl, then

$$S = \frac{l_g}{\mu_0 xl}$$

$$\frac{dS}{dx} = -\frac{l_g}{\mu_0 x^2 l}$$

In this instance, x is measured such that $u = dx/dt$. There is therefore a negative sign in the expression for the force,

$$f_E = -\tfrac{1}{2}\phi^2 . \frac{dS}{dx}$$

$$= -\tfrac{1}{2}B^2 x^2 l^2 . \frac{-l_g}{\mu_0 x^2 l}$$

$$= \frac{B^2 l_g l}{2\mu_0} \qquad (11.16)$$

The polarity of this expression indicates that the force tries to align the poles by increasing the cross-sectional area of the air gap thereby decreasing the reluctance. This expression should be used with great care since it is not a continuous function; for instance it is not immediately obvious that, if the two poles are aligned, then the force drops to zero and reverses in its direction of action thereafter.

Finally it should be noted that relations (11.15) and (11.16) can be applied to the action at any air gap. This is because these expressions are related to the field energy at the gap and this was defined in terms of reluctance and flux. The energy stored in the core of the magnetic circuit, when the core is not ideal, does not affect the validity of the relations.

Example 11.1 An electromagnet is made using a horseshoe core as shown in the diagram below. The core has an effective length of 600 mm and a cross-sectional area of 500 mm^2. A rectangular block of steel is held by the electromagnets force of alignment and a force of 20 N is required to free it. The magnetic circuit through the block

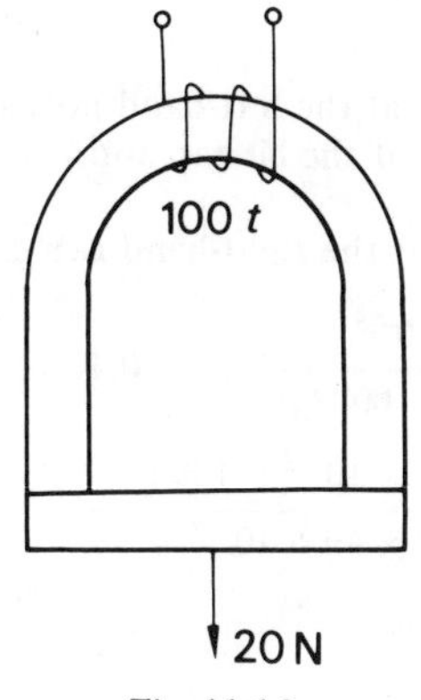

Fig. 11.16

is 200 mm long and the effective cross-sectional area is again 500 mm^2. The relative permeability of both core and block is 700. If the magnet is energised by a coil of 100 turns, estimate the coil current.

There are two air gaps in the magnetic circuit, hence the force to part the circuit members is double that at any one gap.

$$f_M = 2 \cdot \frac{B^2 A}{2\mu_0} = \frac{B^2 A}{\mu_0} = 20 \text{ N}$$

$$B = \left(\frac{20 \times 4\pi \times 10^{-7}}{500 \times 10^{-6}}\right)^{\frac{1}{2}} = 0{\cdot}222 \text{ T}$$

$$H = \frac{B}{\mu_0 \mu_r} = \frac{0{\cdot}222}{4\pi \times 10^{-7} \times 700} = 250 \text{ At/m}$$

$$F = Hl = 250 \times (600 + 200) \times 10^{-3} = 200 \text{ At}$$

$$= NI = 100I$$

$$I = \frac{200}{100} = \underline{2{\cdot}0 \text{ A}}$$

Example 11.2 The poles of an electromagnet are shown relative to a steel bar. If the effects of leakage flux may be neglected, estimate the forces of alignment that act laterally on the steel bar due to each pole.

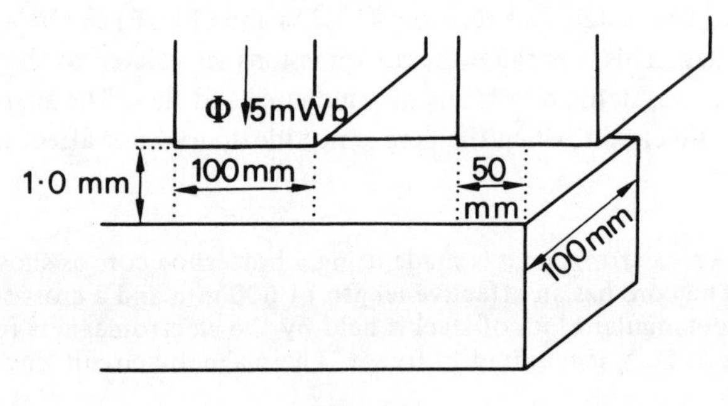

Fig. 11.17

There is no force of alignment at the left-hand pole since motion in either direction would not change the reluctance of the air gap and hence there can be no change in the field energy stored in that gap.

There is a force of alignment at the right-hand gap given as follows,

$$B = \frac{\Phi}{A} = \frac{5 \times 10^{-3}}{50 \times 10^{-3} \times 100 \times 10^{-3}} = 1{\cdot}0 \text{ T}$$

$$f_E = \frac{B^2 l_g l}{2\mu_0} = \frac{1{\cdot}0^2 \times 1 \times 10^{-3} \times 100 \times 10^{-3}}{2 \times 4\pi \times 10^{-7}}$$

$$= \underline{39{\cdot}7 \text{ N}}$$

The positive polarity indicates that it is a force of attraction.

11.8 Solenoid relay and plunger

The solenoid relay and the plunger are two practical applications of the arrangement whereby two magnetised surfaces are attracted together. The solenoid relay was introduced in para. 11.1 and shown in Fig. 11.2. It constructed so that the pole or armature is generally pivotted. Due to the rotational movement about the pivot, the gap length is not uniform over the cross section of the gap—due to the short travel, the distortion thus caused is sufficiently small to be negligible. The pivot hinge is made in a manner that minimises the discontinuity in the magnetic circuit core. In this way, only the air gap at the pole need be considered and the energy stored in the core may be neglected.

The plunger armature has a sliding fit in a bushing in the case of the plunger, which is again used for relay purposes, usually in cases where greater lengths of travel are required. The air gap at the hole through which the plunger armature moves may be neglected for the purposes of analysis and only the air gap between the poles need be considered. The plunger relay is shown in Fig. 11.18.

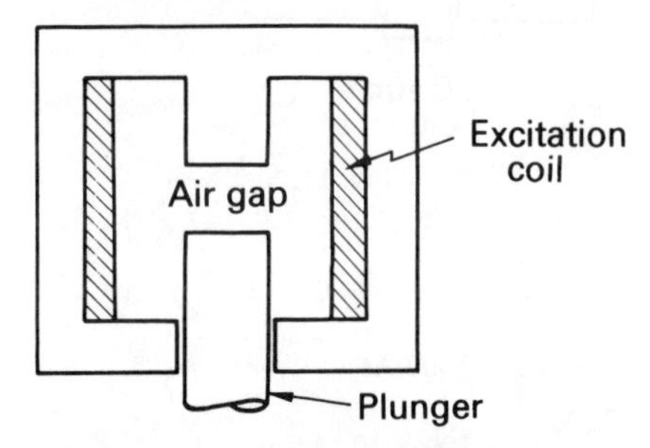

Fig. 11.18. Simple plunger relay

The effects of the current and applied voltage on the energy transfer processes in such relays can best be described with the aid of a flux/m.m.f. characteristic. Considering the air gap(s) only, the appropriate characteristics are linear ones and are shown in Fig. 11.19.

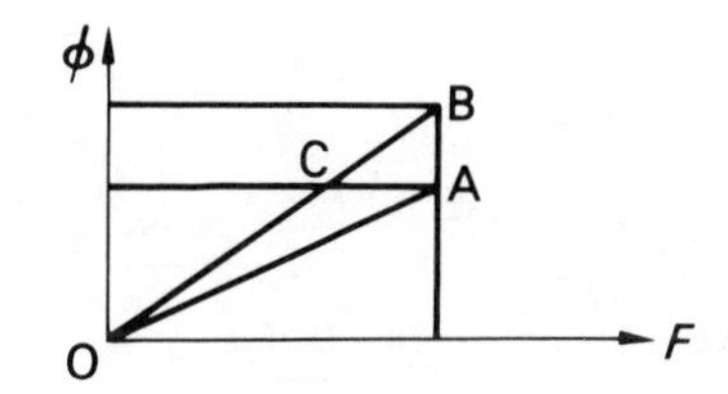

Fig. 11.19. Flux/m.m.f. characteristic for idealised relay operation

If the current remains constant throughout the movement of the armature, the m.m.f. also remains constant but the flux rises from ϕ_A to ϕ_B. The energy supplied by the source is $F(\phi_B - \phi_A)$. Half the energy is absorbed into the field and the other half appears as mechanical energy including the mechanical storages. This is represented by the area ABO and the intermediate conditions are shown by the line AB.

If the applied voltage remains constant, it can be assumed that the flux linkages will also remain constant provided that the energising circuit has no resistance. The changing conditions are now represented by the line AC. The relay moves quickly and completes its movement before the flux can change in value. Thus the m.m.f. will have the value F_C when the plunger strikes the end stop, the flux increasing thereafter. The mechanical energy available is represented by the area ACO.

Example 11.3 A plunger relay is shown in the diagram below. When it is energised, the current remains constant at 2·0 A and the plunger moves such that the air gap is reduced in length from 20 mm to 10 mm. If losses may be neglected, calculate the average force experienced by the plunger and the energy taken from the source.

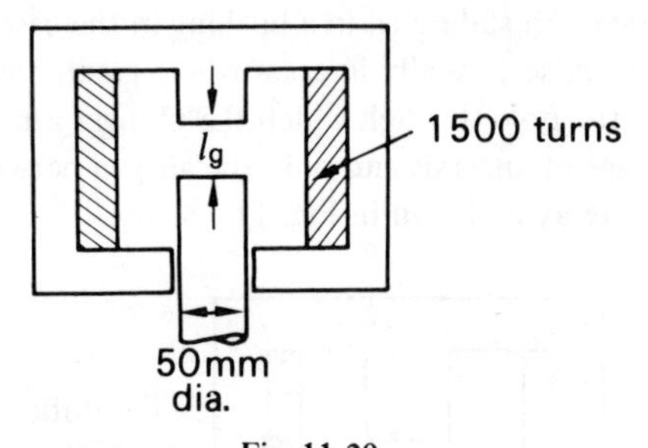

Fig. 11.20

For l_g = 10 mm:

$$F = NI = 1\,500 \times 2 = 3000 \text{ At}$$

$$H = \frac{F}{l} = \frac{3000}{10 \times 10^{-3}} = 300 \times 10^3 \text{ At/m}$$

$$B = \mu_0 H = 4\pi \times 10^{-7} \times 300 \times 10^3 = 0{\cdot}377 \text{ T}$$

$$\Phi = BA = 0{\cdot}377 \times 252 \times \pi \times 10^{-6} = 0{\cdot}74 \times 10^{-3} \text{ Wb}$$

$$L = \frac{N\Phi}{I} = \frac{1\,500 \times 0{\cdot}74 \times 10^{-3}}{2} = 0{\cdot}56 \text{ H}$$

For l_g = 20 mm:

$$L = 0{\cdot}56 \times \tfrac{1}{2} = 0{\cdot}28 \text{ H}$$

$$f_E = -\tfrac{1}{2}I^2 . \frac{\Delta L}{\Delta X} = -\tfrac{1}{2} \times 2^2 \times \frac{0{\cdot}28}{-0{\cdot}01} = \underline{56 \text{ N}}$$

$$\Delta W_f = \tfrac{1}{2}I^2 . \Delta L = \tfrac{1}{2} \times 2^2 \times 0{\cdot}28 = \underline{0{\cdot}56 \text{ J}}$$

$$= -\Delta W_M$$

$$\Delta W_E = -\Delta W_M + \Delta W_f = 0{\cdot}56 + 0{\cdot}56 = \underline{1{\cdot}12 \text{ J}}$$

In this problem, the inductance could have been calculated from $L = N^2/S$. In many problems, however, the intermediate information derived, such as the flux density, is required hence the method of solution demonstrated.

Some difficulty is introduced when the exciting coil has resistance.

The case of constant voltage cannot easily be solved since constant flux linkage requires that the current should change in order that no e.m.f. be induced. This change in current would not produce the required volt drop because of the circuit resistance.

Example 11.4 A solenoid relay is operated from a 110-V, d.c. supply and the 5 000-turn coil resistance is 5·5 kΩ. The core diameter of the relay is 20 mm and the gap length is 1·5 mm, the armature being stationary. The gap faces may be taken as parallel and the permeability of the ferromagnetic parts as very high. Estimate:

(*a*) The gap flux density.
(*b*) The coil inductance.
(*c*) The pull on the armature.

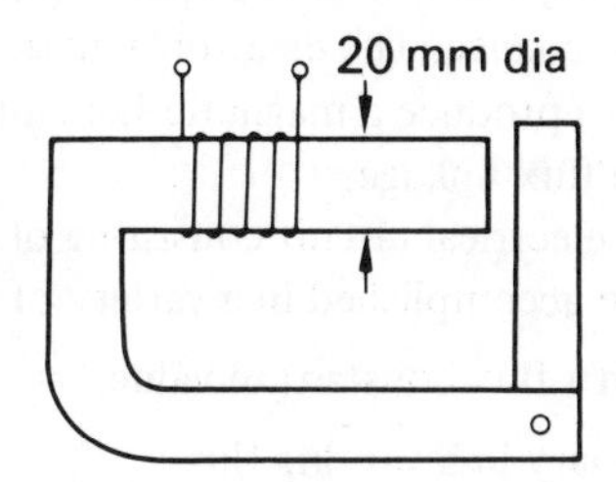

Fig. 11.21

$$I = \frac{V}{R} = \frac{110}{5{\cdot}5 \times 10^3} = 20 \times 10^{-3}\ \text{A}$$

$$F = IN = 20 \times 10^{-3} \times 5000 = \underline{100\ \text{At}}$$

$$H = \frac{F}{l_g} = \frac{100}{1{\cdot}5 \times 10^{-3}} = 0{\cdot}67 \times 10^5\ \text{At/m}$$

$$B = \mu_0 H = 4\pi \times 10^{-7} \times 0{\cdot}67 \times 10^5 = 84 \times 10^{-3}\ \text{T}$$

$$= \underline{84\ \text{mT}}$$

$$\Phi = BA = 84 \times 10^{-3} \times 10^2 \times \pi \times 10^{-6} = 26{\cdot}3 \times 10^{-6}\ \text{Wb}$$

$$L = \frac{\Phi N}{I} = \frac{26{\cdot}3 \times 10^{-6} \times 5000}{20 \times 10^{-3}} = \underline{6{\cdot}56\ \text{H}}$$

The inductance L is inversely proportional to the gap length l_g hence in general

$$L = 6{\cdot}56 \times \frac{1{\cdot}5}{x}$$

$$\frac{dL}{dx} = -\frac{9{\cdot}82}{x^2} = -\frac{9{\cdot}82}{1{\cdot}5^2} = -4{\cdot}37\ \text{H/mm (negative sign as indicated in para. 11.6).}$$

$$= -4370\ \text{H/m}$$

$$f_E = -\tfrac{1}{2}i^2 \cdot \frac{dL}{dx} = -\tfrac{1}{2} \times 20^2 \times 10^{-6} \times 4370 = \underline{0{\cdot}88\ \text{N}}$$

Before proceding to consider the motional condition of the relay, some consideration should be given to the induced e.m.f. Consequent to this, the circuit resistance can then enter into the energy and power balances.

11.9 Induced e.m.f.

The e.m.f. that opposes the flow of current is due to the interlinking of electrical and magnetic circuits. In its most general form:

$$e = \frac{d\psi}{dt}$$

Thus if the electrical circuit were closed on a number of flux linkages due to some externally produced changing magnetic field, the e.m.f. induced in the circuit would produce a current thus generating a self-magnetic field superimposed upon the external field and tending to reduce the change. The electromagnetic method of producing an e.m.f. in a circuit is therefore to produce a magnetic field linked with an electrical circuit and to then change the flux linkage.

For simplicity, consider an electrical circuit consisting of a coil of N turns, then the change of flux linkages may be accomplished in a variety of ways.

1. The coil may move through a flux constant in value.
2. The coil may remain stationary in a varying flux.
3. Both changes may take place simultaneously.

In the first case, the flux-cutting rule can be applied. Hence

$$e_r = Blu$$

This is termed a motional (or rotational) e.m.f. It is always associated with the conversion of energy between the electrical and mechanical forms.

In the second case

$$e_p = N \cdot \frac{d\phi}{dt}$$

This is termed a pulsational e.m.f. or an e.m.f. of transformation. There is no motion involved and hence there is no electromechanical energy conversion.

In the third case, both e.m.f.'s are produced.

To illustrate the motional and pulsational e.m.f.'s, consider again example 11.4. Let the armature be moving with a velocity of 1·0 m/s when the gap length is 1·5 mm. If the current were 20 mA as before, then the motional e.m.f. is

$$e_r = i \cdot \frac{dL}{dx} \cdot u = 20 \times 10^{-3} \times -4370 \times -1 = 87{\cdot}5 \text{ V}$$

It can thus be seen that the current must be less than 20 mA. As movement continues, the current tends to reduce still further, a change opposed by the pulsational e.m.f. $e_p = L \,.\, di/dt$.

An approximate estimate of the transient conditions can be obtained by assuming that the movement of the armature takes place so quickly that the flux does not change sufficiently that the current can remain directly proportional to the reluctance.

Example 11.5 The relay shown below operates at a constant voltage of 50 V. The air gap is initially 2.0 mm long. When the relay is energised, the armature closes passing the point at which the air gap is 1·5 mm long with a velocity of 1·0 m/s. If the relay takes a current of 100 mA when closed, make an estimate of the current when the air gap is 1·5 mm. Hence estimate the power components and the force on the armature.

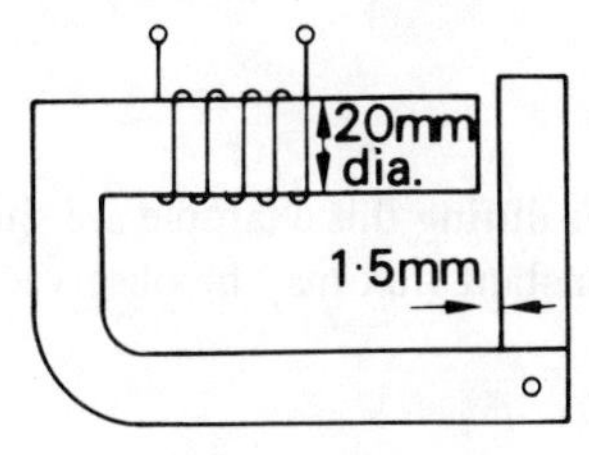

Fig. 11.22

Assume that the flux does not change appreciably.

$$F = \phi S$$

$$\frac{i_{2\cdot 0}}{i_{1\cdot 5}} = \frac{l_{2\cdot 0}}{l_{1\cdot 5}}$$

$$\frac{2\cdot 0}{1\cdot 5} = \frac{100 \times 10^{-3}}{i_{1\cdot 5}}$$

$$i_{1\cdot 5} = 75 \times 10^{-3}\ \text{A} = \underline{75\ \text{mA}}$$

$$V = ir + L\cdot\frac{di}{dt} + i\cdot\frac{dL}{dx}\cdot\frac{dx}{dt}$$

$$L = \frac{N^2 \mu_0 A}{l} = \frac{5000^2 \times 4\pi \times 10^{-7} \times 12\cdot 5^2 \times \pi \times 10^{-6}}{1\cdot 5 \times 10^{-3}} = 10\cdot 25\ \text{H}$$

$$\frac{dL}{dx} = \frac{N^2 \mu_0 A}{l^2} = -\frac{L}{l} = -\frac{10\cdot 25}{1\cdot 5 \times 10^{-3}} = -688\ \text{H/m}$$

$$\frac{dx}{dt} = -1\cdot 0\ \text{m/s}$$

Substituting in the voltage equation.

$$50 = \frac{75}{1000} \times \frac{50}{0\cdot 1} + 10\cdot 25 \times \frac{di}{dt} + \frac{75}{1000} \times -688 \times -1$$

hence $\frac{di}{dt} = -3\cdot 8\ \text{A/s}$

Power from source $Vi = 50 \times 75 \times 10^{-3} = \underline{3\cdot 75\ \text{W}}$

Power to electric loss $= i^2 r = 75^2 \times 10^{-6} \times 500 = \underline{2\cdot 81\ \text{W}}$

Mechanical power $= \frac{1}{2} i^2 \cdot \frac{dL}{dx} \cdot u = \frac{1}{2} \times \frac{75^2}{1000^2} \times -688 \times -1 = \underline{1\cdot 93\ \text{W}}$

Field rate of storage $= iL\cdot\frac{di}{dt} + \frac{1}{2} i^2 \cdot \frac{dL}{dx} \cdot u = \frac{75}{1000} \times 10\cdot 25 \times -3\cdot 8 + 1\cdot 93$

$= \underline{-0\cdot 99\ \text{W}}$

The negative sign means that energy is being released from the field. The field power could also have been estimated from the power balance relation. The armature force is given by

$$p_M = -f_E u = 1{\cdot}93 \text{ W}$$

$$f_E = \frac{1{\cdot}93}{1{\cdot}0} = 1{\cdot}93 \text{ N}$$

The change in the conditions during this example are shown in Fig. 11.23. The limitation to the concept of constant flux may be observed from this flux/m.m.f. chart.

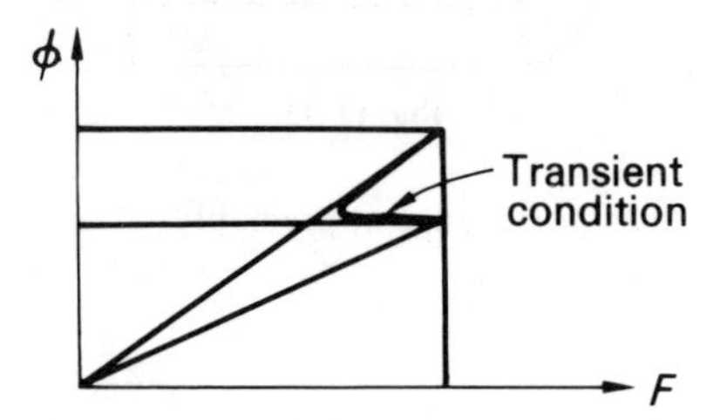

Fig. 11.23. Transient flux/m.m.f. chart

Before leaving these arrangements which are constrained to move in approximately linear modes, the contactor should be noted. This has a movement similar to the solenoid relay but, because it is used to control power circuits, it requires a longer travel. The travel can be sufficiently long to make the condition of parallel pole faces no longer applicable to any degree of accuracy. Nevertheless an estimate of the forces involved can be made using the above principles.

11.10 Force of interaction

The analysis of force as the rate of change of stored field energy with distortion has concerned only the force of alignment. Now consider the principle applied to the force of interaction arrangement, i.e. to a current-carrying conductor.

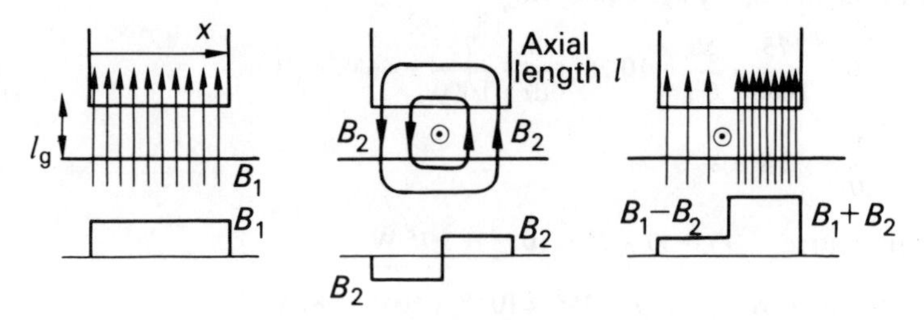

Fig. 11.24. Current-carrying conductor between poles

A uniform field would normally be created between ferromagnetic poles as shown in Fig. 11.24(a); let the resulting flux density be $\mathring{B}_1$. Fig. 11.24(b) shows the same pole arrangement with a conductor carrying a constant current i_2 but B_1 has temporarily been withdrawn from the diagram. Assuming the conductor to belong to a symmetrically constructed circuit and the magnetic circuit also to be symmetrical then the conductor will give rise to a flux of flux density B_2 as shown–note that being a symmetrical arrangement, the conductor lies in the middle of the air gap. Finally Fig. 11.24(*c*) shows the two systems superimposed.

The stored energy density to the left-hand side of the conductor is

$$\omega_f = \frac{(B_1 - B_2)^2}{2\mu_0}$$

whilst on the right-hand side

$$\omega_f = \frac{(B_1 + B_2)^2}{2\mu_0}$$

Movement of the conductor in the direction of x, say through dx, will reduce the stored energy by

$$\mathrm{d}W_f = \left(\frac{(B_1 + B_2)^2}{2u_0} - \frac{(B_1 - B_2)^2}{2u_0}\right) ll_g . \mathrm{d}x$$

$$= \frac{2B_1 B_2}{\mu_0} \cdot ll_g \cdot \mathrm{d}x$$

$$f_E = \left.\frac{\mathrm{d}W_f}{\mathrm{d}x}\right|_{i_2 \text{ constant}}$$

$$= \frac{2B_1 B_2 ll_g}{\mu_0}$$

But

$$B_2 = \mu_0 H_2$$

$$= \frac{\mu_0 i_2}{2l_g}$$

$$f_E = B_1 li_2 \qquad (11.17)$$

This justifies the *Bli* expression introduced in chapter 4. The positive polarity shows that for the given arrangement, a motoring action will take place.

Much emphasis has been placed on the need for symmetry. In most practical rotating, as well as many linear, machines the symmetrical arrangement is realised, hence the importance generally given to the *Bli* relation. However, if the relation is not symmetrical, the *Bli* expression requires to be used with the greatest of caution.

Consider the arrangement shown in Fig. 11.25 in which the core is excited by a 1-turn coil of resistance R_1 through which a constant current I_1 flows–the current is kept constant by variation of the applied voltage v_1. I_1 gives rise to a flux of density B_1.

A single-turn coil of loop resistance R_2 is made to move through the air gap with uniform velocity u and driven by a force f_M. Let one coil side pass between the poles in a time t_b, the other side remaining remote from the pole arrangement.

The movement of the coil side in the air-gap flux will induce a motional e.m.f. E_2 into the coil with the effect of causing a current $I_2 = -E_2/R_2$ to circulate in the coil.

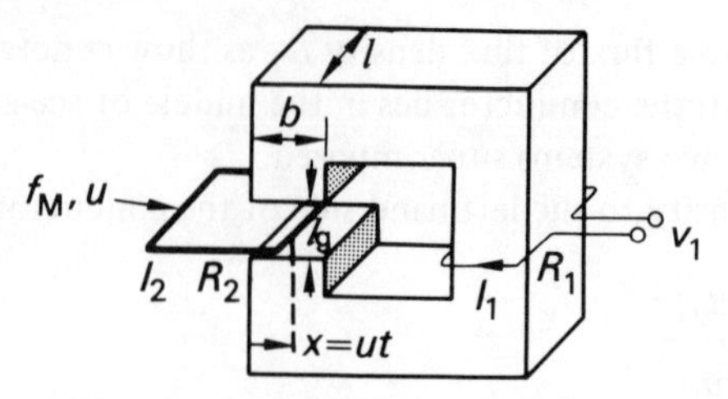

Fig. 11.25. Current-carrying conductor in an unsymmetrical magnetic-circuit air gap

The negative sign appears because the e.m.f. acts as a source e.m.f. and not as an effective volt drop.

Given the above information, it is tempting to suggest that the force of interaction experienced by the coil is

$$f_E = B_1 l I_2$$

This would be wrong! Why?

The presence of I_2 gives rise to a field whereby the gap flux density is effected and the conductor is no longer lying in a field of density B_1. Note that the arrangement is not symmetrical and the second flux does not cross the gap twice–see Fig. 11.26.

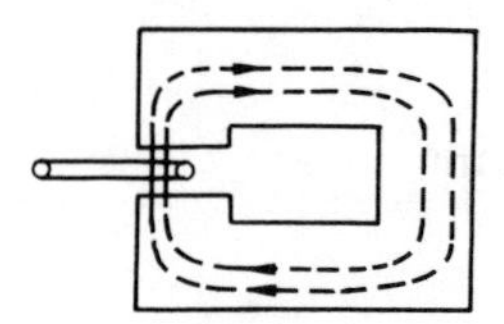

Fig. 11.26. Flux due to closed-coil current

The force of interaction can be found by considering the energy balance. Analytically this can be shown only after extensive manipulation but the following example illustrates the same principle more concisely.

Example 12.6 The device shown in Fig. 11.25 has the following dimensions:

l_g ... 2·5 mm
b ... 200 mm
l ... 500 mm

The excitation coil carries a current of 2 000 A and has a resistance of 0·005 Ω. The closed coil travels with a velocity of 5·0 m/s and has a resistance of 0·005 Ω. Calculate the force of interaction experienced by the closed coil.

In order to evaluate the energy terms in the energy balance it is necessary to calculate the flux densities involved. The initial flux density in the air gap is

$$B_1 = \mu_0 H_1 = \mu_0 \cdot \frac{I_1}{l_g} = \frac{4\pi \times 10^{-7} \times 2000}{2{\cdot}5 \times 10^{-3}} = 1{\cdot}0 \text{ T}$$

Within the coil, the flux density B_2 is given by

$$B_2 = \mu_0 H_2 = \mu_0 \cdot \frac{I_1 + I_2}{l_g}$$

But the flux within the coil is

$$\phi_2 = B_2\, lx = B_2\, lut$$

The flux linking the coil grows at the rate of

$$\frac{d\phi_2}{dt} = B_2\, lu = E_2$$

$$I_2 = -\frac{E_2}{R_2} = -\frac{B_2\, lu}{R_2} = -(I_1 + I_2).\frac{\mu_0\, lu}{l_g R_2}$$

$$= -(I_1 + I_2)k$$

where $$k = \frac{\mu_0\, lu}{l_g R_2} = \frac{4\pi \times 10^{-7} \times 0{\cdot}5 \times 5}{2{\cdot}5 \times 10^{-3} \times 5 \times 10^{-3}} = 0{\cdot}25$$

$$I_2 = -\frac{k}{1+k}\cdot I_1 = -\frac{0{\cdot}25}{1{\cdot}25} \times 2000 = -\,400\ \text{A}$$

$$B_2 = \frac{(I_1 + I_2)\mu_0}{l_g} = \frac{(2000 - 400)4\pi \times 10^{-7}}{2{\cdot}5 \times 10^{-3}} = 0{\cdot}8\ \text{T}$$

The energy terms involved are:

1. The energy input from the electrical system ΔW_E.

Because there is a change in flux in the excitation coil, an e.m.f. is induced in it and the applied voltage has therefore to be modified to keep the current constant. The flux, and therefore the stored energy, is reduced thus stored energy is returned to the electric source. Hence

$$v_1 = I_1\, R_1 - E_1 = I_1\, R_1 - \frac{\Delta\Phi}{\Delta t} = I_1\, R_1 - \frac{(B_2 - B_1)\, lb}{b/u}$$

$$= 2000 \times 0{\cdot}005 - (1{\cdot}0 - 0{\cdot}8)0{\cdot}5 \times 5 = 9{\cdot}5\ \text{V}$$

$$\Delta W_E = v_1\, I_1\, t = v_1\, I_1 \cdot \frac{b}{u} = 9{\cdot}5 \times 2000 \times \frac{0{\cdot}2}{5} = 760\ \text{J}$$

2. The energy input from the mechanical system ΔW_M.

$$\Delta W_M = -f_E\, b = -f_E \times 0{\cdot}2$$

3. The change in field energy ΔW_f.

$$\Delta W_f = \frac{(B_2{}^2 - B_1{}^2)\, lbl_g}{2\mu_0} = \frac{(0{\cdot}8^2 - 1^2)0{\cdot}2 \times 0{\cdot}5 \times 2{\cdot}5 \times 10^{-3}}{2 \times 4\pi \times 10^{67}}$$

$$= -36\ \text{J}$$

4. The i^2R losses $\Delta W_{\cdot\, l}$.

$$\Delta W_l = I_1{}^2\, R_1 . \Delta t + I_2{}^2\, R_2 . \Delta t$$

$$= 2000^2 \times 0{\cdot}005 \times \frac{0{\cdot}2}{5} + 400^2 \times 0{\cdot}005 \times \frac{0{\cdot}2}{5}$$

$$= 832\ \text{J}$$

But $$\Delta W_E + \Delta W_M = \Delta W_f + \Delta W_l$$

$$760 - f_E \times 0{\cdot}2 = -36 + 832$$

$$f_E = \underline{-180\ \text{N}}$$

Hence ignoring any force required for mechanical storages, the mechanical input force is 180 N. Now relate the problem to the initial suggestion that $f_E = B_1 l l_2$. First accept that $I_2 = -400$ A, then

$$f_E = 1{\cdot}0 \times 0{\cdot}5 \times -400 = -200 \text{ N}$$

Alternatively let the flux density have the modified value within the coil, i.e. 0·8 T whence

$$f_E = 0{\cdot}8 \times 0{\cdot}5 \times -400 = -160 \text{ N}$$

It may be observed that in fact f_E is the mean of these two values.

This example is patently abstract. However, the principle illustrated is important in that relation (11.17) only applies to symmetrical arrangements. The mean value is then found to coincide as is illustrated by the above example.

Finally it should be noted that it is quite *incorrect* in the given example to state

$$I_2 = \frac{B_1 lu}{R_2} = -\frac{1 \times 0{\cdot}5 \times 5}{0{\cdot}005} = -500 \text{ A}$$

hence $$f_E = B_1 l I_2 = 1 \times 0{\cdot}5 \times -500 = -250 \text{ N}$$

The moral of this section is to check for symmetry!

PROBLEMS ON UNIFIED MACHINE THEORY

1 A coil of fixed inductance 4·0 H and effective resistance 30 Ω is suddenly connected to a 100-V, d.c. supply. What is the rate of energy storage in the field of the coil at each of the following instants:

(*a*) When the current is 1·0 A.
(*b*) When the current is 2·0 A.
(*c*) When the current is at its final steady value? 70 W; 80 W; 0

2 In question 11.1, the value of the inductance of the coil was not required in the solution. Why was this the case?

3 A simple relay has an air gap of length 1·0 mm and effective cross-sectional area 1 000 mm^2. The magnetising coil consists of 1 000 turns of wire carrying a current of 200 mA. Calculate the energy stored in the air gap. The reluctance of the ferromagnetic part of the magnetic circuit may be neglected. 25 mJ

4 The magnetising coil of a relay carries, at a particular instant, a current of 100 mA which is increasing at the rate 10 A/s. At the same instant, the inductance is 0·5 H rising at 50 H/s. These changes are taking place due to the movement of the relay armature. Calculate the e.m.f. induced in the relay coil.

The effective resistance of the coil is 600 Ω Calculate:

(*a*) The mechanical power input.
(*b*) The rate of field energy storage. 10 V; 0·25 W; 0·75 W

5 Two halves of an iron ring are placed together and a conductor is passed axially through the centre as shown in Fig. 11.27. The ring has a mean diameter of 250 mm and a cross-sectional area of 500 mm^2. When a current is passed through the conductor, a force F of 25 N is required to separate the halves of the ring. For this condition the relative permeability of the iron is 500. Calculate:

(*a*) The flux density in the ring.
(*b*) The mean magnetising force in the ring.
(*c*) The current in the conductor.

0·25 T; 400 At/m; 314 A (SANCAD)

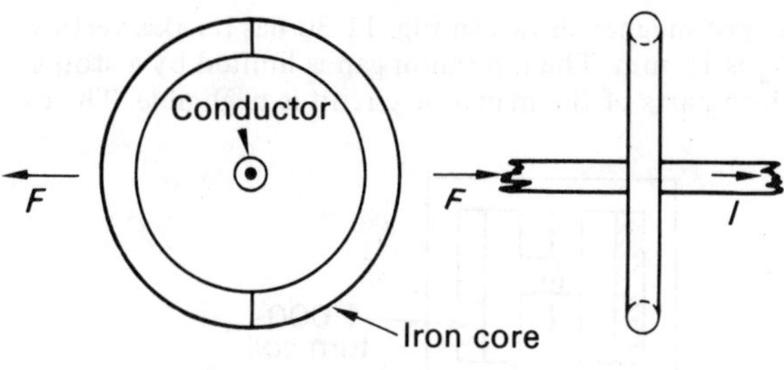

Fig. 11.27

6 An electromagnet is illustrated in Fig. 11.28 and it supports a mass of 11 kg. The cross-section of the magnetic parts of the magnetic circuit is 600 mm², and the excitation current is 2 A. Find the number of turns in the magnetising coil to achieve the required supporting force. The permeability of the core material varies with flux density as in the following table.

B (T)	0·36	0·44	0·48	0·60	0·72
u_r	3 300	3 000	2 900	2 600	2 300

19 turns

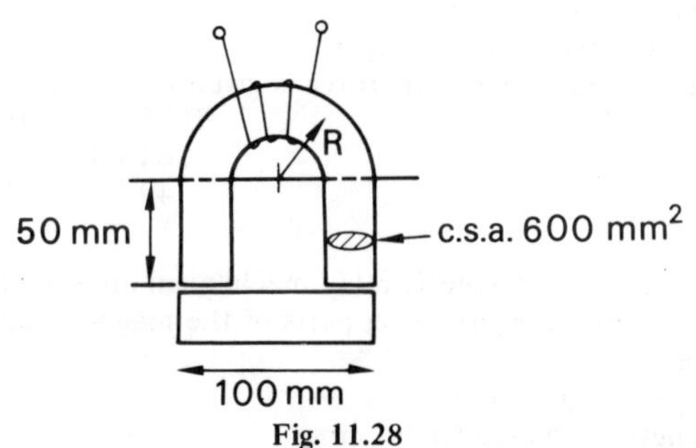

Fig. 11.28

7 An electromagnet is shown in Fig. 11.29 to be in close proximity to a bar of steel. The flux in the magnetic circuit is 2·5 mWb. Assuming that the reluctance of the magnetic circuit is concentrated in the air gap and that fringing is negligible, calculate the horizontal and vertical forces experienced by the steel bar.

79·5 N; 2 985 N (SANCAD)

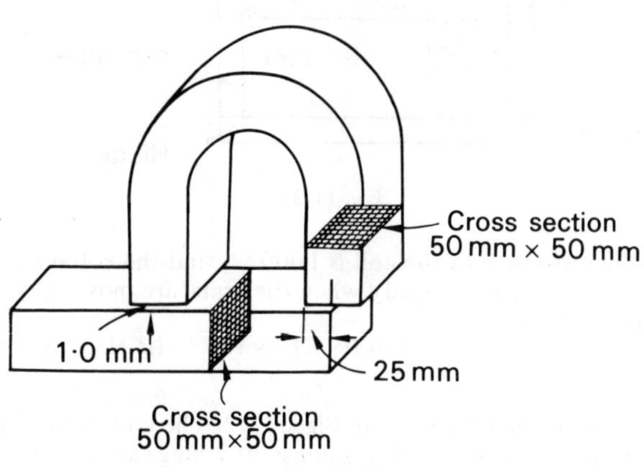

Fig. 11.29

8 The cylindrical pot magnet shown in Fig. 11.30 has its axis vertical. The maximum length of the gap l_g is 15 mm. The minimum gap is limited by a stop to 5 mm. The reluctance of the iron parts of the magnetic circuit is negligible. The exciting coil is

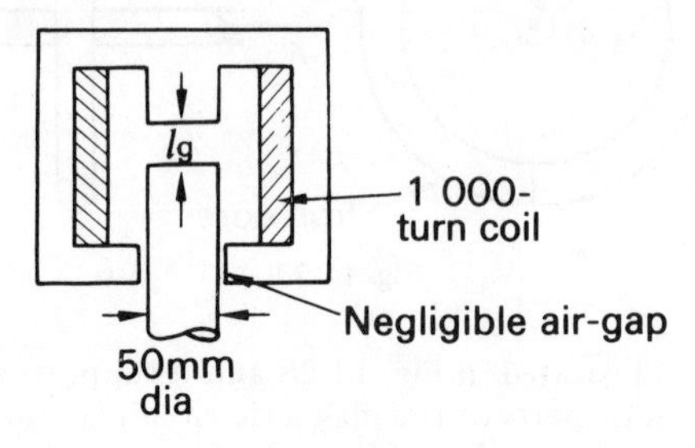

Fig. 11.30

energised by a constant current of 3·0 A from a 60-V d.c. source. For maximum, mid and minimum gap lengths, find:

(*a*) The gap flux density.
(*b*) The coil inductance.
(*c*) The magnetic force on the plunger.
(*d*) Check that the static magnetic force is given by $\frac{1}{2}I^2 \, . \, \mathrm{d}l/\mathrm{d}x$.

0·25 T	0·38 T	0·75 T
164 mH	246 mH	492 mH
49 N	110 N	440 N

9 The essentials of an electromagnetic relay mechanism are shown in Fig. 11.31. Neglecting the m.m.f. required by the iron parts of the magnetic circuit, estimate the inductance of the coil for:

(*a*) The given gap length 3·0 mm.
(*b*) The gap length increased to 3·6 mm.

Hence estimate the average force developed on the armature between these two positions when the coil carries a steady current of 20 mA.

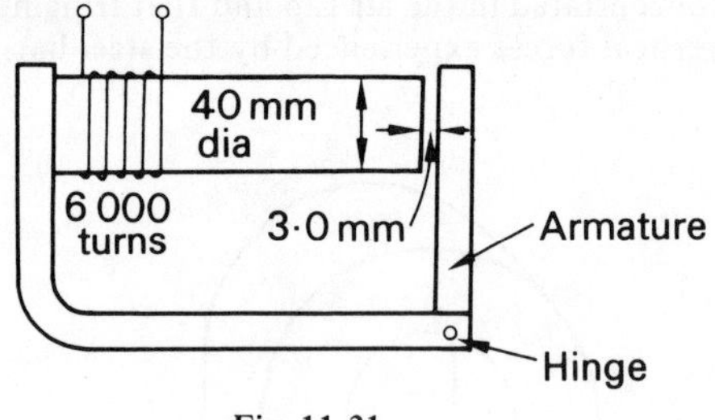

Fig. 11.31

Given that the resistance of the coil is 1 000 Ω, find the voltage required across the coil for the current to remain steady when the armature moves between the two positions in a time of 11 ms.

18·9 H; 15·8 H; 1·04 N; 25·7 V (SANCAD)

10 A cylindrical pot magnet is shown in Fig. 11.32. The coil, which has 1000 turns and a resistance of 20 Ω, is connected to a constant-voltage source of 60 V direct current. The plunger passes the point at which the air-gap is 10 mm travelling upwards with a uniform velocity of 2·0 m/s. Make an estimate of the value of the current at this instant, assuming that the initial air-gap length were 15 mm.

Apportion the instantaneous input power into heat, magnetic-energy storage-rate and mechanical components. Hence estimate the magnetic force on the plunger.

80·0 W; − 78·4 W; 118·4 W; 49·2 N

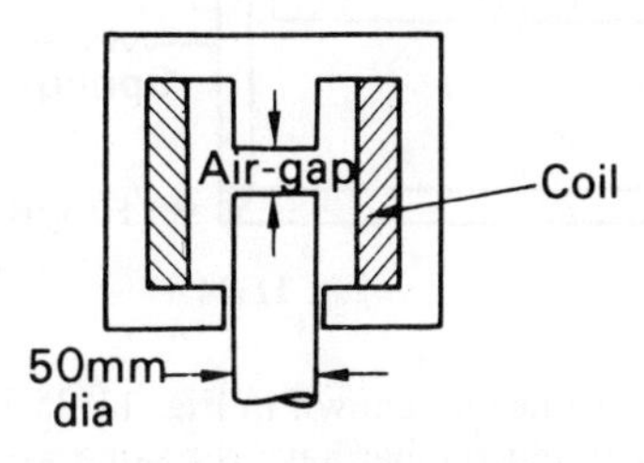

Fig. 11.32

11 A simple electromagnetic relay is shown in Fig. 11.33. The relay coil has a constant resistance R, an inductance L that varies with armature position x, and a constant applied voltage V. Derive an expression relating the converted mechanical power to the rate of change of coil inductance.

Considering the armature to be stationary, and given that $V = 50$ V, $R = 2{\cdot}0$ kΩ, $x = 2{\cdot}0$ mm and the diameter of the core is 20 mm, estimate:

(*a*) The gap flux density.
(*b*) The gap flux.
(*c*) The coil inductance.
(*d*) The rate of change of coil inductance with gap length.
(*e*) The magnetic pull on the armature.

Assume that the gap faces are parallel and that the permeability of the iron parts is very high.

If, at this gap length of 2·0 mm, the armature were moving to close the gap at the rate 1·0 m/s, how would the current be affected? Give a brief explanation.

94 mT; 29·6 μWb; 7·2 H; 3 600 H/m; 1·1 N (SANCAD)

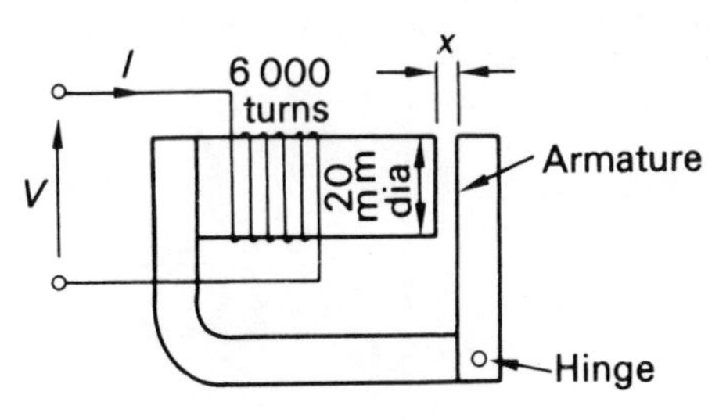

Fig. 11.33

12 The attracted-armature relay shown in Fig. 11.34 has an initial air-gap of 3·0 mm. The reluctance of the iron part is negligible. The coil has 1 000 turns. The armature is held open by a spring, such that an initial force of 0·8 N is required to close it.

Explain what happens following the application of a direct voltage to the coil terminals, stating the energy balances concerned.

Estimate the initial inductance of the coil, the initial rate of change of inductance with gap length and the current required to close the relay.

6 mA (SANCAD)

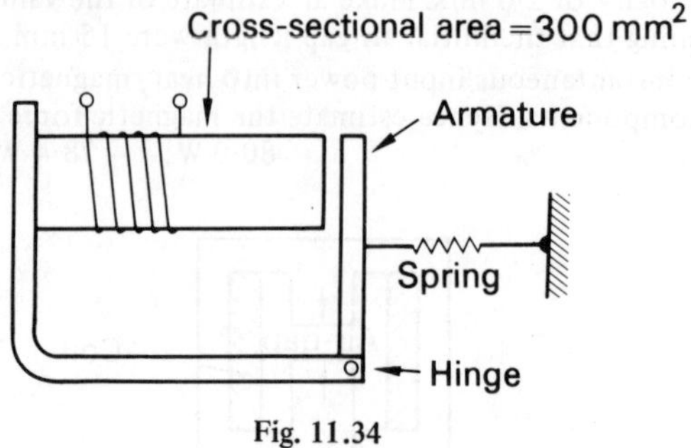

Fig. 11.34

13 The semi-circular electromagnet shown in Fig. 11.35 is to be used to lift a rectangular bar. The electromagnet and the bar have the same square cross-section of $0{\cdot}01\ m^2$. The electromagnet is made of core steel; the bar, of steel plate material, has a bulk

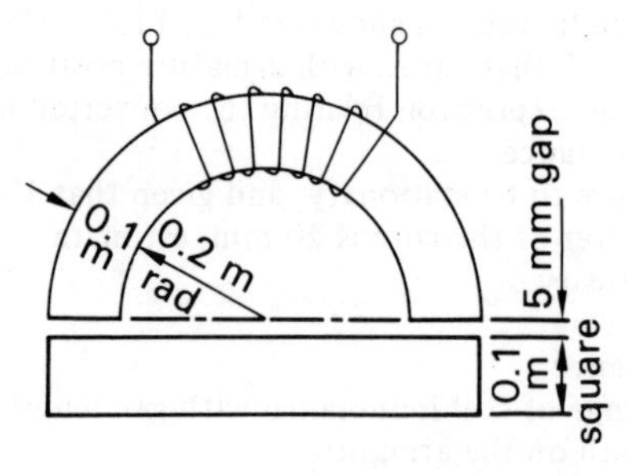

Fig. 11.35

density of 7 500 kg/m^3. The electromagnet is energised by a 500-turn coil connected across a d.c. supply. With the aid of the magnetisation characteristics given on the Chart 5.2 (p. 134), find the current required just to lift the bar. Neglect leakage and fringing in the air-gap. 4·2 A (SANCAD)

12

Rotating machines

In the previous chapter, the linear operation of such common machines as relays has been discussed at some length. These machines are extremely useful for certain modes of operation but they have the drawback that the motion to which they give rise is not continuous. Practical linear motion can only take place over relatively short distances before termination. Rotary motion, on the other hand, can be maintained almost indefinitely.

This quality is required in many walks of present-day life–in the driving of lathes and drills, the propulsion of trains, the operation of cranes and lifts, even in the operation of children's toys. It is therefore appropriate that the reader should now proceed to the study of rotary motion.

12.1 Rotary motion

The analysis of the electromechanical machine has concentrated on the linear machine–however, the principles also apply to the rotating machine by replacing x by λ (the angle of rotational distortion) and u by ω_r (the angular velocity of the rotor).

Angular velocity	Symbol: ω_r	Unit: radian per second (rad/s)

In electrical machines, the speed is often measured in revolutions per second or per minute hence

Angular speed	Symbol: n (or n_r)	Unit: revolution per second (rev/s)

The alternative symbol is used in this book wherever it gives continuity to ω_r, i.e.

$$\omega_r = 2\pi n_r$$

The torque of a rotating machine is given by

$$M_E = \left.\frac{dW_f}{d\lambda}\right|_{I \text{ constant}} \tag{12.1}$$

This compares with relation (11.7).

12.2 Reluctance motor

A simple machine that demonstrates a torque of alignment is shown in Fig. 12.1. If the rotor is displaced through an angle λ, it experiences a torque which tries to align it with the stator poles. The static torque is calculated from relation (12.1).

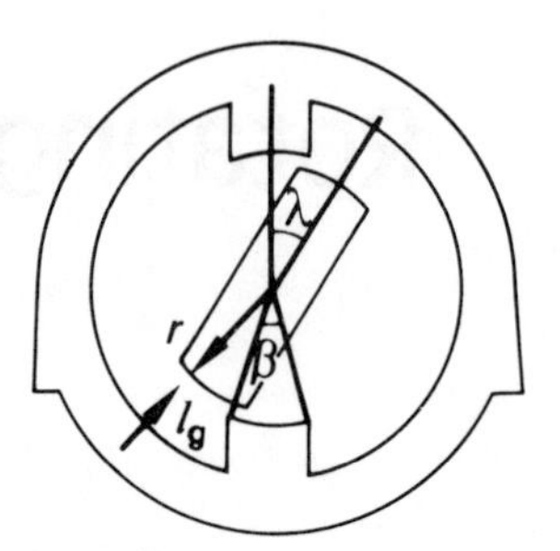

Fig. 12.1. Simple reluctance torque machine

The field energy density in the air gaps is given by

$$w_f = \frac{B^2}{2\mu_0}$$

It will be assumed that only the gap energies need be considered. The total energy is therefore given by

$$W_f = 2\left(\frac{B^2 A l_g}{2\mu_0}\right) = \frac{B^2 lrl_g(\beta - \lambda)}{\mu_0}$$

where A is the gap area through which the pole flux passes and hence is equal to $lr(\beta - \lambda)$

$$M_E = \frac{dW_f}{d\lambda}$$

$$= \frac{B^2 lrl_g}{\mu_0} \tag{12.2}$$

This machine experiences a torque which aligns the rotor with the stator, at which position the torque ceases. It follows that the machine does not produce continuing rotation but by switching the current on and off, e.g. by using alternating current to excite the magnetic system, the machine can be adapted to give continuous rotation. The resulting machine is termed a reluctance motor. It is not particularly important–one common application is clock motors–but its significance should be borne in mind. Any machine with saliency of the type illustrated by the rotor in Fig. 12.1 will produce a significant torque in this way.

For continuous motion, the assumption is required that the reluctance varies sinusoidally with the rotation of the rotor. This compares with practice quite favourably. A sinusoidal voltage is applied to the stator coil giving rise to a flux that may be defined by

$$\phi = \Phi_m \cos \omega t$$

As the rotor rotates, the reluctance will vary. Minimum reluctance occurs when the rotor centre line is coincident with the direct axis of Fig. 12.2. The minimum reluctance is S_d. This occurs when

$$\lambda = 0, \pi, 2\pi, \text{etc.}$$

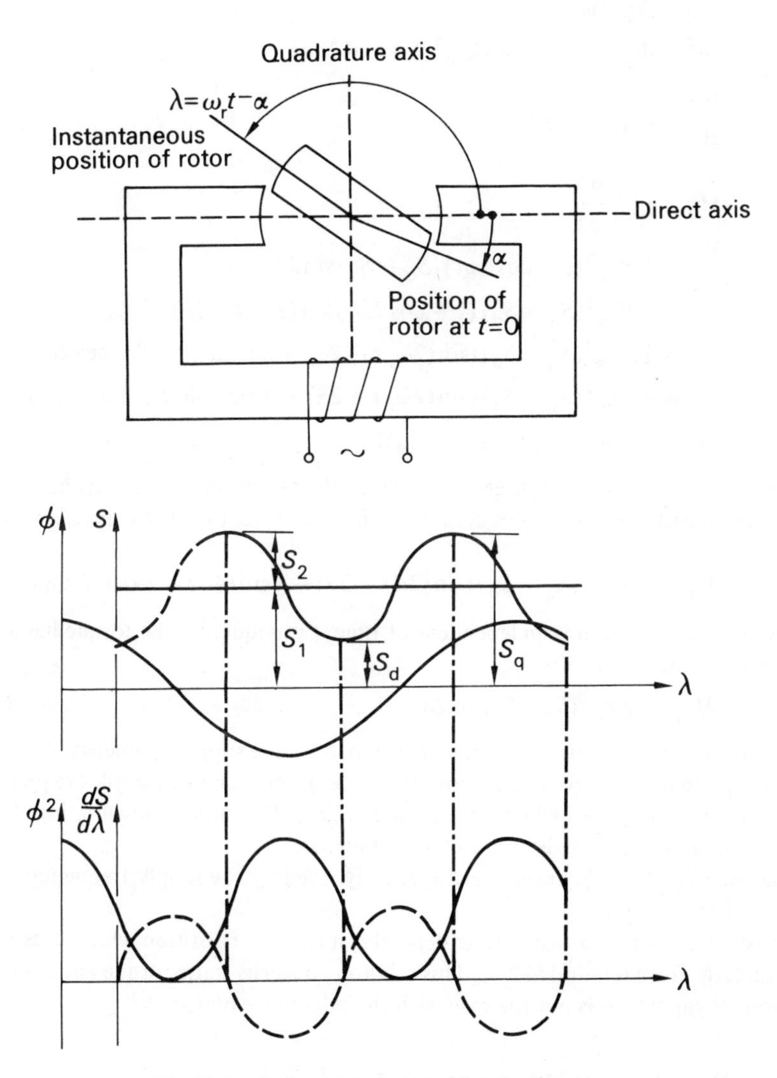

Fig. 12.2. Reluctance motor with operating curves

The maximum reluctance occurs with the rotor centre line coincident with the quadrature axis. The maximum reluctance is S_q. This occurs when

$$\lambda = \frac{\pi}{2}, \frac{3\pi}{2}, \frac{5\pi}{2}, \text{etc.}$$

From the reluctance curves showing the sinusoidal variation between these limits,

$$S = S_1 - S_2 \cos 2\lambda$$
$$= S_d + \tfrac{1}{2}(S_q - S_d) - \tfrac{1}{2}(S_q - S_d)\cos 2\lambda$$
$$= \tfrac{1}{2}(S_d + S_q) - \tfrac{1}{2}(S_q - S_d)\cos 2\lambda$$

but

$$\phi = \Phi_m \cos \omega t$$
$$\phi^2 = \Phi_m{}^2 \cos^2 \omega t = \tfrac{1}{2}\Phi_m{}^2(1 + \cos 2\omega t)$$

also

$$\frac{dS}{d\lambda} = (S_q - S_d)\sin 2\lambda$$

$$M_E = \tfrac{1}{2}\phi^2 . \frac{dS}{d\lambda}$$
$$= \tfrac{1}{4}.\Phi_m{}^2(1 + \cos 2\omega t)(S_q - S_d)\sin 2\lambda$$
$$= \tfrac{1}{4}.\Phi_m{}^2(S_q - S_d)(1 + \cos 2\omega t)\sin(2\omega_r t - 2\alpha)$$
$$= \tfrac{1}{4}.\Phi_m{}^2(S_q - S_d)(\sin(2\omega_r t - 2\alpha) + \sin(2\omega_r t - 2\alpha)\cos 2\omega t)$$
$$= \tfrac{1}{4}.\Phi_m{}^2(S_q - S_d)(\sin(2\omega_r t - 2\alpha) + \tfrac{1}{2}\sin(2\omega_r t + 2\omega t - 2\alpha)$$
$$+ \tfrac{1}{2}\sin(2\omega_r t - 2\omega t - 2\alpha))$$

If ω_r is not equal to ω, all three terms within the last group of brackets become time variables with mean values of zero. If ω_r is equal to ω then the expression becomes

$$M_E = \tfrac{1}{4}.\Phi_m{}^2(S_q - S_d)(\sin(2\omega t - 2\alpha) + \tfrac{1}{2}\sin(4\omega t - 2\alpha) + \tfrac{1}{2}\sin(-2\alpha))$$

Only the last term is now independent of time. Consequently the torque has an average value other than zero.

$$M_{E_{av}} = \tfrac{1}{8}\Phi_m{}^2(S_q - S_d)\sin 2\alpha \qquad (12.3)$$

The rotor can thus rotate at a speed determined by the supply frequency if rotation is to be maintained. In addition, there are double- and quadruple-frequency pulsating torques produced which have no nett effect. The curves shown in Fig. 12.2 are drawn to satisfy the condition of equal frequencies.

A machine in which the rotor speed is exactly fixed by the supply frequency is of the synchronous type.

Finally it will be noted that torque depends on the rotor position which does not follow directly from relation (12.3). This relation concerns static torque and assumes a uniform air-gap which is not the case with the reluctance motor.

12.3 Doubly-excited rotating machines

A machine of the reluctance type discussed above has many disadvantages, the most important of which are the fixed speed and the weak pulsating torque. These can be attributed to the manner in which the torque is produced. The reluctance machine, and also the relays, are singly-excited systems, i.e. only the stator or the rotor is excited by a current-carrying coil. In either case, motion is caused by a movable part changing its position to reduce the reluctance of the magnetic circuit. Because of the physical construction in the rotary case, the axis of the rotor tries to align itself with the axis of the field.

In order to strengthen the attraction towards alignment, both the rotor and the stator can be excited. A simple arrangement is shown in Fig. 12.3. This is termed a doubly-excited system.

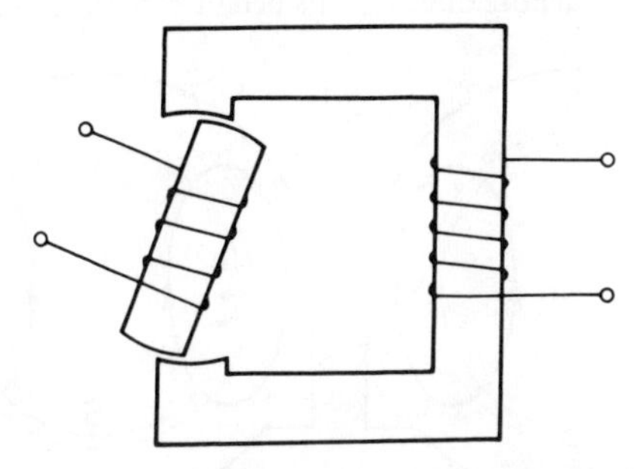

Fig. 12.3. Simple doubly-excited rotary system

With this arrangement, the stator and the rotor each have two magnetic poles. Such a machine is described as a 2-pole machine. Machines can be made with greater numbers of poles but, for the purposes of this introductory chapter, only the 2-pole machine will be considered.

The windings and the magnetic circuit components give rise to clearly defined stator and rotor fields. By the symmetry of the fields, each has an axis. The axes indicate the directions of the mean magnetic field strengths across the air gaps for each of the fields. Since the gap lengths are constant, the field axes can have complexor properties ascribed to them, their magnitudes being related to the respective m.m.f.'s.

It is now possible to describe the torque as being created by the m.m.f. axes trying to align themselves. These axes are functions of the coil constructions and hence become less dependent on the construction of the core members, which only serve to distribute the field although they are of course required to ensure that a sufficient field strength is available and it follows that specially shaped rotors are no longer required as in the reluctance machine. The coil on the rotor ensures that the rotor has a well defined axis and thus cylindrical rotors can be used. This has a considerable advantage with regard to the mechanical design of the rotor, since the cylinder is inherently a very robust structure.

By removing the salient parts of the rotor shape, the reluctance torque is removed, the rotor now offering the same reluctance regardless of its position.

The stator magnetic circuit as shown in Fig. 12.3 is unnecessarily long. Instead the stator can be made in a similar form to the rotor in that it consists of a winding set into a cylinder; however, in the stator, the winding is about the inside surface of a cylinder as shown in Fig. 12.4. It will be noted that in each case, the windings are set into the surface and are not generally wound onto the surface. This gives greater mechanical support to them and it minimises the air gap between the core members.

The current in the rotor winding produces its own m.m.f. and hence a flux that is termed the armature reaction. The total flux in the air gap therefore is the result of the combination of the field due to the rotor winding and the field due to the stator winding. The torque created by the desire for alignment can be analysed by the gap energy method already used in the singly-excited systems. In order to do so, it is necessary to introduce the concept of the current sheet.

The separate conductors around the rotor and stator surfaces create discontinuities locally in the resulting field arrangements, this effect can be avoided by considering

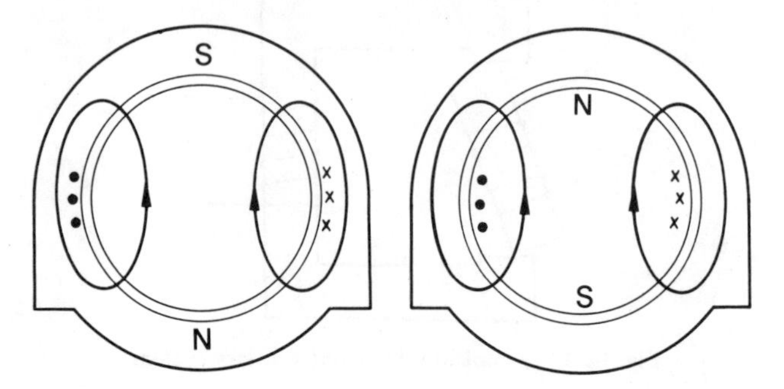

Fig. 12.4. Machine with cylindrical rotor and stator showing typical flux paths

the winding to exist at all points around both surfaces. Viewed end on, this gives the effect of a sheet of current flowing into or out from the stator or the rotor along the air-gap surfaces. In the analysis that follows, current sheets will be considered rather than the details of the winding construction. The construction is a design problem that can be taken into account at a more suitable time. At present, it only interferes with the presentation of the principles involved.

The current sheet is measured either in amperes per metre or in amperes per radian. In the former case, this measurement is taken around the circumference. In the latter case, it is taken about the axis of rotation. The measurement is termed the current surface density.

Current surface density	Symbol: A	Unit: Ampere per radian (A/rad)

For reasons that will become apparent in the subsequent analysis, it is preferable to measure the current surface density in amperes per radian.

A suitable 2-pole machine indicating possible current sheets is shown in Fig. 12.5. The current sheets are indicated by the 'dot and cross' method. Rather than have current sheets suddenly change direction at some points around the surfaces, they are distributed sinusoidally and this is shown by the variations in the sizes of the dots and crosses. This sinusoidal distribution is useful since it will be shown that it compares favourably with many practical machines.

For convenience, the stator quantities are given the subscript 1 whilst the rotor quantities are given the subscript 2. Thus they develop m.m.f.'s F_1 and F_2 respectively and the axes are displaced by the angle λ.

One effect that the sinusoidal current sheets create is a sinusoidal distribution of the flux density around the air gap. Assuming the core to be ideal as usual, apply the work law to any diametral path across both gaps and current sheets, the return being completed through the stator. The resulting m.m.f. causes a flux to be set up around

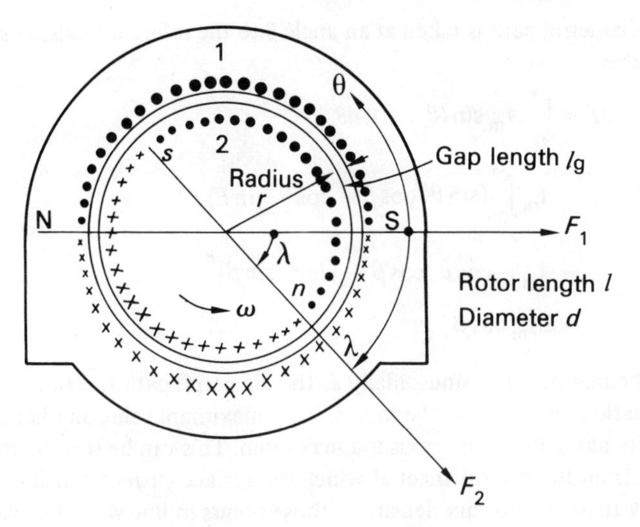

Fig. 12.5. Primitive 2-pole machine

this path and it can be shown that flux density is proportional to the m.m.f., i.e. if the m.m.f. per pole is F, then

$$F = Hl_g$$

$$= \frac{Bl_g}{\mu_0}$$

hence $$B = \frac{\mu_0 F}{l_g} \quad (12.4)$$

The m.m.f. for such a path can be obtained by considering the singly-excited system shown in Fig. 12.6. This incorporates a sinusoidally-distributed current sheet which can be defined by the expression

$$A_m \sin \theta$$

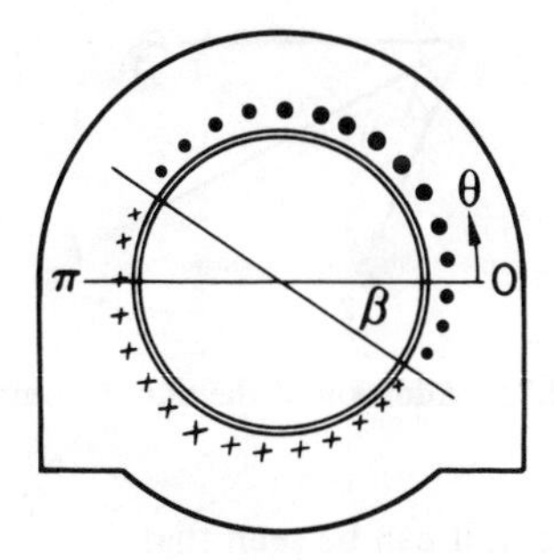

Fig. 12.6. Flux density distribution due to a sinusoidal current sheet

If the diametral path is taken at an angle β to the reference axis, as shown in the diagram, then

$$2F = \int_0^{\pi} A_m \sin(\theta + \beta).d\theta$$

$$= A_m \int_0^{\pi} (\sin\theta.\cos\beta + \cos\theta.\sin\beta)\,d\theta$$

$$= A_m [-\cos\theta.\cos\beta + \sin\theta.\sin\beta]_0^{\pi}$$

$$= 2A_m \cos\beta \qquad (12.5)$$

Thus the m.m.f. varies sinusoidally as the diametral path is rotated about the rotor axis of rotation. When $\beta = 0$, the m.m.f. has a maximum value and hence the resulting flux density along the m.m.f. axis is a maximum. This can be seen to coincide with those points on the current sheet at which the surface current densities are zero. Conversely there is zero flux density at those points in line with the maximum value of current density points.

The m.m.f. of the complete winding is determined by the maximum value of relation (12.5). By determining the path that gives the maximum m.m.f., the axis for any winding arrangement is accurately specified. In this way, the concept of the m.m.f. axis is clarified.

The path along the m.m.f. axis passes through two pole arrangements. The m.m.f. is divided equally between the poles and hence the value derived in expression (12.5) is specified per pole. It is convenient since the total m.m.f. per pole is given by

$$F = A_m \qquad (12.6)$$

In a doubly-excited system, both the stator and the rotor current sheets give rise to fields in the air gap, each having a sinusoidal distribution of the flux density as described above. It has been shown many times that two such sinusoidal quantities having the same angular period combine to give a sinusoidal quantity. Thus the fields combine to give a total field which again has a sinusoidal distribution of flux density. Since the flux density waves can be represented by the m.m.f. complexors, their combined effect can be represented by the complexor addition of the separate m.m.f. complexors. This is illustrated in Fig. 12.7.

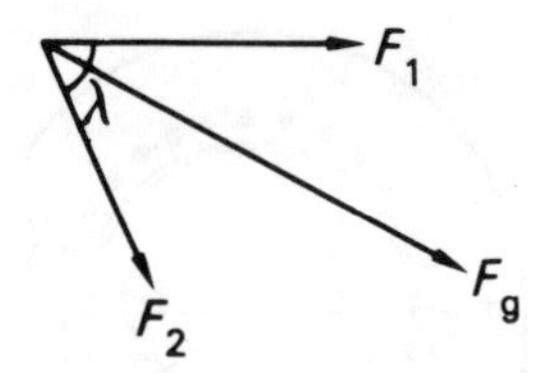

Fig. 12.7. Addition of the m.m.f. complexors

From the complexor diagram, it can be seen that

$$F_g^2 = F_1^2 + F_2^2 + 2F_1 F_2 \cos\lambda$$

It is now possible to analyse the torque of the primitive machine model created by the above discussion. As before, the gap-energy method of analysis can be applied to find an expression for the torque, i.e.

$$M_E = \left.\frac{dW_f}{d\lambda}\right|_I$$

The machine shown in Fig. 12.5 has a constant current-sheet pattern and this is equivalent to fulfilling the condition of constant current. It remains to find an expression for the gap energy. By relation (11.1), the energy density at the point of maximum flux density is given by

$$w_f = \frac{B_m^2}{2\mu_0}$$

However, if F_g is the total m.m.f. per pole, then by relation (12.4)

$$w_f = \frac{\mu_0 F_g^2}{2l_g^2}$$

Both the flux density and the m.m.f. vary sinusoidally around the rotor surface. The average value of the square of a sinusoidal quantity is half of the square of the maximum value, hence

$$w_{fav} = \frac{\mu_0 F_g^2}{4l_g^2}$$

The total gap energy is obtained by multiplying the average gap energy density by the gap volume, i.e.

$$W_f = w_{fav} V$$

$$= \frac{\mu_0 F_g^2}{4l_g^2} \cdot \pi dl . l_g$$

$$= \frac{\mu_0 \pi dl}{4l_g} (F_1^2 + F_2^2 + 2F_1 F_2 \cos\lambda)$$

$$M_E = \frac{dW_f}{d\lambda}$$

$$= -\frac{\mu_0 \pi dl}{2l_g} \cdot F_1 F_2 \sin\lambda \qquad (12.7)$$

The negative sign in this expression arises from the direction in which λ has been measured.

Although this arrangement produces a torque, it does not produce rotation in itself. However it may be assumed that the stator current sheet may be made to rotate about the surface of the stator. The manner in which this is achieved will be explained in para. 12.8. The rotation of the current sheet causes the m.m.f. F_1 to rotate with an angular velocity ω say. If the rotor also is rotating at the same speed, its m.m.f.

axis F_2 will experience the torque defined by relation (12.7). The rotor in consequence moves with respect to a position relative to the m.m.f. axis F_1 whereby it experiences a sufficient torque to maintain its rotation. If the machine were running at a constant speed and had no losses, the axes would align themselves.

However, it is more important to consider the case in which the machine is being used to convert energy. In a motor, a load is connected to the rotor. This load requires a torque to drive it and hence the m.m.f. are displaced by the angle λ in order to produce the required torque. It follows that the angle λ is termed the torque angle. This situation is shown in Fig. 12.5 and it is seen that the rotor axis lags the stator axis. The converse hold true for a generator.

When the machine is operating under steady-state conditions all the energy is transferred from the electrical energy system to the mechanical energy system or vice versa. Thus there is no change in the stored energy in the magnetic field and

$$\frac{\mathrm{d}W_\mathrm{f}}{\mathrm{d}t} = 0$$

hence

$$\frac{\mathrm{d}W_\mathrm{f}}{\mathrm{d}\lambda} \cdot \frac{\mathrm{d}\lambda}{\mathrm{d}t} = 0$$

Since the first term of this expression is used to derive the torque, it correctly follows that $\mathrm{d}\lambda/\mathrm{d}t$ must be zero. This statement leads to an apparent anomaly in that the angular velocity should be zero yet the rotor is rotating due to rotation of the field pattern. The explanation is that there are two discrete angular velocities to be considered.

The angle λ was used to analyse the field patterns with respect to some suitable axis–in this case the stator m.m.f. The angular velocity $\mathrm{d}\lambda/\mathrm{d}t$ therefore measures the angular rate of distortion between the rotor and stator fields. Under steady-state conditions, the field pattern remains constant and thus the rotor and stator m.m.f. axes remain at the same angle λ to each other. If the machine were in a transient condition, i.e. accelerating or decelerating, variations in the torque would cause relative movements between the axes and hence $\mathrm{d}\lambda/\mathrm{d}t$ would be a measure of their relative angular velocity.

Nevertheless both fields are rotating and hence the field pattern is rotating. The velocity of this pattern with respect to some fixed axis in space is ω. There are thus two velocities–one relative within the field pattern and the other spacial.

Finally it should be noted that it is the rotor m.m.f. axis that rotates with the same angular velocity as the stator m.m.f. axis. This does not infer that the rotor rotates with the same angular velocity. If the rotor and the rotor m.m.f. axis are arranged such that the axis has a fixed position with respect to the rotor, the angular velocities coincide. For a machine of the constructional type suggested by Fig. 12.5, the machine is said to be synchronous. The m.m.f. axis however may rotate with respect to the rotor and hence the rotor has an angular velocity other than that of the field pattern. Again for the type of construction being considered, the machine in this case is said to be asynchronous.

This relative motion of the rotor to the rotor m.m.f. axis may raise some doubts about the validity of the derivation of the torque expression in that the derivation required the condition of constant current. However, it should be remembered that it was the field pattern that was analysed and the pattern maintains a fixed position with respect to a current sheet of steady value. Thus the condition of constant current remains valid.

Example 12.1 The primitive 2-pole machine shown below (Fig. 12.8) comprises concentric cores of high permeability material. The rotor diameter is 0·25 m, its effective axial length is 0·20 m and the radial length of the air gap is 5 mm.

(*a*) The stator has sine-distributed currents giving a maximum surface current density of 2 000 A/rad. What peak gap flux density is produced at the rotor surface?

(*b*) The rotor is now magnetised by sine-distributed currents to give a maximum surface current density of 1 500 A/rad on an axis displaced by 60° from that of the stator. Calculate the rotor torque.

(*c*) If the rotor has an angular velocity of $2\pi 50$ rad/s, what mechanical power is converted?

(SANCAD)

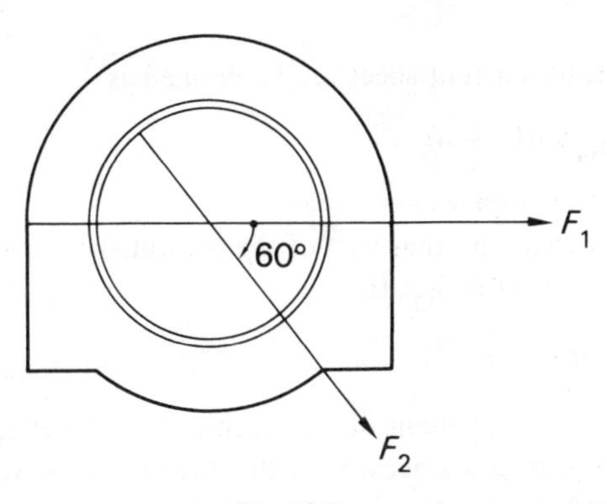

Fig. 12.8

By relation (12.7), it has been shown that provided the surface current density is measured in amperes per radian,

$$F_1 = A_{1m}$$

but
$$F_1 = \frac{B_{1m} l_g}{\mu_0}$$

hence
$$B_{1m} = \frac{\mu_0 F_1}{l_g} = \frac{\mu_0 A_{1m}}{l_g} = \frac{4\pi \times 10^{-7} \times 2000}{5 \times 10^{-3}} = \underline{0{\cdot}5 \text{ T}}$$

$$M_E = \frac{\mu_0 \pi d l}{2 l_g} . F_1 F_2 \sin \lambda$$

where
$$F_2 = A_{2m}$$

hence
$$M_E = \frac{4\pi \times 10^{-7} \times \pi \times 0{\cdot}25 \times 0{\cdot}2 \times 2000 \times 1\,500 \times 0{\cdot}87}{2 \times 5 \times 10^{-3}}$$

$$= \underline{51 \text{ Nm}}$$

$$P_M = \omega_r M_E = 2\pi 50 \times 51 = 16000 \text{ W}$$
$$= \underline{16 \text{ kW}}$$

12.4 Torque due to the forces of interaction

The analysis of the previous paragraphs has concerned the construction of the magnetic core components. However in the doubly-excited case, the force and torque can be derived from the analysis of the force on the rotor conductors. Consider again the primitive 2-pole machine defined by Fig. 12.5. It has been shown that the distribution of the m.m.f. around the air gap is sinusoidal and also that the flux density is proportional to the m.m.f. It follows that the distribution of the flux density around the rotor surface is sinusoidal. Let the flux density be defined by

$$B = B_{1_m} \cos\theta$$

where $$B_{1_m} = \frac{\mu_0 F_1}{l_g}$$

It follows that the rotor current sheet can be defined as

$$A = A_{2_m} \sin(\theta + \lambda)$$

where A is measured in amperes per radian.

If the force experienced by that part of the current sheet confined by the segment defined by a small angle $d\theta$ is f_{E_θ}, then

$$f_{E_\theta} = -BAl.d\theta$$

Because the overall arrangement of the machine is symmetrical, the relation is valid. The negative sign appears because of the direction in which λ is measured, i.e. in the direction to reduce the stored energy, not to increase it.

The torque due to this small section is given by

$$dM_{E_\theta} = -BAlr.d\theta$$

Integrating over the complete rotor surface, the total torque is given by

$$\begin{aligned}
M_E &= \int_0^{2\pi} -BAlr.\, d\theta \\
&= -B_{1_m} A_{2_m} lr \int_0^{2\pi} (\cos\theta.\sin(\theta+\lambda))\, d\theta \\
&= -B_{1_m} A_{2_m} lr \int_0^{2\pi} (\cos\theta.\sin\theta.\cos\lambda + \cos^2\theta.\sin\lambda)\, d\theta \\
&= -B_{1_m} A_{2_m} lr \int_0^{2\pi} (\tfrac{1}{2}\sin 2\theta.\cos\lambda + \tfrac{1}{2}\sin\lambda + \tfrac{1}{2}\cos 2\theta.\sin\lambda)\, d\theta \\
&= -B_{1_m} A_{2_m} lr[-\cos 2\theta.\cos\lambda + \tfrac{1}{2}\theta.\sin\lambda + \sin 2\theta.\sin\lambda]_0^{2\pi} \\
&= -\pi B_{1_m} A_{2_m} lr.\sin\lambda \qquad (12.8)
\end{aligned}$$

The question of the polarity of the torque in this context requires further consideration. However, this may be left until the principles of phase and commutator windings have been explained.

Since $$B_{1_m} = \frac{F_2 \mu_0}{l_g}$$

and $$A_{2_m} = F_2$$

then $$M_E = -\pi B_{1_m} A_{2_m} lr.\sin\lambda$$

$$= -\frac{\pi F_2 \mu_0 F_2}{l_g} lr.\sin\lambda$$

$$= -\frac{\mu_0 \pi dl}{2l_g} \cdot F_1 F_2 \sin\lambda$$

Thus the torque expressions are shown to be equivalent. In particular, it should be noticed that either expression involves the torque angle λ. This is important since much of the subsequent discussion is based on this angle.

If B_{1_m} and A_{2_m} not only represent the peak values as before but also the peak values of sinusoidally time-varying quantities with a time phase displacement ϕ then the average torque is reduced in value. The torque expression at any instant is given by

$$M_E = -\pi B_{1_m} A_{2_m} lr.\sin\lambda.\sin\omega t.\sin(\omega t + \phi)$$

In this expression, ω is the angular frequency of the time-varying quantities and not the angular velocity of the field pattern. In the analysis of the average power in a sinusoidally-varying a.c. circuit, it was shown that the average of $\sin\omega t \,.\, \sin(\omega t + \phi)$ is $\frac{1}{2}\cos\phi$, hence

$$M_E = -\frac{\pi}{2} B_{1_m} A_{2_m} lr.\cos\theta.\sin\lambda \tag{12.9}$$

It can be seen from any of the expressions for torque that the best torque angle is

$$\lambda = \tfrac{1}{2}\pi$$

This occurs when F_1 and F_2 are at right angles. In this situation, the maximum current density A_{2_m} lies in the maximum flux density B_{1_m}. The torque angle can be fixed by using a suitable winding arrangement–the commutator winding. The construction of such a winding is described in para. 12.9. In all other cases, the torque angle may vary from zero upwards. It is with these cases that this paragraph is presently concerned.

For the round-rotor, cylindrical-stator machine of Fig. 12.5 and the succeeding discussion, the current patterns and the torque angle are such that the torque will be anticlockwise and rotation in the same direction will correspond to motor action. If the torque angle is reversed, the torque also is reversed and anticlockwise rotation corresponds to generator action. All the principle nonsalient-pole machines, i.e. cylindrical stator, that are of interest have this current pattern. Their differences result from the possible methods of setting up the pattern of sinusoidal current sheets.

Example 12.2 An elementary 2-pole, 50-Hz, 60-kW synchronous generator shown below has sinusoidal current sheets around both rotor and stator to give m.m.f. axes displaced by an angle of 60°. The core is ideal and the air-gap accounts for all the magnetic circuit reluctance. The uniform radial air-gap length is 2·5 mm, the rotor is effectively 500 mm long and its radius is 200 mm. The generator operates without losses and the peak surface current density on the rotor is 1 000 A/rad. When the rotor is rotated at $2\pi 50$ rad/s, determine:

(*a*) The peak value of the gap flux density due to the stator alone.
(*b*) The peak current density on the stator.

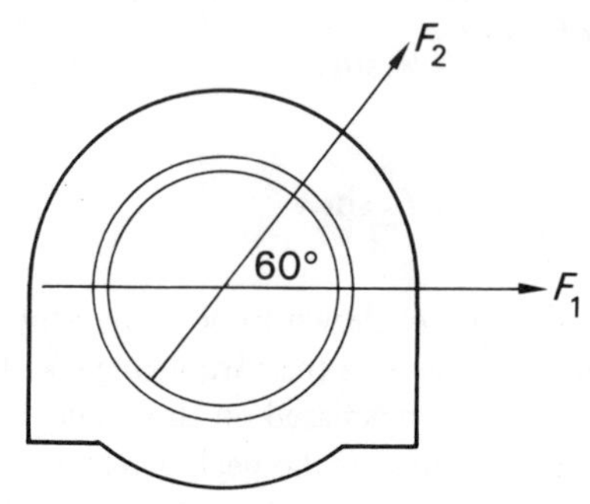

Fig. 12.9

$$M_E = \frac{P_M}{\omega_r} = \frac{60\,000}{2\pi 50} = 191 \text{ Nm}$$

$$= B_{1m} A_{2m} l r \pi . \sin \lambda$$

$$= B_{1m} \times 1\,000 \times 0{\cdot}5 \times 0{\cdot}2 \times \pi \times 0{\cdot}866$$

$$B_{1m} = \underline{0{\cdot}7 \text{ T}}$$

$$F_1 = \frac{B_{1m} l_g}{\mu_0} = \frac{0{\cdot}7 \times 2{\cdot}5 \times 10^{-3}}{4\pi \times 10^{-7}} = 1\,400 \text{ At}$$

but

$$F_1 = A_{1m}$$

$$A_{1m} = \underline{1\,400 \text{ A/rad}}$$

12.5 Salient-pole machines

Not all machines have the uniform air-gap of the cylindrical stator machine, the most common alternative being the case of the salient-pole machine. This arrangement is particularly common when machines are operated from direct current. In such cases, the flux density can be considered uniform between the poles and the rotor. The arrangement is shown in Fig. 12.10.

The flux density in the air-gaps due to the stator excitation windings is uniform over the pole angle β and is represented by B_1. It is convenient to average this flux density over the pole pitch.

The stator field is no longer able to rotate and consequently its m.m.f. axis is fixed in space. It follows that the rotor m.m.f. axis must also be fixed in space if a continuous torque is to be created. However, although the rotor m.m.f. axis is fixed, the rotor must rotate and this is achieved by a commutator winding. One consequence of this method of construction is that the current density A_2 is uniform around the rotor.

Because the m.m.f. axes are fixed, it can be arranged that the torque angle is $\frac{1}{2}\pi$ thus giving the maximum possible torque.

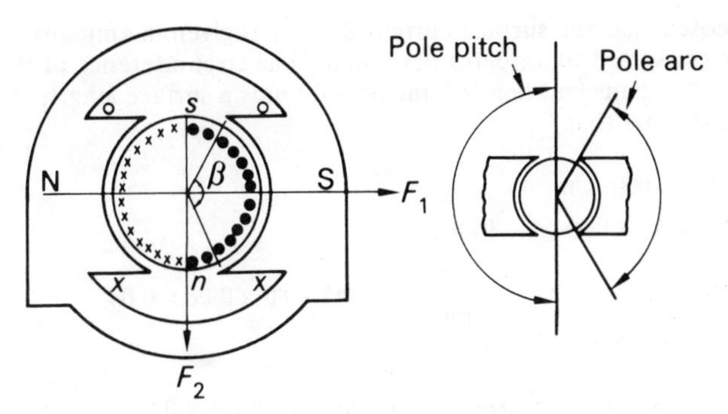

Fig. 12.10. Salient-pole machine

The analysis is carried out using the force on a current-carrying conductor approach, hence

$$dM_E = -B_1 A_2 lr . d\theta$$

$$M_E = -2 \int_0^{\beta} B_1 A_2 lr . d\theta$$

$$= -2\beta B_1 A_2 lr \tag{12.10}$$

Using the average flux density over the pole pitch,

$$M_E = -2 \int_0^{\pi} B_{1\widehat{av}} A_2 lr . d\theta$$

$$= -2\pi B_{1\widehat{av}} A_2 lr \tag{12.10.1}$$

If $B_{1_{av}}$ and A_2 are the peak values of senusoidally-varying quantities of phase difference ϕ_1 it follows as before that

$$M_E = -\pi B_{1\widehat{av}} A_2 lr \cos\phi \tag{12.10.2}$$

Example 12.3 A salient-pole machine with a commutator-type rotor is shown in the diagram below. The radius of the rotor is 40 mm and its effective axial length is 60 mm.

1. The stator winding, acting alone, produces at the rotor surface a uniform radial flux density over the 120° pole arcs of 0·5 T. If the rotor surface current density has a uniform value of 2 000 A/m, calculate the torque on the rotor.
2. Find the magnetic attraction between each pole and the rotor.
3. If the rotor spins at 100 rad/s, what will be the mechanical power developed?

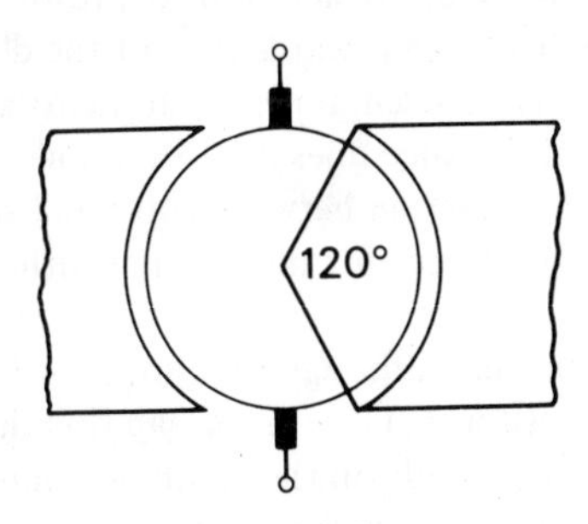

Fig. 12.11

It will be noted that the surface current density is given in amperes per metre. This value must be converted to amperes per radian. The circumference of the rotor is equivalent to 2π radians and hence 1 radian subtends a surface length of r. By proportion, it follows that

$$\begin{aligned} A_2 &= 2000 \text{ A/m} \\ &= 2000 \times 0{\cdot}04 = 80 \text{ A/rad} \\ M_E &= 2\beta B_1 A_2 lr = 2 \times \frac{120}{180} \times \pi \times 0{\cdot}5 \times 80 \times 0{\cdot}06 \times 0{\cdot}04 \\ &= \underline{0{\cdot}4 \text{ Nm}} \\ F &= \frac{B_1{}^2 A}{2\mu_0} = \frac{B_1{}^2 \beta r l}{2\mu_0} = \frac{0{\cdot}5^2 \times 120 \times \pi \times 0{\cdot}04 \times 0{\cdot}06}{2 \times 180 \times 4\pi \times 10^{-7}} \\ &= \underline{500 \text{ N}} \end{aligned}$$

Note that the total force on the rotor conductors is only

$$\begin{aligned} f_E &= \frac{M_E}{r} = \frac{0{\cdot}4}{0{\cdot}04} = 10 \text{ N} \\ P_M &= M_E \omega_r = 0{\cdot}4 \times 100 = \underline{40 \text{ W}} \end{aligned}$$

12.6 Linear machines

The doubly-excited machines considered in the previous paragraphs have all been rotary machines. However, linear machines can also be made operating on the same principles. The change from rotary to linear motion can be considered by unrolling the rotor-stator arrangement of the rotary machine into a straight line. This would give rise to a very short travel and thus either the 'rotor' or 'stator' of the linear machine must be extended over the complete travel required.

It is not proposed to analyse any of the various form of linear machine, because such analyses would duplicate the rotary analyses to a considerable extent but the following example draws attention to the most important difference between rotary and linear machines. In any machine of the doubly-excited type, there are two principle forms of force at work. The first is the lateral force that is used to cause useful motion. The other is the attraction between the air-gap boundary surfaces.

This latter force is generally much greater than the useful lateral force. Fortunately in the rotary machine, this force of attraction of the rotor towards the poles is balanced, there being an equal force towards each of the diametrically opposite poles. The effect therefore never makes itself apparent in the rotary machine.

In the linear machine, this balance does not occur and it follows that there is a problem in dealing with the attraction between rotor and stator. It increases the friction between the bearing surfaces for the machine although for tractive purposes this may be an advantage.

Finally it should be noted that although the torque or force can be derived from the interaction of the current sheet A_2 with the gap flux density B_1 produced by F_1 acting alone, the actual field depends on the combined m.m.f.'s of all the windings on both members. The actual flux is shifted and modified in magnitude by the armature reaction F_2. This is a basic requirement for all machines of the doubly-excited type–otherwise there would be no m.m.f. axes to align.

Example 12.4 The diagram below shows part of a d.c. linear motor. The pole system (stator) is fixed and excited to give a uniform flux density of 0·40 T under the pole shoes.

1. What will be the total force of attraction on the runner in each pole pitch? Would this be enough to lift the runner vertically? The runner is constructed of core plates with a bulk density of 7 500 kg/m³.

The runner is constrained to move horizontally. It carries a winding equivalent to a normal commutator winding, into which currents are fed from fixed brushes to give a uniform current distribution of 10 kA/m.

2. There is a horizontal electromagnetic force on the runner. Why? In what direction and of what value per pole-pitch?

3. If the runner current is switched on with the runner at rest and with no mechanical load attached, what will be its initial acceleration? Neglect friction.

(SANCAD)

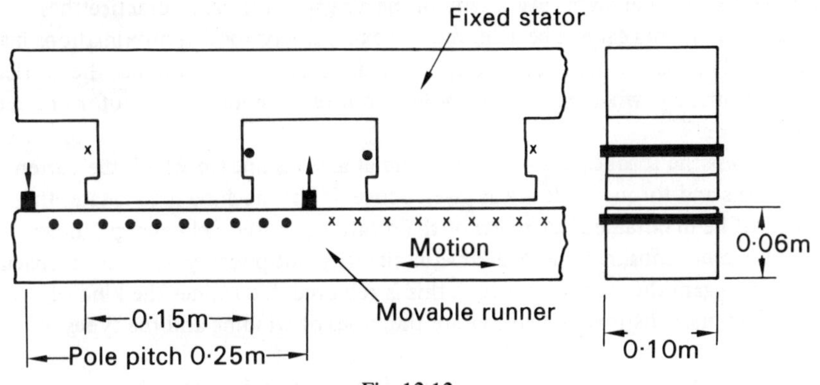

Fig. 12.12

Force of attraction per pole pitch:

$$F_a = \frac{B^2 A}{2\mu_0} = \frac{0{\cdot}4^2 \times 0{\cdot}15 \times 0{\cdot}1}{2 \times 4\pi \times 10^{-7}} = \underline{960 \text{ N}}$$

Mass of runner per pole pitch:

$$m = 0{\cdot}1 \times 0{\cdot}06 \times 0{\cdot}25 \times 7500 = 11{\cdot}3 \text{ kg}$$

Gravitational force per pole pitch F_g = 11·3 x 9·81 = 110 N.
Hence the force of attraction is sufficient to raise the runner.
Active current per pole pitch = 10 × 0·15 = 1·5 kA

= 1 500 A

Lateral force per pole pitch:

$$F_l = BIl = 0{\cdot}4 \times 1500 \times 0{\cdot}1 = \underline{60 \text{ N}}$$

The lateral force that may be explained by the force on a current-carrying conductor principle has a direction given by the left hand rule. The direction is shown in Fig. 12.13.

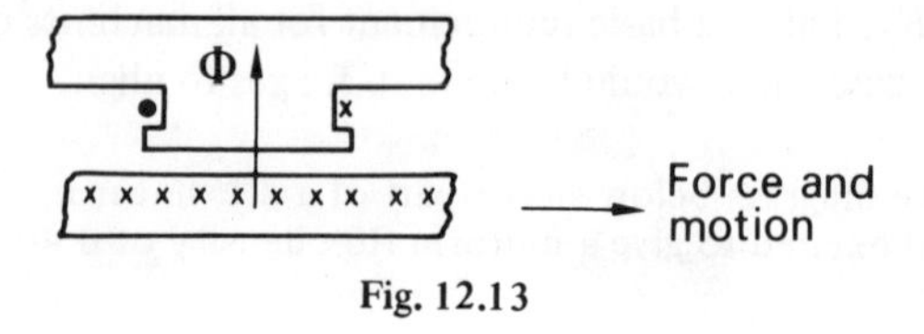

Fig. 12.13

The rotor acceleration is given by:

$$a = \frac{F_1}{m} = \frac{60}{11{\cdot}3} = \underline{5{\cdot}3 \text{ m/s}^2}$$

12.7 Theory into practice

The discussion on doubly-excited machines has purposely made no reference as to the manner in which various machine arrangements may be constructed in practice. In particular, the reader may wonder how a machine may be constructed such that the current has a sinusoidal distribution around the air-gap surfaces. In practice these idealised requirements cannot be achieved but exceedingly good approximations may be arranged. The manner in which the approximation is made determines the particular type of machine, e.g. whether it is, say, a 3-ph squirrel-cage induction motor or a d.c. shunt generator.

Nevertheless the torques derived above hold at any instant for which the current-sheet patterns and torque angle are as given above. If the machine is to rotate, the pattern must be maintained regardless of the rotation, i.e. the field energy, under steady-state conditions, remains the same in quantity but possibly varies in its spacial distribution. Again the manner in which this is achieved determines the kind of machine. The main distinctive features are the types of winding and the types of supply.

The following paragraphs describe the two types of winding–the phase winding and the commutator winding

12.8 Phase windings

It has already been noted that the simple coil of one turn is not sufficient for the purposes of most machines and consequently many turns are used. Partly from physical necessity and partly because of the improved performance, these turns are distributed along the air-gap surfaces as appropriate. Consider the simple arrangement shown in Fig. 12.14.

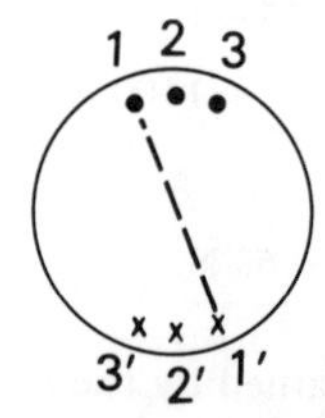

Fig. 12.14. Simple phase winding

Conductor 1 will contribute a torque provided the conductor arrangement lies in a suitable magnetic field, the torque being taken about the centre of the arrangement. Conductor 1′, which is a full pitch away, also gives rise to a similar torque. When 1 and 1′ are connected together, they form a complete turn, as in the simple machine. The number of turns can be increased if 22′ and 33′ are added in series to use the available space. Provided that after as many turns, as have been considered necessary, have been added, there are still two distinct ends to the winding thus formed, the winding is said to be a phase winding.

The phase winding gives two bands of conductors in series. These bands are diametrically opposite and they may cover all or only part of the air-gap surface. The conductors, that make up the winding, may be placed at equal intervals about the surface or at irregular intervals. Regardless of the layout, the m.m.f. axis is fixed with respect to the winding and it will only rotate if the winding rotates.

A simple arrangement is shown in Fig. 12.15. The rotor and the stator have separate external connections, the rotor connections being made through slip rings. A slip ring is a device whereby current may be transferred to the moving rotor from the static body of the machine. The ring, generally made of a copper alloy, is mounted concentrically onto the shaft and one end of the rotor winding is connected to it. A contact is then mounted onto the body of the machine such that it slides on the surface of the ring as it is rotated. The sliding contact is termed a brush and it is made from a block of carbon.

If the stator winding carries a steady current and is constructed to produce a sinusoidally-distributed field, the rotor winding in rotating has a sinusoidal e.m.f. induced in it. It will be noted that a uniform field is not required in this instance because the uniform air gap ensures that the conductors are always cutting the flux at right angles. It follows that a sinusoidal distribution of flux is the only way of inducing a sinusoidal e.m.f. This is the true justification for the concept of sinusoidal distributions of the various patterns. However, it does not follow that all a.c. machines with sinusoidal waveforms must use such phase winding arrangements. In the machine shown in Fig. 12.16, the sinusoidal distribution can be formed by spacing

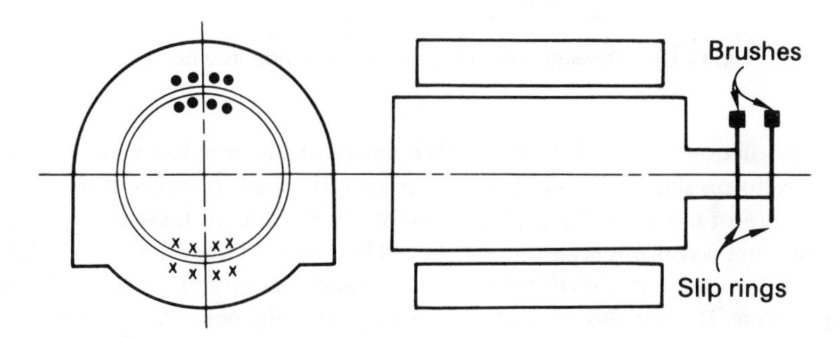

Fig. 12.15. Simple phase-winding machine

the stator winding conductors at appropriate intervals. This necessitates a number of conductors placed close together followed by a gradual increase in the spacing. This is indicated in the diagram.

For such phase windings, the case principally considered in the unified theory was that of the sinusoidally distributed m.m.f. and flux density. These sinusoidal quantities may be more easily observed from a development of the air-gap arrangement. This is shown in Fig. 12.16.

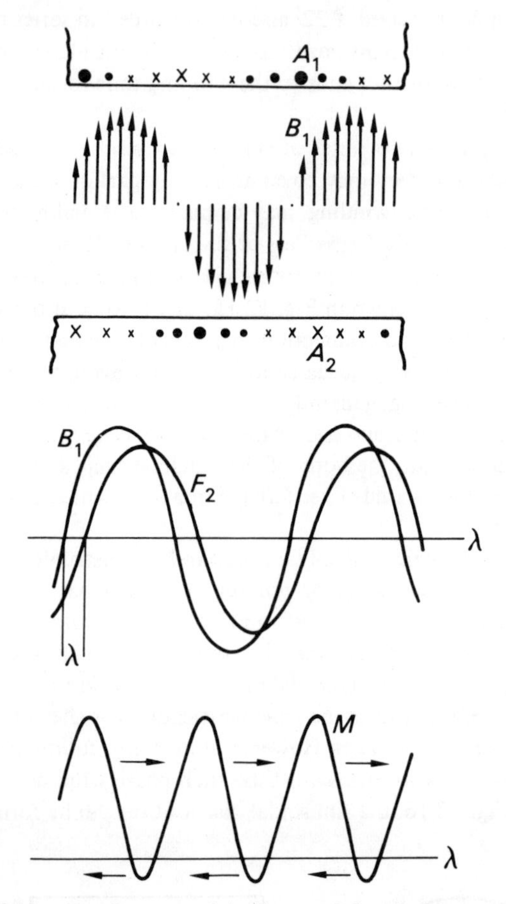

Fig. 12.16. Development of the air-gap of the primitive machine

The construction diagram is a direct development of the primitive machine's air gap and the waveforms shown directly below emphasise the displacements of the stator flux and the rotor m.m.f. The torque depends on the product of these two quantities and this product waveform is also shown. It will be seen that not all of the torque developed acts in the one direction but instead a small quantity of torque opposes the main torque. It is for this reason that difficulty over the polarity of the torque derived by the force on a current-carrying conductor method can be experienced. The effect of sin λ is similar in this analysis to the effect of cos ϕ in the analysis of the power waveform in a sinusoidal a.c. circuit.

The current sheet is made from a finely distributed winding placed in slots. A developed diagram of the rotor or stator surface is shown in Fig. 12.17. The diagram is not to scale but is merely schematic.

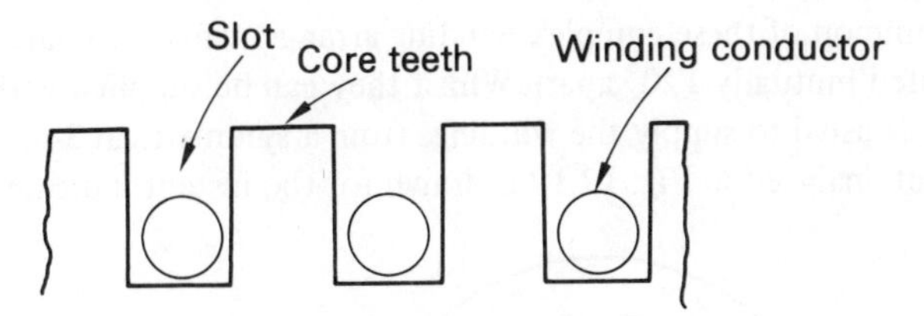

Fig. 12.17. Conductors in slots

Whilst this construction causes a certain amount of discontinuity in the uniformity of the flux, the core teeth reduce the effect at the air gap. Hence the flux appears to be due to a current sheet. Just as the m.m.f. at any point on the surface was found by integrating the current surrounded by the appropriate diametral path in accordance with the work law, a similar analysis can be carried out by integrating the product of current and conductors. The sheet current surface density is therefore measured in ampere-conductors per radian although the conductors are in fact a dimensionless factor in the same way that turns are also a dimensionless factor. The m.m.f. distribution shown in Fig. 12.18 and Fig. 12.19 are derived in this manner.

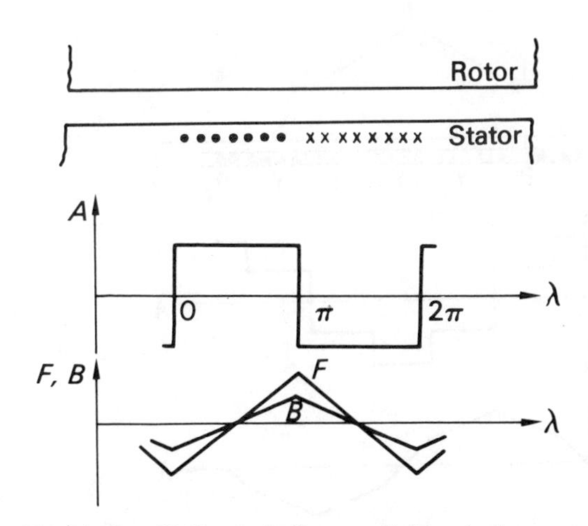

Fig. 12.18. Uniform winding supplied by single source

The most simple phase winding is that fed from a single source. This produces the flux density patterns shown in Fig. 12.18. These approximate very roughly to a sinusoidal waveform and therefore show distinct possibilities in producing the required patterns of the primitive machine.

The winding used for this field pattern analysis is uniform, i.e. all the conductors are equally spaced. As has already been noted, the windings could have been arranged with varying distances between them in order to produce more sinusoidal waveforms.

The most important disadvantage of the 1-ph phase winding or the d.c. phase winding, i.e. regardless of the method of supply, is that the m.m.f. axis remains fixed in space with respect to the winding. This difficulty can be overcome by the use of more complex phase winding arrangements.

The most common of these complex winding arrangements is to have three phase windings mounted mutually 120° apart. Whilst they can be supplied with appropriate d.c. currents, it is usual to supply the windings from a symmetrical 3-ph a.c. source. The arrangement analysed in Fig. 12.19 is drawn for the instant indicated by the

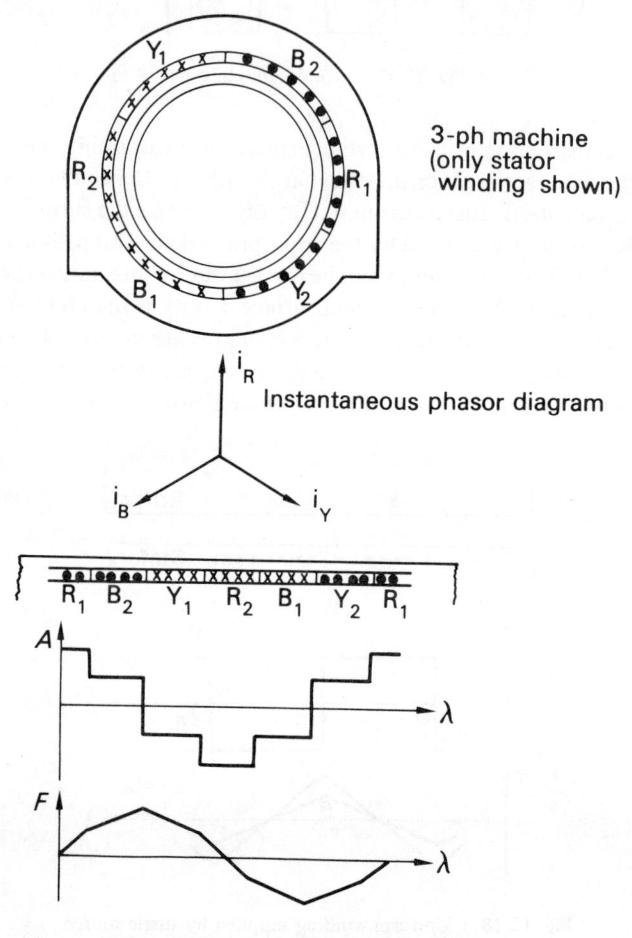

Fig. 12.19. 3-ph winding arrangement

instantaneous phasor diagram. The waveforms are now seen to reasonably approximate to sinusoidal waveforms but it should be noted that the winding waveforms vary in shape according to the instant chosen. Nevertheless the resemblance to a sinusoidal distribution remains.

This arrangement has one important advantage over the single phase winding discussed above. As time progresses, the waveform distribution moves around the air gap. For instance, if the diagram had been drawn for an instant some $2\pi/3\omega$ seconds later, the waveform patterns would be displaced by some $2\pi/3$ radians. Thus three phase windings of simple uniform construction, when supplied with 3-ph currents, give rise to a sinusoidal field pattern which rotates.

Finally a similar situation occurs with two phase windings supplied from a 2-ph source or any other proportional combination. The higher the number of windings involved with a corresponding number of supply phases, the better is the approximation to the sinusoidal condition but the 3-ph winding is generally completely satisfactory.

12.9 Commutator windings

A characteristic of the phase winding is the manner in which the axis of the winding m.m.f. moves with the winding. In order to keep the position of the m.m.f. axis fixed in space whilst its parent winding rotates, it is necessary to use a commutator winding.

A commutator winding is a development of the simple uniform phase winding. The full-pitch windings are continued until the complete rotor surface is covered. The ends of the winding are connected together thus the winding is symmetrical when viewed from any position. The end of each turn (or coil) is brought out to a block of copper; the blocks are so mounted that they form a cylinder but each block is electrically separate from its neighbours. The current is introduced to the windings through carbon brushes that slide on the surface of the cylinder made from the blocks. This arrangement is termed a commutator. The brushes are similar to the slip ring method of current collection but the commutator and the slip ring are quite separate in their actions.

Using a commutator, the m.m.f. can be made to remain at 90°–or any other desired angle–to the m.m.f. axis of the stator field. The generalised representation of the commutator winding is shown in Fig. 12.20.

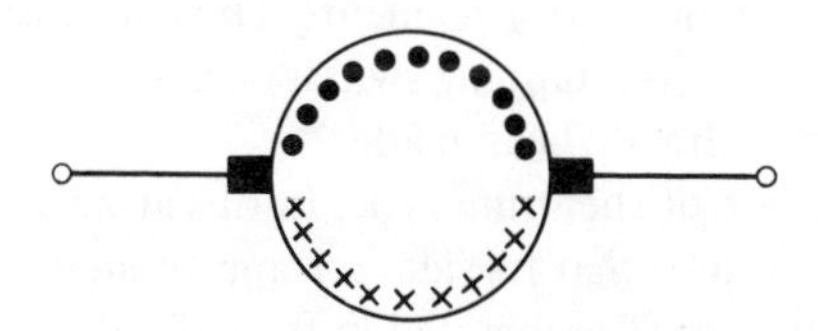

Fig. 12.20. Schematic representation of a commutator

An idea of the commutator action can be obtained by considering the simple one-turn machine, the coil being made to rotate in a uniform field. Unlike the multi-turn coil machines, in this case the turn does not complete a circuit in itself but its ends are brought out to a two-segment commutator–segment is the term used to describe the copper blocks that make up the commutator. The commutator in this case can be thought of as a hollow cylinder of copper, split axially in two and mounted in the shaft supporting the coil. The segments are insulated from one another. When the coil is rotated in the field, the e.m.f. generated in the coil is tapped off using the brushes as shown in Fig. 12.21.

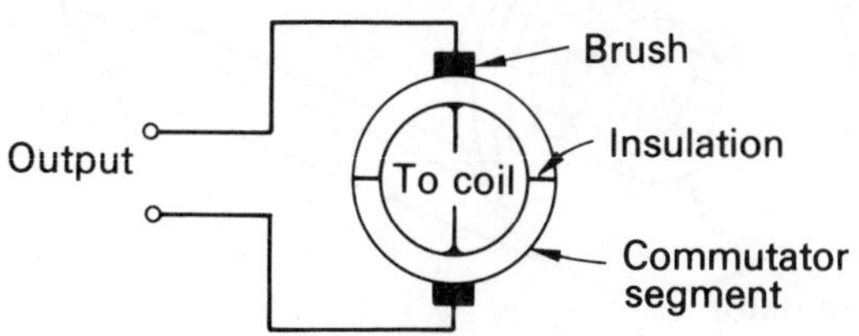

Fig. 12.21. Single commutator

The commutator can be thought of as an automatic reversing switch, although there are other purposes that it can serve. In the case shown in Fig. 12.21, the coil connections change every half revolution. Thus the coil side connected to the top brush is always passing the same field pole. The e.m.f. induced in the coil varies sinusoidally as in any coil rotated in a uniform field but due to the switching action of the commutator, the resulting potential difference between the brushes always has the same polarity. In this way the alternating e.m.f. is converted to a direct e.m.f. The resulting e.m.f. picked up by the brushes is shown in Fig. 12.22. The waveform is similar to that of rectified alternating current–see Chapter 14–which it is in reality.

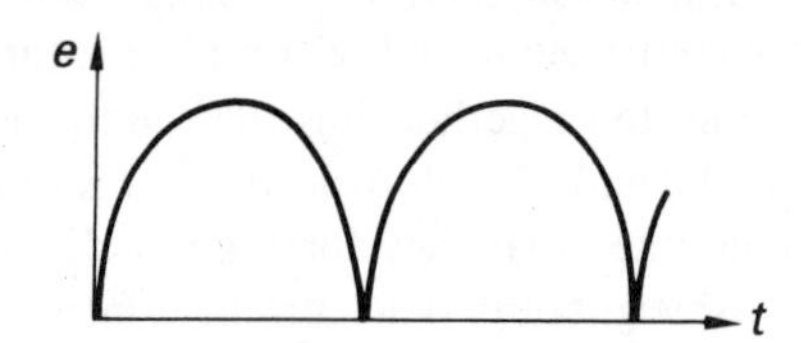

Fig. 12.22. Brush e.m.f. for simple machine

It should be noted that the brushes are so placed that when the two segments are in contact with the same brush, the coil is not cutting any appreciable flux and hence it is not generating an e.m.f. If it did a considerable short-circuit current would flow around the coil and through the brushes. However, practical machines have many coils and there are many related commutator segments. The coils form one continuous winding and the commutator serves as a tapping switch necessary to ensure the rectifier action to which allusion has already been made.

The armature windings are of the drum type. In this arrangement, each coil consists of one or more turns arranged so that its sides are approximately one pole pitch apart. This ensures that the e.m.f.'s generated in the coil sides are additive. All the coils are connected in series, each junction being brought out to a commutator segment. The construction is shown in principle in Fig. 12.23, a 2-pole winding being illustrated.

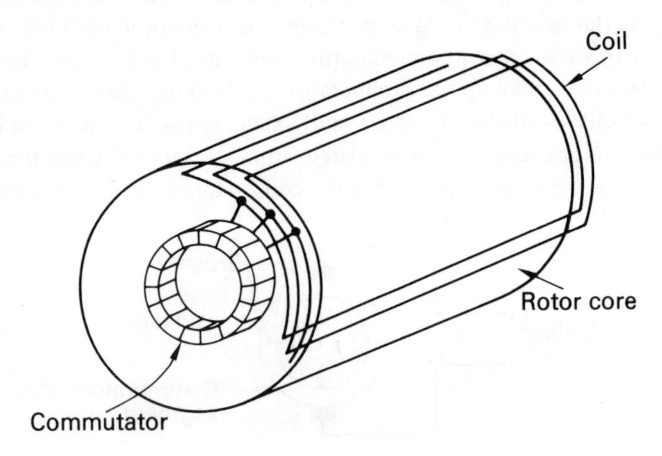

Fig. 12.23. Construction of rotor winding coil

The resulting winding can be shown schematically in the form illustrated in Fig. 12.24. Here it can be seen that all the coils have been connected in series to form a complete loop. The loop can be tapped by the brushes, in this case giving rise to two current paths through the winding.

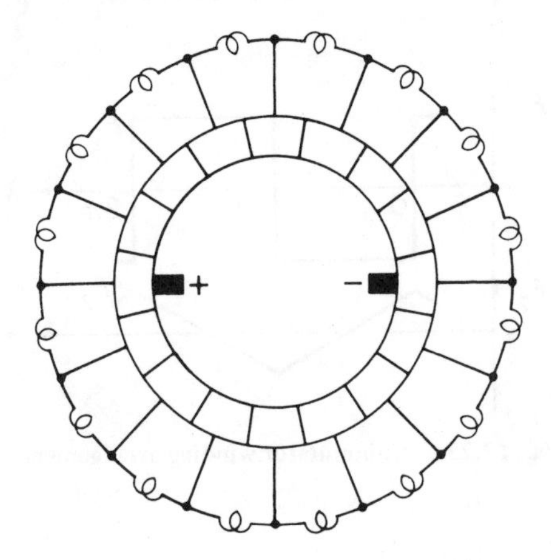

Fig. 12.24. Schematic 2-pole commutator winding

The winding forms a complete or continuous loop and is sometimes termed a continuous winding instead of a commutator winding. Either term is acceptable except in the case of the simple 1-turn machine. Although each coil is shown schematically to have only one turn, it should again be noted that a greater number of turns may be involved.

With the commutator winding, it is possible to change to points at which current is fed into the winding without changing the magnetic field pattern set up. If the winding moves forwards or backwards by one segment, the current pattern in the winding remains apparently unchanged although the currents in several of the coils have been reversed or commutated. In this way the torque angle remains unchanged by the rotor motion and continuous rotation is consequently possible.

In arranging a continuous winding around the rotor surface, by the symmetry of the construction it is found that two coil sides coincide in each slot. For the sake of symmetry and the resulting advantages, each coil has one side placed at the top of a slot and the other side of the coil at the bottom of a slot.

To produce the uniform field in which the commutator winding was shown in para. 12.5, a salient-pole stator is used. The resulting m.m.f. distribution is shown in Fig. 12.25.

An interesting alternative to the sliding contact arrangement is the use of thyristors– a semiconductor switch that can be opened and closed by pulses of electricity. These are being used in many machines for the purposes of speed control.

Both commutator and phase windings can be excited by direct or alternating current. The commutator winding with direct current excitation produces a steady m.m.f. directed along the axis of the brushes whilst alternating current excitation produces a pulsating m.m.f. The phase winding produces similar m.m.f.'s which are directed along the winding axis.

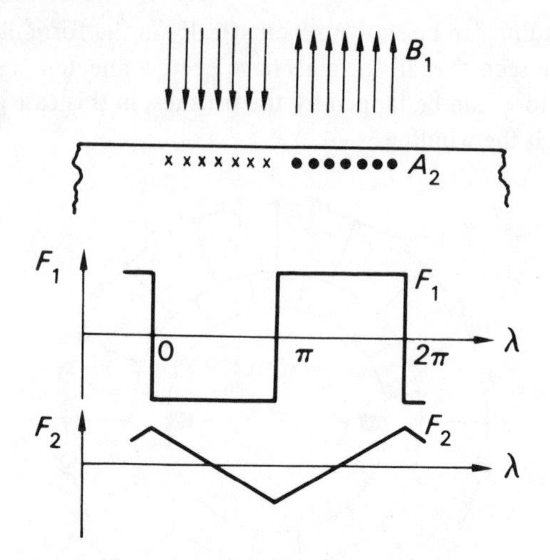

Fig. 12.25. Commutator winding arrangement

12.10 Electrical machine families

It should be noted that the physical arrangements of the windings in relation to the fields determine the types of machines. In the light of the types of windings, it is possible to create six families of electrical machines.

Family	*Flux*	*Coils*	*Connections*	*Type*
1	Constant	Moving	Commutator	D.C. commutator machine
2	Constant	Moving	Phase	Synchronous machine
3	Alternating	Fixed	Commutator	—
4	Alternating	Fixed	Phase	Transformer
5	Alternating	Moving	Commutator	A.C. commutator machine
6	Alternating	Moving	Phase	Induction machine

Family 3 is not important although a machine called the transverter is a practical example. The remaining families are represented by their most common examples in practice. It is not the purpose of an elementary text such as this to describe these in detail but some explanation of the synchronous, induction and d.c. commutator machines would be appropriate.

12.11 Synchronous and induction machines

The most common form of synchronous machines is the 3-ph turbo-generator to be found in electricity supply and also in industrial generating stations. The latter, although smaller in rating than their electricity board counterparts, generate almost as much electrical energy for industry as do the electricity boards.

The synchronous machine has a 3-ph winding on the stator whilst the rotor is supplied with direct current. The stator field therefore 'rotates' whilst the rotor field remains fixed spacially to the rotor. Torque can only be maintained if the fields rotate in synchronism, and this forms the most simple practical example of the primitive machine illustrated in Fig. 12.5. The torque is a pure interaction torque.

The stator field 'rotates' with synchronous angular velocity ω_1 whilst the angular velocity of the rotor is ω_r. For a synchronous machine $\omega_1 = \omega_r$. The machine can function either as a motor or as a generator depending on whether the stator field leads or lags the rotor field.

Some synchronous machines have salient poles, but unlike that shown in Fig. 12.10, the salient poles are on the rotor whilst the stator is cylindrical. The effect of saliency is that the rotor inductance and also the mutual inductance vary with rotation of the fields relative to one another. There is a torque due to alignment as well as the torque of interaction as a consequence.

The problems of employing machines, particularly motors, that operate at constant speed are considerable and the induction machine is more versatile in its applications. It always takes the basic form of the machine shown in Fig. 12.5 but there is no external supply to the rotor. Instead the rotor rotates at a speed other than synchronous speed and the resulting motion relative to the stator field induces a pattern of currents in the rotor winding which is a closed winding. The relative speeds determine the frequency of the induced e.m.f.'s in the rotor and also the rotor impedance. It therefore follows that the rotor speed determines the magnitude of the rotor current sheet and finally this determines the torque that the machine produces.

It should be noted that the rotor and stator field patterns must remain in synchronism if a torque is to be maintained. The basic current patterns are present at all times and the torque is constant without fluctuation, except in the case of the 1-ph machine.

Synchronous machines are mostly constructed to operate with 3-ph supplies, the most important exception being the 1-ph reluctance motor discussed in para. 12.2. Induction machines for ratings of 1 kW and over are also constructed for operation on 3-ph supplies but machines for lesser ratings may operate on 1-ph or, in control applications, on 2-ph supplies.

12.12 D.C. commutator machines

These are always constructed in the form of the machine shown in Fig. 12.10. The rotor winding is supplied through a commutator arrangement and the brushes are so positioned that the m.m.f. axis of the rotor field is always at right angles to the stator field m.m.f. axis, i.e. the torque angle is optimised.

This type of machine develops a torque at any speed and the torque is both unidirectional and unfluctuating. The same form of machine may be operated by alternating current but the torque then becomes pulsational whilst remaining unidirectional. It has already been shown that the e.m.f. induced in the winding is rectified by the action of the commutator but this action is further complicated by the use of alternating current with the result that only very small power may be developed by a.c. commutator motors, except for a few special machines, e.g. traction motors.

The d.c. commutator motor is easily controlable both from the torque and also the speed points of view. For this reason, it finds much use where performance control is of the essence. This function, however, is being replaced by controlled induction motors.

PROBLEMS ON ROTATING MACHINES

1 An elementary 2-pole machine has primary (stator) and secondary (rotor) currents distributed sinusoidally to give m.m.f. axes displaced by an angle of 60°. The uniform radial air-gap length is 5 mm. The peak surface current densities are A_{1m} = 1 500 A/rad and A_{2m} = 1 000 A/rad. The axial length of the machine is 150 mm and the rotor diameter is 200 mm.

(*a*) How could such current distributions be produced (to a reasonable approximation) in a practical machine?

(*b*) Neglecting the m.m.f. required by the iron parts, find the peak value of the gap flux density, and its direction, due to
 (i) the stator alone,
 (ii) the two members in combination.

(*c*) Determine the torque exerted on the rotor, and the mechanical power converted when the rotor spins at $2\pi \times 50$ rad/s.

0·38 T; 0·55 T; 15·7 Nm

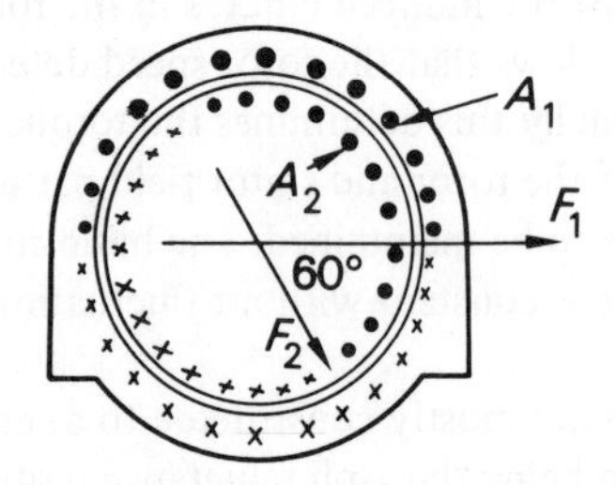

Fig. 12.26

2 A primitive 2-pole machine has current distributions around both rotor and stator surfaces of the sinusoidal variety. The concentric cores are of high permeability materials. The rotor diameter is 200 mm and its effective axial length is 360 mm, the radial length of the air-gap being 0·5 mm. The stator has sine-distributed currents giving a maximum surface current density of 2 kA/m. The rotor is magnetised similarly on an axis displaced by 50°. Calculate the torque on the rotor.

17·1 Nm

3 An elementary 2-pole, 50-Hz, 10-kW synchronous motor, as shown in Fig. 12.27 has primary (stator) and secondary (rotor) currents distributed sinusoidally to give m.m.f. axes displaced by an angle of 60°. The air-gap accounts for all the magnetic circuit reluctance. The peak surface current density on the rotor is 1 000 A/rad. The uniform radial air-gap length is 5 mm, the rotor is 200 mm long and has a radius of 100 mm. The motor operates on full load without losses. For a rotor speed of $2\pi 50$ rad/s, determine:

(*a*) The torque developed.

(*b*) The peak value of the gap flux density due to the stator alone.

(*c*) The peak current density on the stator.

32 Nm; 0·59 T; 2 340 A/rad (SANCAD)

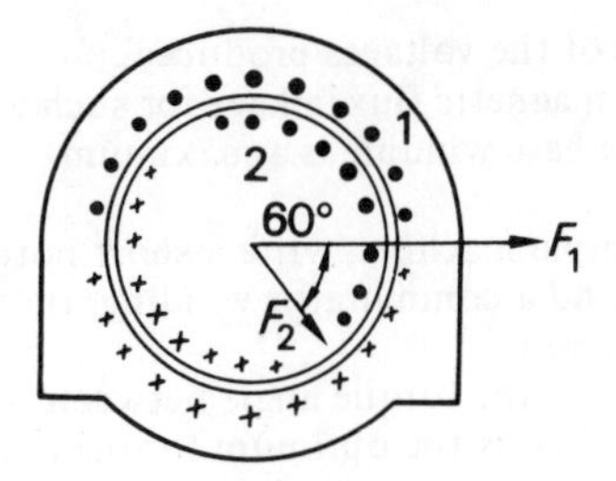

Fig. 12.27

4 A 2-ph, 2-pole motor comprises fixed concentric cores of axial length l forming an annular air-gap across which there is a sinusoidally distributed radial rotating field of peak density B_m and angular velocity ω. A copper cylinder of radial thickness b and mean radius r is arranged to occupy the gap. Show that the torque on the copper cylinder (resistivity ρ) when stationary is given by

$$M = B_m{}^2 \pi \omega r^3 l \cdot \frac{b}{\rho}$$

(SANCAD)

5 Figure 12.28 shows the essentials of a machine with a commutator-type rotor winding. The radius of the rotor is 40 mm and its active axial core length is 60 mm.

(*a*) Explain why torque will be exerted on the rotor, and indicate the direction in which it will act.

(*b*) The stator winding, acting alone, produces at the rotor surface a uniform radial flux density over the 120° arcs of 0·5 T. Calculate the torque on the rotor if its surface current density is 2 000 A/m of circumference.

(*c*) Find the magnetic attraction between each pole and the rotor.

(*d*) If the rotor spins at 100 rad/s, what will be the mechanical power developed, and how can the corresponding electrical power be supplied?

(*e*) The currents in both stator and rotor windings alternate such that the rotor surface current density is given by 2 000 sin ωt amperes per metre and the gap flux density over the polar arcs is 0·5 sin $(\omega t - a)$, teslas. Find the maximum, minimum and mean torques developed for:

(i) $\alpha = 30°$,
(ii) $\alpha = 90°$.

415 mN m; 239 N; 41·5 W
386 mN m; −27·2 mN m; 180 mN m
207 mN m; −207 mN m; 0 (SANCAD)

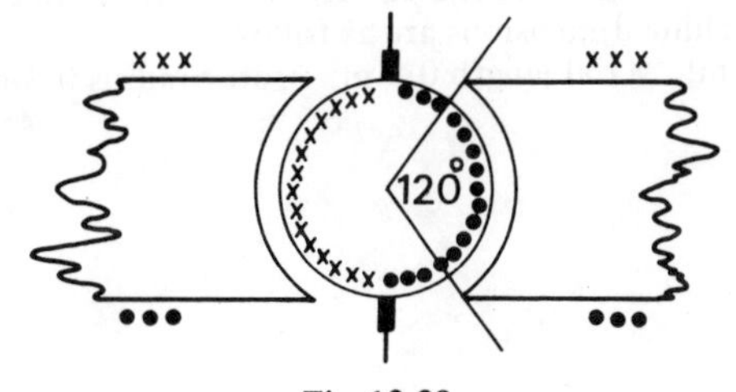

Fig. 12.28

6 Write illustrated notes describing how a 3-ph system of voltages may be produced in an electrical machine by a rotating field. Give diagrams of:

(*a*) The cross-section of such a machine, clearly indicating the direction of rotation of the field relative to the windings.

(*b*) The waveforms of the voltages produced.

Draw a diagram of the magnetic flux linkage for such a machine at the instant when the e.m.f. induced in one phase winding is a maximum.

7 With reference to a simple machine, write a short note differentiating clearly between a phase winding and a commutator winding. Indicate two methods whereby commutation may be achieved.

Explain the importance of the torque angle between the field m.m.f. and the armature reaction m.m.f. What is the optimum torque angle?

Machines can be placed into six groups depending on whether they incorporate phase or commutator windings and the manner whereby the change of flux in the winding is achieved. Discuss the applications of these groups.

8 A long, horizontal rectangular channel of insulating material contains a conducting liquid. In one region, electrodes are let into the walls and a current is passed through the liquid. The liquid path represents a resistance of 0·5 mΩ. Find the current and voltage to be applied to develop a fluid flow of 0·005 m^3/s against an inlet–outlet pressure difference of 100 kN/m^2. Find also the efficiency of this electromagnetic pump device Make the simplifying assumptions that the current, magnetic field and liquid flow directions are uniformly distributed in the duct and are mutually at right angles. Also assume that the fluid friction is zero.

4 kA; 2·125 V; 0·059 (SANCAD)

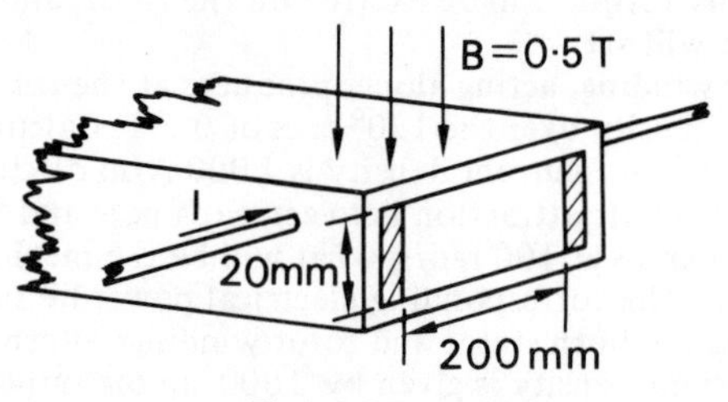

Fig. 12.29

9 A 2-pole rotating machine, energised so that the current distribution patterns in the stator and rotor windings are sinusoidal and have axes of magnetisation displaced by 45°, produces a torque of 12 Nm.

If the flux density distribution in the air-gap is also sinusoidal, having a maximum value at a point midway between the axes of magnetisation of the stator and rotor windings, find the maximum value of the current distribution on each member in amperes per radian. Machine dimensions are as follows:

radial air-gap 8 mm; axial length 0·5 m; rotor radius 0·3 m.

480 A/rad (SANCAD)

13

Electrical physics

Underlying all the aspects of electrical science that have been given consideration is the action that takes place within the constituent materials. This can be a complex study and the electrical engineer may be well satisfied by observing certain physical actions and applying them to his work. The electronics engineer cannot afford such a cursory explanation of his work and must therefore of necessity give consideration to atomic physics.

13.1 Atomic structure and energy levels

It was stated in chapter 1 that all atoms contain protons, neutrons and electrons. The number of protons in an atom is a characteristic unique to the element of which it is an example and, since, under normal conditions, the atom is electrically neutral, it follows that the number of electrons must equal the number of protons. Because these electrons are moving and are considered to have mass, they must possess kinetic energy. A force of attraction exists between each electron and the nucleus and the electrons will possess potential energy due to their positions in the atom. The total energy of any electron is the sum of its kinetic energy and potential energy.

For any atom, its electron energies must be such that they are one of a finite number of permissible energies which will vary for different elements. Moreover, the maximum number of electrons that can exist at any allowable energy is two, and the total electron energy will tend to be as small as possible. Thus the lowest permissible energies, which correspond to the orbits nearest the nucleus, tend to be filled first, leaving unfilled permissible energy levels above the highest occupied level.

The permissible energies can be grouped into shells. These shells are designated, from the nucleus outwards, as the K-, L-, M-, N-, etc. shells. There is only one energy level in the K-shell, four in the L-shell, nine in the M-shell and sixteen in the N-shell. The maximum number of electrons that can be thus accommodated in the K-shell is two, in the L-shell eight, in the M-shell eighteen and in the N-shell thirty two. The electron distribution for a few elements is indicated in Table 13.1.

Table 13.1. Distribution of electrons in shells

	Total no. of electrons	K *Shell*	*No. of electrons in* L *Shell*	M *Shell*	N *Shell*
Hydrogen	1	1	–	–	–
Helium	2	2	–	–	–
Lithium	3	2	1	–	–
Carbon	6	2	4	–	–
Oxygen	8	2	6	–	–
Aluminium	13	2	8	3	–
Silicon	14	2	8	4	–
Phosphorus	15	2	8	5	–
Copper	29	2	8	18	1
Gallium	31	2	8	18	3
Germanium	32	2	8	18	4
Arsenic	33	2	8	18	5

The chemical and electrical properties of an element are dependent on its higher energy electrons. These are known as its valence electrons. For example, the elements silicon and germanium, which are used extensively in the devices considered in chapters 14 and 15, solidify into a crystaline structure in such a way that each atom is equidistant from its four neighbours and shares its four outer electrons with them. There is said to be co-valent bonds established between the atoms. Figure 13.1 illustrates the arrangement of the atoms of silicon and germanium.

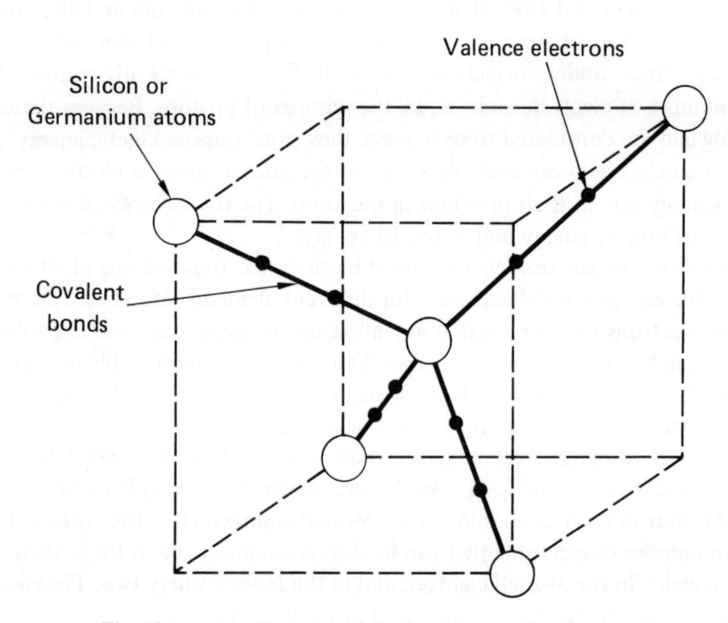

Fig. 13.1. Atomic arrangement for silicon and germanium

13.2 Conductors, insulators and semiconductors

The picture of the atomic structure in the previous section has been made with the underlying assumption that the atom is completely isolated. This assumption is valid for gases since their atoms remain relatively far apart, but, with solids however, the atoms are packed closely together and the available energy levels are influenced by the presence of the other atoms. The general effect is to broaden the higher energy levels into bands containing large numbers of very closely spaced energy levels. These levels are so close in practice that it is usual to consider a continuous range of available energies within the band. The valence electrons exist within a valence band, and above this is another energy band, known as the conduction band, which will be unfilled. Both these bands are associated with all the atoms rather than with individual ones; this introduces the possibility of electrons moving throughout the solid either within the valence or conduction band, and hence for current to flow. The resistance that the material presents to current flow is dependent on the magnitude of the forbidden energy gap between the valence and conduction bands.

For metals, the range of energies associated with the valence and conduction bands overlap (i.e. there is no energy gap between them). One effect of this is that the valence electrons are relatively free to move throughout the material and hence metals (e.g. silver, copper and aluminium) tend to be good conductors.

At first glance, it would appear that the elements silicon and germanium must be perfect insulators since their electrons are either bound closely to the nucleus or are in the valence band which is completely filled. In both cases they are not available to move freely through the crystal as would be the case if current were to flow. If,

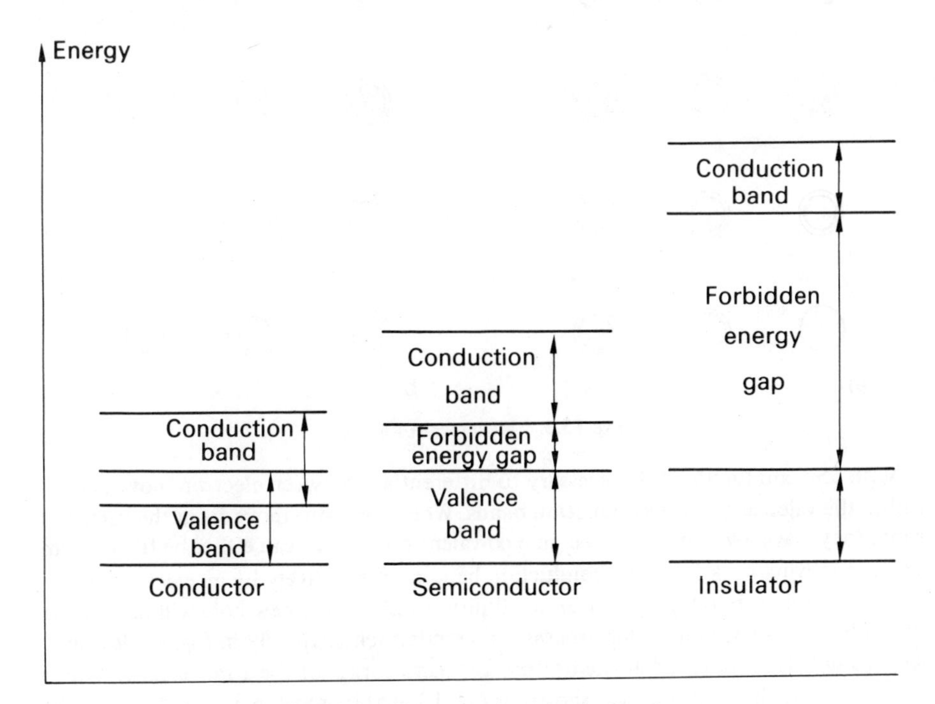

Fig. 13.2. Valence and conduction bands

however, electrons could be raised into the conduction band, they would then become free to move through the crystal. For electrons to reach the conduction band, it is necessary to supply energy (e.g. heat) to the crystal, and in this way the valence electrons can acquire sufficient energy to cross the range of forbidden energies. Even at normal room temperatures, sufficient energy is available in the form of heat for some of the silicon and germanium valence electrons to cross into the conduction band, and thus these materials will conduct to some extent. Their resistivities are such that they fall within the group of materials known as semiconductors which means that they are neither good conductors nor good insulators.

For insulators, the forbidden energy gap is a large one and, at normal temperatures, very few electrons have sufficient energy to cross into the conduction band. Figure 13.2 shows a schematic arrangement of the valence and conduction bands for conductors, semiconductors and insulators.

13.3 Conduction in metals and semiconductors

Current is carried by electrons moving in the valence and conduction bands. With metals, since the energies associated with both bands are of the same order, there is no need to differentiate between them when considering electron movement. Thus when a potential difference is established across the conductor, the picture of current flow is simply a drift of valence electrons towards the end with the more positive potential, the direction of conventional current being in the opposite direction.

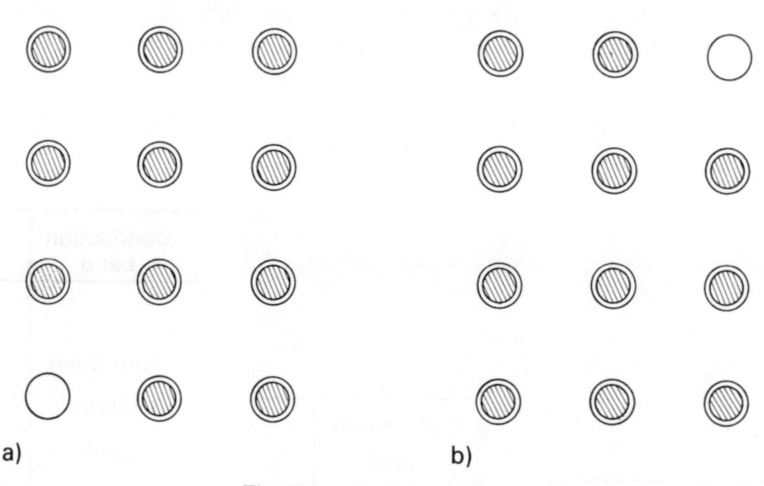

Fig. 13.3. Movement of holes

With semiconductors, it is necessary to differentiate between electron movement within the valence and the conduction bands. When electrons move from the valence band, they leave a vacancy or 'hole' in a co-valent bond. This can either be filled by an electron moving back from the conduction band or alternatively by an electron from another co-valent bond. If the latter possibility occurs then a new hole will be created in another co-valent bond. This process is illustrated schematically in Fig. 13.3(*a*) in which a hole is represented at the bottom left-hand corner of the arrangement. After a period of time the situation is as shown in Fig. 13.3(*b*) in which a hole is shown at the top right-hand corner. During the interval between these two arrangements there may have been a great many movements of the electrons; nevertheless the nett effect is still the movement of the hole from its position at (*a*) to its position at (*b*). As a result, it may be more convenient to describe the movement of charge within the valence band as a movement of holes. These must be considered as having a charge equal in magnitude to that on the electron but of opposite sign (i.e. positive). When a potential difference is therefore applied across a semiconductor, electrons in the conduction band (i.e. free electrons) drift towards the end with the more positive potential, while the holes in the valence band drift in the opposite direction. The total current will be the sum of the current due to the free electrons and that due to the holes. This is illustrated in Fig. 13.4. Note that there will be recombination of electrons from the conducting leads and holes in the semiconductor near the connection shown at A. At the other connection at B, electrons are removed from the semiconductor into the conducting lead. The rate at which electrons enter the semiconductor at A equals the rate at which they leave at B and this constitutes the current from the battery.

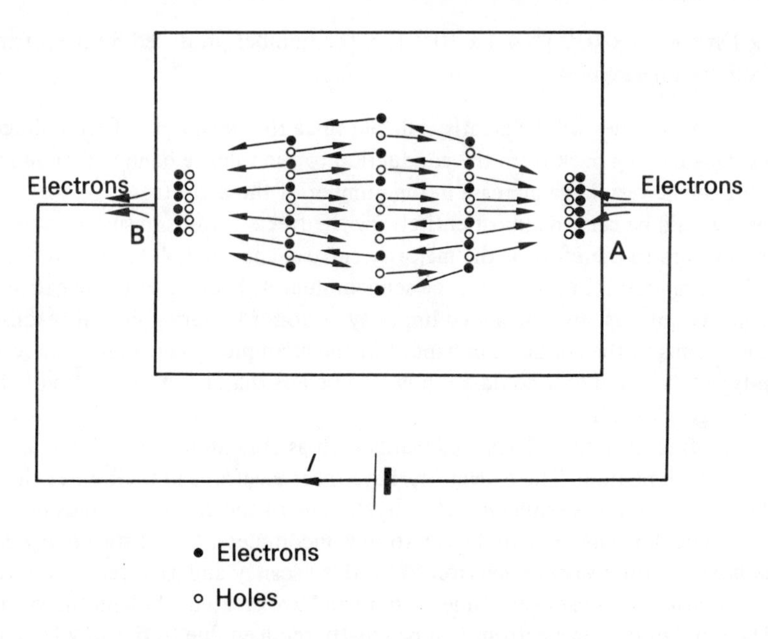

Fig. 13.4. Conduction in a semiconductor

13.4 p- and n-type semiconductors

In para. 13.3 above the conduction process described for semiconductors is applicable to pure (i.e. intrinsic) materials. The number of current carriers available, either electrons or holes, can be modified greatly by the addition of small amounts of certain other materials.

Elements, such as phosphorus and arsenic, can be seen from Table 13.1 to have five electrons in their outer shells. When these elements are added to pure germanium or silicon during the formation of crystals, their atoms form up in the same way as the germanium and silicon atoms. Co-valent bonds are created between them and the four neighbouring semiconductor atoms. There remains, however, a fifth electron which is not required for the co-valent bonds and would tend therefore to remain associated with the added impurity atom. The energy level of this electron is however just below that of the bottom level of the conduction band so that, at normal room temperatures, the great majority of these electrons are in the conduction band. The extent of the change in the number of carriers is illustrated below:

For intrinsic germanium at 300 K the number of atoms per cubic metre is $4{\cdot}4 \times 10^{28}$,
The number of free electrons = the number of holes = $2{\cdot}4 \times 10^{19}/\text{m}^3$.
If this germanium is 'doped' with arsenic to the extent of 1 atom per 10^6 atoms of germanium then:

$$\text{Number of arsenic atoms} = \frac{4{\cdot}4 \times 10^{28}}{10^6} = 4{\cdot}4 \times 10^{22}/\text{m}^3.$$

If each of these atoms has released its extra electron to the conduction band then the number of free electrons will now be:

$(4{\cdot}4 \times 10^{22} + 2{\cdot}4 \times 10^{19}) = 4{\cdot}4 \times 10^{22}$ (i.e. the number produced by the intrinsic germanium is negligible).

The number of holes will be greatly reduced since the possibility of recombination with electrons moving back from the conduction to the valence band will be much greater due to the very large increase in the number of these electrons.

Current would be carried almost entirely by the free electrons in such a material. They are referred to therefore as the majority carriers and the holes as the minority carriers. The material is known as n-type semiconductor, because of the negative charge on its majority carriers, and the added impurity as donor impurity since it readily donates electrons to the conduction band. For the example shown above, the resistivity of the n-type semiconductor could be shown to be less than one thousandth of that of the intrinsic germanium.

A similar effect is obtained when elements such as aluminium and gallium are added to silicon and germanium. These electrons have three electrons in their outer shells and when they form up in the semiconductor crystal one of the co-valent bonds between them and a neighbouring semiconductor atom is incomplete. Very little energy is required however for a valence electron to fill this vacancy and as a result, at normal room temperatures, almost every added atom will have created a hole in the valence band. The number of free electrons will be greatly reduced due to the very large increase in the number of holes.

In this type of material, current will be carried almost entirely by holes. These then become the majority carriers and the free electrons the minority carriers. The material is known as p-type semiconductor, because of the effective positive charge on its majority carriers, and the added impurity as acceptor impurity since it readily accepts electrons from the valence band.

n-type and p-type semiconductors are known as extrinsic semiconductors and semiconductor devices, some of which are considered in chapters 14 and 15, make use of both types. The degree of impurity added in practice is up to about only 1 in 10^5.

13.5 Effect of heat, light and strain on resistance

If the atomic structure of a material were perfect, the current carriers could move through the material perfectly freely and the material would therefore be an ideal conductor. In normal practice however, collisions take place between the carriers and the atoms, the positions of which are very much restricted. This results in an interchange of energy between carriers and atoms and hence a resistance to the flow of current.

At temperatures above 0 K, for instance, the atoms vibrate about their mean positions, the amplitude of the vibration increasing with increasing temperature. This has the effect of increasing the likelihood of collisions. With metals therefore their resistivities increase with increasing temperature. The same effect takes place with semiconductors and insulators but, at the same time, the number of carriers increase due to the supply of further energy. This latter effect predominates and thus their resistivities decrease with increasing temperature. In the case of a metal, there is no further increase in the number of carriers since, as stated in para. 13.2, all the valence electrons are already free to move through the material. The effect of temperature on semiconductors will be more pronounced in intrinsic semiconductor materials where the number of carriers available will be dependent solely on the energy available to lift electrons from the valence to the conduction band.

Electromagnetic radiation (e.g. visible light) can supply energy in the same way as heat to raise electrons to the conduction band. Thus the resistivity of semiconductors will decrease with the absorption of this radiation. Semiconductors can be used therefore as detectors of electromagnetic radiation. As with heat, the effect is most pronounced with intrinsic semiconductors, and by cooling the semiconductor to very low temperatures the number of thermally-produced carriers can be decreased greatly and, as a result, extremely high sensitivity to electromagnetic radiation can be obtained.

The effect of mechanical stress on a material is to alter its physical dimensions (i.e. the material is said to be strained). Thus its resistance will alter as described in para. 2.5. Over and above this however, there will be a change in resistivity due to the re-orientation of the atoms, so producing a variation in the rate of collisions with carriers. Use is made of these phenomena in certain types of strain gauges which are used to measure small changes in dimensions of materials to which they are attached. Recent developments in strain gauges using semiconductors have produced very much greater sensitivities than those obtained with the more conventional metal wire types.

13.6 Atomic radiation

The condition in an atom whereby electrons exist with energies greater than the minimum allowable is an unstable one. The electrons will, after a short period of time, drop back to the lower allowable energy levels. The energy lost by the electron will be radiated from the atom in the form of electromagnetic radiation. The frequency of this radiation is related to the difference in energy between the two levels as specified by the equation:

$$\Delta W = hf \qquad (13{\cdot}1)$$

where ΔW = energy difference (J) and,
h = Planck's constant = $6{\cdot}624 \times 10^{-34}$ (J s).

This effect can be observed in many devices in which conduction takes place in a gas at low pressure. Energy is supplied to the gas atoms in the main by collision with fast moving electrons which lose some of their kinetic energy in the process. Gas atom electrons are raised to higher energies and when they drop back to the lower levels the gas emits light, the difference between the energy levels involved being such that the emission lies within the visible spectrum. The particular gas used gives rise to a characteristic colour of emitted light. Hydrogen, for instance, emits light in the orange, blue and violet regions of the visible spectrum; mercury vapour emits light in the yellow, green and blue regions and also in the ultra-violet range. Radiation from sodium vapour falls within a very narrow band in the yellow region.

The difference in levels near the nucleus is in practice much greater than the difference at the outer levels. When high kinetic-energy electrons are made to bombard a metal plate, some of these lower energy electrons can be removed completely from the atom. When other electrons drop down to take their places, the higher energy differences involved produce radiation of frequencies beyond the visible spectrum. These waves are known as X-rays and they have the property of being able to penetrate materials which are opaque to visible light. X-rays are used in industry for the examination of materials for internal flaws, and in medicine.

PROBLEMS ON ELECTRICAL PHYSICS

1 Two coils connected in parallel have resistances of 600 Ω and 300 Ω and temperature coefficients of 0·001/K and 0·004/K respectively at 20°C. Find the resistance of the combination at 50°C. What is the effective temperature coefficient of the combination?

217·5 Ω; 0·002 92/K

2 The field winding of a shunt motor has a resistance of 98 Ω at 16°C. Later the resistance was found to be steady at 114 Ω. Find the rise in temperature of the winding. Temperature coefficient of copper = 0·004/K.

43·7 K

3 The filament of a 60-W, 230-V lamp has a normal working temperature of 2 000°C. Take the temperature coefficient of the material to be 0·005/K. Find the approximate current which flows at the instant of switching on the supply to the cold lamp.

2·87 A

4 Define the temperature coefficient of resistance.
The field winding of a d.c. motor has a mean temperature of 15°C prior to the motor being energised. When the field is connected to a 240-V, d.c. supply, the initial steady current is 2·5 A, but after a period of operation, the field current falls to 2·3 A. The winding is made of copper, the temperature coefficient of resistance of which is 0·004/K. Calculate the final mean temperature of the winding.

65°C (SANCAD)

5 A light-sensitive circuit is shown in Fig. 13.5. The resistance of the photo-electric device is inversely proportional to the illumination falling on it. For a certain illumination, the ammeter indicates a current of 1·0 mA. What current will the ammeter indicate when the illumination is doubled?

1·84 mA (SANCAD)

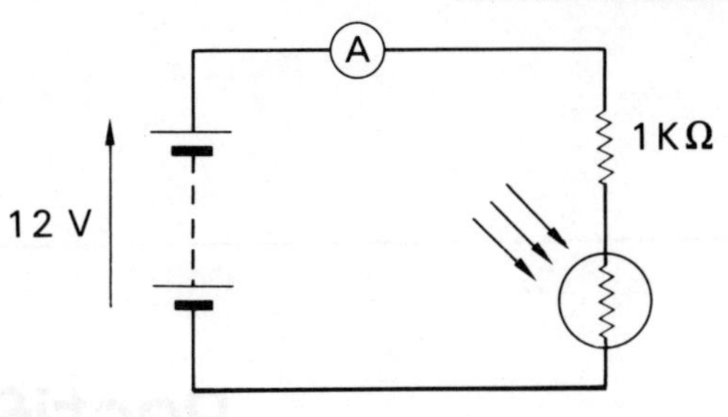

Fig. 13.5.

5 Discuss the factors that affect the resistance of a metallic conductor. State the effect of strain on the resistance of a conductor.
When copper is strained, the extension per unit length in the direction of the applied force is 6×10^{-9} N whilst the contraction per unit length at right angles to the direction of the applied force is $1{\cdot}8 \times 10^{-9}$ N. For a copper conductor having an unstrained resistance of $0{\cdot}124\ \Omega$, calculate, to three significant figures, its resistance when tensioned by a force of 160 kN, assuming the resistivity of the copper to remain unchanged.
(SANCAD)

14

Rectification

The most simple application of semiconductor materials is to be found in diodes and diode networks. Because diode networks permit the conversion of alternating current to direct current, the study of diodes has practical implications as well as being the means of introducing simple electronic devices. The conversion process is known as rectification.

14.1 p-n junction

When a junction is formed between p-type and n-type semiconductors, there is an immediate migration of majority carriers across the junction. This is due to a phenomenon known as diffusion whereby the carriers start to redistribute themselves in such a way that their densities tend to become uniform throughout the material. It is analagous to the distribution of gas molecules in a closed vessel producing uniform pressure within the vessel.

Since the regions on either side of the junction were originally electrically neutral, the passage of the negatively-charged electrons from the n-type to the p-type semiconductor, and the corresponding passage of effectively positively-charged holes from the p-type to the n-type semiconductor will result in a potential difference being built up between the two semiconductor types (since electrons now have to move from a region of high potential to one of low potential and vice-versa for the holes). The polarity of the potential difference is such that the n-type is positive with respect to the p-type and will thus tend to oppose the diffusion. Eventually a condition will be reached whereby the potential difference so built up will be sufficient to stop the diffusion and a state of equilibrium will exist. There will have been created a narrow layer about the junction in which there is a reduction in carrier densities due to the recombination of carriers within the layer. The electric field established across the layer will be relatively large in magnitude due to the small thickness of the layer. It is known as the depletion layer and the conditions existing within and on either side of it are illustrated in Fig. 14.1.

Consider now what happens when an external potential difference is applied across the p-n junction just described. If the polarity of this voltage is such that the p-type

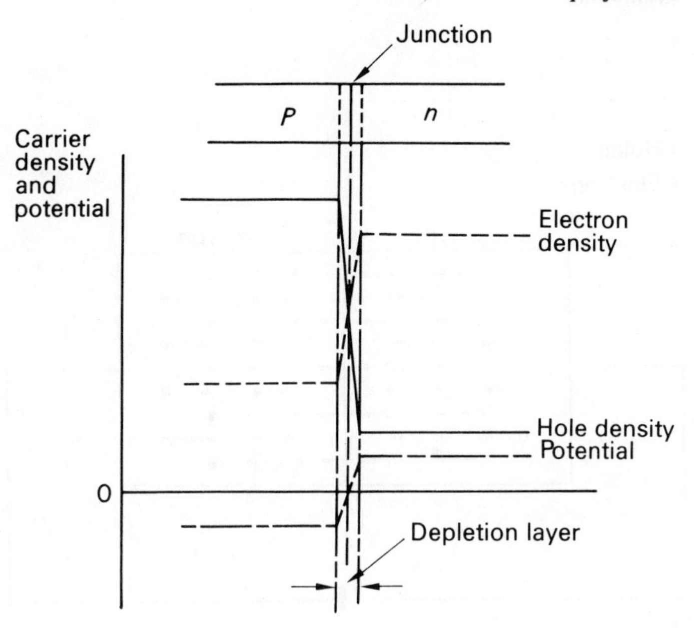

Fig. 14.1. p-n junction

region is positive with respect to the n-type region then it will oppose the potential difference built up internally at the junction. This will destroy the equilibrium established and majority carriers will commence crossing the junction again. The source of the applied voltage injects electrons into the n-type region and collects them from the p-type region, so creating further holes, and thus a current is established. The p-n junction is said to be forward biased.

If the polarity of the applied voltage is reversed, it will assist the internally-developed potential difference. Thus there is even less likelihood of majority carriers crossing the junction. A small current does flow however due to the presence of minority carriers on both sides of the junction. The depletion layer electric field actually assists their crossing of the junction. Initially this small current increases with applied voltage but eventually becomes essentially independent of it. This is a saturation effect whereby the minority carriers are crossing the junction at the rate at which they are being created which is, of course, a function of temperature. Still further increase of applied voltage eventually produces a rapid increase in current. This is due to electron-hole pairs being created in large numbers within the depletion layer and, due to the electric field existing there, being immediately swept out. An explanation of the mechanisms producing the electron-hole pairs is beyond the scope of this book. The effect is known as reverse breakdown although it is basically non-destructive. Care must be taken however to ensure that the maximum power rating of the device is not exceeded when it is operating in this mode. With the polarity of applied voltage as described in this paragraph the p-n junction is said to be reverse biased. Figure 14.2 illustrates the conduction processes in the forward and reverse biased conditions.

The p-n junction does therefore have the useful property of conducting very much better in one direction than in the other. It has similar characteristics to a thermionic vacuum diode which it has to a large extent replaced in electronic circuits. The term

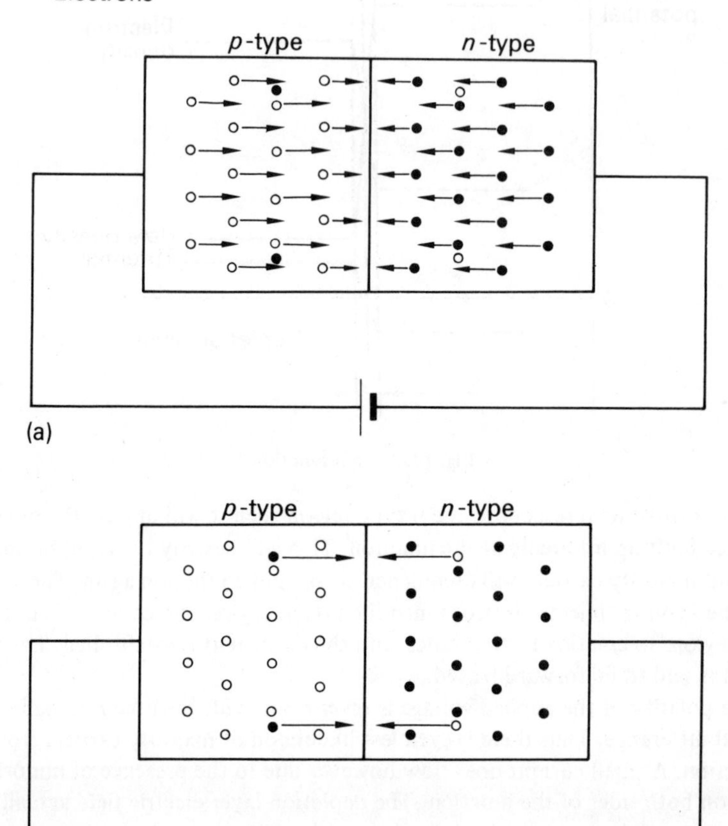

Fig. 14.2. Conduction across p-n junction (*a*) Forward biased, (*b*) Reverse biased

'diode' has been retained and such devices are known as semiconductor diodes. Various processes are employed in practice to produce the junction described, it being sufficient at this stage to say that it is not possible to produce an effective p-n junction by simply bringing into contact p and n-type semiconductors. The impurity content is controlled during manufacture and a p-n junction is formed in a single piece of semiconductor.

Figure 14.3 shows a typical characteristic for a silicon diode. The reader's attention is drawn to the different scales used in the display of the forward and reverse characteristics.

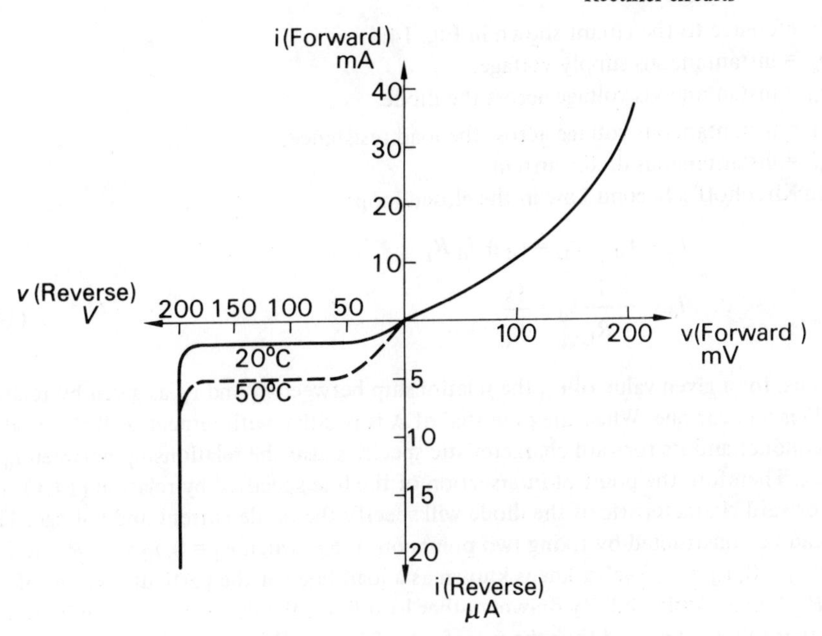

Fig. 14.3. **Silicon diode characteristic**

14.2 Rectifier circuits

Since a diode has the characteristic of having a much greater conductivity in one direction than in the other, it will produce a direct component of current when connected in series with an alternating voltage and a load. This process is known as rectification and is the main use to which diodes are put. There are numerous applications for rectification e.g. driving a d.c. motor from a.c. mains and the production of direct-voltage supplies for electronic amplifiers.

14.2.1 Half-wave rectifier (with resistive load)

The basic half-wave rectifier circuit is shown in Fig. 14.4. The arrowhead on the diode symbol indicates the direction of conventional current when the diode is forward biased.

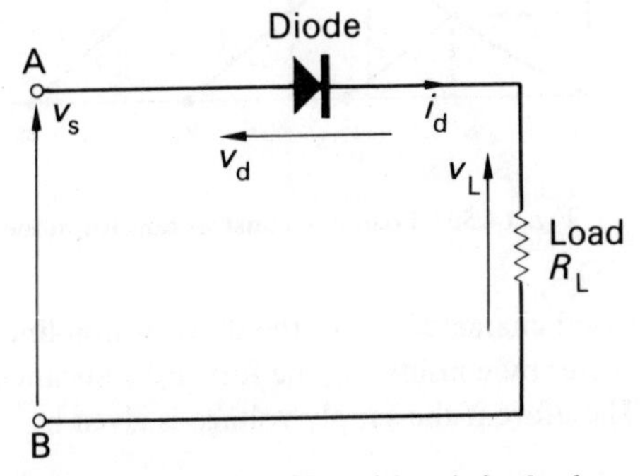

Fig. 14.4. **Half-wave rectifier with resistive load**

With reference to the circuit shown in Fig. 14.4.

v_s = instantaneous supply voltage,
v_d = instantaneous voltage across the diode,
v_L = instantaneous voltage across the load resistance,
i_d = instantaneous diode current.

Using Kirchhoff's Second Law in the closed loop:

$$v_s = v_d + v_L = v_d + i_d R_L$$

$$\therefore \quad i_d = -\frac{1}{R_L} v_d + \frac{v_s}{R_L} \tag{14.1}$$

Thus, for a given value of v_s, the relationship between i_d and v_d as given by relation (14.1) is a linear one. When the potential of A is positive with respect to B the diode will conduct and its forward characteristic specifies also the relationship between i_d and v_d. Therefore the point of intersection of the line specified by relation (14.1) and the forward characteristic of the diode will specify the diode current and voltage. The line can be constructed by fixing two points on it, e.g. when $v_d = 0$, $i_d = v_s/R_L$ and when $i_d = 0$, $v_d = v_s$. Such a line is known as a load line for the particular values of v_s and R_L being considered. By drawing other load lines, the diode current can be found for other values of v_s, and thus the waveform of i_d over the half cycle can be determined. The construction is illustrated in Fig. 14.5.

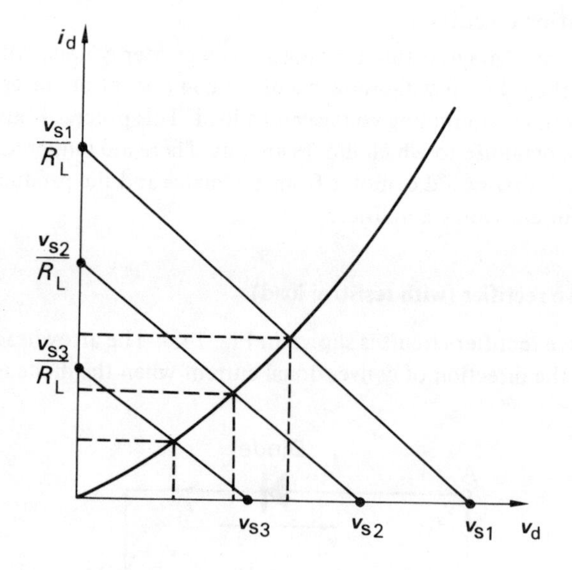

Fig. 14.5. Load line construction for diode

Although the forward characteristic of the diode is non-linear, sufficient accuracy is obtained in many cases by considering the forward resistance r_d of the diode constant or even zero. Therefore if the supply voltage is given by:

$$v_s = V_m \sin \omega t$$

then

$$i_d = \frac{v_s}{r_d + R_L} = \frac{V_m \sin \omega t}{r_d + R_L} = I_m \sin \omega t.$$

Thus if the mean value of the current (neglecting the reverse current) $= I_{dc}$ then:

$$I_{dc} = \frac{1}{2\pi}\int_0^{\pi} I_m \sin \omega t \, d(\omega t) = \frac{I_m}{2\pi}[-\cos \omega t]_0^{\pi}$$

$$= \frac{I_m}{2\pi}[-\cos \pi + \cos 0] = \frac{I_m}{2\pi}[1+1]$$

$$\therefore \quad I_{dc} = \frac{I_m}{\pi}. \tag{14.2}$$

Similarly the r.m.s. value of the current $= I_{rms}$ then:

$$I_{rms} = \sqrt{\frac{1}{2\pi}\int_0^{\pi} I_m{}^2 \sin^2 \omega t \, d(\omega t)}$$

$$= \sqrt{\frac{I_m{}^2}{2\pi}\int_0^{\pi} \tfrac{1}{2}(1 - \cos 2\omega t)\, d(\omega t)}$$

$$= \sqrt{\frac{I_m{}^2}{2\pi} \times \tfrac{1}{2}[\omega t + \tfrac{1}{2}\sin 2\omega t]_0^{\pi}} = \sqrt{\frac{I_m{}^2}{4\pi}[\pi]}$$

$$\therefore \quad I_{rms} = \frac{I_m}{2} \tag{14.3}$$

The voltage across the load is given by:

$$v_L = i_d R_L = I_m \sin \omega t \, R_L = V_{Lm} \sin \omega t.$$

Therefore in the same way as derived for the current:

$$\text{Mean value of the load voltage} = V_{dc} = \frac{V_{Lm}}{\pi} \tag{14.4}$$

$$\text{R.M.S. value of the load voltage} = V_{rms} = \frac{V_{Lm}}{2} \tag{14.5}$$

The maximum voltage, which occurs across the diode in the reverse direction, is known as the peak inverse voltage (P.I.V.). This must be less than the breakdown voltage of the diode if it is not to conduct appreciably in the reverse direction. The peak inverse voltage for the diode in this circuit occurs when the potential of B is positive with respect to A by its maximum amount. The reverse resistance of the diode will, in the great majority of practical cases, be very much greater than the load resistance and most of the applied voltage will appear across the diode. Thus the peak inverse voltage equals approximately the peak value of the supply voltage.

Since the production of direct current from an alternating supply is the object of the circuit, the useful power output is that produced in the load by the direct component of the current. The efficiency of rectifier circuits is defined therefore as:

$$\text{Efficiency } (\eta) = \frac{\text{Power in the load by direct component of current}}{\text{Total power dissipated in the complete circuit}}.$$

Therefore:

$$\eta = \frac{I_{dc}{}^2 R_L}{I_{rms}^2(r_d + R_L)} = \frac{\left(\frac{I_m}{\pi}\right)^2 R_L}{\left(\frac{I_m}{2}\right)^2 (r_d + R_L)}$$

$$\therefore \quad \eta = \frac{4R_L}{\pi^2(r_d + R_L)} \tag{14.6}$$

If $r_d = 0$, the efficiency will be a maximum.

i.e.

$$\eta_m = \frac{4}{\pi^2} = 0{\cdot}405 \text{ (i.e. } 40{\cdot}5\%).$$

Waveforms for the half-wave rectifier circuit are shown in Fig. 14.6.

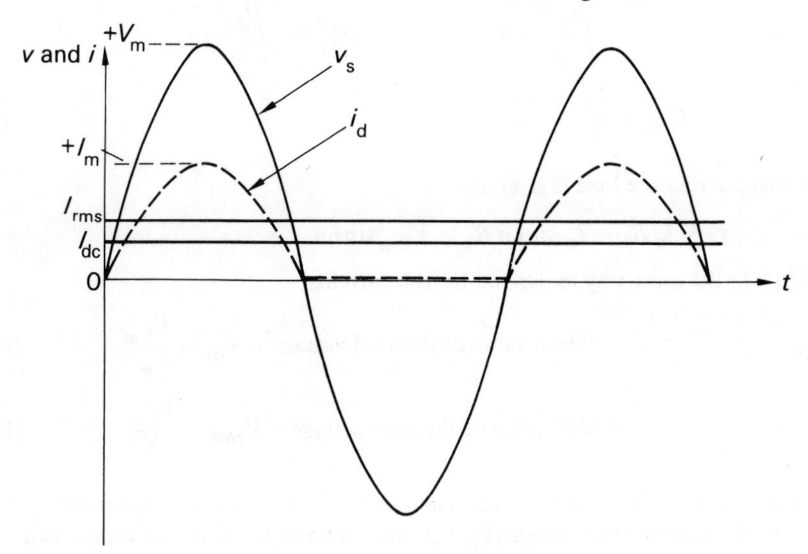

Fig. 14.6. Waveforms for half-wave rectifier circuit

14.2.2 Full-wave rectifier network (with resistive load)

The half-wave rectifier gave rise to an output which had a unidirectional current but the resulting current does not compare favourably with the direct current that one would expect from, say, a battery. One reason is that the half-wave rectifier is not making use of the other half of the supply waveform. It would be technically and economically advantageous if both halves were rectified and this may be achieved by a full-wave rectifier network.

The basic full-wave rectifier network is shown in Fig. 14.7. C is a centre tap on the secondary of the transformer, thus the e.m.f.'s induced in each section of the secondary are equal, and when the potential of A is positive with respect to C, so is that of C positive with respect to B. With these polarities, diode D_1 will conduct while diode D_2 is non-conducting. When these polarities reverse diode D_2 will conduct and D_1 will be non-conducting. In this way, each diode conducts on alternate half cycles, passing current through the load in the same direction.

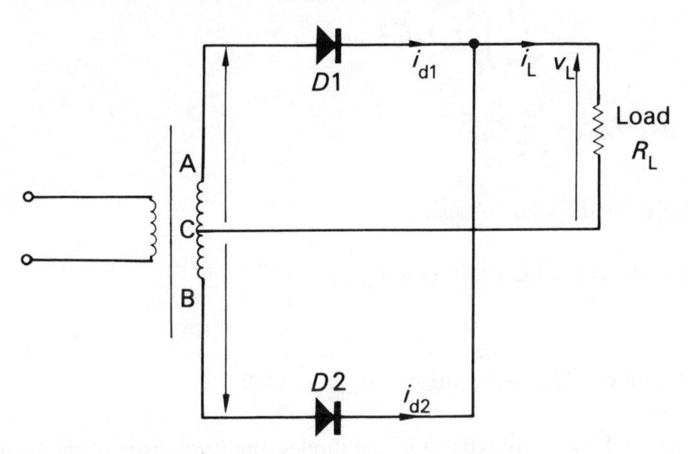

Fig. 14.7. Full-wave rectifier with resistive load

For $v_{AC} = v_s = V_m \sin \omega t$

then $v_{BC} = -v_s = -V_m \sin \omega t$.

For identical diodes with a forward resistance r_d which can be considered constant and an infinite reverse resistance, then during the period that v_s is positive:

$$\text{Diode 1 current} = i_{d1} = \frac{V_m \sin \omega t}{r_d + R_L} = I_m \sin \omega t$$

$$\text{Diode 2 current} = i_{d2} = 0$$

During the period that v_s is negative:

$$\text{Diode 1 current} = i_{d1} = 0$$

$$\text{Diode 2 current} = i_{d2} = \frac{-V_m \sin \omega t}{r_d + R_L} = -I_m \sin \omega t$$

At any instant the load current is given by:

$$i_L = i_{d1} + i_{d2}$$

Thus in this circuit the current will repeat itself twice every cycle of the supply voltage, therefore:

Mean value of the load current = I_{dc}

$$= \frac{1}{\pi}\int_0^{\pi} I_m \sin\omega t \, d(\omega t)$$

$$\therefore \quad I_{dc} = \frac{2I_m}{\pi} \tag{14·7}$$

R.M.S. value of the load current = I_{rms}

$$= \sqrt{\frac{1}{\pi}\int_0^{\pi} I_m^2 \sin^2 \omega t \, d(\omega t)}$$

$$\therefore \quad I_{rms} = \frac{I_m}{\sqrt{2}} \tag{14.8}$$

Similarly for the load voltage:

$$\text{Mean value of the load voltage} = V_{dc} = \frac{2V_{Lm}}{\pi} \tag{14.9}$$

$$\text{R.M.S. value of the load voltage} = V_{rms} = \frac{V_{Lm}}{\sqrt{2}} \tag{14.10}$$

Neglecting the forward resistance of the diodes, the peak value of the load voltage is equal to the peak value of the half secondary voltage. At this instant therefore, one end of the non-conducting diode is + V_m while the other end is – V_m with respect to the centre tap C and the peak inverse voltage is $2V_m$.

The efficiency of the circuit is given by:

$$\eta = \frac{I_{dc}^2 R_L}{I_{rms}^2 (r_d + R_L)} = \frac{\left(\frac{2I_m}{\pi}\right)^2 R_L}{\left(\frac{I_m}{\sqrt{2}}\right)^2 (r_d + R_L)}$$

$$\therefore \quad \eta = \frac{8R_L}{\pi^2 (r_d + R_L)} \tag{14.11}$$

If $r_d = 0$, the efficiency will be a maximum.

i.e.

$$\eta_m = \frac{8}{\pi^2} = 0{\cdot}810 \text{ (i.e. } 81{\cdot}0\%)$$

Waveforms for the full-wave rectifier network are shown in Fig. 14.8.

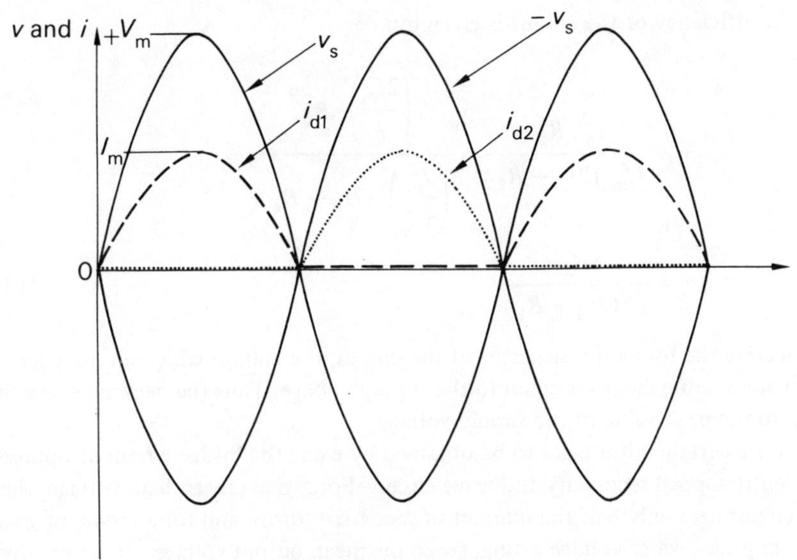

Fig. 14.8. Waveforms for full-wave rectifier circuit

14.2.3 Bridge rectifier network (with resistive load)

The full-wave rectifier only makes use of each half of the transformer winding for half of the time. The winding can be used all of the time, or the transformer can be omitted in certain cases, by the application of the bridge rectifier network.

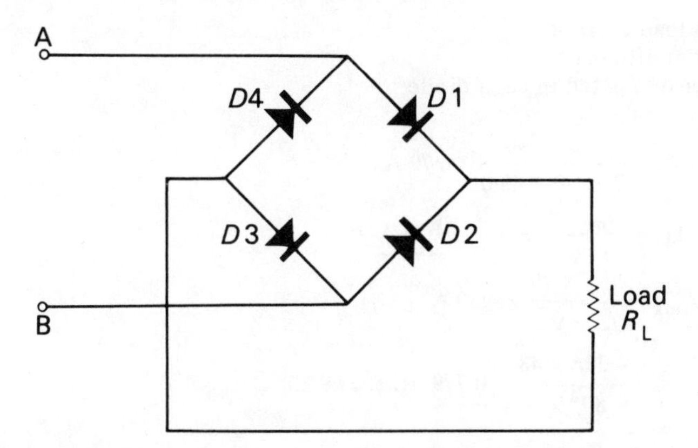

Fig. 14.9. Bridge rectifier network with resistive load

When the potential of A is positive with respect to B, diodes D_1 and D_3 conduct and a current flows in the load. When the potential of B is positive with respect to A diodes D_2 and D_4 conduct and the current in the load is in the same direction as before. Thus a full-wave type of output is obtained. The expressions derived for current and load voltage for the full-wave circuit will be applicable here also. Note should be taken however that, at any instant, two diodes are conducting so that the total resistance in circuit is $2r_d + R_L$.

The efficiency of the circuit is given by:

$$\eta = \frac{I_{dc}^2 R_L}{I_{rms}^2(2r_d + R_L)} = \frac{\left(\frac{2I_m}{\pi}\right)^2 R_L}{\left(\frac{I_m}{\sqrt{2}}\right)^2 (2r_d + R_L)}$$

$$\therefore \quad \eta = \frac{8R_L}{\pi^2(2r_d + R_L)} \qquad (14.12)$$

Neglecting the forward resistances of the diodes, the voltage which appears across the non-conducting diodes is equal to the supply voltage. Thus the peak inverse voltage is equal to the peak value of the supply voltage.

There are certain advantages to be obtained by using the bridge circuit as opposed to the centre tapped secondary full-wave circuit. For a given mean load voltage, the bridge circuit uses only half the number of secondary turns, and for a diode, of given maximum peak-inverse-voltage rating, twice the mean output voltage can be obtained from it since the peak inverse voltage encountered is V_m compared to $2V_m$ for the full-wave circuit.

Example 14.1 The four diodes used in a bridge rectifier circuit have forward resistances which may be considered constant at 1·0 Ω and an infinite reverse resistance. The alternating supply voltage is 240 V r.m.s. and the resistive load is 480 Ω. Calculate:

(*a*) The mean load current.
(*b*) The rectifier efficiency.
(*c*) The power dissipated in each diode.

(*a*)
$$I_m = \frac{\sqrt{2} \times 240}{2 \times 1{\cdot}0 + 48{\cdot}0} = 6{\cdot}79 \text{ A}$$

$$I_{dc} = \frac{2I_m}{\pi} = \frac{2 \times 6{\cdot}79}{\pi} = \underline{4{\cdot}32 \text{ A}}$$

(*b*)
$$I_{rms} = \frac{I_m}{\sqrt{2}} = \frac{6{\cdot}79}{\sqrt{2}} = 4{\cdot}80 \text{ A}$$

$$\therefore \quad \eta = \frac{4{\cdot}32^2 \times 48}{4{\cdot}80^2 \times 50} = \underline{0{\cdot}779} \quad \text{(i.e. 77·9\%)}$$

(*c*) Since each diode conducts for only half a cycle then diode r.m.s. current

$$= \frac{I_m}{2} = \frac{6{\cdot}79}{2} = 3{\cdot}39 \text{ A.}$$

∴ Power dissipated in each diode = $3{\cdot}39^2 \times 1{\cdot}0$ = <u>11·5 W.</u>

Example 14.2 A battery charging circuit is shown in Fig. 14.10. The forward resistance of the diode can be considered constant at 2·0 Ω and the reverse resistance infinite. Calculate the necessary value of the variable resistance R so that the battery charging current is 1·0 A. The internal resistance of the battery is negligible.

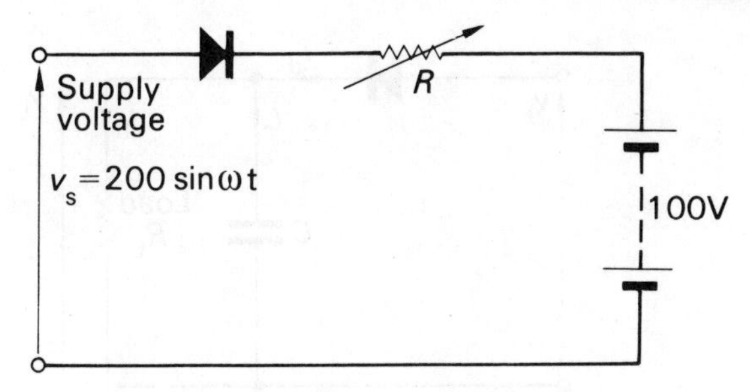

Fig. 14.10. Circuit for example 14.2

Diode conducts during the period $v_s > 100$ V

$200 \sin \omega t = 100$ i.e. $\sin \omega t = 0{\cdot}5$ i.e. when $\omega t = \dfrac{\pi}{6}$ and $\dfrac{5\pi}{6}$

$\therefore$ diode conducts when $\dfrac{\pi}{6} < \omega t < \dfrac{5\pi}{6}$

During conduction $i = \dfrac{v_s - 100}{r_d + R} = \dfrac{200 \sin \omega t - 100}{2{\cdot}0 + R}$

$$\text{Mean value of current} = 1{\cdot}0 = \frac{1}{2\pi}\int_{\frac{\pi}{6}}^{\frac{5\pi}{6}} \frac{200 \sin \omega t - 100}{2{\cdot}0 + \mathrm{R}}\, d(\omega t)$$

$$= \frac{1}{2\pi(2{\cdot}0 + R)}\left[-200 \cos \omega t - 100\,\omega t\right]_{\frac{\pi}{6}}^{\frac{5\pi}{6}}$$

$$= \frac{1}{2\pi(2{\cdot}0 + R)}\left[-200 \cos\frac{5\pi}{6} - 100 \times \frac{5\pi}{6} + 200 \cos\frac{\pi}{6} + 100 \times \frac{\pi}{6}\right]$$

$$= \frac{1}{2\pi(2{\cdot}0 + R)}\left[-200\left(-\frac{\sqrt{3}}{2}\right) - \frac{500\pi}{6} + 200\left(\frac{\sqrt{3}}{2}\right) + \frac{100\pi}{6}\right]$$

$$= \frac{1}{2\pi(2{\cdot}0 + R)}\left[200\sqrt{3} - \frac{400\pi}{6}\right]$$

$$2{\cdot}0 + R = \frac{200\sqrt{3}}{2\pi} - \frac{400\pi}{2\pi \times 6}$$

$$\therefore \quad R = 55{\cdot}1 - 33{\cdot}3 - 2{\cdot}0 = \underline{19{\cdot}8\ \Omega}$$

14.3 Smoothing

The rectifier circuits so far described have produced, as required a direct component of current in the load. There remains, however a large alternating component. In a large number of applications it is desirable to keep this latter component small. This can be accomplished by the use of smoothing circuits, the simplest of which consists of a capacitor in parallel with the load. Figure 14.11 shows such an arrangement.

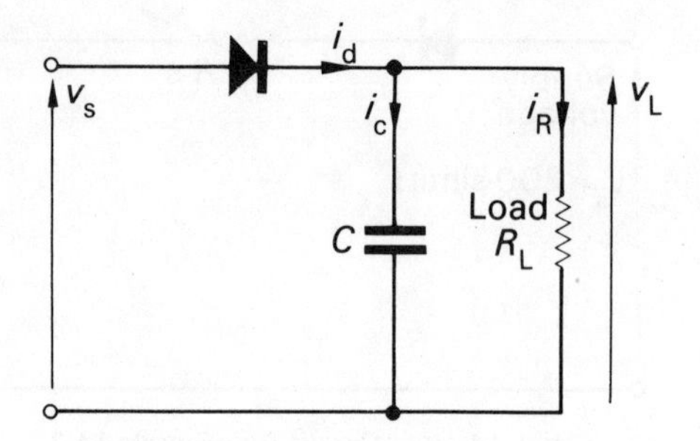

Fig. 14.11. Half-wave rectifier with capacitor input filter

The diode conducts when the supply voltage v_s is more positive than the load voltage v_L. During this conduction period, if the diode forward resistance is neglected, then the load voltage is equal to the supply voltage. Therefore if:

$$v_s = V_m \sin \omega t$$

$$i_R = \frac{v_s}{R_L} = \frac{V_m \sin \omega t}{R_L}$$

$$i_c = C\frac{dv_s}{dt} = C\omega V_m \cos \omega t.$$

$$\text{Diode current} = i_d = i_R + i_c = \frac{V_m}{R_L}\sin \omega t + \omega C V_m \cos \omega t$$

$$= \sqrt{\left(\frac{V_m}{R_L}\right)^2 + (\omega C V_m)^2}\,\sin(\omega t + \Phi)$$

$$\therefore \quad i_d = \frac{V_m}{R_L}\sqrt{1 + \omega^2 C^2 R_L{}^2}\,\sin(\omega t + \Phi) \tag{14.13}$$

where $$\tan \Phi = \frac{\omega C V_m}{\dfrac{V_m}{R_L}} = \omega C R_L$$

If $i_d = 0$, i.e. diode cuts-off, when $t = t_2$ then

$$\sin(\omega t_2 + \Phi) = 0$$

$$\therefore \quad \omega t_2 + \Phi = \pi$$

$$\therefore \quad \omega t_2 = \pi - \Phi \tag{14.14}$$

If $\omega C R_L \gg 1$ then $\Phi \doteqdot \dfrac{\pi}{2}$

and $\omega t_2 \doteqdot \pi - \dfrac{\pi}{2} = \dfrac{\pi}{2}$

i.e. the diode ceases to conduct near the instant at which v_s has its positive maximum value.

While the diode is non-conducting, C will discharge through R_L and the load voltage will be given by:

$$v_L = Ve - \frac{(t-t_2)}{CR_L}$$

where V = the capacitor voltage at the instant the diode cuts off. This will equal approximately the peak value of the supply voltage if $\omega CR_L \gg 1$. The diode will start to conduct again during the period that the supply voltage is positive and increasing. This instant can be determined by equating $Ve^{-(t-t_2)/CR_L}$ and $V_m \sin \omega t$; one solution of which will give $T + t_1$ where:

T = period of the supply voltage, and
t_1 = instant of time at which the diode starts to conduct.

In order to keep the variation in load voltage down during the period when the diode is non-conducting a long time constant CR_L compared to the period of the supply voltage is required. Note, however, from relation (14.13) that the peak value of the diode current increases with C and therefore care must be taken to ensure that the maximum allowable diode peak current is not exceeded. Waveforms for the circuit are shown in Fig. 14.12.

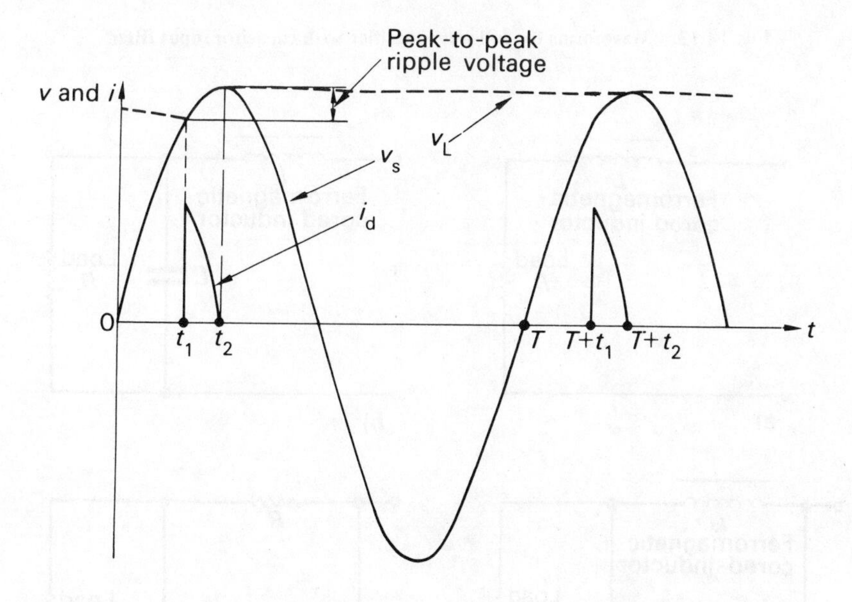

Fig. 14.12. Waveforms for half-wave rectifier with capacitor input filter

A similar analysis could be carried out for the full-wave circuit. The operation is identical to that of the half-wave circuit during the charging period, but the capacitor discharges into the load resistance for a shorter period, giving less amplitude of ripple for a given time constant during the non-conducting period. Figure 14.13 shows waveforms for a full-wave circuit with a capacitor input filter.

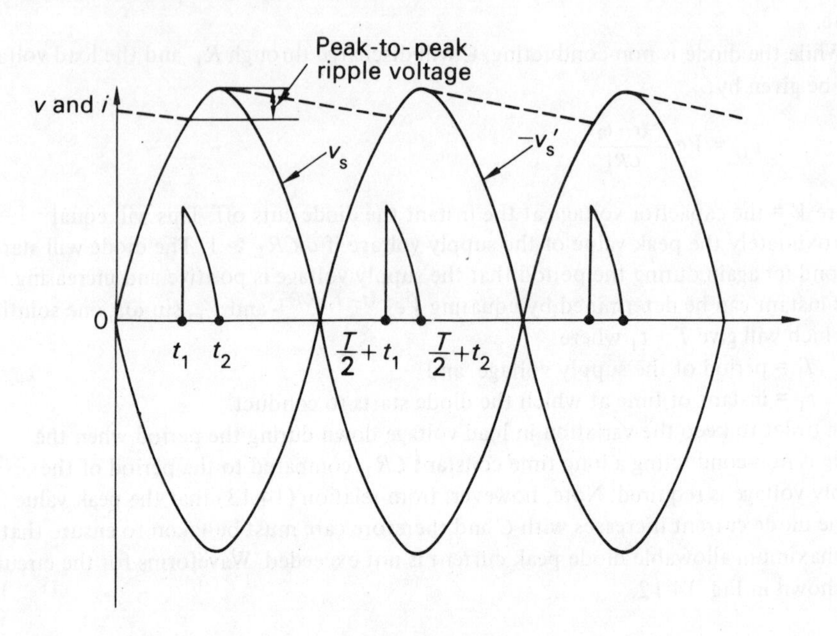

Fig. 14.13. Waveforms for full-wave rectifier with capacitor input filter

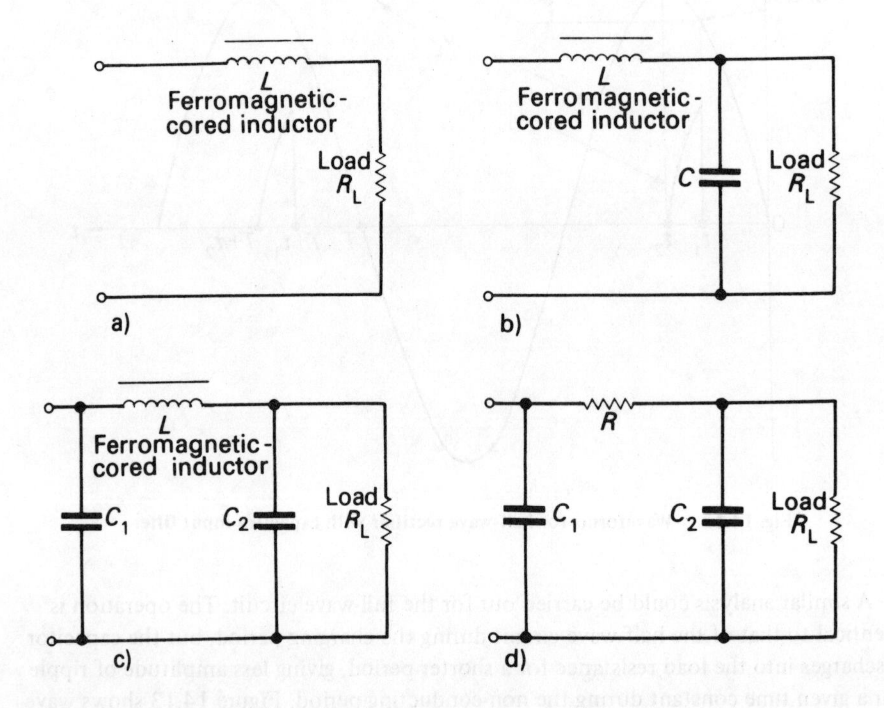

Fig. 14.14. Smoothing circuits (*a*) Series inductor filter (*b*) L-C filter (*c*) π-Filter (*d*) Resistive filter

Figure 14.14 shows some other types of filter circuits used in practice for smoothing purposes. The ideal arrangement is to connect a low valued impedance at the frequency being used and which presents a high resistance to direct current, e.g. a capacitor, in parallel with the load, and/or a high valued impedance at the frequency being used and which presents a low resistance to direct current, e.g. an inductor, in series with the load. This has the effect of minimising the ripple across the load while having little effect on the direct voltage developed at the input to the filter circuit. The circuit of Fig. 14.14 (*d*) does not meet these requirements fully in so much as the resistor R will reduce the direct voltage as well as the ripple. The circuit is useful however in low current applications where the direct voltage drop across R can be kept small; a resistor has an economic advantage over a ferromagnetic-cored inductor.

14.4 3-phase rectifier circuits

Diodes are used with polyphase supplies to give a rectified output which has basically less ripple than rectifiers operating from single phase supplies. Figure 14.15 shows a simple 3-ph rectifier circuit. Each diode will conduct for one third of a cycle, the conduction path changing instantaneously from one diode to another as one phase voltage becomes more positive than another. For example when v_{RN} becomes less positive than v_{YN} diode D_1 ceases to conduct while D_2 starts to conduct.

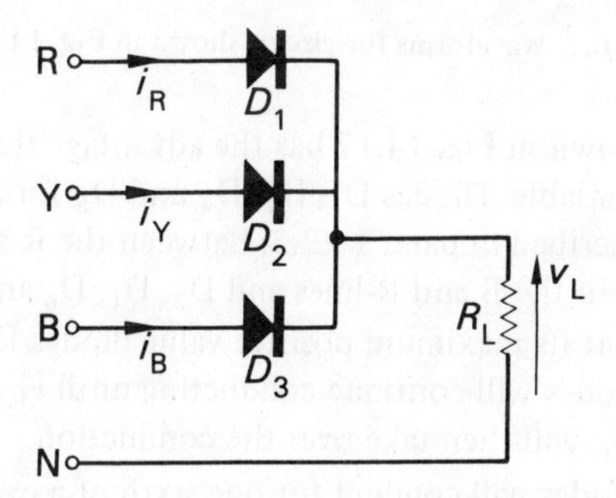

Fig. 14.15. Simple 3-ph rectifier circuit

The mean load current can be determined as follows:

$$I_{dc} = \frac{1}{2\pi/3}\int_{\frac{\pi}{6}}^{\frac{5\pi}{6}} I_m \sin \omega t \, d(\omega t)$$

$$= \frac{3I_m}{2\pi}\Big[-\cos \omega t\Big]_{\frac{\pi}{6}}^{\frac{5\pi}{6}} = \frac{3I_m}{2\pi}\left[-\cos\frac{5\pi}{6} + \cos\frac{\pi}{6}\right]$$

$$= \frac{3I_m}{2\pi}\left[\frac{\sqrt{3}}{2} + \frac{\sqrt{3}}{2}\right]$$

$$I_{dc} = \frac{3\sqrt{3}\,I_m}{2\pi} \qquad (14.15)$$

The mean load voltage will be given by:

$$V_{dc} = I_{dc} R_L$$

$$\therefore \quad V_{dc} = \frac{3\sqrt{3}\, I_m R_L}{2\pi} \tag{14.16}$$

where $I_m R_L = V_m$, the peak value of the phase voltage if the diode forward resistance is neglected. Figure 14.16 shows waveforms for the circuit.

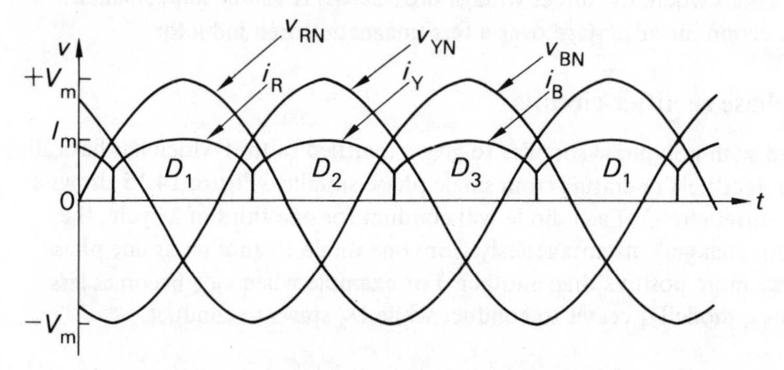

Fig. 14.16. Waveforms for circuit shown in Fig. 14.15

The rectifier circuit shown in Fig. 14.17 has the advantage that there is no necessity to have a neutral point available. Diodes D_1, D_2, D_4 and D_5 for a bridge rectifier circuit, similar to that described in para. 14.2.3, between the R and Y lines, similarly D_2, D_3, D_5 and D_6 between the B and R lines and D_3, D_1, D_6 and D_4 between the B and R lines. When v_{RY} is at its maximum positive value diodes D_1 and D_5 will be conducting. These two diodes will continue conducting until v_{RB} is more positive than v_{RY} and diodes D_1 and D_6 will then take over the conduction.

In this way a pair of diodes will conduct for one sixth of a cycle at a time, each diode conducting for one third of a cycle. The waveforms for the circuit are shown in

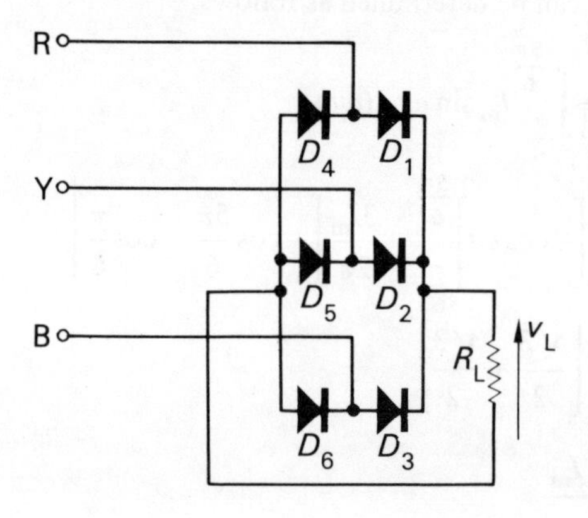

Fig. 14.17. 3-ph bridge rectifier circuit

Fig. 14.18 where it can be seen that the ripple frequency is twice that obtained in the previous circuit. The amplitude of ripple is decreased also since the conduction periods about the peaks of the supply is decreased.

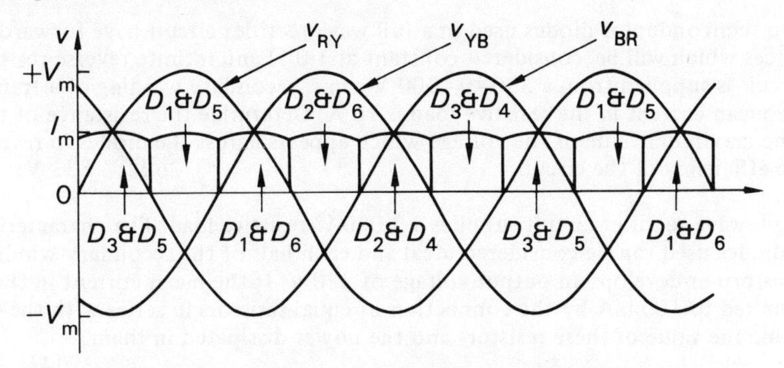

Fig. 14.18. Waveforms for circuit shown in Fig. 14.17

The mean load current can be determined as follows:

$$I_{dc} = \frac{1}{\pi/3} \int_{\frac{\pi}{3}}^{\frac{2\pi}{3}} I_m \sin \omega t \, d(\omega t)$$

$$= \frac{3I_m}{\pi} \Big[-\cos \omega t \Big]_{\frac{\pi}{3}}^{\frac{2\pi}{3}}$$

$$= \frac{3I_m}{\pi} \left[-\cos \frac{2\pi}{3} + \cos \frac{\pi}{3} \right] = \frac{3I_m}{\pi} \left[\tfrac{1}{2} + \tfrac{1}{2}\right]$$

$$\therefore \quad I_{dc} = \frac{3I_m}{\pi} \qquad (14.17)$$

PROBLEMS ON RECTIFICATION

1 A silicon diode has forward characteristics given in the following table:

V (V)	0	0·50	0·75	1·00	1·25	1·50	1·75	2·00
I (A)	0	0·1	0·3	1·5	3·3	5·0	6·7	8·5

The diode is connected in series with a 1·0-Ω resistor across an alternating voltage supply of peak value 10·0 V. Determine the peak value of the diode forward current and the value of the series resistor which would limit the peak current to 5·0 A.

8·1 A; 1·7 Ω

2 A semiconductor diode the forward and reverse characteristics of which can be considered ideal is used in a half wave rectifier circuit supplying a resistive load of 1 000 Ω. If the r.m.s. value of the sinusoidal supply voltage is 250 V determine:

(*a*) The peak diode current.
(*b*) The mean diode current.

(*c*) The r.m.s. diode current and
(*d*) The power dissipated in the load.

(*a*) 354 mA; (*b*) 112 mA; (*c*) 177 mA; (*d*) 31·3 W

3 Two semiconductor diodes used in a full wave rectifier circuit have forward resistances which will be considered constant at 1·0 Ω and infinite reverse resistances. The circuit is supplied from a 300–0–300 V r.m.s. secondary winding of a transformer and the mean current in the resistive load is 10 A. Determine the resistance of the load, the maximum value of the voltage which appears across the diodes in reverse and the efficiency of the circuit. 26 Ω; 833 V; 78·2 %

4 A full wave rectifier circuit supplies a 2 000-Ω resistive load. The characteristics of the diodes used can be considered ideal and each half of the secondary winding of the transformer develops an output voltage of 250 V. If the mean current in the load is to be limited to 100 mA by the connection of equal resistors in series with the diodes determine the value of these resistors and the power dissipated in them.

250 Ω; 1·54 W

5 The four semiconductor diodes used in a bridge rectifier circuit have forward resistances which can be considered constant at 0·1 Ω and infinite reverse resistances. They supply a mean current of 10 A to a resistive load from a sinusoidally varying alternating supply of 20 V r.m.s. Determine the resistance of the load and the efficiency of the circuit. 1·6 Ω; 72 %

6 A half-wave rectifier circuit is used to charge a battery of e.m.f. 12 V and negligible internal resistance. The sinusoidally varying alternating supply voltage is 24 V peak. Determine the value of resistance to be connected in series with the battery to limit the charging current to 1·0 A. What peak current would flow in the diode if the battery were reversed? 2·61 Ω; 13·8 A

15

Transistor circuits

This chapter deals with the operation and use of transistors in amplifier circuits. Both graphical and analytical methods are used in the analysis of these circuits. In the main a simplified h-parameter equivalent circuit has been adapted for gain calculations although a section on T-parameters is included.

15.1 n-p-n and p-n-p transistor action

The transistor is a three-layer semiconductor device consisting of either p-type material sandwiched between two layers of n-type material giving an n-p-n transistor, or alternatively n-type material sandwiched between two layers of p-type material giving a p-n-p transistor. For both types the middle region is known as the base and the outer

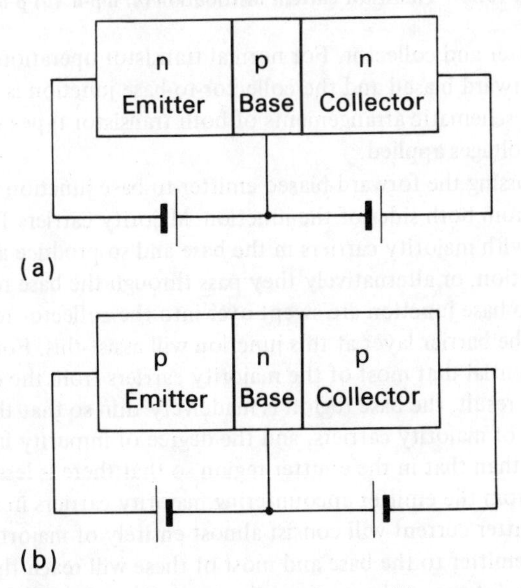

Fig. 15.1. Transistor types (*a*) n-p-n (*b*) p-n-p

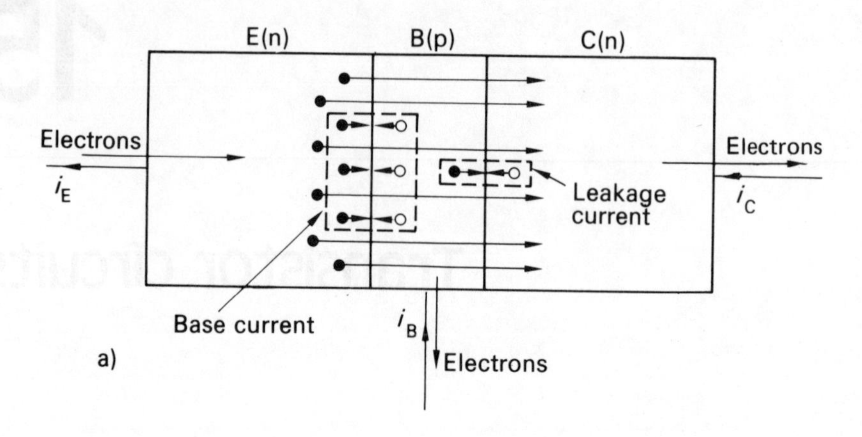

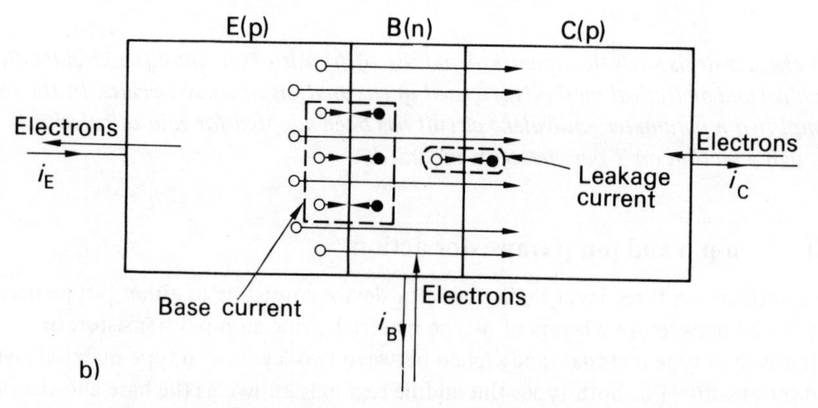

Fig. 15.2. Transistor current distribution (*a*) n-p-n (*b*) p-n-p

regions as the emitter and collector. For normal transistor operation, the emitter-to-base junction is forward biased and the collector-to-base junction is reverse biased. Figure 15.1 shows schematic arrangements of both transistor types with the correct polarities of bias voltages applied.

The current crossing the forward-biased emitter-to-base junction is carried by majority carriers from both sides of the junction. Majority carriers from the emitter either recombine with majority carriers in the base and so produce an external current in the base connection, or alternatively they pass through the base region and on reaching the collector-to-base junction are swept over into the collector region; the potential difference across the barrier layer at this junction will assist this. For normal transistor operation, it is essential that most of the majority carriers from the emitter reach the collector. As a result, the base region is made very thin so that there is little space for recombination of majority carriers, and the degree of impurity in the base region is made much less than that in the emitter region so that there is less likelihood of majority carriers from the emitter encountering majority carriers in the base. This means that the emitter current will consist almost entirely of majority carriers crossing from the emitter to the base and most of these will reach the collector. The normal leakage current across the reverse biased collector to base junction completes

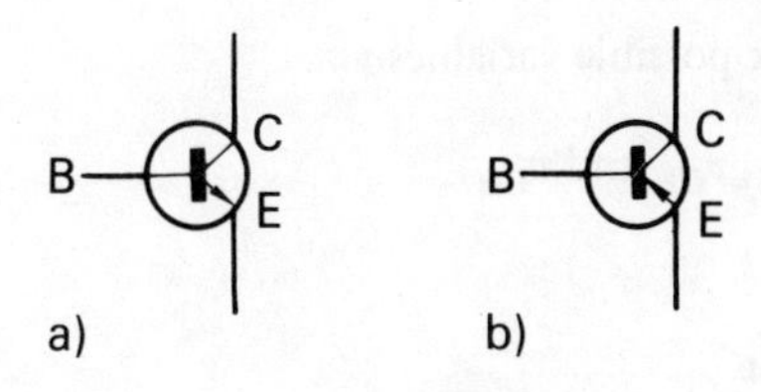

Fig. 15.3. Transistor symbols (*a*) n-p-n (*b*) p-n-p

the collector current. Figure 15.2 shows the current distribution in both the n-p-n-type and p-n-p type transistors.

The collector current = A fraction α (just less than unity e.g. 0·98) of the emitter current + The reverse biased collector to base junction leakage current.

i.e. $$i_C = \alpha i_E + I_{CBO} \qquad (15.1)$$

also $$i_E = i_C + i_B$$

$$\therefore \quad i_C = \alpha(i_C + i_B) + I_{CBO}$$

$$\therefore \quad i_C(1-\alpha) = \alpha i_B + I_{CBO}$$

$$i_C = \frac{\alpha}{1-\alpha} i_B + \frac{I_{CBO}}{1-\alpha} \qquad (15.2)$$

Relations (15.1) and (15.2) show that the collector current of a transistor can be controlled by either the emitter or base currents respectively.

The symbols used for n-p-n and p-n-p transistors are shown in Fig. 15.3. The arrow on the emitter connection indicates the direction of conventional current in the device.

15.2 Transistor characteristics

There are three basic ways of connecting a transistor to obtain gain:

1. Common (or grounded) emitter.
2. Common base.
3. Common collector.

The name in each case is derived from the layer which is common to both the input and output circuits.

15.2.1. Common emitter

Figure 15.4 shows an n-p-n transistor connected in the common emitter configuration with the possible variable quantities shown on it.

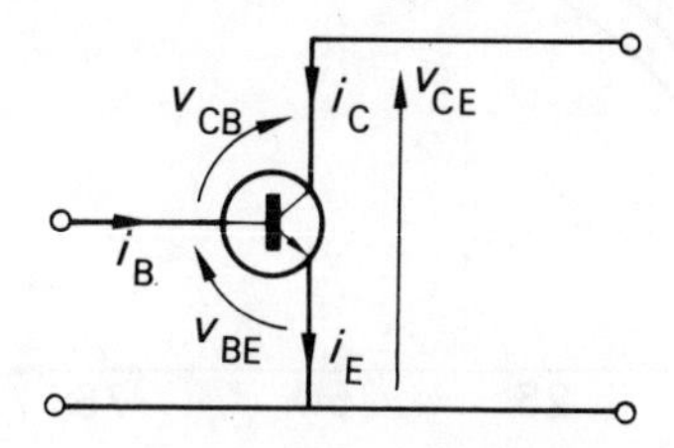

Fig. 15.4. n-p-n transistor in common emitter configuration

It is seen that there are six possible variables viz.

$$i_B,\ i_C,\ i_E,\ v_{BE},\ v_{CE} \text{ and } v_{CB}$$

Since, however

$$i_B + i_C = i_E$$

and

$$v_{CB} + v_{BE} = v_{CE},$$

knowing any two of the voltages and any two of the currents means that the third quantity in each case can be found. The operation of a transistor can thus be predicted given a set of characteristics showing the inter-dependence of two of the voltages and two of the currents. It is most convenient to use the voltages and currents which are the input and output quantities. These, for the common-emitter configuration, are v_{BE}, v_{CE}, i_B and i_C. Any two of these quantities can be considered as independent variables and the other two as dependent variables. Thus, for example, the characteristics could be expressed mathematically as:

$$v_{BE} = f_1(i_B, v_{CE}) \qquad (15.3)$$

$$i_C = f_2(i_B, v_{CE}) \qquad (15.4)$$

Here i_B and v_{CE} have been chosen as the independent variables and v_{BE} and i_C as the dependent variables. Relation (15.3) tells us that v_{BE} is a function of (i.e. is dependent on) i_B and v_{CE}. To show this relationship graphically, it is necessary to keep one of the independent variables constant while the relationship between the dependent variable and the other independent variable is illustrated. For example Fig. 15.5 shows the variation of v_{BE} against i_B for several different values of v_{CE}. Such a characteristic is known as the input characteristic for the transistor concerned in the common emitter configuration.

It can be seen from Fig. 15.5 that for a constant value of v_{BE}, i_B decreases with increasing values of v_{CE}. This is due to the fact that as v_{CE}, and hence v_{BE}, is increased, the thickness of the barrier layer at the reverse-biased collector-to-base junction is increased. Thus the thickness of the base region is effectively reduced and hence there

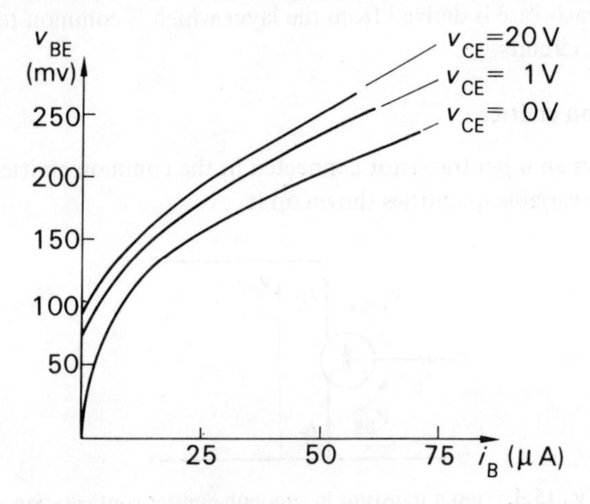

Fig. 15.5. Typical input characteristics for transistor in common emitter configuration

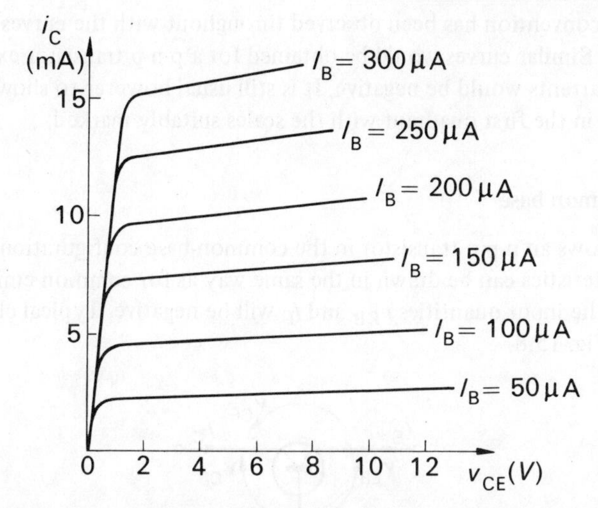

Fig. 15.6. Typical output characteristics for transistor in common emitter configuration

is even less recombination of majority carriers in the base region (i.e. there is a decrease in base current).

Figure 15.6 shows the variation of i_C against v_{CE} for several different values of i_B. Such characteristics are known as the output characteristics for the transistor concerned in the common-emitter configuration. From relation (15.2), it would appear that i_C should remain constant with constant i_B. Due, however, to the effective reduction in base thickness with increasing v_{CE}, as described previously, an increased fraction of the emitter current reaches the collector (i.e. α increases). Although this increase in α tends to be small, $1/(1-\alpha)$ increases rapidly with increasing α as α approaches unity.

e.g. $\alpha = 0{\cdot}98 \qquad \dfrac{1}{1-\alpha} = \dfrac{1}{1-0{\cdot}98} = 50$

$\alpha = 0{\cdot}981$ (approximately 0·1% increase)

$$\frac{1}{1-\alpha} = \frac{1}{1-0{\cdot}981} = 52{\cdot}6 \quad (5{\cdot}2\% \text{ increase})$$

This explains the appreciable increase in i_C with increasing v_{CE} for constant i_B.

The output characteristics show also that as v_{CE} approaches zero, there is a rapid decrease in i_C. This is due to the fact that the collector-to-base junction will become forward biased before $v_{CE} = 0$, since v_{BE} itself is of the same polarity as v_{CE}. When this junction does become forward biased, the collector current will reverse.

Characteristics showing the variation of i_C against i_B for constant values of v_{CE}, and the variation of v_{BE}–against v_{CE} for constant values of i_B could be shown in the same way. Since, however, these are not used to the same extent as the two sets shown they have been omitted here.

As has been specified, the typical characteristics shown in Figs. 15.5 and 15.6 are for n-p-n transistors. Thus the voltages v_{BE} and v_{CE} are of positive polarity. Similarly, using the convention that currents flowing *into* a transistor are positive and currents flowing *away* from it are negative, the currents i_B and i_C are positive also. Therefore mathematical convention has been observed throughout with the curves drawn in the first quadrant. Similar curves would be obtained for a p-n-p transistor except that the voltages and currents would be negative, It is still usual however to show these characteristics in the first quadrant with the scales suitably marked.

15.2.2. Common base

Figure 15.7 shows an n-p-n transistor in the common-base configuration. Input and output characteristics can be drawn in the same way as for common emitter. Note however that the input quantities v_{EB} and i_E will be negative. Typical characteristics are shown in Fig. 15.8.

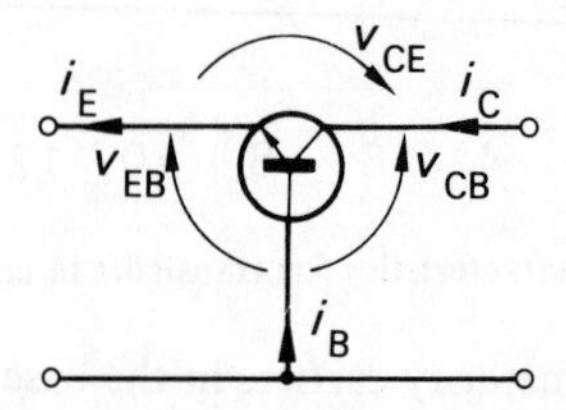

Fig. 15.7. n-p-n transistor in common base configuration

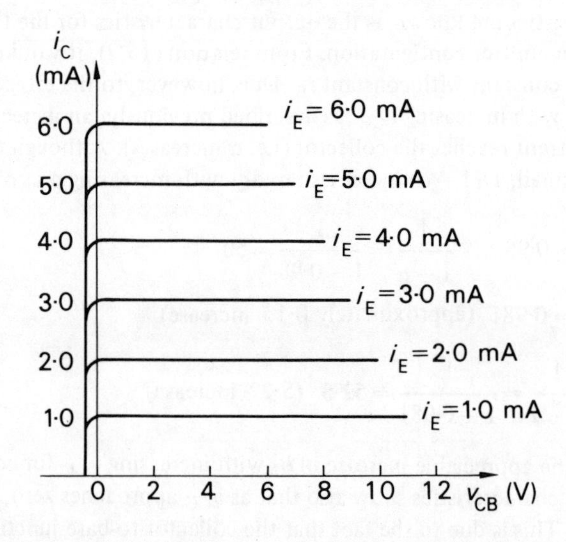

Fig. 15.8. Typical output characteristic for transistor in common base configuration

15.2.3. Common collector

Figure 15.9 shows an n-p-n transistor in the common collector configuration. Input and output characteristics could be drawn in the same way as for the common-emitter and common-base configurations. These are relatively uncommon, however, and are not included here.

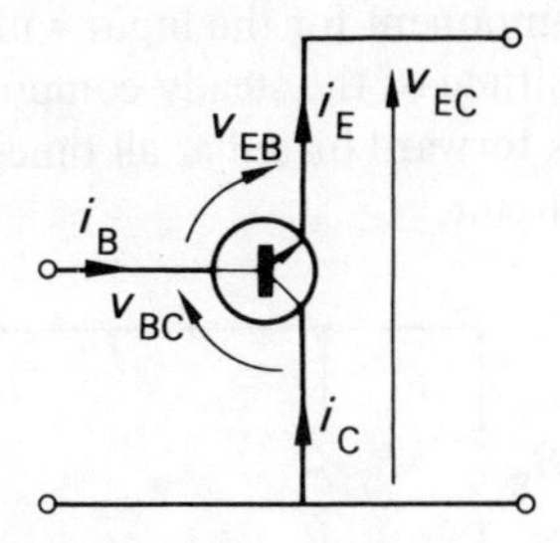

Fig. 15.9. n-p-n transistor in common collector configuration

15.3 Transistor as an amplifier

The basic operation of an amplifier is illustrated in Fig. 15.10. The input signal controls the amount of power that the amplifier takes from the power source (e.g. rectified supply, battery) and converts into power in the load. In practical amplifiers, this convertion is not done with 100% efficiency, some of the power from the source being dissipated in the amplifier as unwanted heat. The amplifier produces gain when the power fed to the load is greater than the input controlling power.

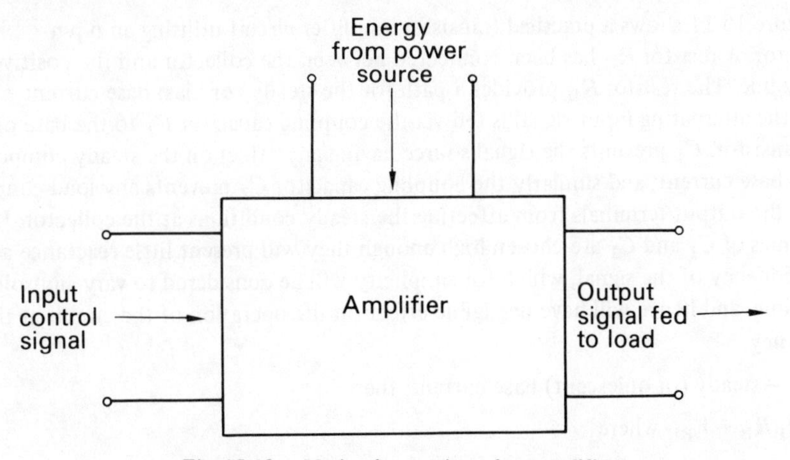

Fig. 15.10. Mode of operation of an amplifier

Relations (15.1) and (15.2) show that the collector current of the transistor can be controlled by the emitter and base currents respectively. By connecting a load effectively between the collector and the common terminal, the transistor can produce gain. In the type of amplifier to be considered here there is another requirement which has to be met. The waveform of the input signal must be maintained to a high degree in the output signal otherwise the 'information' conveyed by the waveform will be distorted to some extent. For example, the tone of a musical note can be altered by the use of an amplifier in a sound reproduction system.

In very many applications the input signal to an amplifier is an alternating quantity. When using transistors, however, it is not possible to have a true alternating input signal since this would entail the emitter-to-base junction being reverse biased during part of the cycle and normal transistor action would not occur. As a result it is necessary to provide a steady component for the input with the alternating component superimposed on to it, the magnitude of the steady component being such that the emitter-to-base junction remains forward biased at all times. A similar arrangement has to be made for the output circuit.

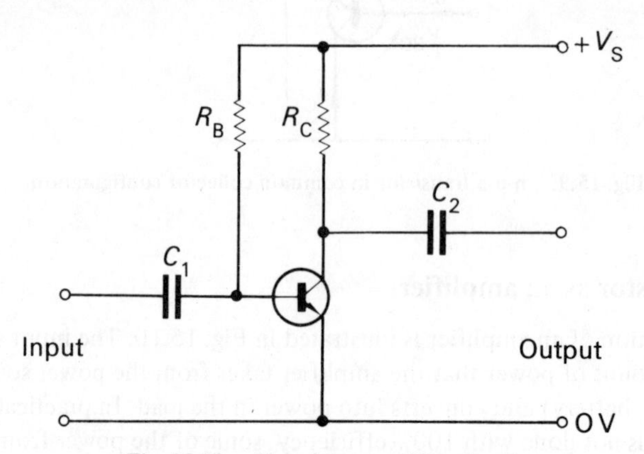

Fig. 15.11. Transistor amplifier circuit

Figure 15.11 shows a practical transistor amplifier circuit utilising an n-p-n transistor. A resistor R_C has been connected between the collector and the positive supply line. The resistor R_B provides a path for the steady (or bias) base current, while the alternating input signal is fed via the coupling capacitor C_1 to the base of the transistor. C_1 prevents the signal source having any effect on the steady component of the base current, and similarly the coupling capacitor C_2 prevents any load connected across the output terminals from affecting the steady conditions at the collector. If the values of C_1 and C_2 are chosen high enough they will present little reactance at the frequency of the signal, which for simplicity will be considered to vary sinusoidally with time, and hence will have negligible effect on the operation of the circuit at this frequency.

Let I_B = steady (or quiescent) base current, then

$V_S = I_B R_B + V_{BE}$ where

V_{BE} = quiescent base-to-emitter voltage.

Therefore $R_B = \dfrac{V_S - V_{BE}}{I_B}$

In most cases however $V_{BE} \ll V_S$ therefore:

$$R_B \doteqdot \frac{V_S}{I_B} \qquad (15.5)$$

If the applied signal produces an alternating base current given by:

$$i_b = I_{bm} \sin \omega t$$

then the total base current is:

$$i_B = I_B + I_{bm} \sin \omega t$$

At any instant:

$$v_{CE} = V_S - i_C R_C$$

therefore

$$i_C = -\frac{1}{R_C} v_{CE} + \frac{V_S}{R_C} \tag{15.6}$$

Comparing relation (15.6) to relation (14.1), it can be seen that a load line can be drawn for the value of R_C being used and the supply voltage V_S. Figure 15.12 shows such a load line superimposed on to a set of output characteristics for the transistor being used. i_C and v_{CE} for any value of base current can be found from the intersection of the characteristic for that base current and the load line.

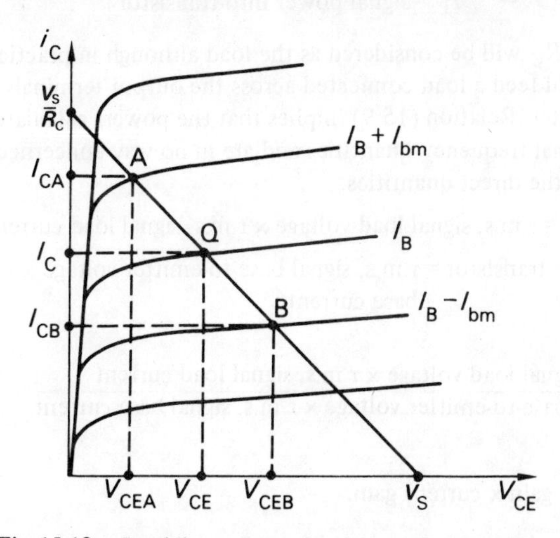

Fig. 15.12. Load line construction for transistor amplifier stage

The points A, Q and B on the load line correspond to the maximum, mean and minimum values of the base current, which are $I_B + I_{bm}$, I_B and $I_B - I_{bm}$ respectively. The corresponding values of the collector current are I_{CA}, I_C and I_{CB} respectively, and of the collector-to-emitter voltage V_{CEA}, V_{CE} and V_{CEB} respectively. The point Q is known as the quiescent (or standing) point.

From the characteristics described, it is possible to derive the gain which is the ratio of output to input. The ratio may be one of current, voltage or power.

Gain	Symbol: G	Unit: none

For example, the current gain G_i is given by

$$G_i = \frac{\Delta I_o}{\Delta I_i} = \frac{I_{CA} - I_{CB}}{2I_{bm}} \tag{15.7}$$

Similarly the voltage gain G_V is given by

$$G_V = \frac{\Delta V_o}{\Delta V_i} \qquad (15.8)$$

To calculate the voltage gain, information must be available concerning the change in input voltage that has produced the change in base current. This could be estimated if a set of input characteristics similar to those shown in Fig. 15.5 were available. Fundamentally it would be fairly complex to carry out since the collector-to-emitter voltage is continuously varying. However, little error is introduced if the collector-to-emitter voltage is assumed constant when calculating the alternating base-to-emitter voltage producing the alternating base current. A further simplification can be introduced by considering the input characteristic linear. The power gain is given by

$$G_p = \frac{P_o}{P_i} = \frac{\text{signal power in load}}{\text{signal power into transistor}} \qquad (15.9)$$

For the moment R_C will be considered as the load although in practice the transistor stage would feed a load connected across the output terminals. Such a load will be considered later. Relation (15.9) implies that the powers calculated are those produced by the signal frequency quantities and are in no way concerned with the power produced by the direct quantities.

signal power in load = r.m.s. signal load voltage x r.m.s. signal load current.

signal power into the transistor = r.m.s. signal base-to-emitter voltage x r.m.s. signal base current.

Therefore:

$$G_p = \frac{\text{R.M.S. signal load voltage} \times \text{r.m.s. signal load current}}{\text{R.M.S. signal base-to-emitter voltage} \times \text{r.m.s. signal base current}}$$

Thus:

power gain = voltage gain x current gain.

i.e. $$G_p = G_V G_i \qquad (15.10)$$

If from the load line construction illustrated in Fig. 15.12 the relationship between i_c and i_b was plotted then a curve similar to that shown in Fig. 15.13 would be obtained. Such a curve is known as a dynamic characteristic for the values of R_C and V_S being used with the particular transistor. The value of such a characteristic is that if it is linear over the range being used it follows that the waveform of i_c will be identical to that of i_b. This does not mean necessarily that the transistor does not distort the input signal waveform since examination of Fig. 15.5 shows that the input characteristics of the transistor are themselves not linear. Thus the waveform of the signal base current is not an exact replica of the waveform of the applied signal. The amount of distortion produced in this way is not usually very large, however, since the source resistance of the signal is usually large enough to swamp the effect of the variation in input resistance. Therefore a linear i_c/i_b characteristic is a good indication of little distortion.

A resistance R_L (i.e. a practical load) will now be considered connected across the

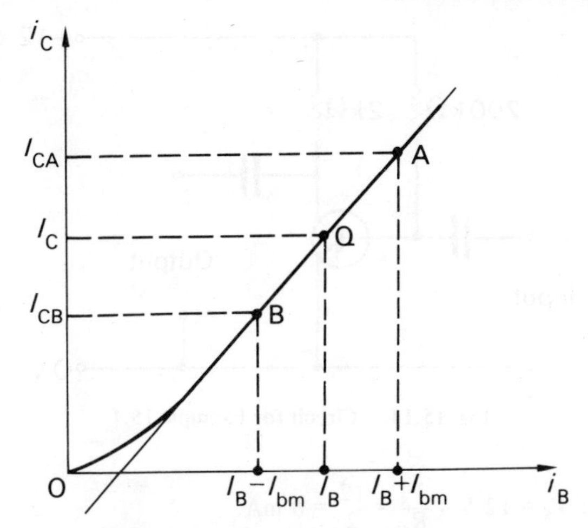

Fig. 15.13. Dynamic characteristic

output terminals. This could, for example, be the input resistance of another amplifier stage which could provide further gain. If the reactance of C_2 is small then the load presented to the *transistor* is now R_C and R_L in parallel, since the impedance of the d.c. supply will be very low to the signal frequency. If therefore

$$R_P = \frac{R_C R_L}{R_C + R_L}$$

then the performance of the circuit can be determined by drawing a load line for R_P. Such a load line will have a slope of $-1/R_P$ in the same way that the load line for R_C had a slope of $-1/R_C$. The quiescent point Q will be a point on the load line for R_P as well as on that for R_C, since the collector to emitter voltage will not be affected by R_L when the input signal is zero. Thus knowing one point on the line and its slope the line can be drawn in. The procedure will be to draw a load line for R_C as before and to mark the quiescent point on it. Through the quiescent point is then drawn the load line for R_P. This new load line is used in exactly the same way as before to calculate gain. The load line for R_C is known as the d.c. load line (since R_C is the d.c. load on the transistor), while that for R_P is known as the a.c. load line. The use of an a.c. load line is illustrated in Example 15.1.

Example 15.1 The transistor used in the amplifier stage shown in Fig. 15.14 has characteristics that may be considered linear within the range given in the table below.

A pure resistive load of 2 kΩ is connected across the output terminals, and the stage is supplied from a signal source of e.m.f. 0·50 V peak and internal resistance 10 kΩ. The input resistance of the transistor may be considered constant at 2·5 kΩ. Determine the current, voltage and power gains of the amplifier stage.

The reactances of the coupling capacitors C_1 and C_2 may be considered zero.

	$i_B = 0\mu A$		$i_B = 20\mu A$		$i_B = 40\mu A$		$i_B = 60\mu A$		$i_B = 80\mu A$		$i_B = 100\mu A$		$i_B = 120\mu A$	
v_{CE}(V)	1	10	1	10	1	10	1	10	1	10	1	10	1	10
i_C(mA)	0·20	0·20	0·88	1·40	1·70	2·38	2·69	3·27	3·70	4·39	4·65	5·39	5·56	6·86

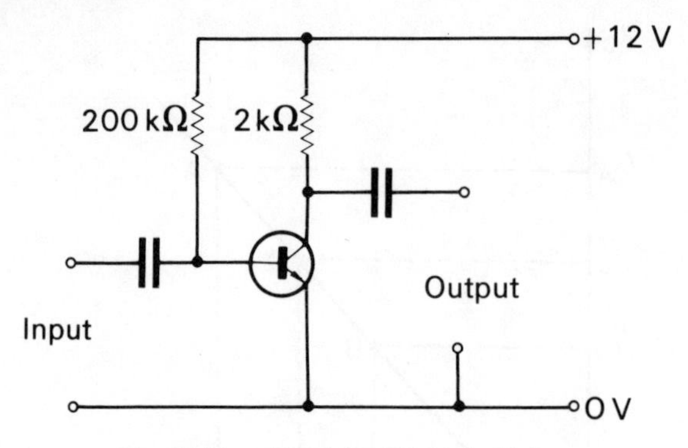

Fig. 15.14. Circuit for Example 15.1

For d.c. load line: $V_s = 12$ V $\dfrac{V_s}{R_c} = \dfrac{12}{2} = 6$ mA.

For a.c. load line: a.c. load $= \dfrac{2 \times 2}{2 + 2} = 1{\cdot}0$ kΩ

∴ Slope of a.c. load line $= -\dfrac{1}{1{\cdot}0} = -1{\cdot}0$ mA/V

Quiescent base current $= I_B = \dfrac{12}{200} = 0{\cdot}06$ mA $= 60$ μA.

(neglecting small voltage drop between base and emitter)

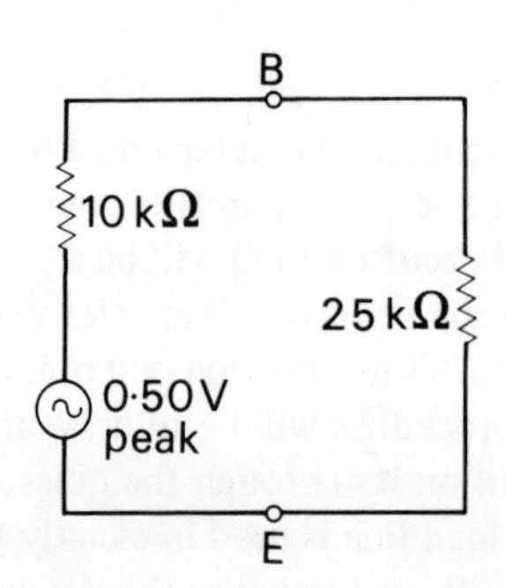

Fig. 15.15. Part of Example 15.1

The input circuit is shown in Fig. 15.15. This neglects the effect of the 200-kΩ bias resistor, which is effectively in parallel with the 2·5-kΩ input resistance, assuming that the 12-V supply has negligible impedance.

∴ Peak value of signal base current $= \dfrac{0{\cdot}50}{10 + 2{\cdot}5} = 0{\cdot}04$ mA $= 40$ μA.

∴ Maximum base current = 60 + 40 = 100 μA.

Minimum base current = 60 − 40 = 20 μA.

The graphical construction is shown in Fig. 15.16.

From the a.c. load line:

Change in collector current = 4·9 − 1·2 = 3·7 mA peak-to-peak.

This change in collector current is shared between the 2·0-kΩ collector resistor and the 2·0-kΩ load resistor i.e. in this case it is shared equally:

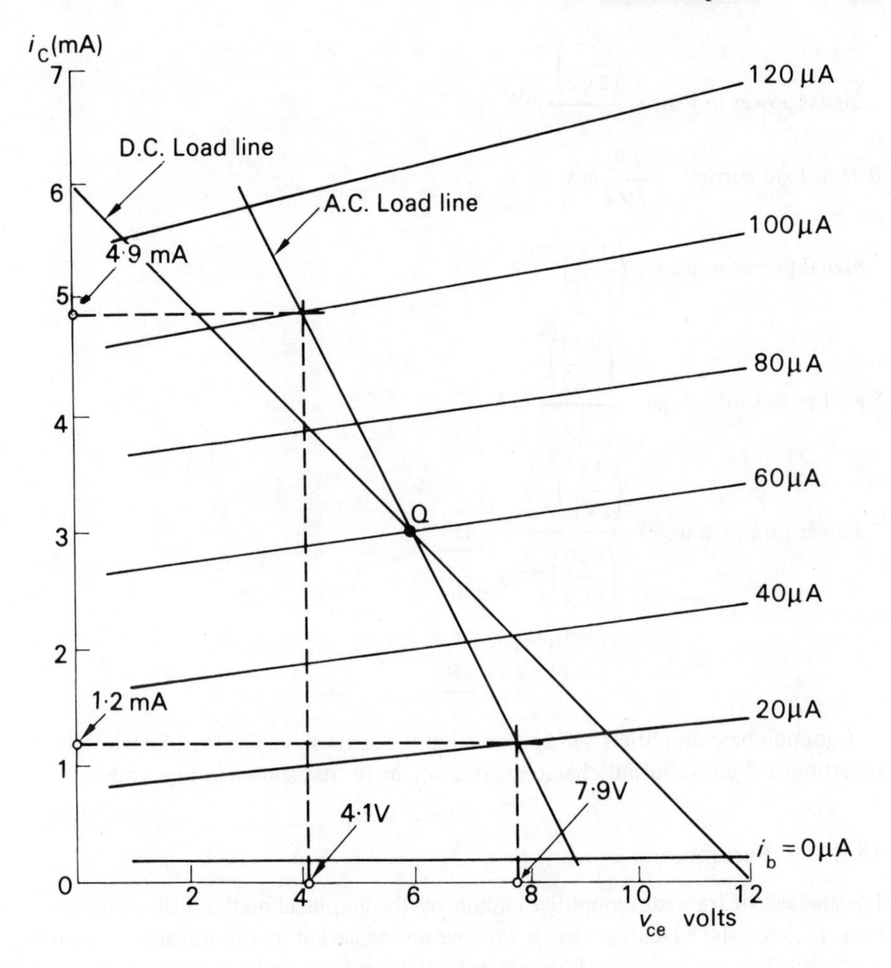

Fig. 15.16. Graphical construction for example 15.1

$\therefore$ Change in load current $= \dfrac{3{\cdot}7}{2} = 1{\cdot}9$ mA peak-to-peak.

$\therefore$ Current gain of the stage $= \dfrac{1{\cdot}9}{0{\cdot}10 - 0{\cdot}02} = \dfrac{1{\cdot}9}{0{\cdot}08} = \underline{24}$.

Note: The current gain of the *transistor* in this example will be twice that of the *stage*, since the transistor is feeding two 2·0-kΩ resistors in parallel. Only half of its current gain is being utilised in the load.

Change in collector to emitter voltage
= change in load voltage = 7·9 – 4·1 = 3·8 V peak-to-peak.

Change in base to emitter voltage = 0·08 x 2·5 = 0·20 V peak-to-peak.

$\therefore G_v = \dfrac{3{\cdot}8}{0{\cdot}20} = \underline{19}$.

R.M.S. load voltage $= \dfrac{3{\cdot}8}{2\sqrt{2}}$ V

$$\therefore \text{Signal power in load} = \frac{\left(\frac{3\cdot8}{2\sqrt{2}}\right)^2}{2} \text{ mW}$$

or

$$\text{R.M.S. load current} = \frac{1\cdot9}{2\sqrt{2}} \text{ mA}$$

$$\therefore \text{Signal power in load} = \left(\frac{1\cdot9}{2\sqrt{2}}\right)^2 \times 2$$

$$\text{Signal power into stage} = \frac{\left(\frac{0\cdot20}{2\sqrt{2}}\right)^2}{2\cdot5} \text{ mW}$$

$$\therefore \text{Power gain of stage} = \frac{\left(\frac{3\cdot8}{2\sqrt{2}}\right)^2 \Big/ 2}{\left(\frac{0\cdot20}{2\sqrt{2}}\right)^2 \Big/ 2\cdot5} = \underline{456}$$

or

$$G_p = G_i\, G_v$$
$$= 24 \times 19 = \underline{456}$$

Common-base amplifiers can be treated in the same way. The load lines will be superimposed on to output characteristics similar to that shown in Fig. 15.8.

15.4 Parameters

The analysis of transistor amplifier circuits by the graphical methods illustrated in para. 15.3 has the advantage that it will give an indication of any waveform distortion produced. The method does, however, tend to be tedious and moreover characteristics for the transistor are not always readily available. As a result analytical methods are employed making use of sets of parameters for the transistor which specify its operation. The derivation of one such set of parameters for the common emitter configuration is illustrated below.

It was shown in para. 15.2 that the characteristics for a transistor in the common emitter configuration could be expressed as:

$$v_{BE} = f_1(i_B, v_{CE})$$
$$i_C = f_2(i_B, v_{CE})$$

If the independent variables i_B and v_{CE} change by δi_B and δv_{CE} respectively there would be corresponding changes δv_{BE} and δi_C produced in the dependent variables. Although the curves relating the variables are non-linear, if the changes are kept relatively small then as a reasonable approximation it could be said that:

$$\delta v_{BE} = \left.\frac{\partial v_{BE}}{\partial i_B}\right|_{v_C \text{ constant}} \times \delta i_B + \left.\frac{\partial v_{BE}}{\partial v_{CE}}\right|_{i_B \text{ constant}} \times \delta v_{CE}.$$

and

$$\delta i_C = \left.\frac{\partial i_C}{\partial i_B}\right|_{v_{CE\,\text{constant}}} \times \delta i_B + \left.\frac{\partial i_C}{\partial v_{CE}}\right|_{i_{B\,\text{constant}}} \times \delta v_{CE}$$

where the partial derivatives will be the slopes (assumed constant) of the relative characteristics.

e.g. $$\left.\frac{\partial v_{BE}}{\partial i_B}\right|_{v_{CE\text{ constant}}} = \text{slope of the } v_{BE}/i_B \text{ characteristic}$$

If therefore:

$$h_{ie} = \left.\frac{\partial v_{BE}}{\partial i_B}\right|_{v_{CE\text{ constant}}}$$

$$h_{re} = \left.\frac{\partial v_{BE}}{\partial v_{CE}}\right|_{i_{B\text{ constant}}}$$

$$h_{fe} = \left.\frac{\partial i_C}{\partial i_B}\right|_{v_{CE\text{ constant}}}$$

$$h_{oe} = \left.\frac{\partial i_C}{\partial v_{CE}}\right|_{i_{B\text{ constant}}}$$

then

$$\delta v_{BE} = h_{ie}\,\delta i_B + h_{re}\,\delta v_{CE} \tag{15.11}$$

$$\delta i_C = h_{fe}\,\delta i_B + h_{oe}\,\delta v_{CE} \tag{15.12}$$

If the changes that take place are sinusoidal with time then δv_{BE}, δv_{CE}, δi_E and δi_c can be represented by V_{be}, V_{ce}, I_b and I_c respectively, where these terms represent the r.m.s. (or peak) values of the sinusoidal changes. Under these conditions therefore:

$$V_{be} = h_{ie}\,I_b + h_{re}\,V_{ce} \tag{15.11.1}$$

$$I_c = h_{fe}\,I_b + h_{oe}\,V_{ce} \tag{15.12.1}$$

It should be noticed that h_{ie} has the dimensions of resistance (ohms) while h_{oe} has the dimensions of conductance (siemens). The other two h_{re} and h_{fe} are pure numbers. All of them are derived with v_{ce} zero (i.e. output short circuited to a.c.) or i_b zero (i.e. input open circuited to a.c.) and are thus termed:

h_{ie} = short circuit input impedance in common emitter,
h_{oe} = open circuit reverse voltage ratio in common emitter,
h_{fe} = short circuit forward current ratio in common emitter and
h_{oe} = open circuit output admittance in common emitter.

This set of parameters are known as the small signal hybrid parameters. The suffix 'e' is used throughout to indicate that the common-emitter configuration is being considered.

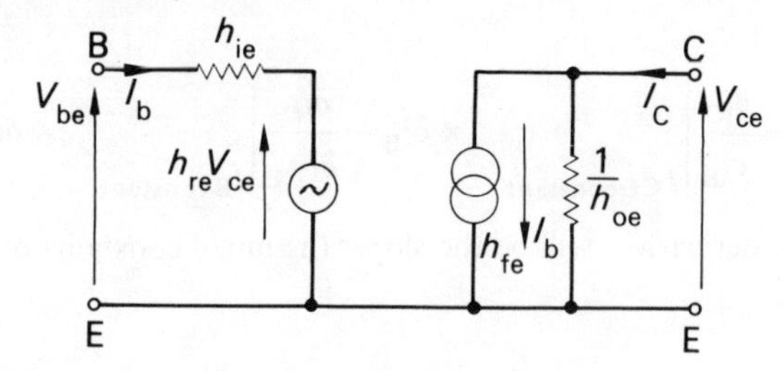

Fig. 15.17. Equivalent circuit for transistor in the common emitter configuration

Relations (15.11.1) and (15.12.1) can be represented by the circuit shown in Fig. 15.17. This can be used therefore as an equivalent circuit for the transistor. The conditions specified in its derivation should always be borne in mind when using it however. These are that the signals involved should be relatively small so that the assumption that the characteristics are linear is reasonable, and that the circuit is applicable only to changes in the quantities involved.

The voltage generator $h_{re}V_{ce}$ represents a feedback characteristic of the transistor whereby the output signal voltage affects, to some extent, the input signal current. Although V_{ce} can be very much greater than V_{be} due to the voltage gain obtainable from the transistor, h_{re} is very small (typically 5×10^{-4}) and $h_{re}V_{ce}$ is small in comparison to V_{be}. Therefore this voltage generator can be left out of the equivalent circuit without introducing large errors. This would certainly be justifiable in practice due to the wide tolerances allowed on parameter values for a transistor of a given type.

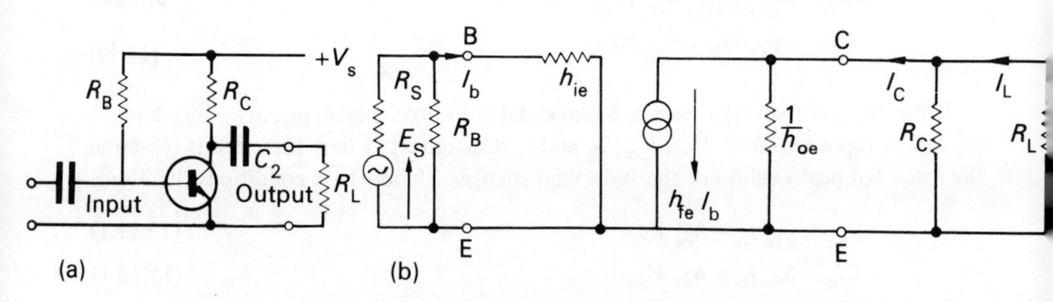

Fig. 15.18. **(*a*) Common emitter amplifier stage (*b*) Equivalent circuit for (*a*)**

Figure 15.8 shows a basic common-emitter amplifier stage with its equivalent circuit. The circuit will be analysed and expressions derived for the important properties.

Note

(i) R_B, the bias resistor, is effectively in parallel with the base-to-emitter input of the transistor, since the source of the supply voltage V_S will appear as a near short circuit to the signal. Similarly R_C is connected effectively between the collector and the emitter for the signal.

(ii) The source of signal to the stage has been considered as a source of e.m.f. E_s in series with a resistance R_s.

(iii) The reactances of the coupling capacitors C_1 and C_2 have been considered small enough to be neglected.

From the equivalent circuit for the transistor

$$G_i = \frac{I_c}{I_b}$$

$$I_c = \frac{\dfrac{1}{h_{oe}}}{\dfrac{1}{h_{oe}} + R_p} h_{fe} I_b \quad \text{where} \quad R_p = \frac{R_C R_L}{R_C + R_L}$$

$$G_i = \frac{h_{fe}}{1 + h_{oe} R_p} \qquad (15.13)$$

$$\frac{I_L}{I_C} = \frac{R_C}{R_C + R_L}$$

Thus the current gain of the stage $= \dfrac{R_C}{R_C + R_L} G_i$ (15.13.1)

$V_{ce} = -I_c R_P$ (minus sign because the polarity of the current has been taken in the direction of emitter to collector)

$$V_{be} = I_b h_{ie}$$

Voltage gain (G_v) of the transistor $= \dfrac{V_{ce}}{V_{be}} = \dfrac{-I_c R_p}{I_b h_{ie}}$

$$G_v = \frac{-h_{fe} R_p}{h_{ie}(1 + h_{oe} R_p)} \qquad (15.14)$$

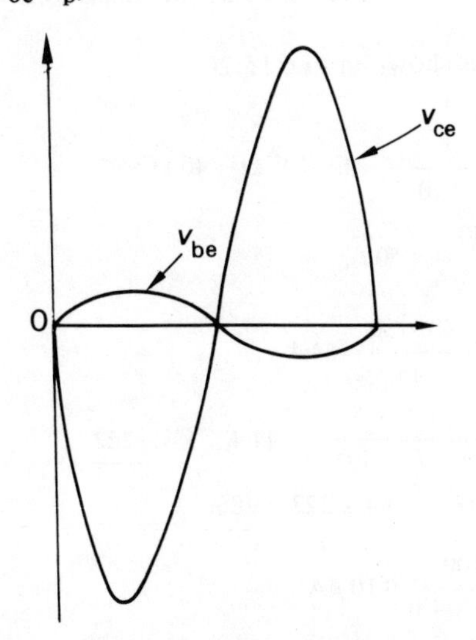

Fig. 15.19. Waveform diagram showing v_{be} and v_{ce}

Since the same voltage appears across the load as appears across the output of the transistor it follows that the voltage gain of the stage is identical to that of the transistor. The minus sign in equation (15.14) indicates that there is 180° phase shift between V_{be} and V_{ce}. This is illustrated in Fig. 15.19.

The power gain of the transistor G_p

$$= \frac{\text{Signal power in } R_P}{\text{Signal power fed into the transistor}} = \frac{V_{ce} I_c}{V_{be} I_b}$$

$= G_i G_v$ (magnitudes)

$$\therefore \quad G_p = \frac{h_{fe}^2 R_P^2}{h_{ie}(1 + h_{oe} R_P)^2} \tag{15.15}$$

Example 15.2 A transistor has the following small-signal hybrid parameters: $h_{ie} = 1{\cdot}0$ kΩ, $h_{fe} = 50$, $h_{oe} = 25$ μS and $h_{re} = 0$.

The total effective load connected between the collector and emitter can be considered as a 5-kΩ resistor. The signal is supplied from a source which can be considered as an e.m.f. of 100 mV in series with a resistance of 9 kΩ. Determine the current, voltage and power gains of the transistor and the signal power fed to the load. The effect of the bias components and the reactances of the coupling capacitors can be neglected.

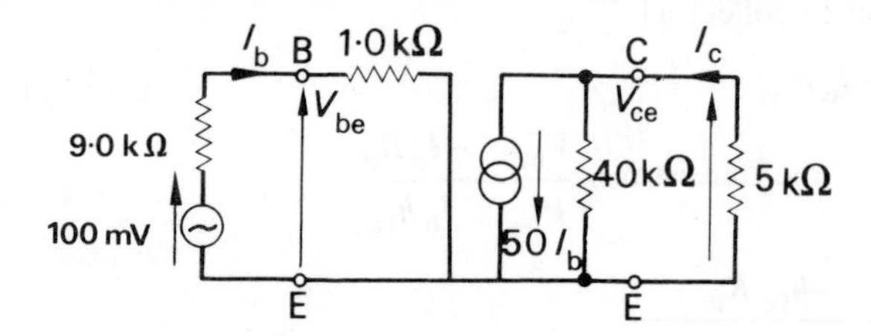

Fig. 15.20. Part of example 15.2

The equivalent circuit is shown in Fig. 15.20.

$$\frac{1}{h_{oe}} = \frac{1}{25 \times 10^{-6}} = 4 \times 10^4 \ \Omega = 40 \text{ k}\Omega$$

$$I_c = \frac{40}{40 + 5} \times 50 \, I_b$$

$$G_i = \frac{I_c}{I_b} = \frac{40 \times 50}{45} = 44{\cdot}4$$

$$G_v = \frac{V_{cb}}{V_{eb}} = \frac{-I_c \times 5{\cdot}0}{I_b \times 1{\cdot}0} = -44{\cdot}4 \times 5 = \underline{-222}$$

$$G_p = G_i \, G_v = 44{\cdot}4 \times 222 = \underline{9860}$$

$$I_b = \frac{100}{9{\cdot}0 + 1{\cdot}0} = 10 \ \mu\text{A}$$

$$\therefore \quad I_c = 44{\cdot}4 \times 10 = 444 \ \mu\text{A}$$

Signal power fed to load $= (4{\cdot}44 \times 10^{-4})^2 \times 5 \times 10^3$
$= 9{\cdot}85 \times 10^{-4}\ \text{W} = \underline{0{\cdot}985\ \text{mW}}$

Example 15.3 The transistor used in the cirucit shown in Fig. 15.21 has the following small signal parameters:

$h_{ie} = 2{\cdot}0\ \text{k}\Omega$, $h_{fe} = 60$, $h_{oe} = 20\ \mu\text{S}$ and h_{re} is negligible.

The input signal is supplied from a source of e.m.f. E in series with a resistance of 10 kΩ. Determine the necessary value of E to give a signal power of 10 mW in a 10-kΩ resistive load connected across the output terminals. The reactances of the coupling capacitors can be neglected.

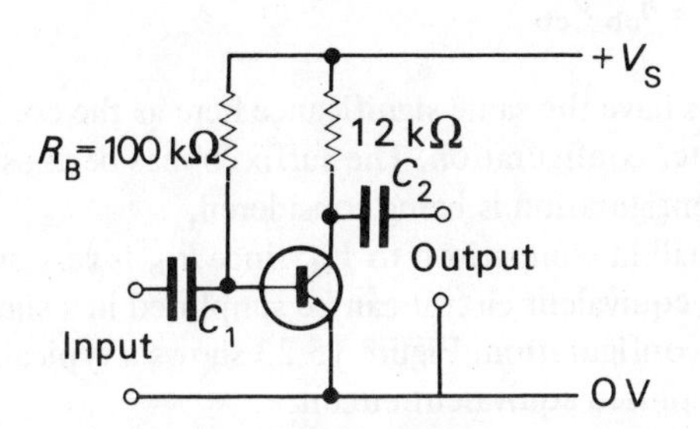

Fig. 15.21. Circuit for example 15.3

The equivalent circuit is shown in Fig. 15.22.

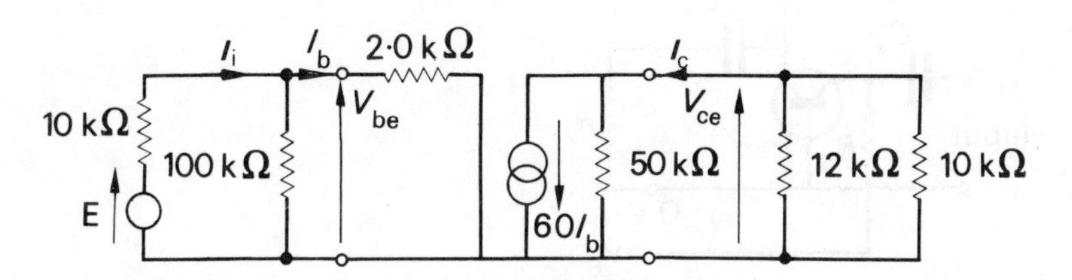

Fig. 15.22. Part of example 15.3

$$\frac{1}{h_{oe}} = \frac{1}{20 \times 10^{-6}} = 5 \times 10^4\ \Omega = 50\ \text{k}\Omega$$

$$\frac{V_{ce}{}^2}{10} = 10 \quad \therefore \quad V_{ce} = 10\ \text{V}$$

$$\therefore \quad I_c = \frac{10}{\dfrac{12 \times 10}{12 + 10}} = \frac{10 \times 22}{120} = 1{\cdot}83\ \text{mA}$$

$$1{\cdot}83 = \frac{10}{50} + 60\, I_b$$

$$\therefore \quad I_b = \frac{1{\cdot}63}{60} = 0{\cdot}027\ \text{mA}$$

$$\therefore \quad V_{be} = 0{\cdot}027 \times 2{\cdot}0 = 0{\cdot}054\ \text{V}$$

$$\therefore \text{Input current to stage} = I_i = 0{\cdot}027 + \frac{0{\cdot}054}{100}$$

$$= 0{\cdot}028 \text{ mA}$$

$$E = 0{\cdot}028 \times 10 + 0{\cdot}054$$
$$= 0{\cdot}334 \text{ V}$$

In a similar way a set of small-signal hybrid parameters can be derived for the transistor in the common-base configuration. The corresponding equations are:

$$V_{eb} = h_{ib} I_e + h_{rb} V_{cb} \quad (15.16)$$

$$Ic = h_{fb} I_e + h_{ob} V_{cb} \quad (15.17)$$

The hybrid parameters have the same significance here as the corresponding ones had in the common-emitter configuration. The suffix 'b' has been used to indicate that the common-base configuration is being considered.

The term $h_{rb} V_{cb}$ is small in comparison to V_{eb} since h_{rb} is very small (typically 5×10^{-4}). Therefore the equivalent circuit can be simplified in a similar way to that for the common emitter configuration. Figure 15.23 shows a typical common-base amplifier stage with a simplified equivalent circuit.

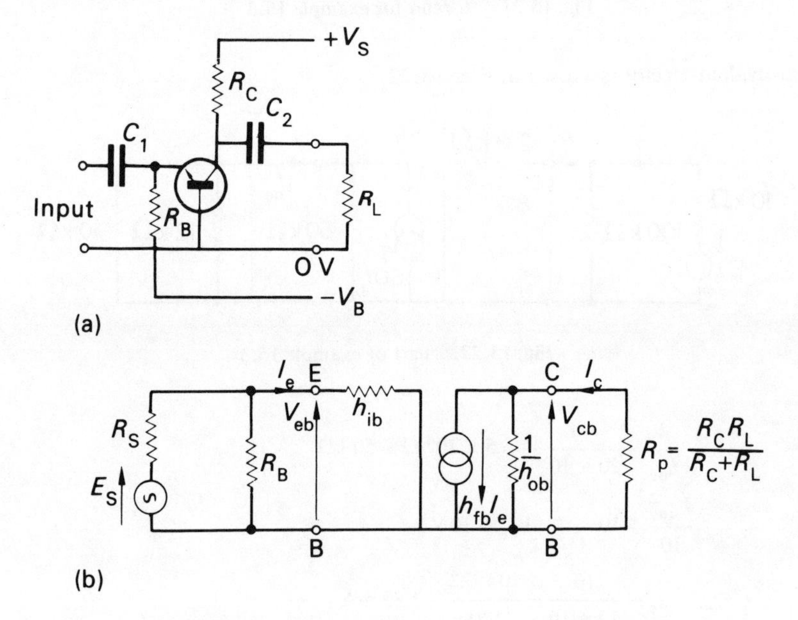

Fig. 15.23. (*a*) Common base amplifier stage (*b*) Equivalent circuit for (*a*)

The expressions for the gains will be identical to those derived for the common emitter configuration with the appropriate parameters substituted.

$$G_i = \frac{I_c}{I_e} = \frac{h_{fb}}{1 + h_{ob} R_P} \quad (15.18)$$

$$G_v = \frac{V_{cb}}{V_{eb}} = \frac{-h_{fb} R_P}{h_{ib}(1 + h_{ob} R_P)} \tag{15.19}$$

$$G_p = G_i\, G_v = \frac{h_{fb}^2 R_P}{h_{ib}(1 + h_{ob} R_P)^2} \tag{15.20}$$

The magnitude of h_{fb} is less than unity (typically 0·98) and is negative i.e. an increase in emitter current in the direction shown produces an increase in collector current in the opposite direction to that shown. This means that the current gain is less than unity and that there is 180° phase shift between I_e and I_c with the polarities as shown, and that there is zero phase shift between V_{eb} and V_{cb}.

Example 15.4 The transistor used in the common base circuit shown in Fig. 15.24 has the following small signal parameters:

$h_{ib} = 50\ \Omega$, $h_{fb} = -0{\cdot}96$, $h_{ob} = 1{\cdot}0\ \mu S$ and h_{rb} is negligible.

The signal source can be considered as a constant-current generator of 100 μA and an internal resistance of 450 Ω. Calculate the voltage and power gains of the stage when the load is a 1·0-kΩ resistor connected across the output terminals. The effect of the bias resistor and the reactance of the coupling capacitors can be neglected.

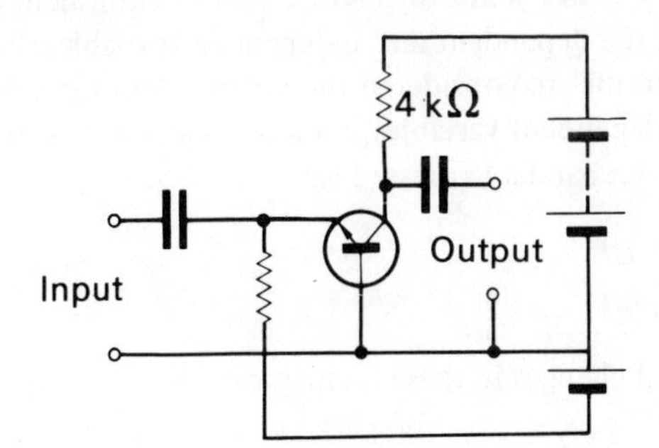

Fig. 15.24. Circuit for example 15.4

The equivalent circuit is shown in Fig. 15.25.

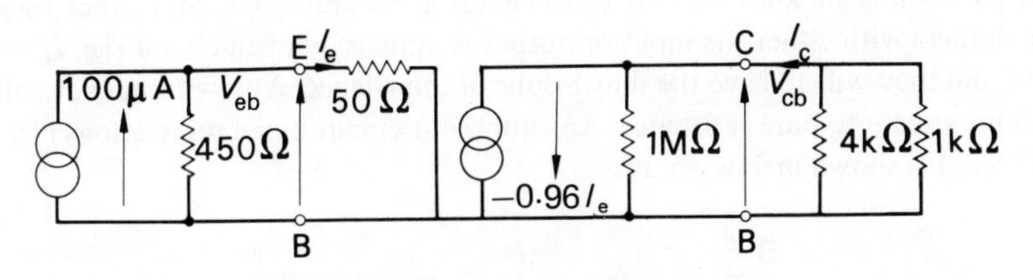

Fig. 15.25. Part of example 15.4

$$\frac{1}{h_{ob}} = \frac{1}{10^{-6}} = 10^6\ \Omega = 1\ \text{M}\Omega$$

Thus the effect of $1/h_{ob}$ will be negligible in comparison to the 4 kΩ in parallel with the 1 kΩ

$$I_e = \frac{450}{450 + 50} \times 100 = 90\ \mu A$$

$$V_{eb} = 0{\cdot}09 \times 50 = 4{\cdot}5\ mV$$

$$I_c = -0{\cdot}96 \times 90 = -86\ \mu A$$

$$V_{cb} = -(-86) \times \frac{4{\cdot}0 \times 1{\cdot}0}{4{\cdot}0 + 1{\cdot}0} = 69\ mV$$

$$\therefore \quad G_v = \frac{69}{4{\cdot}5} = \underline{15}$$

$$\text{Signal power in load} = \frac{(69 \times 10^{-3})^2}{10^3} = 4{\cdot}8 \times 10^{-6}\ W$$

$$= 4{\cdot}8\ \mu W$$

$$\text{Signal power into stage} = (90 \times 10^{-6})^2 \times 50 = 4{\cdot}1 \times 10^{-7}\ W$$

$$= 0{\cdot}41\ \mu W$$

$$\therefore \text{Power gain } (G_p) = \frac{4{\cdot}8}{0{\cdot}41} = \underline{12}$$

Note: The power gain of the transistor = 15 x 0·96 = 14·4.

Other sets of parameters, and hence different forms of equivalent circuit, can be derived by interchanging the dependent and independent variables. For example, if the input and output currents are considered the independent variables and the input and output voltages the dependent variables, then for the common emitter configuration the transistor characteristics can be expressed as:

$$v_{BE} = f_1(i_B, i_C)$$

$$v_{CE} = f_2(i_B, i_C)$$

Hence for small sinusoidal changes in these quantities:

$$V_{be} = z_{ie} I_b + z_{re} I_c \tag{15.21}$$

$$V_{ce} = z_{fe} I_b + z_{oe} I_c \tag{15.22}$$

These parameters are known as the open circuit impedance parameters since they will all be defined with either the input or output terminals on open-circuit (i.e. $I_b = 0$ or $I_c = 0$) and they will all have the dimensions of impedance. At low frequencies they will approximate to pure resistance. An equivalent circuit based on relations (15.21) and (15.22) is shown in Fig. 15.26.

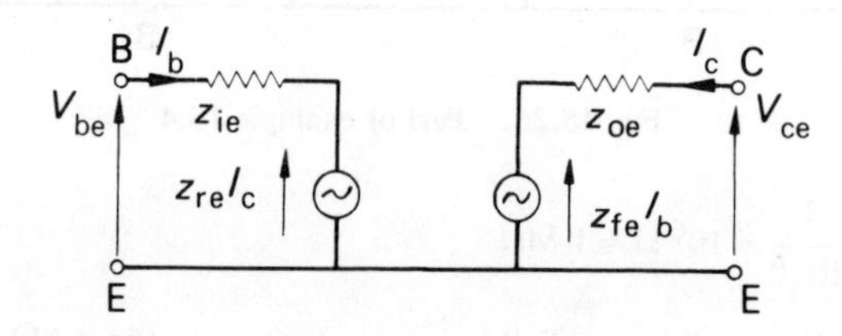

Fig. 15.26. Equivalent circuit based on *z*-parameters

For low frequencies relations (15.21) and (15.22) can be written as

$$V_{be} = r_{ie} I_b + r_{re} I_c \tag{15.21}$$

$$V_{ce} = r_{fe} I_b + r_{oe} I_c \tag{15.22}$$

Similarly for the common-base configuration

$$V_{eb} = r_{ib} I_e + r_{rb} I_c \tag{15.23}$$

$$V_{cb} = r_{fb} I_e + r_{ob} I_c \tag{15.24}$$

Adding and subtracting $r_{rb}I_e$ to relation (15.23) gives

$$V_{eb} = r_{ib} I_e + r_{rb} I_c + r_{rb} I_e - r_{rb} I_e$$

$$V_{eb} = (r_{ib} - r_{rb}) I_e + r_{rb}(I_e + I_c) \tag{15.25}$$

Similarly adding and subtracting $r_{rb}I_e$ and $r_{rb}I_c$ to relation (15.24) gives

$$V_{cb} = r_{fb} I_e + r_{ob} I_c + r_{rb} I_e - r_{rb} I_e + r_{rb} I_c - r_{rb} I_c$$

$$V_{cb} = (r_{fb} - r_{rb}) I_e + (r_{ob} - r_{rb}) I_c + r_{rb}(I_e + I_c) \tag{15.26}$$

Let $r_{ib} - r_{rb} = r_e =$ the emitter resistance

$r_{rb} = r_b =$ the base resistance

$r_{fb} - r_{rb} = r_m =$ the mutual resistance

and $r_{ob} - r_{rb} = r_c =$ the collector resistance.

Therefore relations (15.25) and (15.26) can be written as

$$V_{eb} = r_e I_e + r_b(I_e + I_c) \tag{15.27}$$

$$V_{cb} = r_m I_e + r_c I_c + r_b(I_e + I_c) \tag{15.28}$$

Using relations (15.27) and (15.28) gives an equivalent circuit which has only one generator. This is shown in Fig. 15.27 and is known as the T-equivalent circuit for the common-base configuration. It has the advantage that it gives a somewhat better physical picture of the transistor.

15.4.1. T-parameters

It may be that the student reader does not require the derivation but is prepared to accept the T-equivalent circuit shown in Fig. 15.27. In this case, he need only apply Kirchhoff's Second Law to the network circuits in order to obtain relations (15.27) and (15.28.)

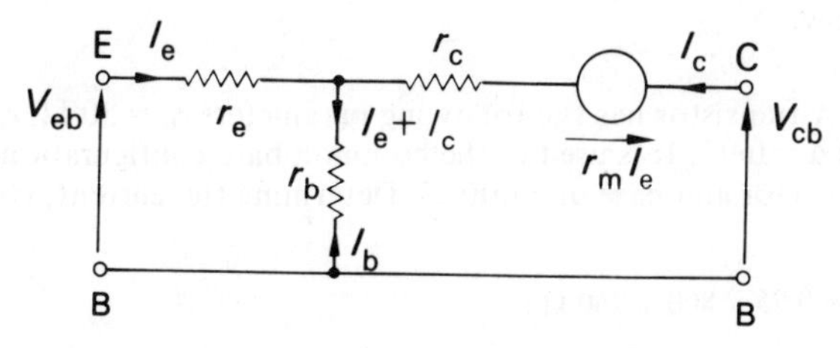

Fig. 15.27. T-equivalent circuit for common base configuration (constant voltage form)

If the output terminals are short circuited then $V_{cb} = 0$ and from relation 15.28

$$0 = r_m I_e + r_c I_c + r_b(I_e + I_c)$$

The short-circuit current gain in common base, which is usually given the symbol α when using the T-equivalent circuit, is defined as the ratio of the magnitudes of I_c and I_e with the output on short circuit. Therefore

$$\alpha = \left|\frac{I_c}{I_e}\right| = \rightarrow \frac{r_m + r_b}{r_c + r_b} \tag{15.29}$$

In practice $r_m \gg r_b$ and $r_c \gg r_b$

$$\therefore \quad \alpha \doteqdot \frac{r_m}{r_c}$$

$$\text{i.e.} \quad r_m \doteqdot \alpha r_c \tag{15.30}$$

Since α is typically 0·98 it follows that

$$r_m \doteqdot r_c$$

The T-equivalent circuit shown in Fig. 15.27 can be converted to the constant-current form by converting the branch containing the voltage generator. This gives a current generator of magnitude

$$\frac{r_m I_e}{r_c} = \alpha I_e$$

The circuit is given in Fig. 15.28.

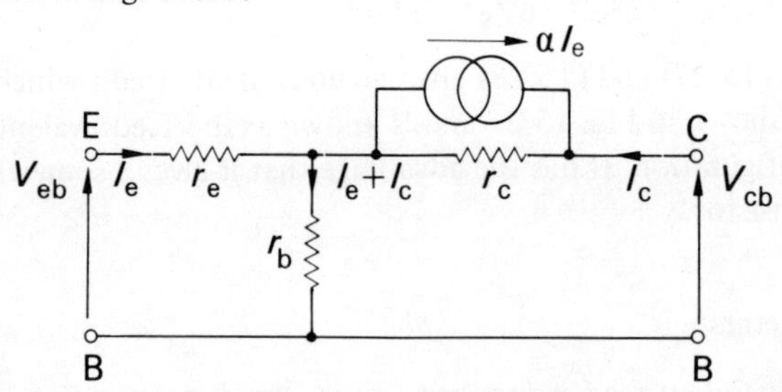

Fig. 15.28. T-equivalent circuit for common base configuration (constant current form)

The use of the T-equivalent circuit for a common-base amplifier is illustrated in the example below.

Example 15.5 A transistor has the following parameters $r_b = 30\ \Omega$, $r_e = 400\ \Omega$, $r_c = 800\ \text{k}\Omega$ and $\alpha = 0{\cdot}95$. It is used in the common-base configuration with an effective load between collector and base of 5 000 Ω. Determine the current, voltage and power gains.

$$r_m = 0{\cdot}95 \times 800 = 760\ \text{k}\Omega$$

The equivalent circuit is shown in Fig. 15.29

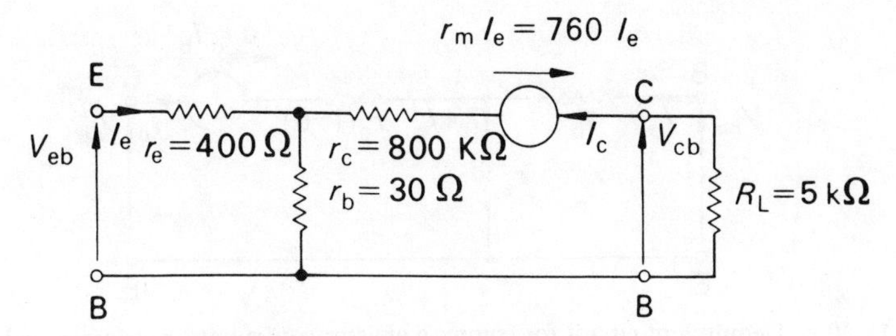

Fig. 15.29. Part of example 15.5

$$V_{eb} = I_e \times 0{\cdot}4 + (I_e + I_c)0{\cdot}03$$

$$\therefore \quad V_{eb} = 0{\cdot}43\ I_e + 0{\cdot}03\ I_c \quad [1]$$

$$760\ I_e = -I_c \times 5{\cdot}0 - (I_c + I_e)0{\cdot}03 - I_c 800$$

$$\therefore \quad 760\ I_e \doteqdot -805\ I_c \quad [2]$$

$$G_i = \frac{I_c}{I_e} = -\frac{760}{805} = \underline{-0{\cdot}94}$$

Note that there is virtually no difference between the magnitude of the current gain and the short-circuit current gain. This is due to the fact that the load resistance R_L is very small in comparison to r_c. Because of the inherent high value of r_c this will be true in most practical circuits.

$$\text{From [1]} \quad V_{eb} = 0{\cdot}43\left(-\frac{I_c}{0{\cdot}94}\right) + 0{\cdot}03\ I_c$$

$$= -0{\cdot}46\ I_c + 0{\cdot}03\ I_c$$

$$= -0{\cdot}43\ I_c$$

$$V_{cb} = -I_c \times 5{\cdot}0 = -\left(-\frac{V_{eb}}{0{\cdot}43}\right)5{\cdot}0$$

$$G_v = \frac{V_{cb}}{V_{eb}} = \frac{5{\cdot}0}{0{\cdot}43} = \underline{12}$$

Note that the calculated value of the voltage gain is positive indicating that V_{cb} is in phase with V_{eb}.

$$G_p = \frac{V_{cb} I_c}{V_{eb} I_e} = G_v\, G_i = 12 \times 0{\cdot}94 = \underline{11{\cdot}3}$$

The T-equivalent circuit for common-base operation can be modified for common emitter operation as follows.

From Fig. 15.27

$$I_e = -(I_b + I_c)$$

$$r_m I_e = -\alpha r_c (I_b + I_c)$$

$$= -\alpha r_c I_b - \alpha r_c I_c$$

Therefore the voltage generator $r_m I_e$ can be replaced by another voltage generator $-\alpha r_c I_b$, which is more convenient since I_b is now the input current, with an additional voltage drop $-\alpha r_c I_c$. This gives the T-equivalent circuit as shown in Fig. 15.30. The constant-current form is shown in Fig. 15.31.

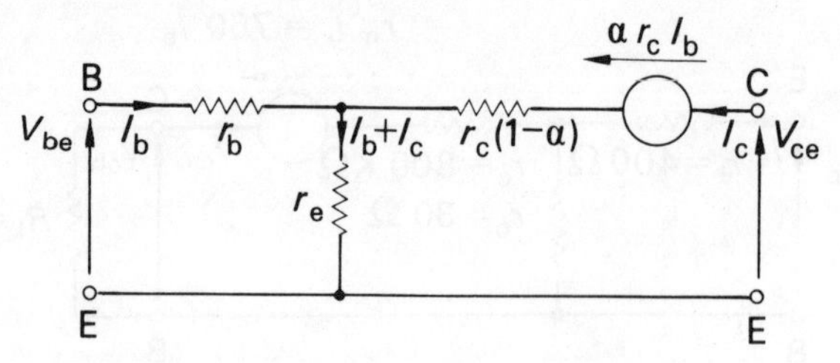

Fig. 15.30. T-equivalent circuit for common emitter configuration (constant voltage form)

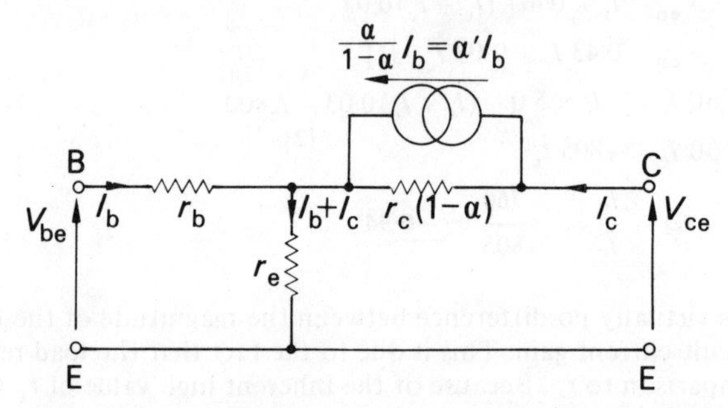

Fig. 15.31. T-equivalent circuit for common emitter configuration (constant current form)

Example 15.6 The transistor specified in Example 15.5 is used in the common-emitter configuration with an effective load of 5 000 Ω between collector and emitter. Determine the current, voltage and power gains.

$$r_c(1-\alpha) = 800\ (1 - 0{\cdot}95) = 40\ \text{k}\Omega$$

The equivalent circuit is shown in Fig. 15.32.

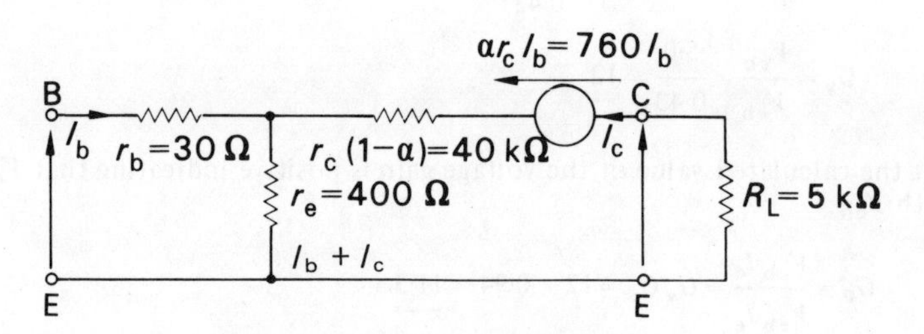

Fig. 15.32. Part of example 15.6

$$V_{be} = I_b \times 0{\cdot}03 + (I_b + I_c)0{\cdot}4$$

$$V_{be} = 0{\cdot}43\ I_b + 0{\cdot}4\ I_c \qquad [1]$$

$$760\ I_b = I_c \times 40 + (I_b + I_c)\ 0{\cdot}4 + I_c \times 5{\cdot}0$$

$$\therefore \quad 760 I_b = 45{\cdot}4\ I_c \qquad [2]$$

$$\therefore \quad G_i = \frac{I_c}{I_b} = \frac{760}{45{\cdot}4} = \underline{17}$$

Note that current gains greater than unity are obtainable.

From [1] $V_{be} = 0{\cdot}43 \times \frac{I_c}{17} + 0{\cdot}4\, I_c$

$$= 0{\cdot}02\, I_c + 0{\cdot}4\, I_c = \underline{0{\cdot}42\, I_c}$$

$$V_{ce} = -I_c \times 5{\cdot}0 = -\frac{V_{be}}{0{\cdot}42} \times 5{\cdot}0$$

$$G_v = \frac{V_{ce}}{V_{be}} = -\frac{5{\cdot}0}{0{\cdot}42} = \underline{-12}$$

Note that there is 180° phase shift between V_{ce} and V_{be}.

$$G_p = \frac{V_{ce} \times I_c}{V_{be} \times I_b} = A_v \times A_c = 12 \times 17$$

$$= \underline{204}$$

The large power gains obtainable with the common-emitter configuration is one of the main reasons why this configuration is used so extensively in practice.

15.5 Low- and high-frequency response

In the preceding analysis the reactances of the coupling capacitors have been neglected. There will however be frequencies at which appreciable voltage drops will be produced across these capacitors. There will therefore be a decrease in signal voltage across the load and hence a loss in gain. There is a corresponding loss of gain at high frequencies also. This is due to two main reasons. Firstly, there is a decrease in h_{fe} at high frequencies due to the transit time of carriers in the base region being comparable to the period of the signal. Secondly the presence of stray shunt capacitance across the load reduces the effective load across the transistor at high frequencies. Figure 15.33 shows a typical gain/frequency characteristic for a transistor amplifier stage. The frequency scale is logarithmic. It is seen that there is a wide frequency range over which the gain remains essentially constant. The difference in frequency between the two frequencies at which the gain has fallen to $1/\sqrt{2}$ of this constant value, is known as the bandwidth of the stage. The two frequencies are known as the lower and upper half-power frequencies since the signal power in the load will be reduced to one half of the signal power at the mid-band frequencies for a constant amplitude of input signal.

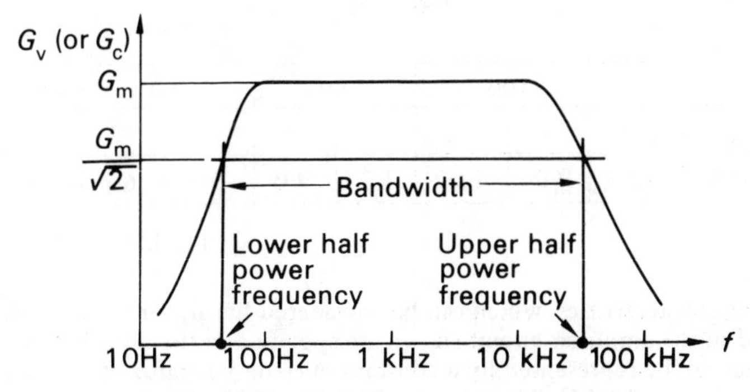

Fig. 15.33. Gain/frequency characteristic

PROBLEMS ON TRANSISTOR CIRCUITS

1 A n-p-n transistor, the characteristics of which can be considered linear between the limits shown in the table, is used in an amplifier circuit. A 2·0-kΩ resistor is connected between its collector and the positive terminal of the 9 V d.c. supply and its emitter is connected directly to the negative terminal. Given that the quiescent base current is 40 μA determine the quiescent collector-to-emitter voltage and the quiescent collector current.

If when a signal is applied to the circuit the base current varies sinusoidally with time with a peak alternating component of 20 μA, determine the alternating component of the collector current and hence the current gain of the stage. The load across the output terminals of the circuit can be considered very high in comparison to 2·0 kΩ.

	I_b = 60 μA		I_b = 40 μA		I_b = 20 μA	
	V_{ce} = 1 V	V_{ce} = 10 V	V_{ce} = 1 V	V_{ce} = 10 V	V_{ce} = 1 V	V_{ce} = 10 V
I_c mA	2·7	3·2	1·8	2·1	0·9	1·1

5·1 V; 1·9 mA; 0·9 mA; 45

2 The transistor used in the circuit shown has characteristics which can be considered linear between the limits shown in the table. Determine the value of R_B to give a quiescent base current of 80 μA.

If the signal base current varies sinusoidally with time and has a peak value of 40 μA determine the r.m.s. value of the signal voltage across the load and hence the signal power in the load. The reactances of the coupling capacitors can be considered zero.

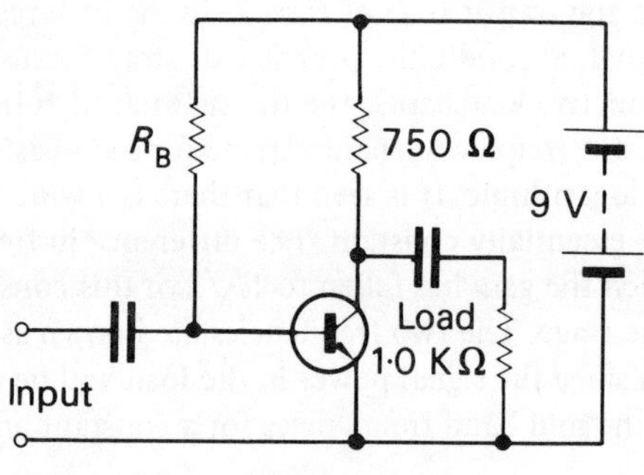

Fig. 15.34

I_b (μA)	120		100		80		60		40	
V_{ce} (V)	1	12	1	12	1	12	1	12	1	12
I_c (mA)	10·8	13·4	9·0	11·2	7·2	9·0	5·3	6·7	3·7	4·5

113 kΩ; 1·1 V; 1·2 mW

3 The output characteristics, which can be considered linear, for the n-p-n silicon transistor used in the amplifier circuit shown are specified in the table below. The source of signal can be represented by a constant current generator of 24 μA peak and internal resistance 3·0 kΩ. The stage feeds an identical one. Determine the current,

voltage and power gains of the stage. The input resistances of the transistors can be considered constant at 6·0 kΩ and the reactances of the coupling capacitors are negligible.

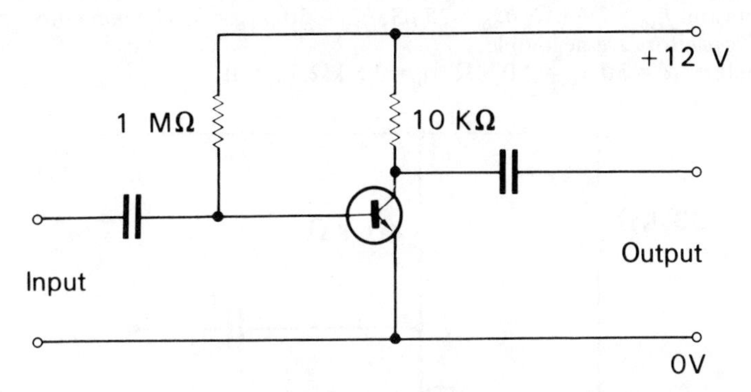

Fig. 15.35. Circuit for 15.3

I_b (μA)	0		4		8		12		16		20	
V_{ce} (V)	1	10	1	10	1	10	1	10	1	10	1	10
I_c (mA)	0·0	0·0	0·14	0·18	0·33	0·39	0·49	0·61	0·68	0·83	0·86	1·04

30; 30; 900

4 Determine the current gain (I_o/I_i) and the voltage gain (V_o/V_i) for the amplifier circuit shown. The effects of R_B, C_1 and C_2 on the calculation can be neglected. The transistor's small signal hybrid parameters are: h_{fe} = 50, h_{oe} = 10 μS, h_{ie} = 5·0 kΩ, h_{re} = 0.
(T-parameters: α' = 50, r_c = 5·0 MΩ, r_b = 5·0 kΩ, r_c = 0).

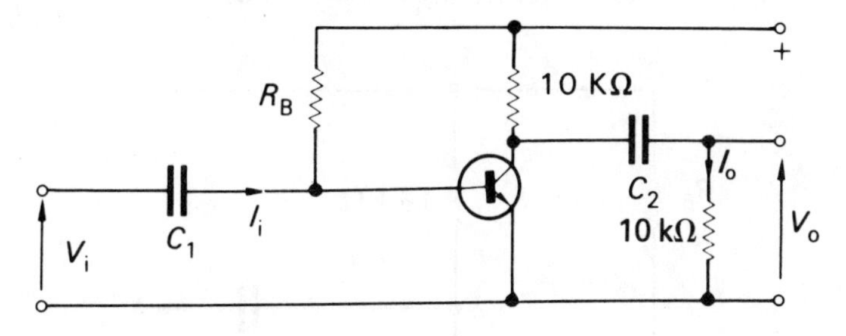

Fig. 15.36

– 23·8; – 47·6

5 The small signal hybrid parameters for a transistor used in an amplifier circuit are: h_{ie} = 2·0 kΩ, h_{fe} = 60, h_{oe} = 2·0 x 10^{-5} S and h_{re} = 0. The total collector to emitter load is 10 k resistance and the transistor is supplied from a source of signal of e.m.f. 100 mV r.m.s. and internal resistance 3·0 kΩ. Determine the current, voltage and power gains of the transistor and the signal power developed in the load.
(T-parameters: α' = 60, r_c = 3·12 MΩ, r_b = 2·0 kΩ, r_e = 0).

50; 250; 12 500; 10 mW

6 The amplifier shown in the circuit diagram feeds an identical stage. If the input signal source has an internal resistance of 1·0 kΩ determine the power gain of the stage (i.e. signal power into second stage/signal power into first stage). The transistor parameters are: h_{ie} = 2·5 kΩ, h_{oe} = 25 μS, h_{fe} = 50, h_{re} = 0. The reactances of the coupling capacitors are negligible.
(T-parameters: α' = 50, r_c = 2·0 MΩ, r_b = 2·5 kΩ, r_e = 0).

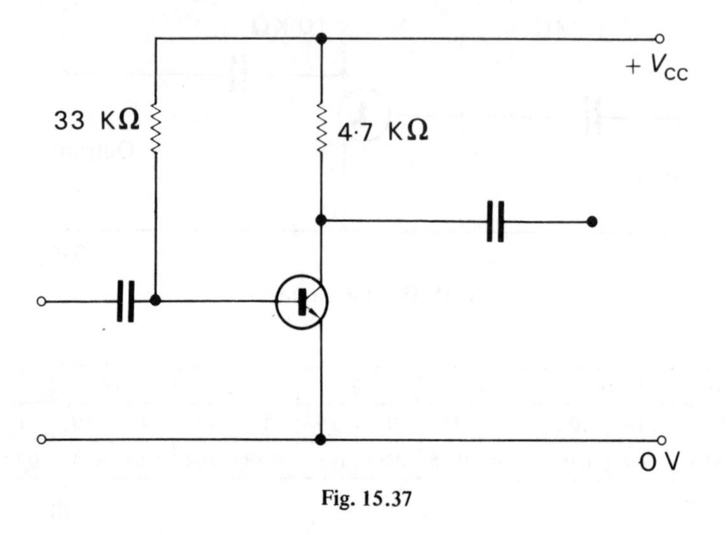

Fig. 15.37

900

7 Given that the input signal source to the circuit shown can be represented by a constant current generator I μA and internal resistance 10 kΩ, determine the magnitude of I so that the signal power dissipated in a 4·0 kΩ resistor connected across the output terminals is 1·0 mW. The small signal hybrid parameters for the transistor are: h_{ie} = 2·0 kΩ, h_{fe} = 100, h_{oe} = 50 μS and h_{re} = 0. The reactances of the coupling capacitors are negligible and the signal current taken by R_B is very small.
(T-parameters: α' = 100, r_c = 2·0 MΩ, r_b = 2·0 kΩ, r_e = 0).

8·8 μA r.m.s.

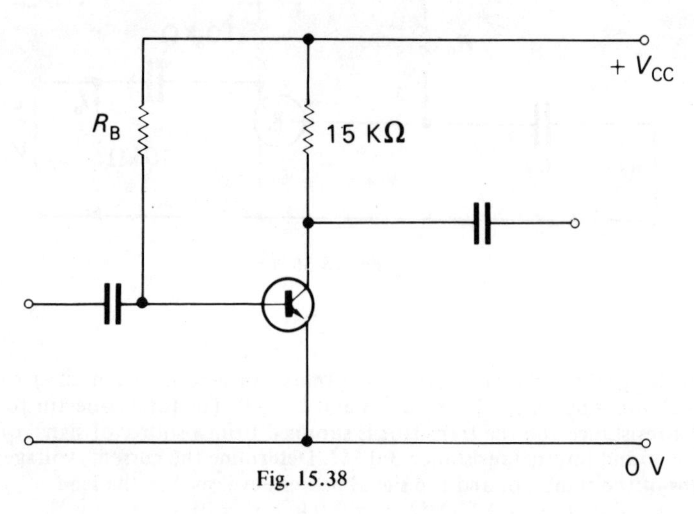

Fig. 15.38

8 A transistor used in the common base configuration has the following small signal parameters: $h_{ib} = 75\ \Omega$, $h_{fb} = -0{\cdot}95$, $h_{ob} = 0$ S and $h_{rb} = 0$. The effect of the applied sinusoidal signal can be represented by a source of e.m.f. of 30 mV and internal resistance 75 Ω connected between the emitter and base, and the total load connected between the collector and base is 5·0 kΩ resistance. Determine the collector-to-base signal voltage in magnitude and phase relative to the source of e.m.f., and the power gain of the transistor.
(T-parameters: $\alpha = 0{\cdot}95$, $r_e = 75\ \Omega$, effects of r_c and r_b negligible).

0·95 V, 0°; 60

16

Electronic systems

Chapter 15 has dealt with the use of transistors in amplifier applications. Other devices exist, e.g. thermionic valves and field-effect transistors, which are used in somewhat similar ways to provide gain. To include a treatment of such devices in comparable detail to that adopted for the transistor would entail a considerable extension to the physics contained in Chapter 13 in order that the operation of the devices could be understood. While this would be undoubtedly advantageous, the extra time involved could be obtained only by the exclusion of other topics in a fixed time period course of study. There is therefore much to be said in favour of a 'systems' approach to the teaching of electronics whereby the amplifier is treated as a 'block' with input and output terminals, and little or nothing said about the contents of the block; all the time available being devoted to its characteristics. The advent and very extensive use of the 'integrated circuit' where the components are formed on one small piece of silicon has increased the desirability of a systems approach, since the design of the circuitry does not in any case follow closely the techniques adopted using discrete components, and there is certainly no necessity to know the exact circuit details to be able to use them in electronic systems. This chapter is therefore an introduction to the systems approach to the teaching of electronics. It is complete in itself and requires no reference to the theory of bipolar transistors and can be considered as an alternative approach to the introduction of electronics.

16.1 Basic amplifier principles

The purpose of an amplifier is to produce gain. That is to say, a small input signal power controls a larger output signal power. Certain devices not normally referred to as amplifiers do nevertheless come into this category by definition. In a relay, for example, the power required by the coil to close the contacts can be considerably less than that involved in the circuit switched by the contacts. Another example is the separately-excited d.c. generator being driven at constant speed. Here the power being fed to a load connected across the output terminals can be controlled by a relatively small power fed to its field winding.

In the type of amplifier to be considered here, there is a further characteristic that it must exhibit. That is the waveform of the input signal voltage or current must be maintained to a fairly high degree of accuracy in the output signal, since the waveform carries some of the information represented by the signal. This can be illustrated by

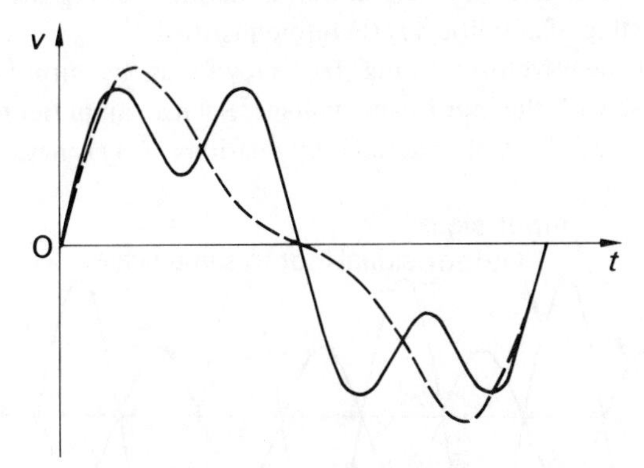

Fig. 16.1. Signal waveforms

comparing two signals representing the same note played by two different musical instruments. They are both of the same fundamental frequency representing the pitch but the waveforms differ representing different tones. This is illustrated in Fig. 16.1.

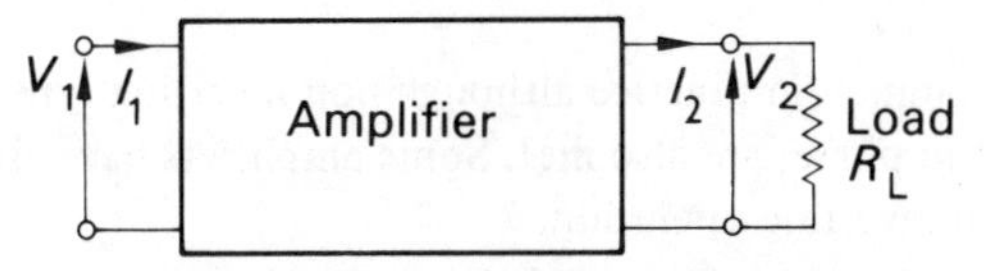

Fig. 16.2. Amplifier block diagram

Figure 16.2 shows an amplifier with a resistive load R_L connected across the output terminals. The basic parameters of the amplifier are

$$\text{Voltage gain } (G_v) = \frac{\text{Output signal voltage}}{\text{Input signal voltage}}$$

$$G_v = \frac{V_2}{V_1} \tag{16.1}$$

$$\text{Current gain } (G_i) = \frac{\text{Output signal current}}{\text{Input signal current}}$$

$$G_i = \frac{I_2}{I_1} \tag{16.2}$$

$$\text{Power gain } (G_p) = \frac{\text{Output signal power}}{\text{Input signal power}}$$

$$G_p = \frac{V_2 \times I_2}{V_1 \times I_1}$$

$$G_p = G_v G_i \tag{16.3}$$

Figure 16.3 shows typical waveforms where the signals are assumed to vary sinusoidally with time. The waveforms met in practice tend to be more complex, as illustrated in Fig. 16.1, but it can be shown that such waves are formed from a series of pure sine waves the frequencies of which are exact multiples of the basic or fundamental frequency. These sine waves are known as harmonics. The use of sine waves in the analysis and testing of amplifiers is therefore justified.

Examination of the waveforms in Fig. 16.3 shows that the output signal voltage is 180° out of phase with the input signal voltage. Such an amplifier is known as a *phase-inverting* one and the gain as defined by relation (16.1) is negative. These

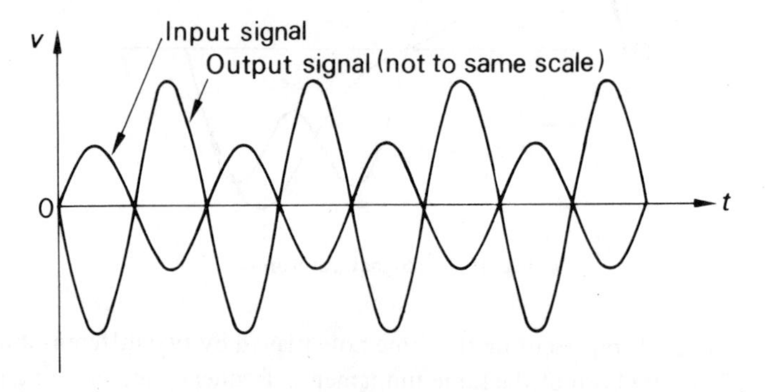

Fig. 16.3. Amplifier signal waveforms

amplifiers are very common in practice although non-inverting types, where the output and input signals are in phase, are also met. Some amplifiers have alternative inputs for inverting and non-inverting operation.

So far only the input and output signals have been considered in the operation of the amplifier. It is necessary however to provide a source of power from which is obtained the output signal power fed to the load; the magnitude of this signal power being controlled by the magnitude of the input signal. The source itself has to be a direct current one and could be a dry battery or a rectified supply as described in chapter 14. The magnitude of this supply voltage depends on the type of device used in the amplifier. For thermionic valves it will be of the order of hundreds of volts while for transistors it is often in the range 6 to 30 volts.

The voltage range within which the output signal voltage can vary is limited, being very much dependent on the value of the supply voltage. This means therefore that there is a maximum value of input signal that will produce an output signal the waveform of which remains an acceptable replica of the input signal waveform. Increasing the input signal beyond this level will produce 'clipping' of the input signal waveform as illustrated in Fig. 16.4, although the clipping levels need not be symmetrical about the zero level.

Examination of the gain of an amplifier will show that it does not remain constant with the frequency of the input signal. Some amplifiers exhibit a reduction in gain at both high and low frequencies while others have a reduction at high frequencies only. In both cases there is a considerable frequency range over which the gain remains essentially constant. It is within this frequency range that the amplifier is designed to

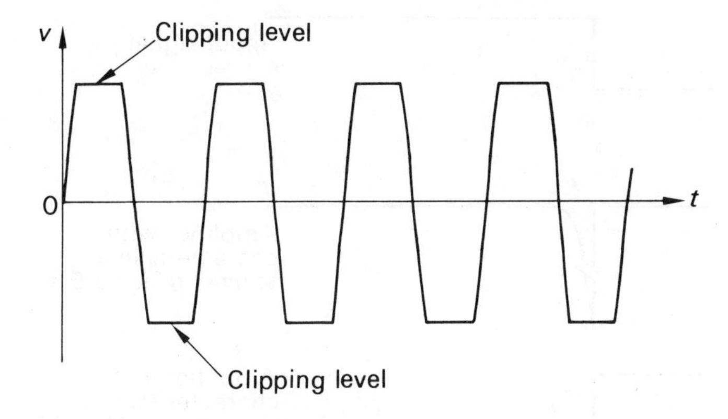

Fig. 16.4. Clipping of output signal waveform

operate. The effect of the unequal gain is to produce another form of waveform distortion since the harmonics present in a complex input signal waveform may not be amplified by the same amount. Fig. 16.5 shows typical gain/frequency characteristics where it should be noted that the frequency scales are logarithmic. The advantage of an amplifier with a characteristic as illustrated in Fig. 16.5 (*b*) is that it is capable of

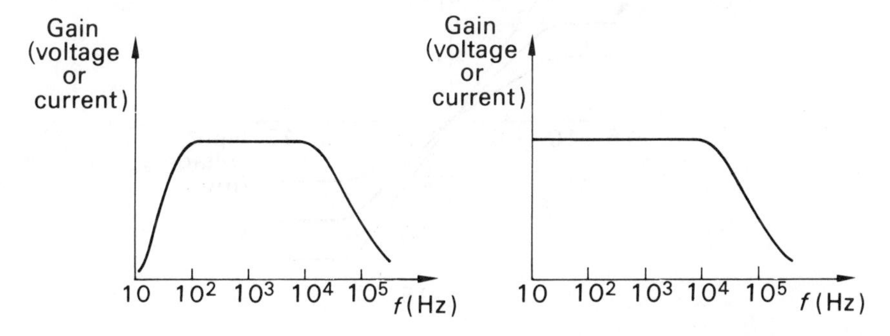

Fig. 16.5. Gain frequency characteristics (*a*) Capacitance coupled amplifier (*b*) Direct coupled amplifier

amplifying signals at the very low frequencies met with in many industrial applications. It is known as a direct-coupled (d.-c.) amplifier and integrated-circuit amplifiers are of this type. Figure 16.6 shows the responses obtained from both types of amplifier for an input signal which changes instantaneously from one level to another. Lack of low frequency gain in the first type prevents faithful reproduction of the steady parts of the input signal, the output being merely a 'blip' as the input signal changes from one level to the other. The characteristics showing output voltage against input voltage for several d.c. supply voltages as given in Fig. 16.7 are representative of integrated-circuit d.-c. amplifiers. Output voltage variation on either side of zero volts is obtained by the use of power supply voltages which are spaced on both sides of zero e.g. + 12 V and − 12 V as opposed to + 24 V and 0 V.

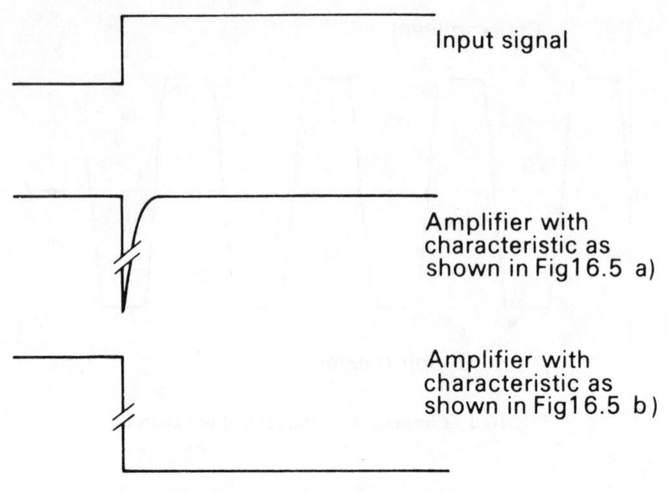

Fig. 16.6. Response to step change

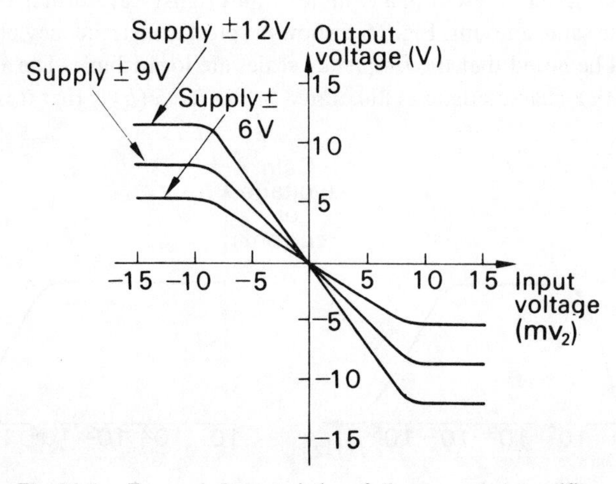

Fig. 16.7. Forward characteristics of direct-coupled amplifier

16.2 Logarithmic units

It is sometimes found convenient to express the ratio of two powers P_1 and P_2 in logarithmic units known as *bels* as follows

$$\text{Power ratio in bels} = \log\frac{P_2}{P_1} \tag{16.4}$$

It is found that the *bel* is rather a large unit and as a result the *decibel* (one tenth of a bel) is more common so that

$$\text{Power ratio in decibels (dB)} = 10\log\frac{P_2}{P_1} \tag{16.5}$$

If the two powers are developed in the same resistance or equal resistances then

$$P_1 = \frac{V_1^2}{R} = I_1^2 R \quad \text{and} \quad P_2 = \frac{V_2^2}{R} = I_2^2 R$$

where V_1, I_1, V_2 and I_2 are the voltages across and the currents in the resistance. Therefore

$$10 \log \frac{P_2}{P_1} = 10 \log \frac{V_2^2/R}{V_1^2/R} = 10 \log \frac{V_2^2}{V_1^2}$$

$$\text{Power ratio in dB} = 20 \log \frac{V_2}{V_1} \qquad (16.6)$$

Similarly

$$\text{Power ratio in dB} = 20 \log \frac{I_2}{I_1} \qquad (16.7)$$

The relationships defined by relations 16.6 and 16.7 although expressed by ratios of voltages and currents respectively still represent power ratios. They are however used extensively, although by fundamental definition erroneously, to express voltage and current ratios where common resistance values are not involved. For example, the voltage gain of an amplifier is expressed often in decibels. Care should be taken in this use of decibels since the expression of power gain in decibels would be in complete agreement with the basic definition.

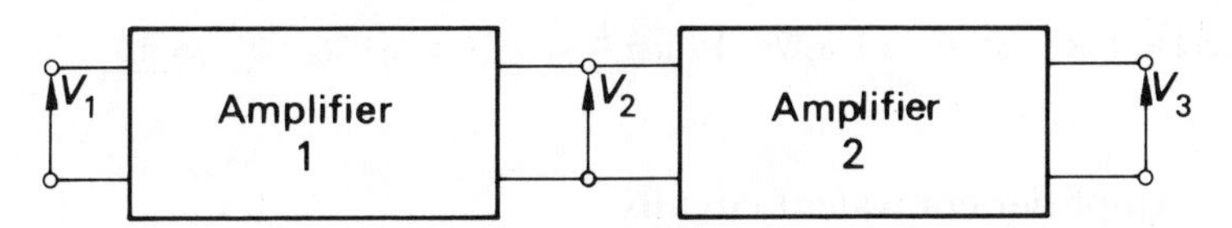

Fig. 16.8. Amplifiers in cascade

Figure 16.8 shows two amplifiers connected in cascade in which case the input of the second amplifier is the load on the first one. The overall voltage gain is given by

$$G_v = \frac{V_3}{V_2} \times \frac{V_2}{V_1}$$

Expressing this in decibels

$$\text{Voltage gain in dB} = 20 \log \frac{V_3}{V_2} \times \frac{V_2}{V_1}$$

$$= 20 \log \frac{V_3}{V_2} + 20 \log \frac{V_2}{V_1}$$

Thus the overall voltage gain in decibels is equal to the sum of the voltage gains in decibels of the individual amplifiers. This is a most useful result and can be extended to any number of amplifiers in cascade. It is obviously applicable to current and power gains.

The use of decibels gives a representation of one power (or voltage) with reference to another. If P_2 is greater than P_1 then P_2 is said to be $10 \log P_2/P_1$ dB *up* on P_1. For P_2 less than P_1, $\log P_2/P_1$ is negative. Since $\log P_1/P_2 = -\log P_2/P_1$, it is usual to determine the ratio greater than unity and P_2 is said to be $10 \log P_1/P_2$ *down* on P_1.

Example 16.1 The voltage gain of an amplifier when it feeds a resistive load of 1·0 kΩ is 40 dB. Determine the magnitude of the output signal voltage and the signal power in the load when the input signal is 10 mV.

$$20 \log \frac{V_2}{V_1} = 40$$

$$\therefore \quad \log \frac{V_2}{V_1} = 2{\cdot}0$$

$$\therefore \quad \frac{V_2}{V_1} = 100$$

$$\therefore \quad V_2 = 100 \times 10 = 1000 \text{ mV} = \underline{1{\cdot}0 \text{ V}}$$

$$P_2 = \frac{{V_2}^2}{R_L} = \frac{1{\cdot}0^2}{1000} = \frac{1}{1000} \text{ W} = \underline{1 \text{ mW}}$$

Example 16.2 Express the power dissipated in a 15-Ω resistor in decibels relative to 1·0 mW when the voltage across the resistor is 1·5 V r.m.s.

$$P_2 = \frac{1{\cdot}5^2}{15} \text{ W} = 150 \text{ mW}$$

Power level in DB relative to 1 mW $= 10 \log \frac{150}{1} = 10 \times 2{\cdot}176 = \underline{21{\cdot}76 \text{ dB}}$

16.3 Amplifier equivalent circuits

An equivalent circuit, as the name implies, is one that can replace the actual circuit for the purpose of analysis. The circuit shown in Fig. 16.9 (*a*) can be used as an

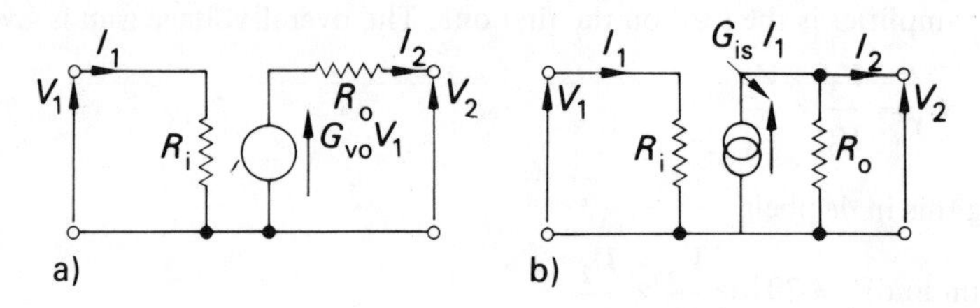

Fig. 16.9. Amplifier equivalent circuits (*a*) Constant-voltage form (*b*) Constant-current form

equivalent circuit for an amplifier. The input resistance of the amplifier is R_i. This represents the property whereby the input circuit of the amplifier loads the source of signal. It can be determined from

$$R_i = \frac{V_1}{I_1} \tag{16.8}$$

The capability of the amplifier to produce gain is represented by the voltage generator $G_{vo}V_1$ where G_{vo} is the voltage gain that would be obtained if the output terminals were on open circuit. The output resistance is R_o, a finite value of which will mean that the voltage gain G_v obtained when the amplifier is loaded is less than the open-circuit voltage gain G_{vo}, due to the loss of signal in R_o.

An alternative form of the equivalent circuit is shown in Fig. 16.9 (*b*). Here the output circuit is in a constant current form with the generator current $G_{is}I_1$ where G_{is} is the current gain that would be obtained with the output terminals on short circuit.

Since the circuits are themselves equivalent to each other it follows that

$$G_{vo}V_1 = G_{is}I_1R_o = G_{is}\frac{V_1}{R_i}R_o$$

$$\therefore \quad G_{vo} = \frac{R_o}{R_i}G_{is} \tag{16.9}$$

The use of such equivalent circuits is restricted to the signal quantities only. They should not be used in calculations concerned with the direct quantities associated with the amplifier. Moreover their use assumes an exact linear relationship between input and output signals i.e. the amplifier produces no waveform distortion.

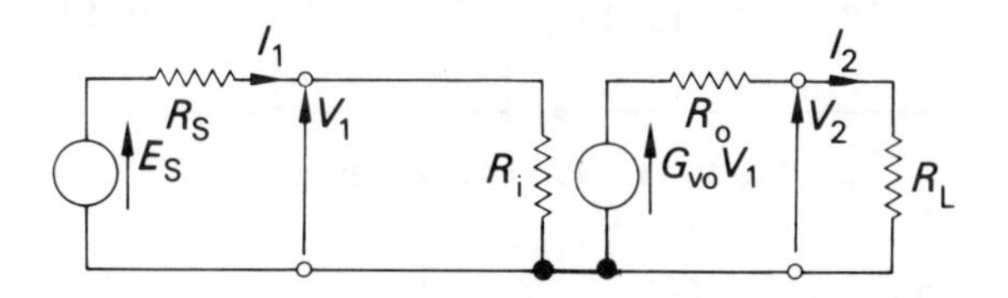

Fig. 16.10. Amplifier with signal source and load

Figure 16.10 shows an amplifier, represented by its equivalent circuit, being supplied by a source of signal E_s volts and source resistance R_s, and feeding a load resistance R_L. Expressions for the important parameters are derived.

$$\text{Input signal voltage to amplifier} = V_1 = \frac{R_i E_s}{R_s + R_i} \tag{16.10}$$

$$\text{Output signal voltage} = V_2 = \frac{R_L}{R_o + R_L}G_{vo}V_1$$

$$\therefore \text{Voltage gain} = G_v = \frac{V_2}{V_1} = \frac{G_{vo}R_L}{R_o + R_L} \tag{16.11}$$

$$\text{Output signal current} = I_2 = \frac{G_{vo}V_1}{R_o + R_L} = \frac{G_{vo}I_1R_i}{R_o + R_L}$$

$$\therefore \text{Current gain} = G_i = \frac{I_2}{I_1} = \frac{G_{vo}R_i}{R_o + R_L} \tag{16.12}$$

Substituting equation (16.9) gives

$$G_{\mathrm{i}} = \frac{\dfrac{R_{\mathrm{o}}}{R_{\mathrm{i}}} G_{\mathrm{is}} R_{\mathrm{i}}}{R_{\mathrm{o}} + R_{\mathrm{L}}} = \frac{G_{\mathrm{is}} R_{\mathrm{o}}}{R_{\mathrm{o}} + R_{\mathrm{L}}} \tag{16.13}$$

$$\text{Power gain} = G_P = \frac{\text{Signal power into load}}{\text{Signal power into amplifier}}$$

$$\therefore \quad G_P = \frac{V_2^2/R_L}{V_1^2/R_i} = \frac{I_2^2 R_L}{I_1^2 R_i} = \frac{VI}{V_1^2 I_1^2} = G_v G_i \tag{16.14}$$

Example 16.3 An integrated-circuit amplifier has an open-circuit voltage gain of 5 000, an input resistance of 15 kΩ and an output resistance of 25 Ω. It is supplied from a signal source of internal resistance 5·0 kΩ and it feeds a resistive load of 175 Ω. Determine the magnitude of the signal source voltage to produce an output signal voltage of 1·0 V. What value of load resistance would half the signal voltage output for the same input?

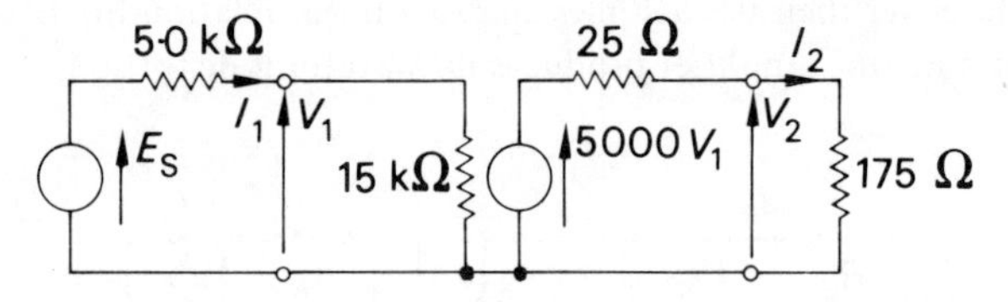

Fig. 16.11. Part of Example 16.3

The circuit is shown in Fig. 16.11

$$V_2 = \frac{175}{25 + 175} \times 5000\ V_1$$

$$\text{Voltage gain} \quad = G_{\mathrm{v}} = \frac{V_2}{V_1} = \frac{175 \times 5000}{200} = 4350$$

$$\therefore \quad V_1 = \frac{1 \cdot 0}{4350} = 2 \cdot 30 \times 10^{-4}\ \mathrm{V} = 230\ \mu\mathrm{V}$$

$$V_1 = \frac{15}{5 + 15} E_{\mathrm{s}} \qquad \therefore \quad E_{\mathrm{s}} = \frac{20 \times 230}{15} = 307\ \mu\mathrm{V}$$

For half the signal output the voltage gain must be halved.

$$\therefore \quad G_{\mathrm{v}} = \frac{4350}{2} = 2175 = \frac{R_{\mathrm{L}} \times 5000}{25 + R_{\mathrm{L}}}$$

$$\therefore \quad 2175 \times 25 + 2175\ R_{\mathrm{L}} = 5000\ R_{\mathrm{L}}$$

$$\therefore \quad R_{\mathrm{L}} = \frac{2175 \times 25}{2825} = 19 \cdot 2\ \Omega$$

Example 16.4 An amplifier has an open-circuit voltage gain of 70 dB and an output resistance of 1·5 kΩ. Determine the minimum value of load resistance so that the voltage gain is not more than 3·0 dB down on the open-circuit value. With this value of load resistance determine the magnitude of the output signal voltage when the input signal is 1·0 mV.

$$20 \log G_{vo} = 70 \qquad \therefore \quad \log G_{vo} = 3{\cdot}50$$

$$\therefore \quad G_{vo} = 3160$$

$$20 \log G_{v} = 70 - 3 = 67 \qquad \therefore \quad \log G_{v} = 3{\cdot}35$$

$$\therefore \quad G_{v} = 2240$$

$$\therefore \quad \frac{R_L}{R_o + R_L} 3160 = 2240$$

$$\therefore \quad 3160\, R_L = 2240 \times 1{\cdot}5 + 2240\, R_L$$

$$\therefore \quad R_L = \frac{2240 \times 1{\cdot}5}{920} = \underline{3{\cdot}65\ \text{k}\Omega}$$

Alternatively

since $\quad 20 \log G_{vo} - 20 \log G_{v} = 3{\cdot}0$

$$\therefore \quad 20 \log \frac{G_{vo}}{G_{v}} = 3{\cdot}0$$

$$\therefore \quad \frac{G_{vo}}{G_{v}} = 1{\cdot}41$$

$$\therefore \quad \frac{R_L}{R_o + R_L} = \frac{1}{1{\cdot}41} \qquad \therefore \quad R_L = \frac{1{\cdot}5}{0{\cdot}41} = \underline{3{\cdot}65\ \text{k}\Omega}$$

$$V_2 = 2240 \times 1{\cdot}0 = 2240\ \text{mV} = \underline{2{\cdot}24\ \text{V}}$$

16.4 Frequency response

It has already been stated in para. 16.1 that amplifier gain decreases at high frequencies and in some cases at low frequencies also. The equivalent circuits, as have been used so far, gives no indication of this since they contain pure resistance only. They have therefore to be modified for the frequency ranges in which the gain decreases.

One cause of loss of gain at high frequencies is the presence of shunt capacitance across the load. This can be due to stray capacitance in the external circuit and, what is usually more important, capacitance within the amplifier itself. The effective load on the amplifier tends to zero as the frequency tends to infinity. Thus the voltage gain decreases because of the decrease in load impedance, and the current gain decreases because the shunt capacitance path drains current away from the load. Another cause of loss of gain at high frequencies is the inherent decrease of available gain in the amplifier i.e. G_{vo} decreases with frequency. The manner in which the gain decreases due to this effect is similar to that produced by shunt capacitance and both effects can be represented on the equivalent circuit by the connection of capacitance across the output terminals assuming that, at the frequency being considered, only one of the effects is appreciable. In practice this is often a reasonable assumption over a considerable frequency range. Figure 16.12 shows the output section of the equivalent circuit modified for high frequency operation. Associated with the loss of gain will be a shift of phase between input and output signals from the nominal value. A more complex circuit would be required if both the effects considered above had to be taken into account simultaneously.

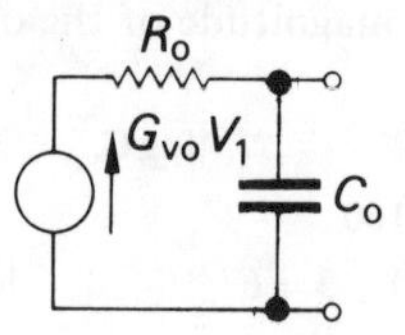

Fig. 16.12. Equivalent circuit for high-frequency operation

Loss of gain at low frequencies is due to the use of certain capacitors in the circuit. Their values are chosen such that the reactances are very small at the frequencies being used, so that little of the signal voltage is developed across them. At low frequencies however appreciable signal is developed across them resulting in a loss of signal at the load. As with the high frequency response there is a corresponding shift of phase between input and output signals.

The *bandwidth* of an amplifier is defined as the difference in frequency between the lower and upper frequencies, f_1 and f_2 respectively, at which the gain is 3·0 dB down on its maximum value. For a direct coupled amplifier the bandwidth will be simply f_2 since the gain extends down to zero frequency. The *passband* or working frequency

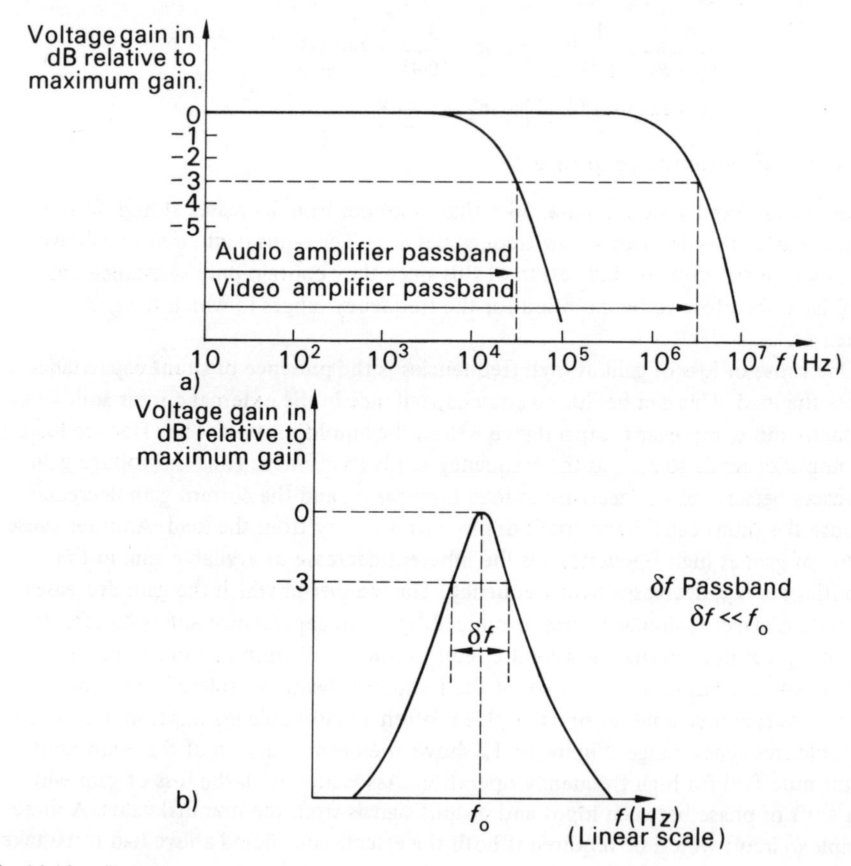

Fig. 16.13. Frequency response characteristics (*a*) Broad-band amplifiers (*b*) Narrow-band amplifier

range is that bounded by f_1 and f_2. If this lies within the range of frequencies normally audible to the ear as sound waves e.g. 30 Hz to 15 kHz, the amplifier is referred to as an *audio amplifier.* Such amplifiers are used in sound reproduction systems. The signals applied to the cathode ray tube in television receivers require greater passbands extending from 0 to several megahertz and are known as *video amplifiers.* In both audio and video amplifiers approximately constant gain over a fairly wide range of frequencies is required and they are collectively known as *broad-band amplifiers.* Other types known as *narrow-band amplifiers* are used where the bandwidth is considerably less than the centre frequency. This type of amplifier provides selectivity between signals of different frequency. Figure 16.13 shows typical frequency response curves for different types of amplifiers.

16.5 Feedback

Feedback is the process whereby a signal derived in the output section of the amplifier is fed back into the input section. In this way the amplifier can be used to provide characteristics which differ from those of the basic amplifier. The signal fed back can be either a voltage or a current, being applied in series or shunt respectively with the input signal. Moreover the feedback signal, whether voltage or current can be directly proportional to the output signal voltage or current. This gives rise to four basic types of feedback, i.e. series-voltage, series-current, shunt-voltage and shunt-current. The characteristics produced by these four types of feedback are similar in some respects and differ in others. Series-voltage feedback will be considered here as a representative type and Fig. 16.14 shows such a feedback amplifier.

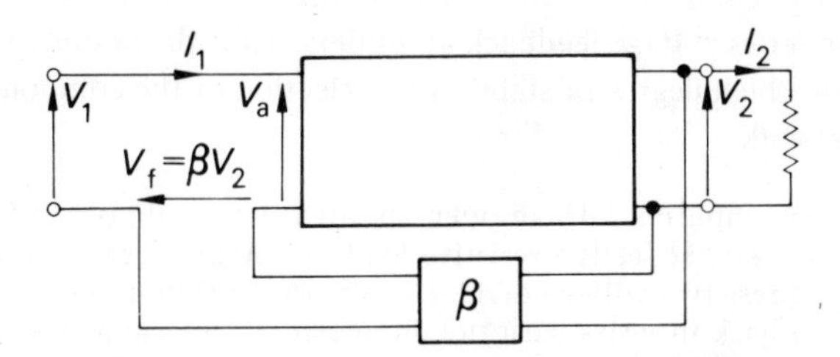

Fig. 16.14. Series voltage feedback amplifier

The block marked β is that part of the circuit which provides the feedback voltage $V_f = \beta V_2$. In one of its simplest forms it could consist of two resistors connected across the output to form a voltage divider, the feedback voltage being the signal developed across one of the resistors. The voltage gain G_v will be dependent on the load which the β network presents to the amplifier, although in many practical applications the values of the components used in the network are such as to present negligible loading.

From Fig. 16.14

$$\text{Input signal to basic amplifier} = V_a = V_1 + V_f = V_1 + \beta V_2$$

$$V_2 = G_v V_a = G_v(V_1 + \beta V_2)$$

$$V_2(1 - \beta G_v) = G_v V_1$$

Voltage Gain with Feedback $= G_{vf} = \dfrac{V_2}{V_1}$

$$G_{vf} = \frac{G_v}{1 - \beta G_v} \tag{16·15}$$

If the magnitude of $1 - \beta G_v$ is greater than unity then the magnitude of G_{vf} is less than that of G_v and the feedback is said to be *negative* or degenerative. The simplest means of accomplishing this is to provide a phase inverting amplifier, in which case G_v is negative, and for β to be a positive fraction, as would be obtained with a simple resistive voltage divider. For this particular case relation (16.10) can be written as

$$|G_{vf}| = \frac{|G_v|}{1 + \beta|G_v|}$$

where $|G_{vf}|$ and $|G_v|$ represent the magnitudes of the quantities.

Thus if $\beta\,|G_v| \gg 1$ then

$$G_{vf} \doteqdot \frac{|G_v|}{\beta\,|G_v|}$$

i.e. $$|G_{vf}| \doteqdot \frac{1}{\beta} \tag{16.16}$$

Relation (16.16) illustrates that the voltage gain with negative feedback is relatively independent of the voltage gain of the basic amplifier, provided that the product $\beta\,|G_v|$ remains large compared to unity. This is one of the most important characteristics of negative series voltage-feedback amplifiers. Thus the desired voltage gain can be obtained with a high degree of stability by selection of the component values in the feedback network.

Example 16.5 An amplifier with an open-circuit voltage gain of – 1 000 and an output resistance of 100 Ω feeds a resistive load of 900 Ω. Negative feedback is provided by connection of a resistive voltage divider across the output and one fiftieth of the output voltage fed back in series with the input signal. Determine the voltage gain with feedback. What percentage change in the voltage gain with feedback would be produced by a 50% change in the voltage gain of the basic amplifier due to a change in the load.
The loading effect of the feedback network can be neglected.

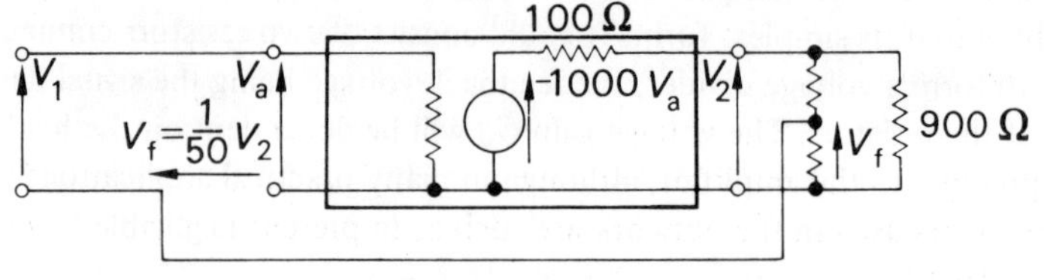

Fig. 16.15. Part of Example 16.5

The circuit is shown in Fig. 16.15

$$G_v = \frac{900}{100 + 900}(-1000) = -900$$

Voltage gain with feedback $= G_{vf} = \dfrac{G_v}{1 - \beta G_v}$

$$= \frac{-900}{1 - (\frac{1}{50})(-900)} = \frac{-900}{1 + 18} = \underline{-47{\cdot}4}$$

For $\quad G_v = -450$

$$\therefore \quad G_{vf} = \frac{-450}{1 - (\frac{1}{50})(-450)} = \frac{-450}{1 + 9} = \underline{-45{\cdot}0}$$

$$\Delta G_{vf} = 47{\cdot}4 - 45{\cdot}0 = 2{\cdot}4$$

Percentage change in $G_{vf} = \dfrac{2{\cdot}4}{47{\cdot}4} \times 100 = \underline{5{\cdot}1\ \%}$

Example 16.6 An amplifier is required with an overall voltage gain of 100 and which does not vary by more than 1·0%. If it is to use negative feedback with a basic amplifier the voltage gain of which can vary by 20%, determine the minimum voltage gain required and the feedback factor.

$$100 = \frac{G_v}{1 + \beta G_v} \qquad [1]$$

$$99 = \frac{0{\cdot}8\ G_v}{1 + \beta 0{\cdot}8\ G_v} \qquad [2]$$

$$\therefore \quad 100 + 100\ \beta G_v = G_v \qquad [3]$$

$$99 + 79{\cdot}2\ \beta G_v = 0{\cdot}8\ G_v \qquad [4]$$

$$[3] \times 0{\cdot}792$$

$$\therefore \quad 79{\cdot}2 + 79{\cdot}2\ \beta G_v = 0{\cdot}792\ G_v \qquad [5]$$

$$[4] - [5]$$

$$19{\cdot}8 = 0{\cdot}008\ G_v$$

$$\therefore \quad G_v = \frac{19{\cdot}8}{0{\cdot}008} = \underline{2475}$$

Substitute in [3]

$$100 + 100\ \beta \times 2475 = 2475$$

$$\therefore \quad \beta = \frac{2375}{100 \times 2475} = \underline{0{\cdot}00960}$$

If the magnitude of $1 - \beta G_v$ is less than unity then from relation (16.15) the magnitude of G_{vf} is greater than that of G_v. The feedback is then said to be *positive* or regenerative. It is not very common however to use positive feedback to increase gain since a positive feedback amplifier has opposite characteristics to that of a negative feedback amplifier. Thus the stability of the voltage gain with feedback will be worse than that of the basic amplifier.

There is one case of the use of positive feedback that is of considerable practical importance. That is the case where $\beta G_v = 1$, which gives from relation (16.10)

$$G_{vf} = \frac{G_v}{0} = \infty$$

An amplifier with infinite gain is one that can produce an output signal with no externally applied input signal. It provides its own input signal via the feedback network. Such a circuit is known as an *oscillator*, and to produce an output signal at a predetermined frequency the circuitry is arranged so that $\beta G_v = 1$ at that frequency only. It must be stressed that the condition $\beta G_v = 1$ must be satisfied in magnitude and phase.

It is necessary to consider the condition $\beta G_v > 1$. This can only be a transient condition in a feedback amplifier since it means that the output signal amplitude would be increasing with time. This build up of amplitude will eventually entail the amplifier operating in non-linear parts of its characteristics, perhaps even into the cut-off and/or saturation regions. This results in an effective decrease in G_v and the system settles down when $\beta G_v = 1$. Thus the substitution of values of βG_v greater than $+1$ in relation (16.15) has little practical significance.

In para. 16.4 it was stated that loss of gain at high and low frequencies was accompanied by a change of phase shift between input and output signals. It is possible therefore for feedback, which is designed to be negative in the passband, to become positive at high or low frequencies and introduce the possibility of the condition $\beta G_v = 1$ existing. The amplifier would then oscillate at the appropriate frequency and it is said to be unstable. Much of the design work associated with nominally negative feedback amplifiers is concerned with maintaining stability against oscillation.

Example 16.7 An amplifier has a voltage gain of $-1\,000$ within the passband. At a specific frequency f_x outwith the passband the voltage gain is 15 dB down on the passband value and there is zero phase between input and output signal voltages. Determine the maximum amount of nagative feedback that can be used so that the feedback amplifier will be stable.

'Let voltage gain at $f_x = G_{vx}$

$$\therefore \quad 20 \log \frac{G_v}{G_{vx}} = 15$$

$$\therefore \quad \frac{G_v}{G_{vx}} = 5{\cdot}62$$

$$\therefore \quad G_{vx} = \frac{1000}{5{\cdot}62} = 178$$

Oscillation will occur if $\beta G_v = 1$

$$\therefore \text{Maximum value of } \beta = \frac{1}{178} = 0{\cdot}005\,8$$

So far the effect of series voltage feedback on voltage gain has been considered. It is necessary to consider its effect on the other characteristics of the amplifier. While the introduction of series feedback will affect the magnitude of the input current I_1 for a given value of input voltage V_1, the current gain will be unaffected i.e. a given value of I_1 will still produce the same value of I_2 as specified by relation (16.12). It follows therefore that the power gain which is the product of the voltage and current gains will change by the same factor as the voltage gain.

The input resistance with feedback can be determined with reference to Fig. 16.14 as follows

$$\text{Input resistance with feedback} = R_{if} = \frac{V_1}{I_1}$$

$$= \frac{V_a - V_f}{I_1} = \frac{V_a - \beta V_2}{I_1} = \frac{V_a - \beta G_v V_a}{I_1}$$

$$= \frac{V_a(1 - \beta G_v)}{I_1}$$

$$R_{if} = R_i(1 - \beta G_v) \tag{16.17}$$

Thus the input resistance is increased by negative series voltage feedback and hence decreased by positive series-voltage feedback.

The output resistance with feedback can be determined from the ratio of the open circuit output voltage to the short circuit output current.

$$V_{2o/c} = G_{vo} V_a = G_{vo}(V_1 + V_f) = G_{vo}(V_1 + \beta V_{2o/c})$$

$$V_{2o/c} = \frac{G_{vo} V_1}{1 - \beta G_{vo}}$$

$$I_{2s/c} = \frac{G_{vo} V_a}{R_o} = \frac{G_{vo} V_1}{R_o}$$

(there is no signal fed back in this case since there is no output voltage)

$$\text{Output resistance with feedback} = R_{of} = \frac{V_{2o/c}}{I_{2s/c}}$$

$$R_{of} = \frac{R_o}{1 - \beta G_{vo}} \tag{16.18}$$

Thus the output resistance is decreased by negative series voltage feedback and hence increased by positive series voltage feedback.

Example 16.18 An amplifier has an open circuit voltage gain of 1 000, an input resistance of 2 000 Ω and an output resistance of 1·0 Ω. Determine the input signal voltage required to produce an output signal current of 0·5 A in a 4·0-Ω resistor connected across the output terminals. If the amplifier is then used with negative series voltage feedback so that one tenth of the output signal is fed back to the input determine the input signal voltage to supply the same output signal current

From equation (16.12)

$$\frac{I_2}{I_1} = \frac{G_{vo} R_i}{R_o + R_L} = \frac{1000 \times 2000}{1{\cdot}0 + 4{\cdot}0} = 4{\cdot}0 \times 10^5$$

$$I_1 = \frac{0{\cdot}5}{4{\cdot}0 \times 10^5} = 1{\cdot}25 \times 10^{-6}\ \text{A}$$

$$V_1 = I_1 R_i = 1{\cdot}25 \times 10^{-6} \times 2 \times 10^3 = \underline{2{\cdot}5\ \text{mV}}$$

With feedback

$$\frac{I_2}{I_1} = 4{\cdot}0 \times 10^5 \text{ (as before)} \quad \therefore \quad I_1 = 1{\cdot}25 \times 10^{-6} \text{ A.}$$

$$R_{\text{if}} = R_\text{i}(1 + \beta G_\text{v}) = 2000\left(1 + \frac{1}{10} \times \frac{4}{1+4} \times 1000\right)$$

$$= 2000\,(1 + 80) = 162000\ \Omega$$

$$\therefore \quad V_1 = 1{\cdot}25 \times 10^{-6} \times 1{\cdot}62 \times 10^5$$

$$= 0{\cdot}202 \text{ V} = \underline{202 \text{ mV}}$$

16.6 Logic gates

Certain electronic circuits operate with input and output signals which can exist only at one of two possible levels e.g. 0 V and + 6·0 V. In a particular group of such circuits known as *logic gates* the state of the output signal is a function of the states of the input signals, and the basic logic gates used in practice will be considered here. The two possible levels at which the signals can exist will be denoted by 0 and 1, these symbols not having their usual arithmetical significance. For example the level represented by 1 could be less than that represented by 0. If the more positive of the two levels is represented by 1 the system is said to employ *positive logic* while, on the other hand, if the more negative of the two levels is allocated 1 the system is said to employ *negative logic.* Most modern systems employ positive logic and this will be used exclusively here.

The AND gate

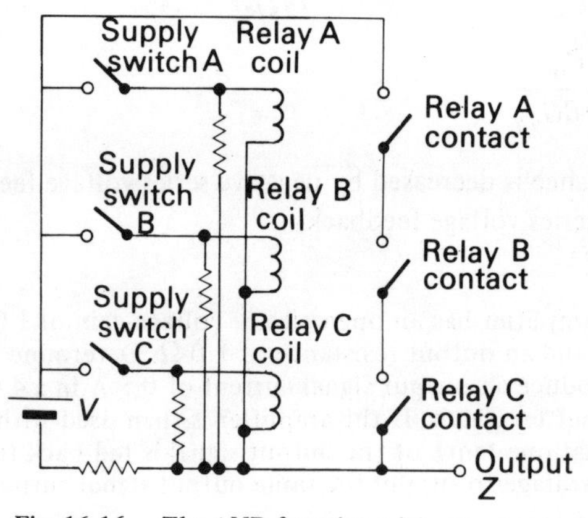

Fig. 16.16. The AND function with relay contacts

Consider the circuit shown in Fig. 16.16 which will be used to illustrate the AND function. The output signal Z will be at the 0 V level i.e. logical 0 unless all the contacts are closed. To close a contact it is necessary to supply the appropriate relay coil with V volts i.e. logical 1. This can be done by closing the relevant supply switch. If therefore the relay supply voltages are considered the inputs then the output will be 1 if, and only if, all the inputs are 1. This statement can be written in Boolean Algebra (i.e. mathematics of logic) form as follows

$$Z = A.B.C \qquad (16.19)$$

i.e. $Z = A \text{ AND } B \text{ AND } C$

The same function can be generated by electronic circuitry. It is necessary however to specify minimum and maximum voltage levels which must be accepted by all gates as a logical 1 and 0 respectively. For example in a circuit with a d.c. supply voltage of 4·5 V, the minimum voltage for a logical 1 could be 2·0 V and the maximum voltage for a logical 0 could be 0·8 V although the nominal levels are 4·5 V and 0 respectively. The symbol used for an AND Gate is shown in Fig. 16.17. The *truth table* shows the

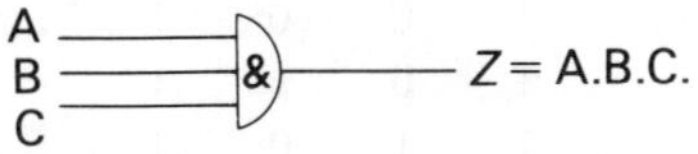

A
B & Z= A.B.C.
C

Fig. 16.17. Symbol for AND gate

state of the output for all possible combinations of the states of the inputs. A three-input gate has been illustrated although in practice there could be many more inputs.

Truth table for AND function

A	B	C	Z
0	0	0	0
0	0	1	0
0	1	0	0
0	1	1	0
1	0	0	0
1	0	1	0
1	1	0	0
1	1	1	1

The OR gate

The output of this gate will be 1 if any one of the inputs is 1. This function is illustrated with relay contacts (coils omitted in diagram) in Fig. 16.18 and the symbol used in Fig. 16.19. The function is written as

$$Z = A + B + C \qquad (16.20)$$

i.e. $Z = A \text{ OR } B \text{ OR } C.$

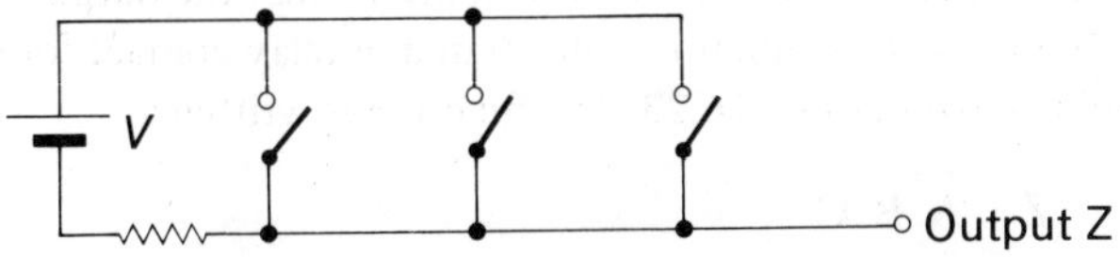

V
Output Z

Fig. 16.18. The OR function with relay contacts

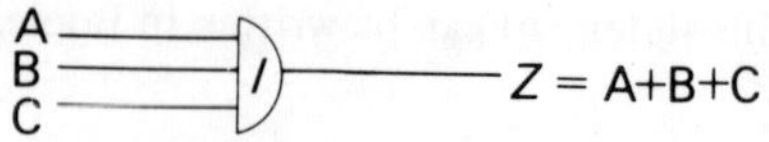

A
B
C
$Z = A+B+C$

Fig. 16.19. Symbol for OR gate

Truth table for OR function

A	B	C	Z
0	0	0	0
0	0	1	1
0	1	0	1
0	1	1	1
1	0	0	1
1	0	1	1
1	1	0	1
1	1	1	1

The NOT gate

This is a one input gate where the output is always the inverse or complement of the input. The function is shown in relay contact form in Fig. 16.20 where the contact is

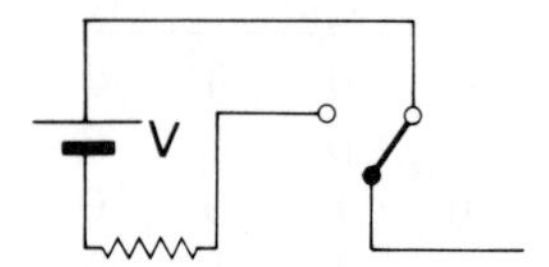

Fig. 16.20. The NOT function with relay contacts

shown in its position corresponding to the coil de-energised i.e. input = 0. The symbol is shown in Fig. 16.21. The function is written as

$$Z = \overline{A} \tag{16.21}$$

i.e. Z = NOT A.

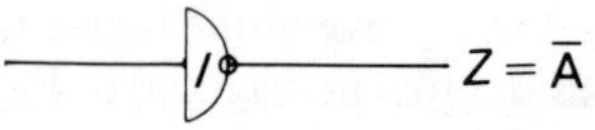

$Z = \overline{A}$

Fig. 16.21. Symbol for NOT gate

The NAND gate

The term NAND is short for NOT AND which means that the output is 0 if, and only if, all the inputs are 1's. This function is illustrated in relay contact form in Fig. 16.22, and the symbol is shown in Fig. 16.23. The function is written as

$$Z = \overline{A.B.C} \tag{16.22}$$

i.e. Z = NOT (A AND B AND C)

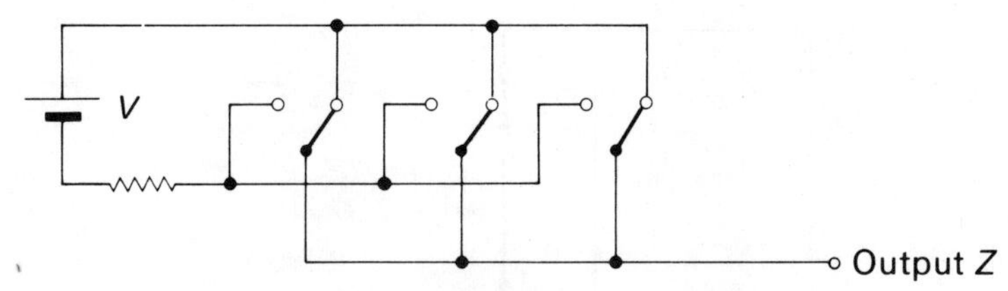

Fig. 16.22. The NAND function with relay contacts

Comparison of the circuit shown in Fig. 16.22 with those in Figs. 16.18 and 16.20 shows that the NAND function is the OR function on the complements of the inputs.

A
B
C
&
$Z = \overline{A.B.C.}$

Fig. 16.23. Symbol for NAND gate

Truth table for NAND function

A	B	C	$\bar{A}$	$\bar{B}$	$\bar{C}$	Z
0	0	0	1	1	1	1
0	0	1	1	1	0	1
0	1	0	1	0	1	1
0	1	1	1	0	0	1
1	0	0	0	1	1	1
1	0	1	0	1	0	1
1	1	0	0	0	1	1
1	1	1	0	0	0	0

This can be verified from the truth table which shows the complements of the inputs. Thus the NAND function can be written in the following alternative form

$$Z = \bar{A} + \bar{B} + \bar{C} \tag{16.23}$$

Comparing relations (16.23) and (16.22) introduces the important relationship that

$$\overline{A.B.C} = \bar{A} + \bar{B} + \bar{C} \tag{16.24}$$

This relationship holds for any number of inputs and is one form of De Morgan's Theorem met with in Boolean Algebra.

The NOR gate

The term NOR is short for NOT OR which means that the output is 0 if any of the inputs are 1. This function is illustrated in relay contact form in Fig. 16.24 and the symbol is shown in Fig. 16.25. The function is written as

$$Z = \overline{A + B + C} \tag{16.25}$$

i.e. Z = NOT (A OR B OR C)

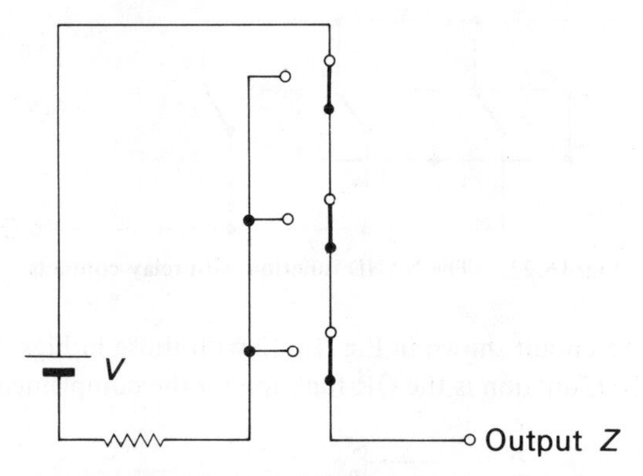

Fig. 16.24. The NOR function with relay contacts

Comparison of the circuit shown in Fig. 16.24 with those in Figs. 16.16 and 16.20 shows that the NOR function is the AND function of the complements of the inputs.

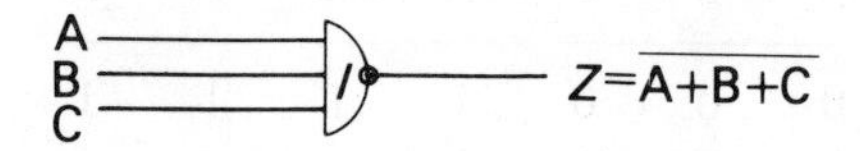

Fig. 16.25. Symbol for NOR gate

Truth table for NOR function

A	B	C	$\bar{A}$	$\bar{B}$	$\bar{C}$	Z
0	0	0	1	1	1	1
0	0	1	1	1	0	0
0	1	0	1	0	1	0
0	1	1	1	0	0	0
1	0	0	0	1	1	0
1	0	1	0	1	0	0
1	1	0	0	0	1	0
1	1	1	0	0	0	0

This can be verified from the truth table. Thus the NOR function can be written in the following alternative form

$$Z = \bar{A}.\bar{B}.\bar{C} \qquad (16.26)$$

Comparing relations (16.26) and (16.25) introduces the important relationship that

$$\bar{A}.\bar{B}.\bar{C} = \overline{A + B + C} \qquad (16.27)$$

This relationship holds for any number of inputs and is the other form of De Morgan's Theorem.

It should be noted that the relationships expressed by relations 16.19 to 16.27 are

Boolean algebra functions and care should be taken to avoid any misunderstanding in their interpretation. For example the signs used (i.e. = + .) do not have the same meaning as they have in basic algebra, the rules of which need not necessarily hold here.

16.7 Combinational logic

Combinations of the logic gates considered in the previous section are used to perform the logical functions met with in industrial switching and computer applications. The rules listed below are useful in the simplification of the logical functions used in the development of the circuitry. Their validity has been justified where necessary by the use of truth tables.

Absorption rules

1. $X + 0 = X$
2. $X + X = X$
3. $X + 1 = 1$
4. $X . 0 = 0$
5. $X . 1 = X$
6. $X . X = X$

Truth table for absorption rules

X	1	0	X + 0	X + X	X + 1	X . 0	X . 1	X . X
0	1	0	0	0	1	0	0	0
1	1	0	1	1	1	0	1	1

Commutative rules

1. $X + Y = Y + X$
2. $X . Y = Y . X$

Associative rules

1. $X + (Y + Z) = (X + Y) + Z$

Truth table for first associative rule

X	Y	Z	Y + Z	X + Y	X + (Y + Z)	(X + Y) + Z
0	0	0	0	0	0	0
0	0	1	1	0	1	1
0	1	0	1	1	1	1
0	1	1	1	1	1	1
1	0	0	0	1	1	1
1	0	1	1	1	1	1
1	1	0	1	1	1	1
1	1	1	1	1	1	1

2. X .(Y . Z) = (X . Y) . Z

Truth table for second associative rule

X	Y	Z	Y . Z	X . Y	X . (Y . Z)	(X . Y) . Z
0	0	0	0	0	0	0
0	0	1	0	0	0	0
0	1	0	0	0	0	0
0	1	1	1	0	0	0
1	0	0	0	0	0	0
1	0	1	0	0	0	0
1	1	0	0	1	0	0
1	1	1	1	1	1	1

Distributive rules

1. X + Y . Z = (X + Y) . (X + Z)

Truth table for first distributive rule

X	Y	Z	Y . Z	X + Y . Z	X + Y	X + Z	(X + Y) . (X + Z)
0	0	0	0	0	0	0	0
0	0	1	0	0	0	1	0
0	1	0	0	0	1	0	0
0	1	1	1	1	1	1	1
1	0	0	0	1	1	1	1
1	0	1	0	1	1	1	1
1	1	0	0	1	1	1	1
1	1	1	1	1	1	1	1

2. X . (Y + Z) = X . Y + X . Z

Truth table for second distributive law

X	Y	Z	Y + Z	X . (Y + Z)	X . Y	X . Z	X . Y + X . Z
0	0	0	0	0	0	0	0
0	0	1	1	0	0	0	0
0	1	0	1	0	0	0	0
0	1	1	1	0	0	0	0
1	0	0	0	0	0	0	0
1	0	1	1	1	0	1	1
1	1	0	1	1	1	0	1
1	1	1	1	1	1	1	1

Rules of complementation

1. $X + \overline{X} = 1$
2. $X . \overline{X} = 0$

Truth table for rules of complementation

X	$\overline{X}$	$X + \overline{X}$	$X . \overline{X}$
0	1	1	0
1	0	1	0

De Morgan's Theorem

1. $\overline{A . B . C} = \overline{A} + \overline{B} + \overline{C}$
2. $\overline{A} . \overline{B} . \overline{C} = \overline{A + B + C}$

These rules have already been dealt with in para. 16.6 in the sections concerned with NAND and NOR gates.

The use of these rules will be illustrated in the examples which follow.

Example 16.9 Determine the relationship between the output Z in the circuit shown in Fig. 16.26 and the inputs A, B and C. Hence construct a circuit using only two gates which will perform the same function.

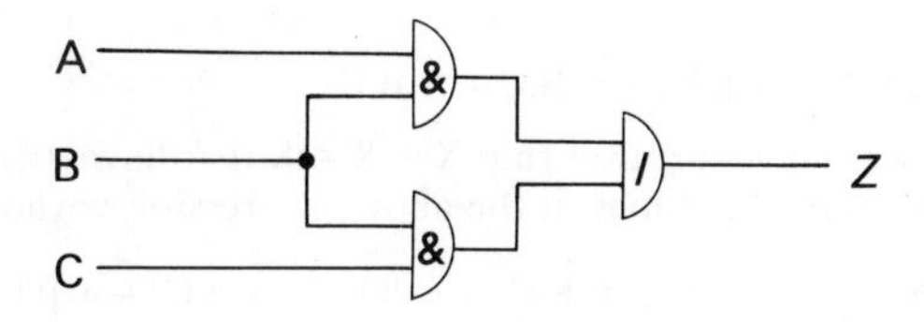

Fig. 16.26

Output from first AND Gate = A . B
Output from second AND Gate = B . C
Therefore output from OR Gate = Z = A . B + B .C

Using second distributive rule

$$Z = A . B + B . C = B . (A + C)$$

Thus the circuit shown in Fig. 16.27 will perform the required function.

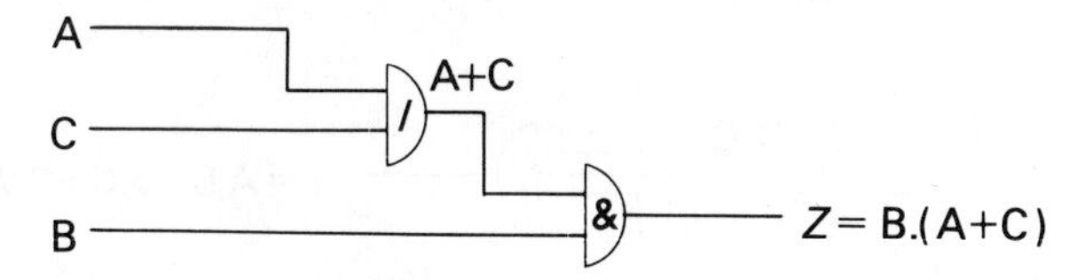

Fig. 16.27. Part of Example 16.9

Example 16.10 Draw a circuit using AND, OR and NOT gates which will produce a 1 output when its two inputs are different.

If the inputs are A and B then the required function is

$$f = A \,.\, \overline{B} + \overline{A} \,.\, B$$

This function can be generated as shown in Fig. 16.28. The circuit is known collectively as NOT-EQUIVALENT Gate.

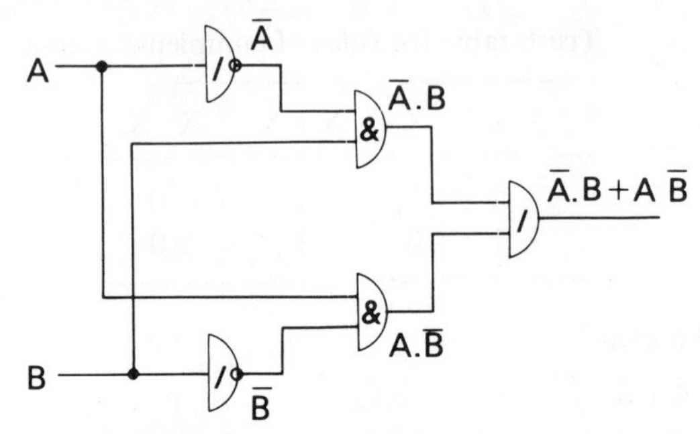

Fig. 16.28. Part of Example 16.10

Example 16.11 An electric furnace uses three identical temperature sensing devices which produce an output equivalent to a logical 1 when the temperature of the furnace rises above a certain threshold value. These devices are to be used in conjunction with a logic circuit comprising AND and OR gates to produce an output which will be used to shut off the furnace when two or more of the devices produce a 1 output. Draw a circuit diagram of a suitable arrangement of the gates.

If the temperature sensing device outputs are A, B, and C then a logical 1 will be obtained if any two or all three of these outputs are logical 1's. Therefore the desired function is:

$$f = A.B.\overline{C} + A.\overline{B}.C. + \overline{A}.B.C + A.B.C$$

Since, from the second absorption rule X + X = X it follows that the term A . B . C can be repeated any number of times in the above expression without affecting its validity.

$$f = A.B.\overline{C} + A.B.C + A.\overline{B}.C. + A.B.C. + \overline{A}.B.C. + A.B.C.$$

Using the second distributive rule

$$f = A.B(C + \overline{C}) + A.C(B + \overline{B}) + B.C(A + \overline{A})$$

From first law of complementation

$$A + \overline{A} = 1 \quad B + \overline{B} = 1 \quad C + \overline{C} = 1.$$

$$f = A.B + B.C + C.A$$

Thus the desired function can be generated using the circuit shown in Fig. 16.29.

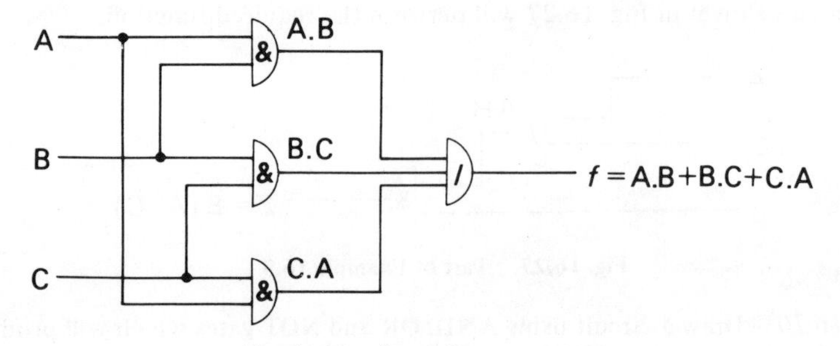

Fig. 16.29. Part of Example 16.11

So far all the problems have been solved using AND, OR and NOT gates. It is however necessary to consider the use of NOR and NAND gates since with integrated circuits these are readily available. Fig. 16.30 shows the generation of the NOT function with a NOR gate, the unused inputs, which are never shown on logic diagrams, being maintained at the 0 level. Examination of the truth table for the NOR function will confirm that under these circumstances the NOT function is being generated. Similarly Figs. 16.31 and 16.32 show the generation of the AND and OR functions respectively. The AND, OR and NOT gates could therefore be replaced directly by the corresponding arrangements of NOR gates. Figs. 16.33, 16.34 and 16.35 show the generation of the basic functions using NAND gates.

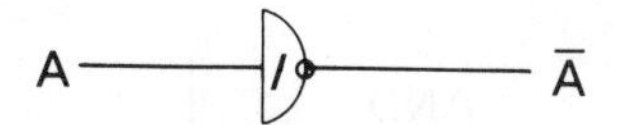

Fig. 16.30. Generation of NOT function with NOR gate

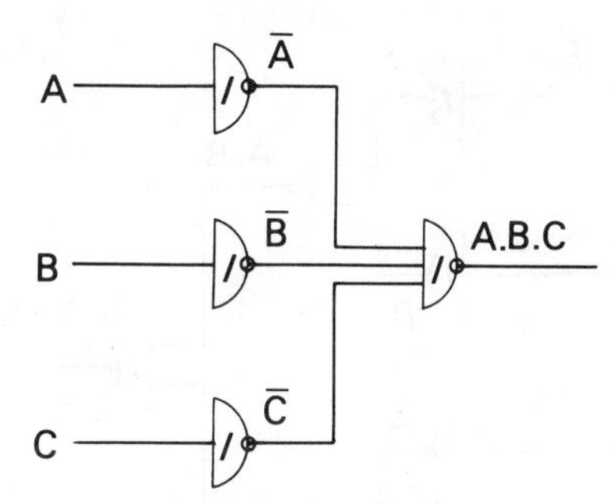

Fig. 16.31. Generation of AND function with NOR gates

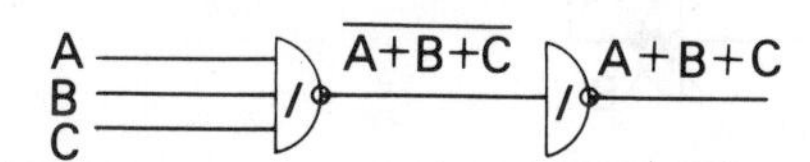

Fig. 16.32. Generation of OR function with NOR gates

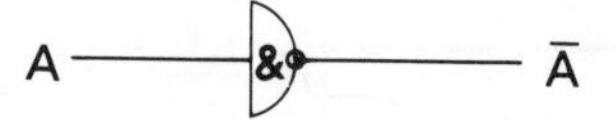

Fig. 16.33. Generation of NOT function with NAND gate

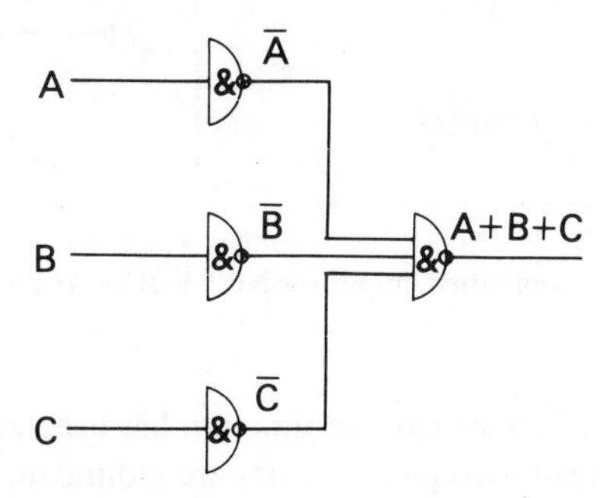

Fig. 16.34. Generation of OR function with NAND gates

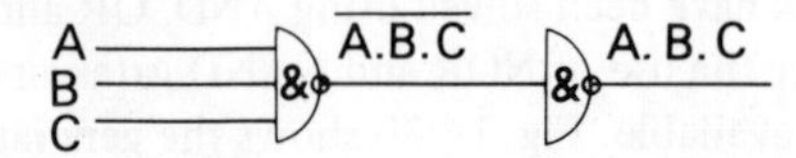

Fig. 16.35. Generation of AND function with NAND gates

Consider the NOT-EQUIVALENT gate as derived in Example 16.10. This has been converted directly into NOR logic gate circuitry as shown in Fig. 16.36. Examination of the circuitry shows that two pairs of NOR gates are redundant since the output of each pair is the same as its input. These gates have been crossed out and the simplified

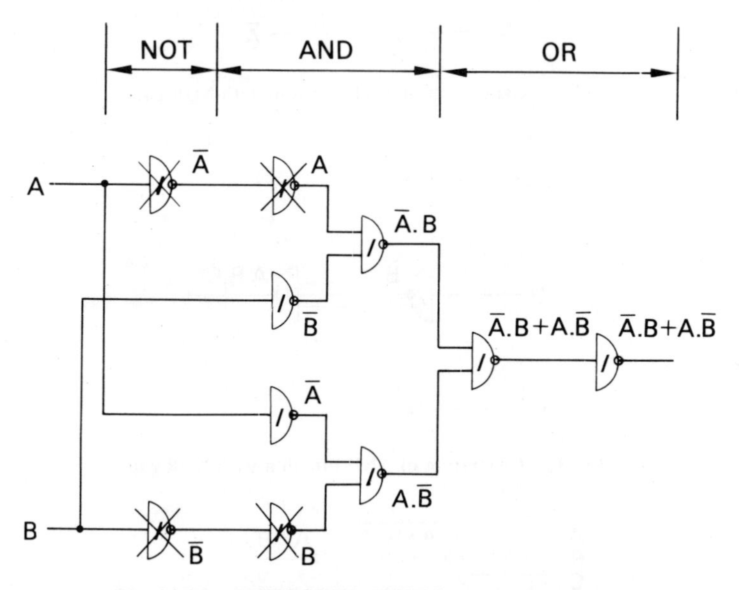

Fig. 16.36. NOT-EQUIVALENT function with NOR gates

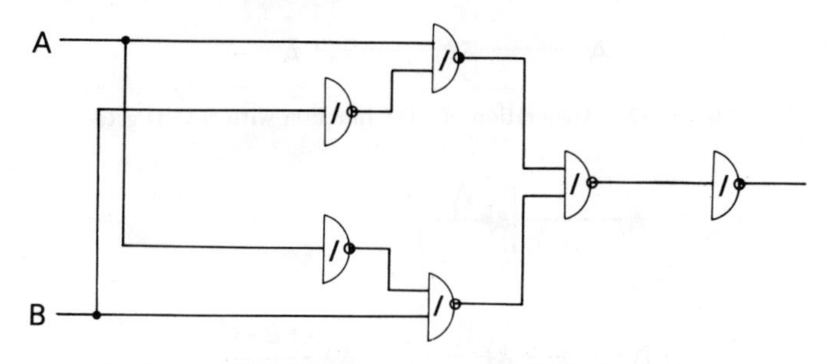

Fig. 16.37. Simplified circuit for NOT-EQUIVALENT function

circuit is shown in Fig. 16.37. Similarly the function has been generated with NAND gates in Fig. 16.38 where again two pairs of gates are redundant giving the simplified circuit in Fig. 16.39.

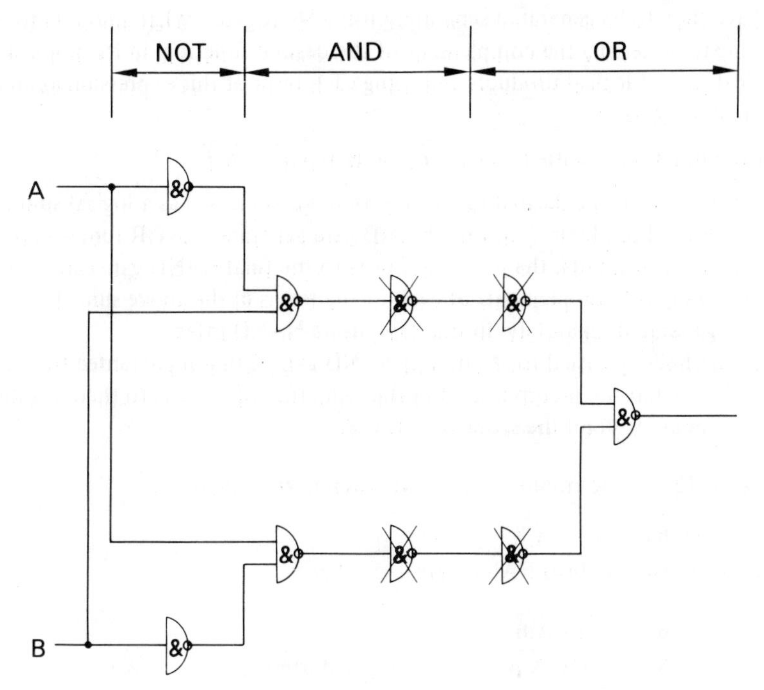

Fig. 16.38. **NOT-EQUIVALENT function with NAND gates**

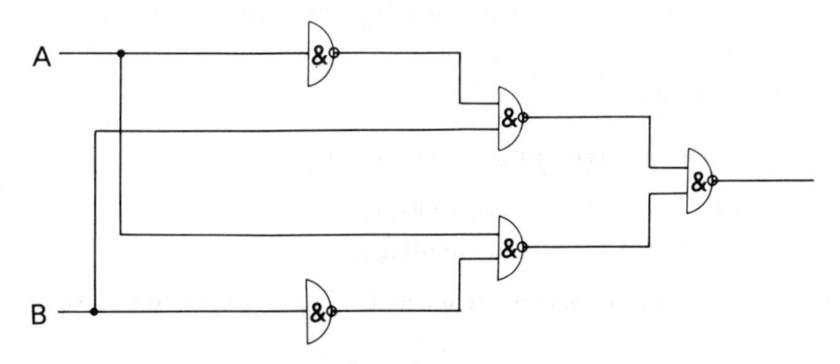

Fig. 16.39. **Simplified circuit for NOT-EQUIVALENT function**

In general if a circuit is to be constructed from NOR gates, the desired function should be expressed, in its simplest form, as a logical product of a number of logical sums.

e.g. $f = (A + B).(A + C).(A + D)$

Since the NOR gate generates the AND function on the complements of its inputs, the required inputs to the final NOR gate can be found by taking the complement of each term of the product.

e.g. for the function specified above $\overline{A + B}$, $\overline{A + C}$ and $\overline{A + D}$

Each of these can be converted into logical products using De Morgan's Theorem

i.e. $\overline{A + B} = \overline{A}.\overline{B}$, $\overline{A + C} = \overline{A}.\overline{C}$ and $\overline{A + D} = \overline{A}.\overline{D}$

These have then to be generated separately using NOR gates. What amounts to the same thing is expressing the complement of the desired function, in its simplest form, as a logical sum of logical products and using each term of this expression as an input to the final NOR gate.

e.g. for the function specified above $\overline{f} = \overline{A}.\overline{B} + \overline{A}.\overline{C} + \overline{A}.\overline{D}$

With NAND gates the desired function should be expressed as a logical sum of a number of logical products. Since the NAND gate generates the OR function on the complements of its inputs, the required inputs to the final NAND gate can be determined by taking the complements of each of the terms in the above sum. These have then to be generated separately, in this case, using NAND gates.

These methods specified for NOR and NAND gates will not guarantee the simplest form of circuit, but the methods used in the reduction of circuits to their absolute minimal form are beyond the scope of this text.

Example 16.12 Draw circuits which will generate the function

$$f = B.(\overline{A} + \overline{C}) + \overline{A}.\overline{B}$$

using (*a*) NOR gates and (*b*) NAND gates.

$$\begin{aligned} f &= B.(\overline{A} + \overline{C}) + \overline{A}.\overline{B} \\ &= B.\overline{A} + B.\overline{C} + \overline{A}.\overline{B} && \text{(second distributive rule)} \\ &= \overline{A}(B + \overline{B}) + B.\overline{C} && \text{(second distributive rule)} \\ &= \overline{A} + B.\overline{C} && \text{(first rule of complementation)} \end{aligned}$$

(*a*) For NOR gates

Complement of function $= \overline{f} = \overline{\overline{A} + B.\overline{C}}$

$$\begin{aligned} &= A.\overline{(B.\overline{C})} && \text{(De Morgan's Theorem)} \\ &= A.(\overline{B} + C) && \text{(De Morgan's Theorem)} \\ &= A.\overline{B} + A.C && \text{(second distributive rule)} \end{aligned}$$

A . $\overline{B}$ and A . C are generated separately giving the circuit shown in Fig. 16.40

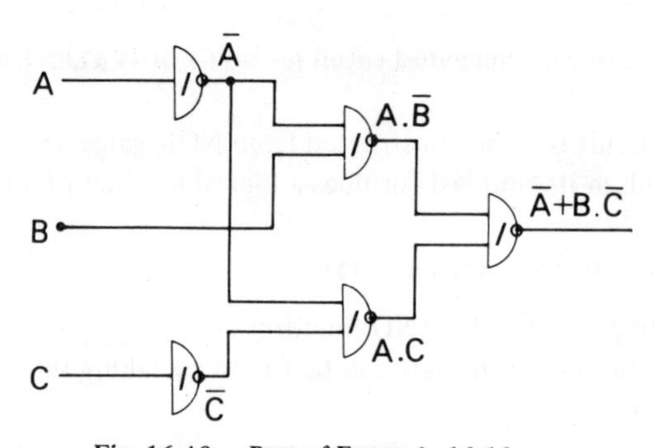

Fig. 16.40. Part of Example 16.12

(*b*) For NAND gates

$$f = \overline{A} + B.\overline{C}$$

Inputs to final NAND gate are

$$\overline{\overline{A}} = A \quad \text{and} \quad \overline{B.\overline{C}} = \overline{B} + C$$

$\overline{B} + C$ has to be generated separately giving the circuit shown in Fig. 16.41

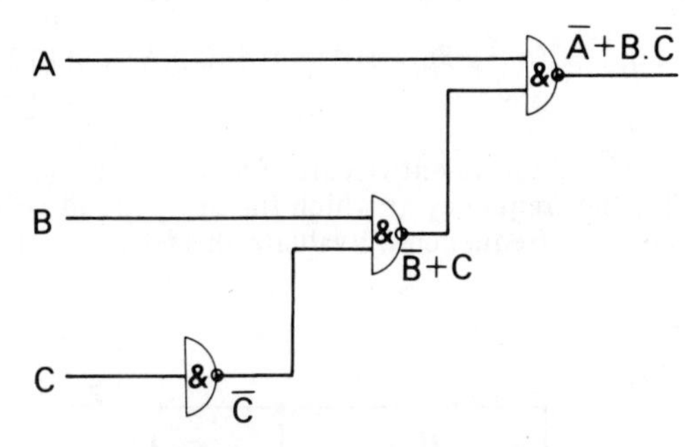

Fig. 16.41. Part of Example 16.12

PROBLEMS ON ELECTRONIC SYSTEMS

1 If an amplifier, when loaded by a 2-kΩ resistor, has a voltage gain of 80 and a current gain of 120, determine the necessary input signal voltage and current to give an output signal voltage of 1·0 V. What is the power gain of the amplifier?
12·5 mV; 4·17 μA; 9 600

2 An amplifier has a voltage gain of 50 dB. Determine the output voltage when the input voltage is 2·0 mV.
632 mV

3 The output of a signal generator is calibrated in dB for a resistive load of 600 Ω connected across its output terminals. Determine the terminal voltage to give
(*a*) 0 dB corresponding to 1 mW dissipation in the load,
(*b*) + 10 dB and
(*c*) – 10 dB.
0·775 V; 2·45 V; 0·245 V

4 For testing purposes a voltage amplifier is connected in cascade with an attenuator. If, when the attenuator is set to 70 dB, the magnitude of the output signal voltage is (*a*) equal to and (*b*) twice the magnitude of the input signal voltage to the arrangement, determine the voltage gain of the amplifier. (*a*) 70 dB; (*b*) 76·0 dB

5 An amplifier has an open circuit voltage gain of 80 dB and an output resistance of 100 Ω. If a resistive load of 1·0 kΩ is connected across the output terminals determine the necessary input voltage to give an output voltage of 2·0 V. 0·220 mV

6 An amplifier has an open circuit voltage gain of 1 000, an output resistance of 15 Ω and an input resistance of 7·0 kΩ. It is supplied from a signal source of e.m.f. 10 mV and internal resistance 3·0 kΩ and it feeds a load of 35 Ω. Determine the magnitude of the output signal voltage and the power gain in dB of the amplifier.
4·9 V; 79·9 dB

7 When an amplifier is loaded by a 4·0-kΩ resistor its voltage gain is 72·0 dB. When loaded by a 9·0-kΩ resistor its voltage gain is 73·2 dB. Determine the open-circuit voltage gain and the output resistance of the amplifier. 74·3 dB; 1·23 kΩ

8 An amplifier has a short-circuit current gain of 100, an input resistance of 2·5 kΩ and an output conductance of 25 μS. It is supplied from a signal source that can be considered as a current generator of 12 μA in parallel with a 5·0 kΩ resistor. The amplifier feeds a resistive load of 10 kΩ. Determine the current, voltage and power gains of the amplifier, and hence the current in, the voltage across and the power dissipated in the load.

80; 320; $2{\cdot}56 \times 10^4$; 0·640 mA; 6·40 V; 4·10 mW

9 The output section of the equivalent circuit of an amplifier is shown in the diagram. Derive an expression for the frequency at which the open circuit voltage gain is 3·0 dB down on its value at very low frequencies. Evaluate this frequency for $R_0 = 5{\cdot}0$ kΩ and $C_0 = 20$ pF.

Note: $20 \log \sqrt{2} = 3{\cdot}0$

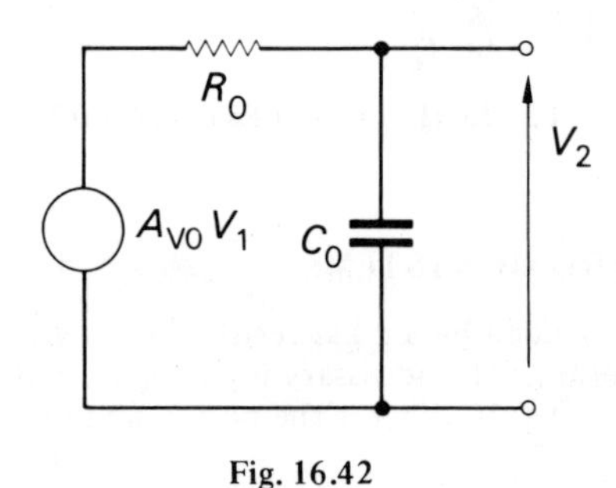

Fig. 16.42

$f = \dfrac{1}{2\pi R_0 C_0}$; 1·59 MHz

10 An amplifier is required with a voltage gain of 100. It is to be constructed from a basic amplifier unit of voltage gain 500. Determine the necessary fraction of the output voltage that must be used as negative series voltage feedback. Hence determine the percentage change in the voltage gain of the feedback amplifier if the voltage gain of the basic amplifier (*a*) decreases by 10% (*b*) increases by 10%.

0·008; (*a*) 2·2%; (*b*) 1·9%

11 A series voltage feedback amplifier has a feedback factor $\beta = 9{\cdot}5 \times 10^{-4}$. If its voltage gain without feedback is 1 000, calculate the voltage gain when the feedback is (*a*) negative and (*b*) positive. What percentage increase in voltage gain without feedback would produce oscillation in the positive feedback case.

(*a*) 513; (*b*) 20 000; 5·3%

12 An integrated circuit amplifier has the following characteristics:

Open-circuit voltage gain = 75 dB,
Input resistance = 40 kΩ,
Output resistance = 1·5 kΩ.

It has to be used with negative series voltage feedback to produce an output resistance of 5·0 Ω. Determine the necessary feedback factor. With this feedback factor and a resistive load of 10 Ω determine the output signal voltage when the feedback amplifier is supplied from a signal source of 10 mV and series resistance 10 kΩ.

0·0532; 115 mV

13 For the circuits illustrated in Fig. 16.43 the switches are shown in their positions represented by logical 0. Hence establish logic equations which will specify the state of the output signal Z in relation to the switch positions for each circuit.

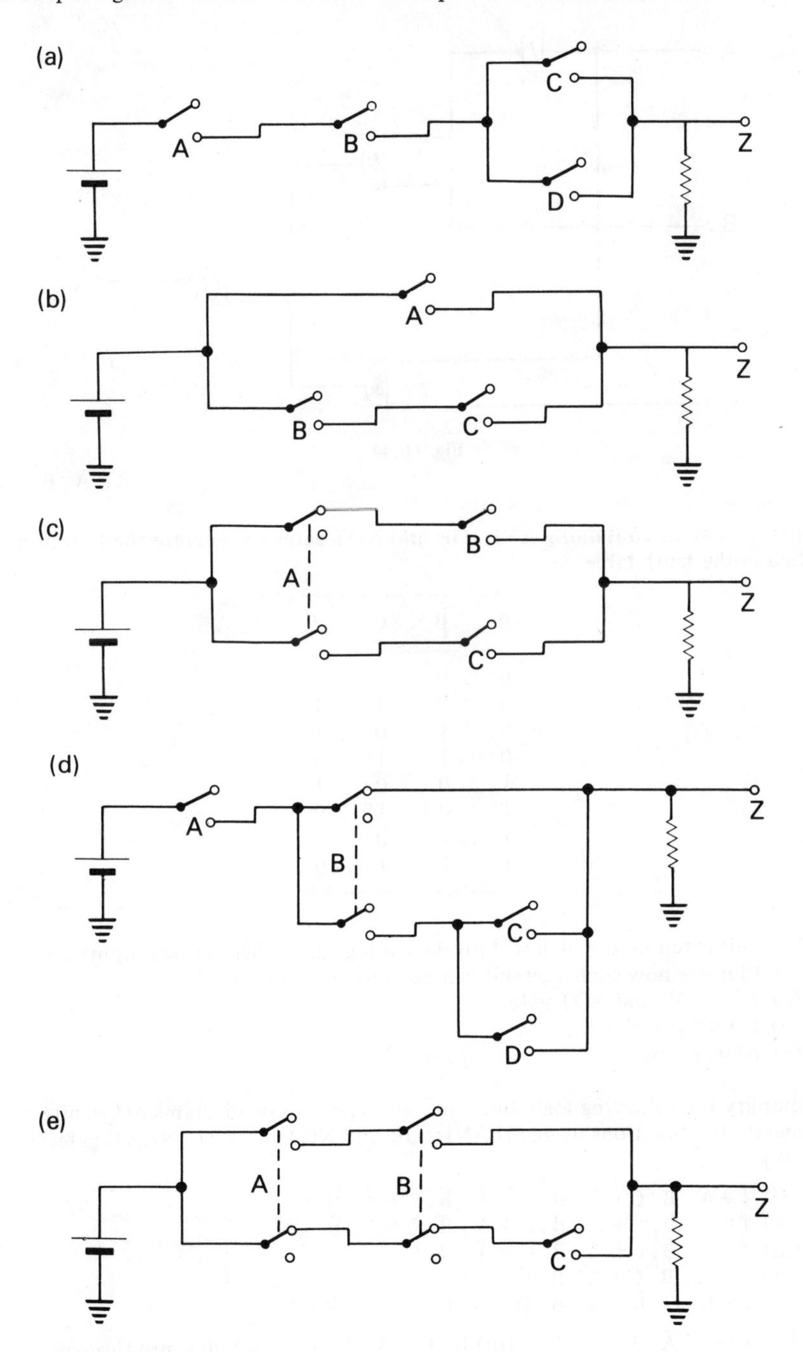

Fig. 16.43. (*a*) $Z = A \cdot B \cdot (C + D)$ (*b*) $Z = A + B \cdot C$ (*c*) $Z = \bar{A} \cdot B + A \cdot C$ (*d*) $Z = A \cdot (\bar{B} + B(C + D))$ (*e*) $Z = A \cdot B + \bar{A} \cdot \bar{B} \cdot C$

14 For the circuit shown in Fig. 16.44, determine the relationship between the output Z and the inputs A, B and C. Construct a truth table for the function.

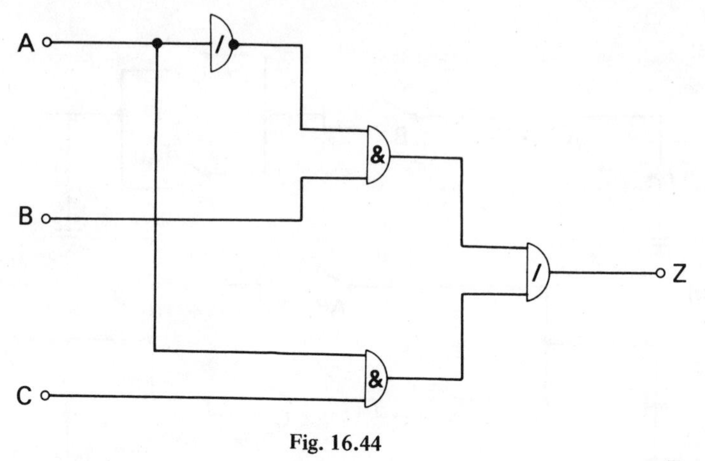

Fig. 16.44

$Z = \overline{A} . B + A . C$

15 Draw a circuit containing AND, OR and NOT gates to generate the function specified in the truth table.

A	B	C	Z
0	0	0	0
0	0	1	1
0	1	0	1
0	1	1	0
1	0	0	1
1	0	1	0
1	1	0	0
1	1	1	0

16 A circuit is required which will produce a logical 1 when its two inputs are identical. Indicate how such a circuit can be constructed using

(*a*) AND, OR and NOT gates
(*b*) NAND gates and
(*c*) NOR gates.

17 Simplify the following logic functions and hence draw diagrams of circuits which will generate the functions using (*a*) AND, OR and NOT gates, (*b*) NAND gates and (*c*) NOR gates.

(i) $f = A . \overline{B} . \overline{C} + A . \overline{B} . C + \overline{A} . \overline{B} . \overline{C} + \overline{A} . \overline{B} . C$
(ii) $f = \overline{A} . B . \overline{C} + \overline{A} . B . C + A . \overline{B} . C + \overline{A} . \overline{B} . C$
(iii) $f = A . \overline{B} . \overline{C} + \overline{A} . \overline{B} . \overline{C} + \overline{A} . B . C$
(iv) $f = A . B . C + \overline{A} . \overline{B} . \overline{C}$
(v) $f = B . \overline{C} . \overline{D} + A . \overline{B} . D + B . C . \overline{D} + \overline{A} . \overline{B} . D$

(i) $f = \overline{B}$; (ii) $f = \overline{A} . B + \overline{B} . C$; (iii) $\overline{B} . \overline{C} + \overline{A} . B . C$; (iv) no simplification; (v) $f = B . \overline{D} + \overline{B} . D$

17

Measurements and measuring instruments

The previous chapters have described and analysed basic electrotechnology. It remains to describe the operation of some of the devices that make these observations possible. Initially, it was sufficient to present a box with a dial on it and to suggest that, connected in the appropriate manner, it would indicate voltage, current and power. Now, by noting the principles of operation of the various devices, it is possible to understand the observations interpreted as they were. It is also possible to note the limitations to the various methods of measurement.

The most reliable, although not the most serviceable, are those devices using comparison methods. Due to their accuracy, they are often used as reference devices whereby other portable instruments may be calibrated. The most common is the potentiometer.

17.1 D.C. potentiometer

The potentiometer is a piece of apparatus used to compare potential differences. Provided that one of the p.d.'s is accurately known, the other may be derived from the comparison. By interpreting the latter p.d., it is also possible to measure current and resistance.

An elementary form of d.c. potentiometer is shown in Fig. 17.1. The slide wire is

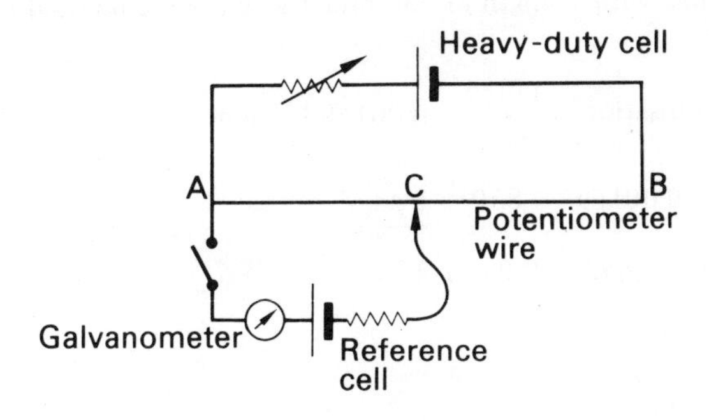

Fig. 17.1. Simple d.c. potentiometer

a uniform wire stretched between two heavy terminal blocks. A heavy-duty cell is connected across the slide wire and the current is regulated if necessary by a rheostat. A cell of known e.m.f. is connected between one terminal and a sliding contact, which makes a sharply defined point of contact with the slide wire. Into the latter circuit are introduced a switch, a resistor of high value and a galvanometer. The switch is closed only when the potentiometer is in use to conserve the reference cell and the resistor prevents the cell from being short-circuited. The galvanometer indicates the presence of current through the cell.

Because the slide wire is uniform, the volt drop will be evenly distributed along the wire. It follows that the potential difference between any two points on the wire is proportional to the distance between them.

When the switch is closed, a current may flow in the galvanometer. If the potential difference between A and C is greater than the e.m.f. of the reference cell, a current will flow through the reference cell from A to C. The converse is also true. However it will be noted that the e.m.f. of the reference cell opposes the passage of current due to the heavy-duty cell. If the p.d. between A and C is equal to the e.m.f. of the reference cell, no current will flow. This is the desired condition and it is for this reason that the galvanometer need only shows whether there is a current flowing or not.

From the balance condition, i.e. zero current flow, it may be concluded that the e.m.f. of the reference cell is proportional to the length AC. By dividing the e.m.f. by the length AC, the potentiometer may be calibrated in volts per metre. This is termed the constant of standardisation.

It is possible to calibrate the wire directly by setting the sliding contact to a point on the scale equivalent to the reference e.m.f. The slide-wire current is then adjusted by the rheostat until balance is obtained.

If there is now an unknown p.d. or e.m.f. to be measured, it is exchanged for the reference cell by the operation of the switch—see Fig. 17.3. The potentiometer is balanced by movement of the sliding contact and the appropriate length AD, say, is noted. The product of this length and the constant of standardisation gives the value of the unknown p.d. or e.m.f.

Example 17.1 Whilst standardising a simple potentiometer, balance was obtained on a length of 600 mm of wire when using a standard cell of 1·018 3 V. The standard cell was replaced by a dry cell and balance was obtained with a length of 850 mm. Calculate the e.m.f. of the cell.

When a 5·0-Ω resistor was connected across the terminals of the dry cell, a balance was obtained with a length of 750 mm. Calculate the internal resistance of the dry cell.

$$\text{Constant of standardisation} = \frac{1{\cdot}0183}{600} = 0{\cdot}001693 \text{ V/mm}$$

$$\text{e.m.f. of dry cell} = 0{\cdot}001693 \times 850 = \underline{1{\cdot}44 \text{ V}}$$

$$\text{p.d. across load} = 0{\cdot}001693 \times 750 = 1{\cdot}27 \text{ V}$$

$$E = V + Ir$$

$$= V + \frac{V}{R} \cdot r$$

$$1{\cdot}44 = 1{\cdot}27 + \frac{1{\cdot}27}{5} \times r$$

$$r = \underline{0{\cdot}67\ \Omega}$$

Certain precautions should be taken in operating the potentiometer. The heavy-duty cell must be sufficiently well charged so that its e.m.f. remains unchanged throughout the measurement. Also the temperature of the circuit should be permitted to stabilise before any measurements are taken.

The reference or standard cell can not be relied upon if it is permitted to pass any considerable current. The most common cell is the Weston cell which has an e.m.f. of 1·018 3 V at 20° C. Usually it is marked with a correction factor to allow for temperature. It is for this reason that the potentiometer must be standardised before use each time.

When the potentiometer is balanced, it takes no current from the source, the e.m.f. or p.d. of which is to be measured. In the case of a battery, it is therefore the open-circuit e.m.f. that is measured.

The accuracy of the potentiometer depends on the length of slide wire used. Most industrial potentiometers obtain the effect of a very long wire by connecting a number of resistors in series with a comparatively short wire. Each resistor has the same resistance as the wire. A simple arrangement giving an increased accuracy of ten times is shown in Fig. 17.2.

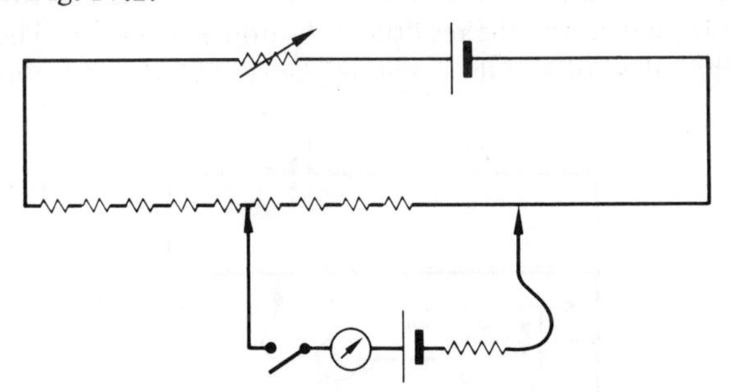

Fig. 17.2. Improved d.c. potentiometer

One important disadvantage about the potentiometer is that it cannot measure a p.d. or e.m.f. greater than the e.m.f. of the heavy-duty cell. If larger voltages are to be measured, a volt ratio box must be incorporated into the circuit as shown in Fig. 17.3.

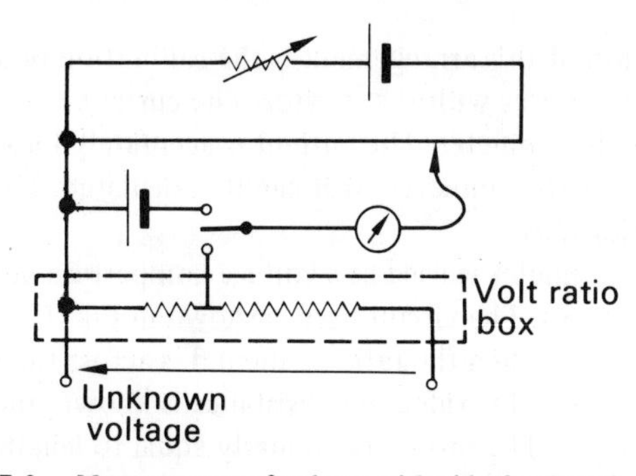

Fig. 17.3. Measurement of voltage with aid of volt ratio box

The volt ratio box consists of a high-value resistance tapped at some known point. The voltage measured is therefore a known fraction of the voltage applied to the box. It should be noted that the potentiometer draws no current from the volt ratio box and therefore the ratio of the box is not affected.

One disadvantage is that the volt ratio box requires a certain current to pass through it. In the case of measuring an e.m.f., the passage of the current will mean a certain internal volt drop and therefore the measured value is somewhat less than the true e.m.f. However, if the resistance of the volt ratio box is high enough, the resulting error should be small. The resistance cannot be too high since it would not then allow enough current to flow in the galvanometer when the potentiometer is off-balance. A similar problem may occur in the measurement of a potential difference.

The potentiometer is a comparatively large piece of apparatus and therefore is not readily portable. It also requires some time to make a measurement. One application, due to the great accuracy possible, is to calibrate portable indicating instruments. A voltmeter can be calibrated by being connected across the volt ratio box of Fig. 17.3 when the box is being fed from a variable-voltage supply. The supply is adjusted until the instrument indicates some desired value. Whilst the supply voltage is held constant, the corresponding p.d. is measured by the potentiometer from which reading the voltmeter error can be derived. This procedure is repeated for various suitable indications on the voltmeter dial.

The potentiometer can also be modified to measure current. The current is passed through a known resistance and the resulting volt drop is measured. The current is the calculated from the values of volt drop and resistance. The circuit is shown in Fig. 17.4.

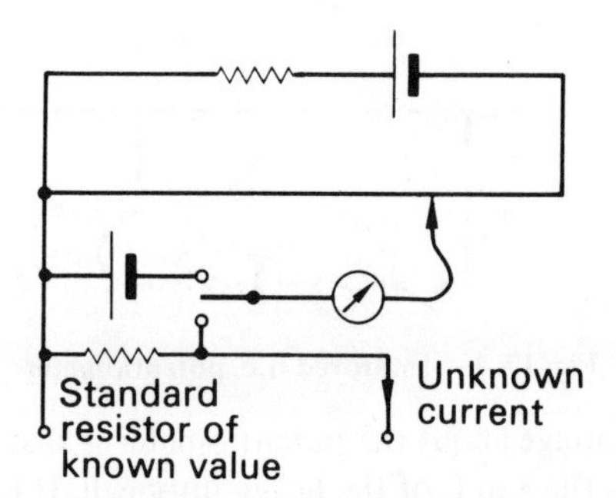

Fig. 17.4. Measurement of current by potentiometer

The usual application of this arrangement is the calibration of ammeters. The ammeter is connected in series with the resistor. The current is adjusted to give a suitable indication by the ammeter. The current is accurately measured by the potentiometer and hence the ammeter error can be calculated. The process is repeated for other suitable indications.

Normally the potentiometer is used as a voltage comparison device but it can be used to compare resistances. The circuit used is shown in Fig. 17.5. If a current is passed through the resistors then the ratio of the p.d.'s across the resistors is equal to the ratio of their resistances. Provided one resistance is known, the other can consequently be measured. The ratios are similarly equal to lengths of measured slide wire at the balance conditions.

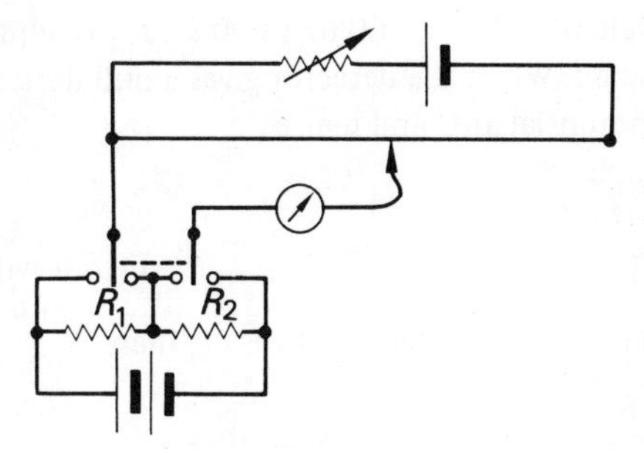

Fig. 17.5. Comparison of resistors by potentiometer

If l_1 and l_2 are the balance lengths for resistances R_1 and R_2, then

$$\frac{R_1}{R_2} = \frac{l_1}{l_2} \tag{17.1}$$

Note that a standard cell is not required since the potentiometer is not standardised but the sources must be constant during the time the comparison is made. The method can be unduly long and it is more simple to compare resistances by means of the Wheatstone bridge.

17.2 Wheatstone bridge

Any circuit made up in the form shown in Fig. 17.6 is termed a bridge circuit provided that the supply is connected to one pair of opposite terminals and the load is connected to the other pair. In a measurement bridge such as the Wheatstone bridge, one side, or arm, is variable. With suitable adjustment, zero deflection on the galvanometer can be achieved, i.e. the bridge is said to be balanced. The Wheatstone bridge is the most elementary using resistors only. It is energised by direct voltage from the terminals A and D, say, and the balance condition is observed from the bridge terminals B and C.

It should be noted that the galvanometer only serves to detect the presence of current in the connection BC and is consequently termed the detector. In order to protect the detector, a variable shunt resistor (not shown) is usually connected across it, thus by-passing the detector to some extent. The shunt resistance is increased as balance is approached and is eventually open-circuited. The protection is necessary because the detector is a delicate instrument being capable of very high sensitivity.

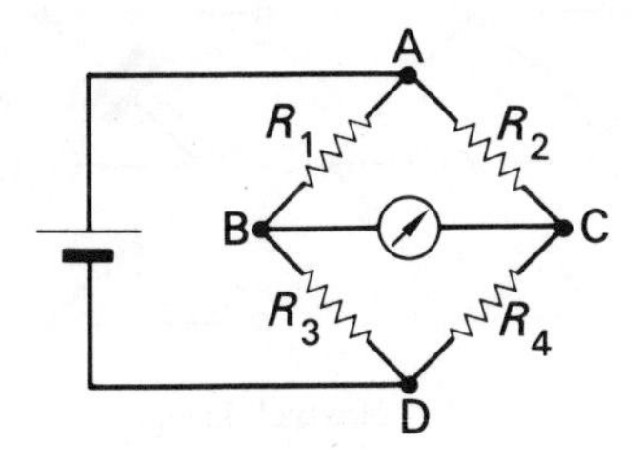

Fig. 17.6. Wheatstone bridge

Let R_1 be the unknown resistance, that is to be measured. R_2 will therefore be a graduated variable resistor (usually a decade resistance box) whilst R_3 and R_4 are resistors with possible values of 1, 10, 100 or 1 000 Ω. R_2 is adjusted until the best possible balance is obtained. When the detector gives a null deflection, the potential at B is the same as the potential at C and hence

$$V_{AB} = V_{AC}$$

also $$V_{BD} = V_{CD}$$

If the current in R_1 is I_1 and the current in R_2 is I_2, then

$$I_1 R_1 = I_2 R_2$$

Since the detector current is zero, the current I_1 also flows in R_3 and I_2 flows in R_4, hence

$$I_1 R_3 = I_2 R_4$$

$$\frac{I_1 R_1}{I_1 R_3} = \frac{I_2 R_2}{I_2 R_4}$$

$$\frac{R_1}{R_3} = \frac{R_2}{R_4}$$

$$R_1 = \frac{R_3}{R_4} \cdot R_2 \qquad (17.2)$$

Note that the sides of the bridge containing R_3 and R_4 are termed the ratio arms. The values are so chosen that all decades of R_2 are brought into use if possible thereby giving the best accuracy available. The bridge measures resistances over 0·1 Ω whilst the potentiometer can be used for lower values.

17.3 Universal bridge

Provided a suitable detector is used, the Wheatstone bridge can be operated by alternating current. Suitable detectors include the vibration galvanometer, the electronic voltmeter and the oscilloscope. This change serves no advantage generally in resistance measurement but it does permit the measurement of inductance and capacitance.

A simple comparison bridge is the Maxwell bridge shown in Fig. 17.7. Here the unknown and variable resistance arms are replaced by capacitive or inductive arms.

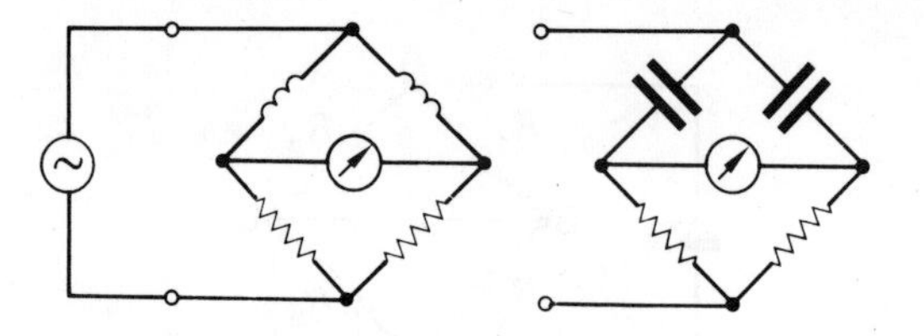

Fig. 17.7. Maxwell bridges

Rather than operate from the industrial supply frequency, these bridges are usually operated from a higher frequency, 1 kHz say, source. This gives a suitable response from both inductors and capacitors.

A common piece of test apparatus incorporates the properties of both Wheatstone and Maxwell bridges in that it can measure resistance, inductance and capacitance. Such a device is termed a universal bridge, a simple form of which is shown in Fig. 17.8.

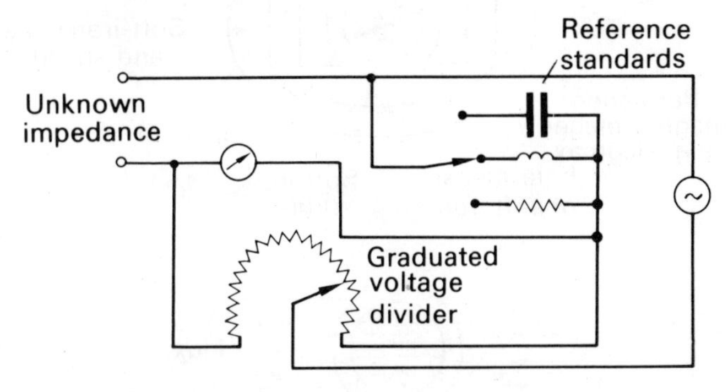

Fig. 17.8. Simple universal bridge

In this form, balance is obtained by tapping a graduated potentiometer device thereby varying the ratio arms. The potentiometer scale permits the ratio to be read off its dial. This affords two advantages:

(*a*) The bridge is easier to operate.
(*b*) Fixed-value reference components can be used which is very much cheaper.

The simple bridge cannot measure the resistance and inductance or capacitance of a component simultaneously. If this is desired, a more sophisticated system of circuits is required. These are usually incorporated into the better commercial universal bridges but the principle of operation remains unchanged.

17.4 Moving-coil instrument

The function of the galvanometer in the d.c. potentiometer and the Wheatstone bridge is to detect the presence of a direct current. Usually it takes the form of a moving-coil instrument.

The moving coil, from which the instrument gets its name, is placed in the magnetic field of a permanent magnet in a manner such that the passage of a current through the coil sets up a torque; the coil is thus displaced rotationally against the restraining torque of a system of control springs to a degree determined by the coil current. This effect can be calibrated and an appropriate indication given on a scale by the movement of a pointer. A typical arrangement is shown in Fig. 17.9.

The coil, usually rectangular, is mounted centrally on a spindle. The spindle is mounted in turn on jewelled bearings or at least needle bearings. These reduce both friction and wear.

So that the coil experiences a reasonably uniform field, it is free to rotate around a fixed cylindrical core and the pole pieces are shaped in a manner that gives rise to a uniform magnetic field in the air gaps. It will be noted that the field is radially uniform as shown in Fig. 17.9(*b*).

The restraint is introduced by two spirally-wound, phosphor-bronze springs, one

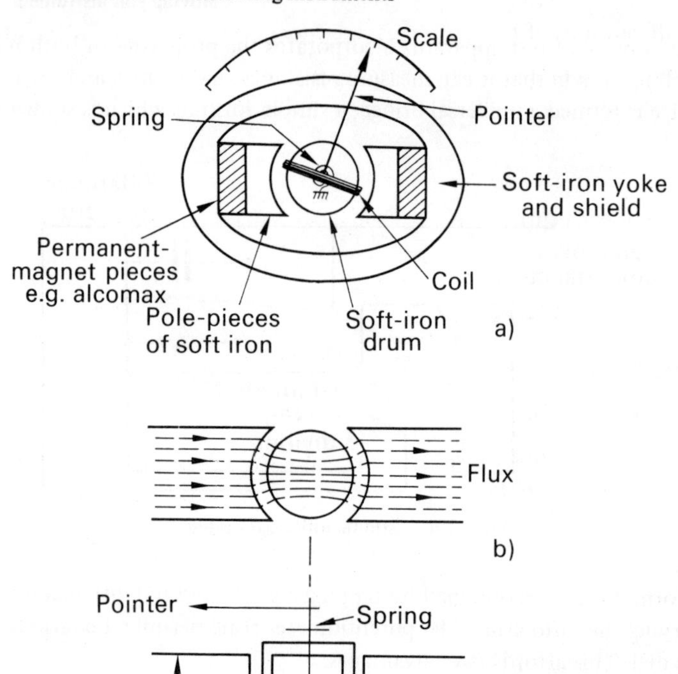

Fig. 17.9. Moving-coil instrument

at each end of the spindle. They are wound in opposition and are adjusted such that they balance when the pointer is at the zero mark on the scale. The torque they produce is proportional to the deflection.

Because conductor connections to the coil would introduce further torque to the moving system, the coil current is introduced through the restraint springs.

To analyse the torque developed by the coil, let the coil current be I and let it produce a deflection θ. Also let the spring factor be K expressed in newton metres per unit deflection. As noted above, this is a constant for the spring assembly. Thus for a deflection θ,

spring restraint torque $= K\theta$

$=$ torque set up by coil when the instrument attains steady-state conditions and is at rest.

The instrument operates on the principle of interaction between a current-carrying conductor. Provided the coil sides are not close to the edges of the poles, the arrangement is symmetrical and the relation $F = Bli$ may be applied using the gap flux density due to the permanent magnet alone.

Using the dimensions of Fig. 17.9(*c*), therefore a 1-turn coil:

$$\text{Force on each coil side} = BlI$$
$$\text{Torque on each coil side} = BlIr = \frac{BlId}{2}$$

Each side of each turn will exert this torque, therefore an *N*-turn coil will experience a torque:

$$M = \frac{2NBlId}{2}$$

since each turn has two sides. Hence

$$M = NBlId \qquad (17.3)$$

It should be noted that the total number of coil sides is an odd number because the electrical connections are made through the springs. This has the curious effect that the total number of turns in the coil is an integer plus a half, e.g. 50·5 turns.

At balance:

$$K\theta = NBlId$$

$$\theta = \frac{NBld}{K} \cdot I$$

The term $NBld/K$ is called the sensitivity of the galvanometer.

Over the majority of the indicating scale, this sensitivity remains constant and the torque is proportional to the current. The scale is therefore linear except at the ends due to the dissymmetry of the magnetic field. This effect is quite small and the instrument is noted for the regular linearity of its scale.

When the instrument is first excited, its movement rotates but, in so doing, it attains a certain inertia. If it were only subjected to the influences of the deflecting and restoring torques, some time would pass before the movement would come to rest. It is possible that continuing changes in the supply to the meter would keep it continually vibrating about the mean indication. For satisfactory operation, the instrument should quickly move to the mean deflection and come to rest. This requires a damping torque which opposes motion but ceases when motion ceases. In the moving-coil instrument, electromagnetic damping is used.

If the coil is wound on a former made of aluminium, say, then the movement of this former through the magnetic field will induce an e.m.f. and hence an eddy current into the former. The effect by Lenz Law is to oppose the change and therefore the eddy current gives rise to a torque opposing the motion. This damping torque is also proportional to the angular velocity and it follows that the torque ceases when the movement stops. The former is designed to permit the movement to overshoot the mean position on the first approach permitting the quickest response and also showing that the instrument is not sticking.

The form of construction described above is reasonably sensitive but a more sensitive form can be made. For instance, the coil can be suspended on a fine fibre. This cannot support a pointer arrangement and the indication is therefore obtained by deflecting a beam of light using a mirror. This form of construction is described in para. 17.15.

Example 17.2 A moving-coil, permanent-magnet instrument is to give a full-scale deflection of 60°, when the coil current is 10 mA. The uniform radial flux density in the air gap is 0·25 T, the rectangular coil has an effective length of 2·5 cm and an effective radius of 1·0 cm. The spring constant is $1{\cdot}0 \times 10^{-6}$ Nm/deg. Calculate the number of turns on the coil.

Using the symbols and construction appropriate to Fig. 17.9,

$$\text{Restoring torque} = K\theta = 1 \times 10^{-6} \times 60$$
$$= 60 \times 10^{-6}\ \text{Nm}$$
$$\text{Deflecting torque} = 2NBIlr$$
$$= 2 \times N \times 0{\cdot}25 \times 2{\cdot}5 \times 10^{-2} \times 10 \times 10^{-3} \times 1 \times 10^{-2}$$
$$= 60 \times 10^{-6}$$
$$N = 48{\cdot}0$$

But number of turns must be integer plus a half. Take number up to next value, hence number of turns required is 48·5 turns.

17.5 Extension of range

The coil of the instrument movement is made from fine wire and therefore it can only pass a very small current. A typical movement passes a current of not more than 20 mA and has a resistance of not more than 5 Ω although higher values are possible. The basic movement can therefore be calibrated for these ranges of current only.

The range can be extended by connecting a shunt R_s in parallel with the movement. This is indicated in Fig. 17.10. The scale is then calibrated according to the total current taken by the arrangement. The shunt is made from manganin or some similar material that has a negligible temperature coefficient of resistance. This introduces an error due to temperature variation since the copper wire of the moving coil will vary in resistance. This is swamped by connecting a resistance in series with the coil made of

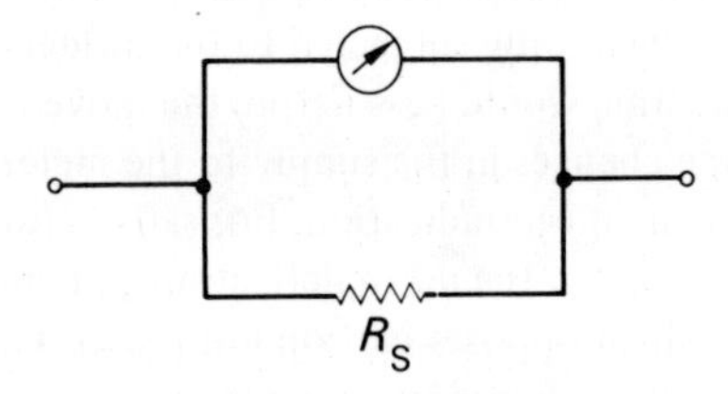

a) Simple ammeter

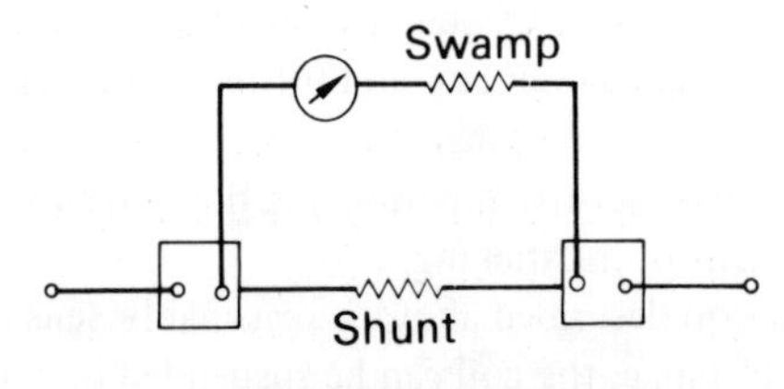

b) Ammeter circuit

Fig. 17.10. Moving-coil ammeter

a material similar to the shunt. Usually it has a resistance about three times that of the coil and reduces the error to about a quarter of what it would otherwise have been.

The shunt is provided with four terminals as indicated in Fig. 17.10(*b*). Because the resistance of the shunt is very small, the terminal resistance may be important; by making separate terminals for the instrument movement, then the resistance that the shunt offers relative to the moving coil and swamp resistor is a constant. It should be noted that terminal resistance of shunts is very important and the terminals should be regularly cleaned. A similar arrangement of four-terminal resistor is used in the potentiometer when measuring resistance.

Example 17.3 A moving-coil instrument movement has a resistance of 5·0 Ω and gives full-scale deflection when passing a current of 5·0 mA. It is to be incorporated into an ammeter with a full-scale deflection of 5·0 A. Calculate the resistance of the required shunt.

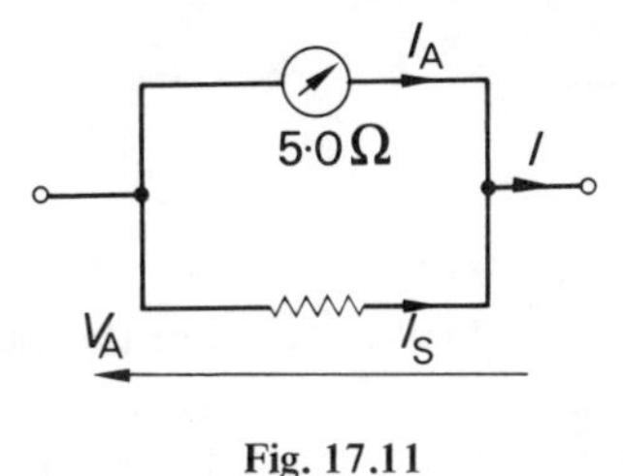

Fig. 17.11

$$I = 5{\cdot}0 \text{ A}$$

$$I_A = 5 \times 10^{-3} \text{ A}$$

$$I_S = I - I_A$$

$$= 5 - 5 \times 10^{-3} = 4{\cdot}995 \text{ A}$$

$$V_A = I_A R_A = 5 \times 10^{-3} \times 5$$

$$= 25 \times 10^{-3} \text{ V}$$

$$R_S = \frac{V_A}{I_S} = \frac{25 \times 10^{-3}}{4{\cdot}995}$$

$$= 0{\cdot}005\ 015\ \Omega$$

For metallic conductors, the voltage across the conductors is proportional to the current. Since the instrument movement circuit is metallic, it follows that the scale can be calibrated in volts. However, with the values of current and resistance quoted for a typical movement above, it will be seen that the measurable voltage is very small. The range can be extended by the use of a series resistor termed a multiplier. The arrangement is shown in Fig. 17.12.

The total resistance of the voltmeter between its terminals is given by the sum of the multiplier and movement resistances. Often this is used to give a figure of merit for the meter by dividing the resistance by the full-scale deflection voltage. This figure of merit is therefore given in ohms per volt.

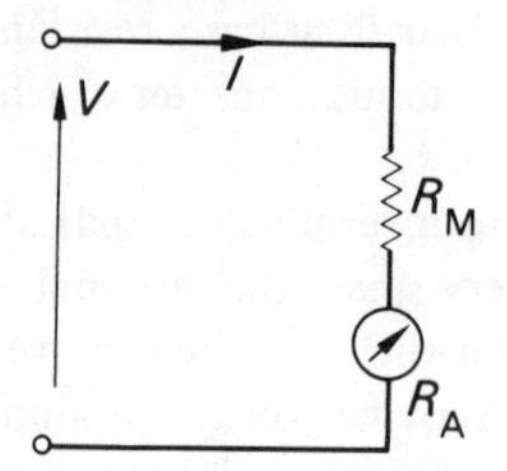

Fig. 17.12. Moving-coil voltmeter

It will be noted that the figure of merit is given by

$$\frac{R_M + R_A}{V} = \frac{1}{I}$$

It follows that a meter requiring 1 mA for full-scale deflection will have a figure of merit given by

$$\frac{1}{I} = \frac{1}{1 \times 10^{-3}}$$
$$= 1000\ \Omega/\text{V}$$

No swamp resistor is required in a voltmeter since the multiplier serves this function.

Example 17.4 The same moving coil instrument of ex. 17.3 is to be adapted to be a voltmeter with a full-scale deflection of 5·0 V. Calculate the resistance of the required multiplier.

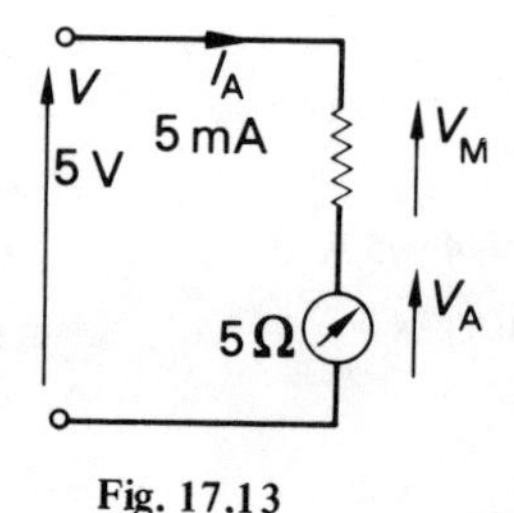

Fig. 17.13

As before, $V_A = I_A R_A = 5 \times 10^{-3} \times 5$
$$= 25 \times 10^{-3}\ \text{V}$$
$$V_M = V - V_A = 5 - (25 \times 10^{-3})$$
$$= 4{\cdot}975\ \text{V}$$
$$R_M = \frac{V_M}{I_A} = \frac{4{\cdot}975}{0{\cdot}005}$$
$$= 995\ \Omega$$

17.6 Rectifier meter

Whilst the moving-coil instrument has the advantages of high sensitivity and linear indication, it has the disadvantages of being operated only by direct current. If the current is reversed, the deflection also is reversed. If alternating current is passed through the instrument, the average torque on the movement is zero, the coil being pulled first one way and then the other. However if the alternating current is rectified by a diode bridge, the resulting direct current will give an appropriate deflection of the instrument. The circuit is shown in Fig. 17.14.

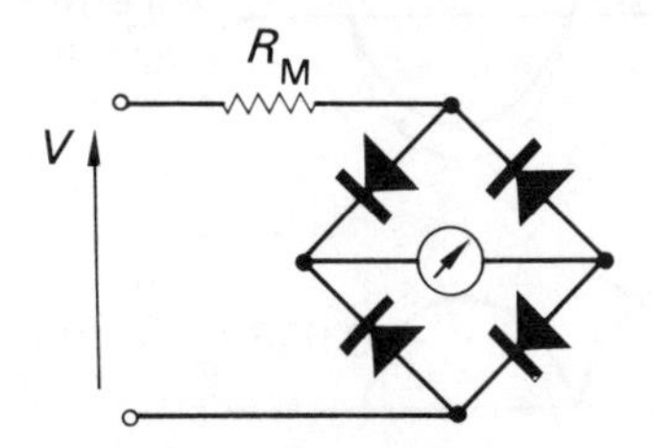

Fig. 17.14. Rectifier instrument

Ideally the rectifiers should offer zero resistance to current flow in one direction and infinite resistance in the other. Whilst the latter condition can almost be fulfilled, the former condition is difficult to satisfy. This problem partly arises because there is a certain forward resistance but mainly because a certain voltage must be attained before appreciable conduction takes place. The most suitable rectifier elements are copper oxide and selenium due to the low voltages at which conduction commences. The appropriate characteristics are shown in Fig. 17.15. During forward conduction, the ratio of voltage to current gives the rectifier resistance and since the characteristic is a curve in either case, the resistance is not constant. This causes the overall resistance of the instrument to be variable; the effect can be offset however by a swamp resistor and thus the rectifier instrument has a more linear scale when operating as a voltmeter than as an ammeter.

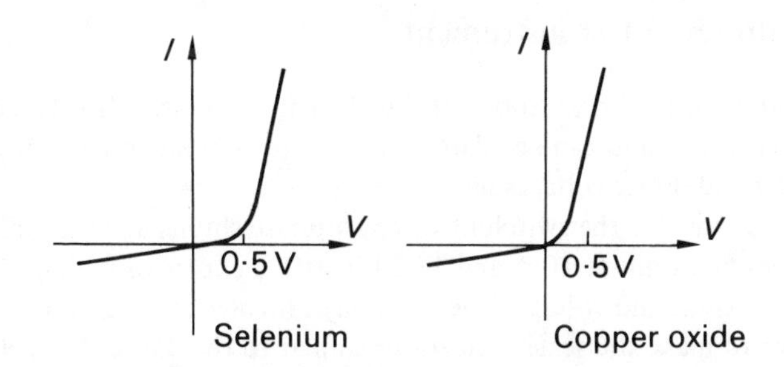

Fig. 17.15. Rectifier element voltage/current characteristics

Due to the lower position of the bend in the copper-oxide rectifier characteristic, this type is preferred to the selenium. The latter can be used in higher power circuits.

Because a rectifier bridge circuit is used, the current in the coil of the meter movement takes the form shown in Fig. 17.16. It has been shown in Chapter 14 that the average value of this current is 0·636 of the maximum and therefore the torque is proportional to 0·636 I_m. However, the meter is being used to measure an alternating quantity and therefore should be calibrated in r.m.s. values. The indication for sinusoidal quantities is therefore 1·11 times the mean value experienced by the movement. The figure 1·11 is the waveform factor for sine waves but there is the consequent drawback to this meter in that the measured quantities must remain sinusoidal.

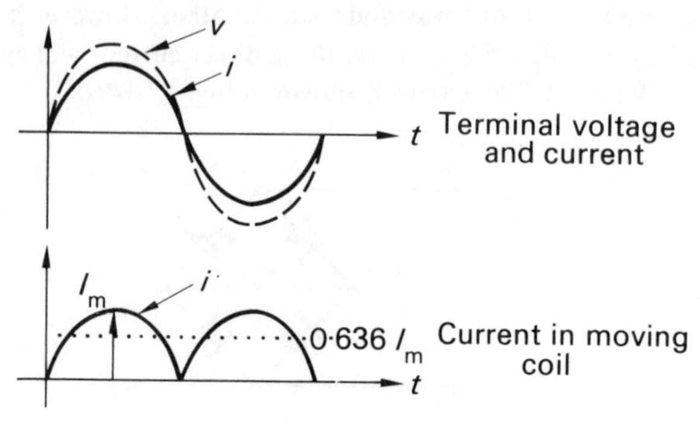

Fig. 17.16. Rectifier instrument waveforms

In the bridge rectifier, the reverse voltage across any single diode element is never greater than the sum of the voltage across the moving-coil instrument plus the forward volt drop across a conducting element. The forward rectifier resistance is only due to two elements at any one instant and the consequent error due to non-linearity can be neglected in the voltmeter application for ranges over 50 V.

In conclusion, the rectifier meter retains the advantages of the moving-coil instrument to a reasonable extent. It also has the advantage that it can operate at quite high frequencies–up to 100 kHz–without appreciable change in error. The limit may be extended to 1 GHz by the use of point-contact silicon diodes. The advantages are offset if the meter is not operated from a sinusoidal source.

17.7 Multi-range test instrument

This is one of the most popular 'tools' of the electrical engineer. It is based on a moving-coil instrument and can be adapted, by the turn of a control switch, to read both alternating and direct voltages and currents.

The d.c. ranges involve the switching of appropriate shunts and multipliers, the maximum values being about 10 A and 1 000 V with a choice of perhaps four or more scales for both current and voltage. The a.c. ranges involve the use of a rectifier bridge. In order to make the scale indications similar to the d.c. scales, yet retaining the ability to read r.m.s. values, different shunts and multipliers are required for the a.c. scales. The choice offered is similar to the choice of d.c. ranges.

In some test meters, a current transformer is incorporated for use when measuring alternating currents. The meters give an indication on alternating current only. They also introduce an element of inductance that can be of importance at higher frequencies.

Many test meters also contain a battery as a source of e.m.f. When this e.m.f. is applied to a circuit, the resulting current is a measure of the circuit resistance and the meter can be appropriately calibrated. It should be noted that the greatest deflection occurs when the circuit resistance is zero and therefore the scale increases in the opposite direction to the voltage and current scales. Also when used in this manner, i.e. as an ohmmeter, the polarity of the terminals is reversed. This is necessary to make the current flow in the direction appropriate to the meter movement.

The test meter, like any other meter, must have the pointer indicating zero before use. This is done by an eccentric screw mechanism that adjusts one of the restraint springs. Appropriate balance of the restraint springs brings the needle to rest at the zero mark when no current is passing through the movement. This adjustment cannot be used to zero the meter when used to measure resistance. Instead, variable resistances are introduced to the battery circuits so that, when the terminals are short-circuited, the pointer indicates zero resistance; this it will be remembered is full-scale deflection.

As far as possible, the multi-range test meter makes use of the same scales on its dial for both voltage and current. This is possible due to the linearity of the moving-coil movement. However the influence of the rectifiers on a.c. measurements may change this linearity and further sets of scales have to be introduced for small a.c. voltages and currents. This leads to test meters having somewhat complex dials. This disadvantage is offset by the compactness of the meter and by the high figures of merit that can be attained, e.g. 20 kΩ/V. In conclusion, the meter is an exceedingly useful instrument that can be used on a wide range of applications with reliability of accuracy.

17.8 Insulation test meter

For the purposes of measuring insulation, in which the resistance may well exceed 1 MΩ, the multi-range test set is of little use due to the low voltage employed in its resistance-measuring circuit. A higher voltage, resulting in sufficient current to give indication on a moving-coil instrument, can be obtained in two ways:

1. From a d.c. generator.
2. From an electronic inverter/rectifier set.

The Megger is typical of the first group. It contains a d.c. generator excited by a permanent magnet. The machine is driven by hand through a controlled clutch which slips at a given speed thereby giving a steady voltage. Typical output voltages of the test meter are 250 V, 500 V, 1 000 V and 2 500 V.

The instrument movement has two coils, one of which is energised directly from the generator as indicated in Fig. 17.17. The insulation resistance to be measured is

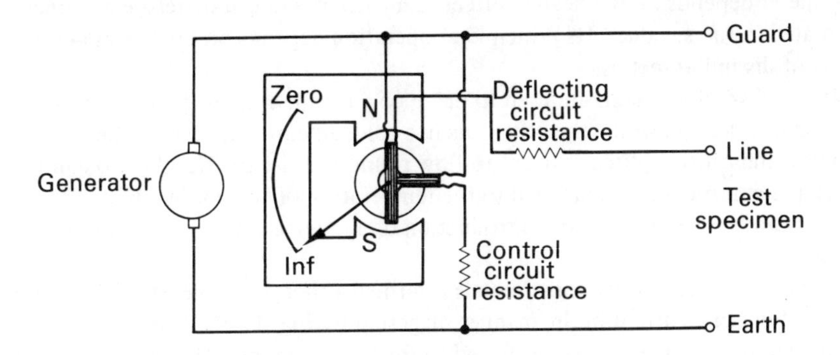

Fig. 17.17. Megger insulation test meter

connected in series with the other coil. These coils are termed the control and deflecting coils respectively.

If the control coil only is energised, the movement comes to rest with the coil axis in line with the field of the permanent magnet. The needle then indicates infinite resistance since this is effectively the resistance inserted into the other circuit. When a resistance of a somewhat lower value is connected in series with the other circuit, a current flows in the deflecting coil and the resulting torque displaces the instrument movement. A position of balance between the torques will be obtained and the scale deflection can be according to the value of the inserted insulation resistance. The calibration is usually in megohms.

The electronic test uses a battery as its source of energy. The direct current is inverted into alternating current by a transistorised chopper circuit and this is fed to a transformer which steps up the voltage. The high voltage signal is then rectified back to direct current usually at 250 V, 500 V or 1 000 V. According to the current taken from the battery by the inverter, an indication of the resistance connected across the output terminals can be obtained. The schematic diagram for such an instrument is shown in Fig. 17.18.

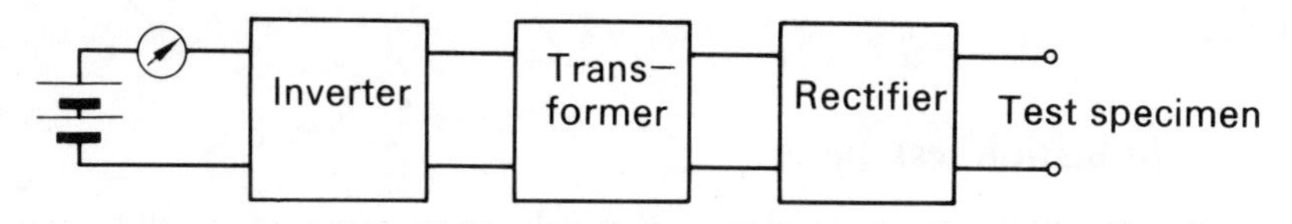

Fig. 17.18. Electronic insulation test meter

The electronic test meter has the advantage of being much smaller than that using a hand generator set and is therefore more readily portable. This advantage is offset by its inability to supply much energy to the apparatus under test. When testing a cable, for instance, the cable must be charged for some time by the test set before the current observed is due only to the insulation resistance and not due to the charging current.

17.9 Thermocouple ammeter

This meter depends on the heating effect of a current and can therefore be directly calibrated in r.m.s. values. Its principle of operation depends on the thermo-electric effect of dissimilar metals.

If two dissimilar metals are used to make up a circuit and one junction between the metals is heated whilst the other remains cold, an e.m.f. appears at the hot junction and causes a direct current to flow round the circuit. The thermo-e.m.f. and hence the current are proportional to the temperature of the hot junction. The current can be measured by the introduction of a moving-coil instrument into the circuit.

The arrangement of dissimilar metals is termed a thermocouple. It can be heated by an alternating current in the manner suggested by Fig. 17.19. The heat is dissipated by the strip and is transferred to the thermocouple. The temperature attained by the thermocouple approximately depends therefore on the square of the

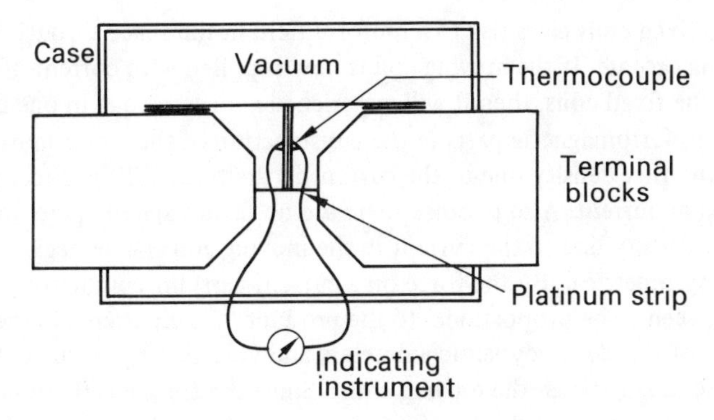

Fig. 17.19. Thermocouple ammeter

r.m.s. current. The scale of the moving-coil instrument, when calibrated in terms of the heating current, is consequently not linear.

The meter has the disadvantages of slow response since the thermocouple temperature varies slowly; also the overload capacity of the meter is low. On the other hand, the meter will operate on both alternating and direct current, indicating r.m.s. values at all times. Note that the r.m.s. and mean values of a constant direct current are one and the same. The meter operates satisfactorily with frequencies up to 1 GHz.

Because the meter is expensive and slow in its response, it is used either in high-frequency circuits or in the measurement of current in distribution systems where instantaneous currents can be ignored and the maximum demand of current over a period is to be measured.

17.10 Electrodynamic instrument

The moving-coil instrument does not respond to alternating currents because the direction of the torque alternates with each half cycle and the average torque is therefore zero. This difficulty can be avoided if both coil current and magnetic field alternate together. This is the principle of the electro-dynamic instrument, once known as the dynamometer instrument.

The basic form of electrodynamic instrument is shown in Fig. 17.20. The two fixed coils are mounted symmetrically about the rotating coil. In this way, the

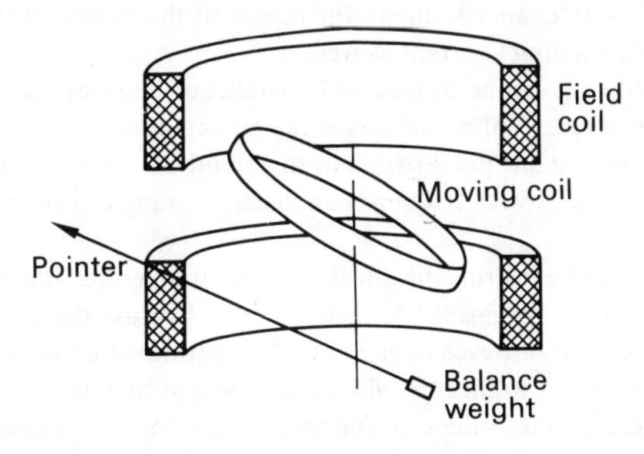

Fig. 17.20. Electrodynamic movement

current in the fixed coils gives rise to a uniform field in the space through which the moving coil may rotate. If the moving coil is now supplied with current of the same frequency as the fixed coils, then it will experience a mean torque in one direction.

There are no ferromagnetic parts in the construction of the meter movement. It follows that the flux density due to the current in the fixed coils is directly proportional to that current. Also because there are no ferromagnetic parts to distort the field, the flux density due to the current in the moving coil can be neglected, the system being symmetrical. By the force on a current-carrying conductor principle the torque can be seen to be proportional to the product of the currents in the coils.

In the case of the electrodynamic ammeters and voltmeters, the currents in the fixed coils also pass through the moving coils. Since the torque is therefore proportional to the mean square of the current, it is possible to calibrate the scale in r.m.s. values. It might be expected that the scale would be exceedingly non-linear due to its square-law characteristic. However the torque depends on the angle of deflection and this helps to make the scale reasonably linear.

The meter movement can also be used as a wattmetric measuring device. In this case, the current in the moving coil is proportional to the voltage whilst the load current is passed through the fixed coils. The torque is proportional therefore to the mean product of voltage and current. It also takes into account the factor that voltage and current need not be in phase since this reduces the resulting torque. The torque is hence proportional to the power in the circuit and the meter acts as a wattmeter. The connections are shown in Fig. 17.21.

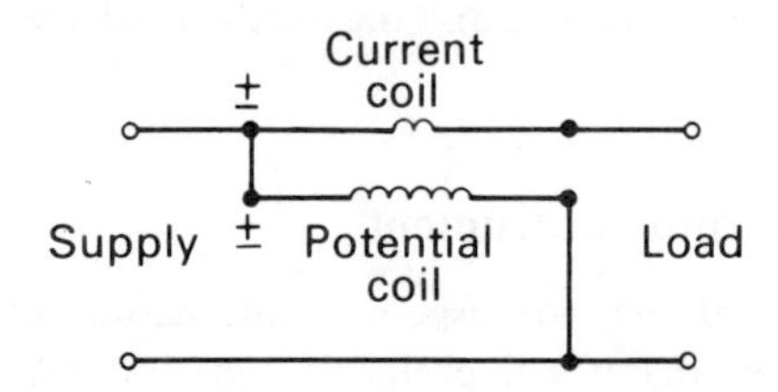

Fig. 17.21. Wattmeter connections

Regardless of the mode of operation, when the electrodynamic instrument is being operated by alternating quantities, the torque continually fluctuates. The indication is proportional to the mean torque and the fluctuations do not appear, except at very low frequencies, due to the inertia of the moving system. The instrument can operate on direct current as well.

The two fixed coils can be connected in parallel or in series, consequently most meters have two ranges. Voltmeter ranges can be extended by the use of multipliers but it is not usual to shunt the instrument to extend the current range. The effect of the shunt would change with frequency and current transformers give more accurate results.

The electrodynamic instrument can be calibrated on either alternating or direct current. The accuracy remains the same with either because there is no hysteresis effect. It is convenient however to calibrate the instrument on direct current and to then use the instrument to measure alternating current quantities. It will be noted that the meter reads r.m.s. values irrespective of waveform or frequency, provided that the latter is not too high.

Electrodynamic instruments are most often found in the form of wattmeters, although electrodynamic voltmeters and ammeters are not uncommon. The most important applications of these meters are as transfer instruments, i.e. they are calibrated on direct current and then used as comparison standards for other a.c. instruments, e.g. integrating energy meters.

The drawback to the electrodynamic instrument is the amount of current required to operate it. The relatively large currents are required because there is no ferromagnetic core to the magnet circuit. The voltmeter can only be used satisfactorily in high-power circuits. The ammeter also has a relatively high resistance. Because of the resulting power loss, it can also be used only in high-power circuits. However the voltmeter and ammeter functions can be achieved just as well by the moving-iron instrument which is cheaper. Only as a wattmeter, is the extra cost justified for the purposes of being a transfer instrument.

17.11 Moving-iron instrument

This instrument can measure both alternating and direct current, although generally it is only used in the former application. The basic movement can be modified in the same manner used with the moving-coil instrument to extend the ranges of operation both as a voltmeter and as an ammeter.

There are two forms of moving-iron instrument–the attracted armature type and the repulsion type.

The attracted-armature type, shown in Fig. 17.22, has a ferromagnetic armature

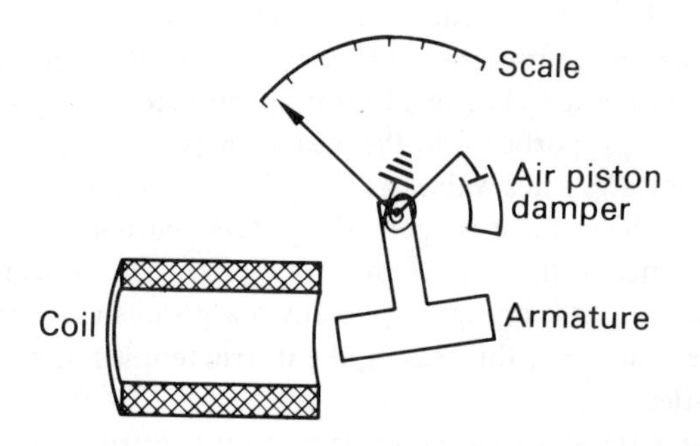

Fig. 17.22. Attracted-armature moving-iron instrument

placed in close proximity to an air-cored coil of wire. The armature is pivotted but to rotate it has to move against the restraint of a system of springs similar to that used in the moving-coil movement.

When a current is passed through the coil, a magnetic field is set up. In order to increase the energy stored in the field, the armature is attracted into the coil. The force of attraction depends on the energy and consequently on the square of the current. It will be noted that the direction of the current does not matter hence the instrument operates on either direct or alternating current. Because the deflection depends on the mean square of the current, the indication of the pointer is proportional to the r.m.s. value of the coil current.

An important difficulty experienced in the design of square law instruments is to obtain a reasonably linear scale. In the moving-iron instrument, the shape of the armature can be varied to give a scale that is almost linear except at the low-value readings.

The other form of moving-iron instrument is the repulsion movement. A typical arrangement is shown in Fig. 17.23. In this case two pieces of ferromagnetic material are placed inside the coil, one being fixed in space and the other attached to the spindle carrying the pointer. When the coil is excited, these pieces are magnetised in a similar manner and therefore repel one another. It is this force which is used to operate the movement.

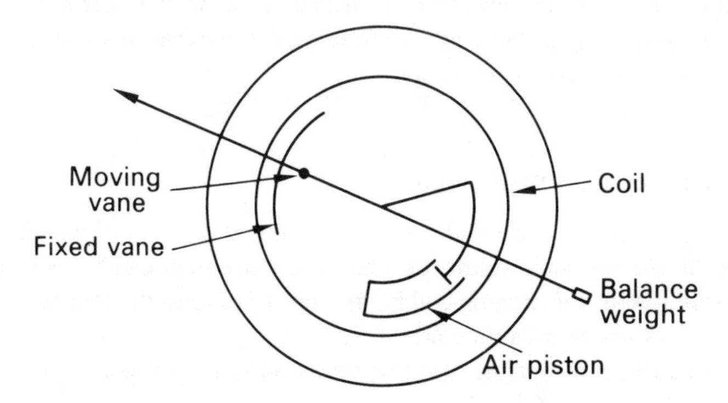

Fig. 17.23. Repulsion-type moving-iron instrument

The deflection depends on the product of the magnetisation of each of the ferromagnetic pieces. It is therefore proportional to the square of the magnetic field strength. However the magnetic field strength is proportional to the coil current and therefore the deflection is proportional to the mean square of the current. The repulsion meter thus indicates r.m.s. values as does the attracted-armature meter.

When the moving-iron instrument is operated by direct current, a hysteresis effect is produced by the presence of the ferromagnetic material. The deflection is therefore not the same when the current is brought up to any given value as when the current is reduced to the same value. For this reason, the instrument is not generally used to measure direct quantities.

Due to the alternating fluxes that are present when the instrument is used to measure alternating quantities, the electromagnetic system of damping is not suitable for use in the moving-iron instrument. Instead a pneumatic system is used. This comprises a piston moving inside a closed cylinder. It does not touch the sides of the cylinder but it does experience an opposition to motion due to the air being forced past the piston. This system of damping is also used in the electrodynamic instrument.

The moving-iron instrument can be used to measure either voltage or current. Because the exciting coil can be made of heavy-gauge wire, it does not require a shunt when operating as an ammeter but a multiplier is required when used as a voltmeter. In either mode, the instrument operates on either direct current or alternating current provided the frequency is not too high, say up to 2 kHz.

The meter has a very low figure of merit and therefore the required current is very high, e.g. a voltmeter may require 0·1 A for full-scale deflection. The moving-iron ammeter has a high resistance relative to its moving-coil equivalent. The moving-iron meter can therefore only be used in power circuits. In this application, it is very popular especially since it is both robust and cheap.

17.12 Electrostatic voltmeter

The attracted-armature moving-iron instrument depends on the magnetic force of attraction between magnetised surfaces. A similar force of attraction exists between the boundary surfaces of an electrostatic field. If these surfaces either move towards one another or present a larger boundary area, then there will be such a force. It is on this principle that the electrostatic voltmeter is founded.

The usual form of instrument is shown in Fig. 17.24. The two sets of plates are free to move between one another, the one being fixed and the other being able to rotate on a spindle. The potential difference to be measured is applied to the plates, one connection being made to the fixed plates and one to the moving plates.

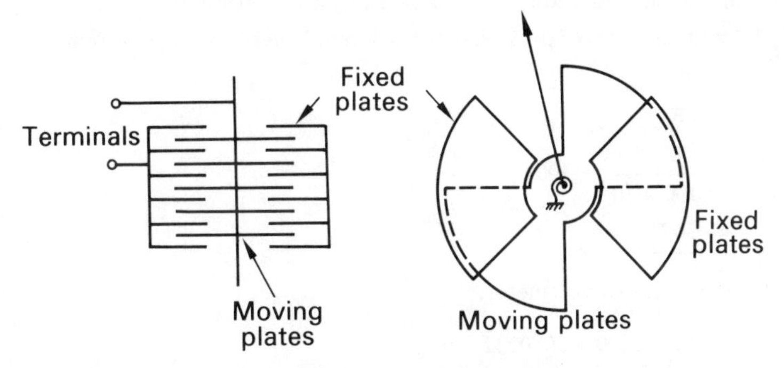

Fig. 17.24. Electrostatic voltmeter

The energy in the electrostatic field is given by $\frac{1}{2}CV^2$ and, since the force is dependent on the rate of change of stored energy the force is therefore proportional to the mean square of the voltage. The deflection is therefore able to indicate r.m.s. values.

Again the torque fluctuates when the voltmeter is measuring an alternating voltage and the indication is the mean square. The voltmeter also operates with direct voltages. As might be expected, the waveform of the supply does not affect the accuracy of the meter and the frequency range is quite high.

The electrostatic voltmeter has the important advantage that it requires little current to make it operate. In particular, it can be used to measure induced static voltages. After taking the initial charge, it takes virtually no further current to continue its deflection and indication.

17.13 Effect of meters on circuits

Reference has been made to the figures of merit and resistance of the various meters discussed in the preceding paragraphs. The importance of these factors requires some further discussion.

Two points should be remembered. First of all, any meter requires a certain amount of power to make it operate. Provided this power is small relative to the power in the measured circuit, then little error will result. However if the meter power is comparable to the power in the circuit a serious error will result. Before giving an example to illustrate this, the other point should be borne in mind; any meter will within the limitations of calibration indicate the conditions at its terminals correctly. Thus a voltmeter will indicate, within the limitations of its normal accuracy, the terminal voltage and similarly an ammeter will indicate the current passing through it.

To illustrate these remarks consider the following example.

Example 17.5 A voltage of 100 V is applied to a circuit comprising two 50-kΩ resistors in series. A voltmeter, with a full-scale deflection of 50 V and a figure of merit of 1 000 Ω/V, is used to measure the voltage across one of the 50-kΩ resistors. Calculate:

(*a*) The voltage across the 50-kΩ resistor.
(*b*) The voltage measured and hence indicated by the voltmeter.

Let V_1 be voltage across 50-kΩ resistor when voltmeter is not in circuit.

$$V_1 = \frac{R}{2R} \times \mathrm{V}$$

$$= \frac{50 \times 10^3}{100 \times 10^3} \times 100$$

$$= \underline{50\ \mathrm{V}}$$

Let R_v be resistance of voltmeter.

$$R_v = 50 \times 1\,000 = 50000\ \Omega$$

$$= 50\ \mathrm{k}\Omega$$

When the voltmeter is connected in circuit, it shunts the 50-kΩ resistor. If R_e is the resistance of the parallel circuit, then

$$R_e = \frac{R \times R_v}{R + R_v} = \frac{50 \times 10^3 \times 50 \times 10^3}{100 \times 10^3}$$

$$= 25 \times 10^3\ \Omega$$

$$= 25\ \mathrm{k}\Omega$$

If V_2 is the voltage across the 50-kΩ resistor as measured by the voltmeter, then

$$V_2 = \frac{R_e V}{R + R_e} = \frac{25}{50 + 25} \times 100$$

$$= \underline{33{\cdot}3\ \mathrm{V}}$$

From this example, it can be seen that the voltage measured by the voltmeter is quite erroneous. The error has been caused by the effect of the voltmeter on the circuit. Because of the values chosen, the voltmeter takes the same current as the load whose voltage is being measured. It can thus be seen that the power taken by the voltmeter is equal to the power in the measured load. Even if the resistance of the voltmeter had been ten times as great, the error would have been almost 2%.

It can therefore be seen that the meter can affect the circuit to which it is being applied. This can be further illustrated by examining the method of resistance measurement using a voltmeter and an ammeter.

If the voltage applied to a resistor is measured by a voltmeter as V and the current is measured by an ammeter as I, then the apparent resistance of the resistor is given by:

$$R_{\text{apparent}} = \frac{V}{I} \tag{17.4}$$

The meters could have been connected in two ways in order to achieve these readings. These are shown in Fig. 17.25.

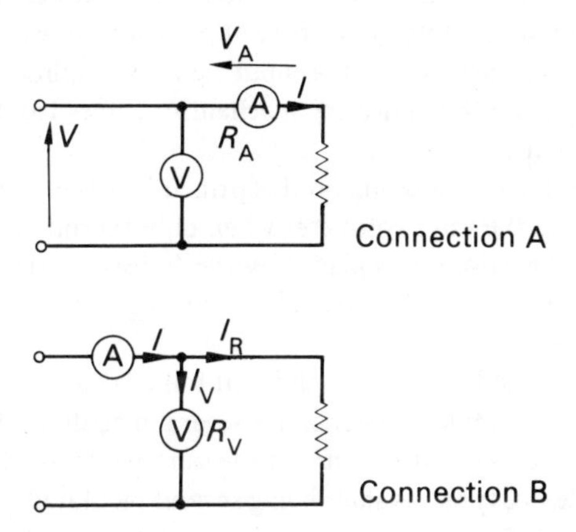

Fig. 17.25. Connections of voltmeter and ammeter to measure resistance

In connection A, the voltmeter indicates the sum of the volt drops across the resistor and the ammeter, i.e. V_R and V_A respectively. The true value of resistance is therefore given by

$$R_{\text{true}} = \frac{V - V_A}{I}$$

$$= \frac{V - IR_A}{I} \tag{17.5}$$

If R_A is much less than R, then V is almost equal to V_R. It follows that this connection is most suitable when measuring resistors of high resistance values.

It will also be observed that the true resistance is equal to the apparent resistance less the resistance of the ammeter.

In connection B, the ammeter indicates the sum of the resistor and voltmeter currents, i.e. I_R and I_V respectively. The true value of resistance is therefore given by

$$R_{\text{true}} = \frac{V}{I_R} = \frac{V}{I - I_V}$$

$$= \frac{V}{I - V/R_V} \tag{17.6}$$

If R_V is much greater than R, then I is almost equal to I_R. It follows that this connection is most suitable when measuring resistors of low value. It will also be observed that the less conductance is equal to the apparent conductance less the conductance of the voltmeter.

17.14 Electronic voltmeter

Although quite high values of input impedance can be had from the electromagnetic and electrostatic meters already described, these values are not sufficiently high for application to electronic and communication circuits. These circuits are unable to contribute enough power for satisfactory operation without material effect to the measured circuits. The meter suitable to these applications must therefore take less power. This can be readily achieved if the input signal is amplified electronically and hence the power to operate the indicating mechanism comes from a source other than the measured circuit.

The most common meter operating on this principle is the electronic voltmeter. This was once known as the valve voltmeter when only thermionic valve amplifiers were available. When the valve was replaced by the transistor, the meter was ubiquitously termed the transistorised valve voltmeter. The term electronic voltmeter covers either method of amplification.

The voltmeter is designed to have a high input impedance–as high as is reasonably possible. Typical values of input resistance are several megohms. When considering the input impedance relative to electronic circuit measurements, the input capacitance should also be considered, typical values being several picofarads.

The circuit design of electronic voltmeters is generally complex and certainly in advance of the electronic circuitry considered in this book. However a primitive circuit is shown in Fig. 17.26. No biassing or voltage stabilisation has been shown in order to keep the circuit as fundamental as possible.

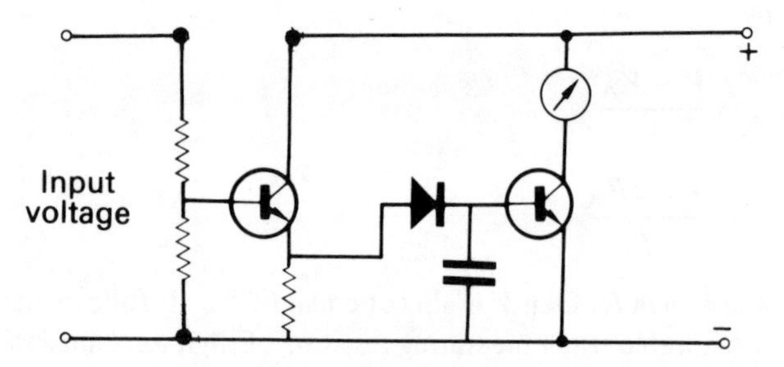

Fig. 17.26. Primitive electronic voltmeter

The input voltage to be measured may be applied to a voltage divider in order to reduce the measured voltage to a reasonable value. The divider may be interchangeable to give variation of measurement range. The resulting signal is supplied via a coupling capacitor to an emitter-follower stage which offers high input impedance and thereby prevents the source of the signal from being overloaded. The output signal is now effectively divorced from its source and can be amplified. To convert the a.c. signal to an equivalent d.c. signal, it is applied to a capacitor through a diode. The charge on the capacitor is proportional to the peak of the a.c. voltage.

The charge attained by the capacitor determines the base current of the following common-emitter amplifier stage and hence the collector current. This current is made to pass through a moving-coil instrument which can therefore be calibrated in terms of the input a.c. voltage.

Such a voltmeter has the drawback that it really measures the peak voltage and not the r.m.s. voltage. The calibration will assume a sinusoidal waveform and hence the voltmeter is susceptible to error.

More complex circuitry will allow the voltmeter to measure r.m.s. values. It will also give better temperature stability and even higher values of input impedance.

The method of indication, i.e. the moving-coil instrument, can be replaced by an illuminated digital display. Although more expensive, it is more easily read and is more robust.

17.15 Oscilloscope

The oscilloscope or cathode-ray oscillograph (C.R.O.) is one of the most commonly used measuring and fault-finding instruments in current practice. By means of a cathode-ray tube, it displays the waveform of any suitable signal—usually an alternating-voltage waveform.

The cathode-ray tube makes use of the phenomenon exhibited by certain materials, known as phosphors, whereby they emit light when bombarded by electrons. When light is emitted at the instant of bombardment, this phenomenon is known as fluorescence; when emitted after bombardment has ceased, it is known as phosphorescence. All phosphors exhibit both phenomena to some extent although one or other will predominate. A phosphor will be chosen for a particular application according to the required degree of persistence of light emission and of the colour of the emitted light. Figure 17.27 shows the construction of a common type of cathode-ray tube.

When a voltage is applied between two plates in a vacuum, there is a tendency for electrons to leave the plate of lower voltage (the electrode) and move through the vacuum to the plate of higher voltage (the anode). This tendency is greatly encouraged

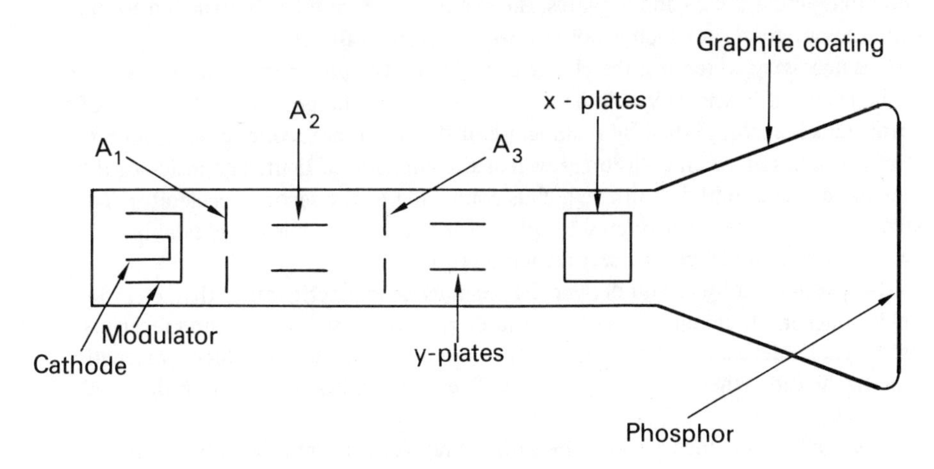

FIg. 17.27. Construction of cathode-ray tube

if the cathode is heated and coated with an appropriate oxide. Thus in the tube shown in Fig. 17.27, electrons are emitted from the indirectly-heated cathode at the left-hand end of the evacuated glass tube.

Around the cathode is a metal cylindrical electrode which has a hole in its flat end and through which the electrons pass. The potential of the cylinder, with respect to that of the cathode, is kept negative and, by varying this electrode potential, the rate of flow of electrons through the hole can be varied. This in turn varies the brightness of the emitted light which depends on the number of electrons striking the phosphor. The controlling electrode is termed the modulator.

After passing through the modulator, the electrons enter a system of electrodes, usually three in number; this system accelerates the electrons. These electrodes are also used to focus the beam of electrons so that they are concentrated and give rise to a small spot of light emitted from the phosphor. Focussing is necessary because the electrons, being negatively charged, tend to repel one another and the beam would otherwise diverge as it passes along the tube. Electrodes A_1 and A_3, known as the first and second accelerators respectively, are metal discs with holes in their centres; A_2, known as the focussing electrode, is an open cylinder. The potential of A_3 is maintained positive with respect to that of A_2 which is, in turn, positive with respect to that of A_1. The electric field configurations between A_1 and A_2, and A_2 and A_3 are such that they have axial and radial components. The axial components accelerate the electrons whilst the radial components have a nett convergent effect—nett effect because the radial component of the field does, at certain points, produce a divergent force on the electrons. Overall convergence is achieved because the beam exists in the sections which tend to converge it for longer periods of time than in the sections which tend to diverge it. The cathode, modulator and the electrodes A_1, A_2 and A_3 are known collectively as the electron gun.

On emerging from the electrode A_3 the beam is passed through two sets of parallel plates at right angles to each other. One set of plates can deflect the beam in the Y-direction and the other set in the X-direction. They are known as the Y- and X-plates respectively. The beam then impinges on the phosphor, which coats the inside of the end of the tube to form the screen, at a point dependent on the potentials across the X- and Y-plates; the beam in each case being attracted to the plate the potential on which is positive with respect to the other.

It is necessary to remove the charge acquired by the phosphor from the electrons of the beam, otherwise it would become so negatively charged that it would repel the beam. This is accomplished by making use of the fact that secondary emission of electrons takes place on collision as well as the emission of light. The inside of the tube has a thin graphite coating which is connected to the second accelerator. The secondary electrons are attracted to this coating and thus a current is established between the cathode and the second accelerator.

It is possible to focus and deflect the beam using magnetic rather than electric fields. Magnetic focussing is rare but cathode ray tubes used in television receivers use magnetic deflection. Two sets of coils mounted externally produce a magnetic field in the tube when they carry current. The electron beam situated in this field is deflected.

The oscilloscope incorporates the cathode-ray tube in the following manner. The Y-plates are controlled by the signal that is to be observed. This signal is not directly applied to the plates but is electronically amplified for two reasons. First of all, there is the problem of attaining a high input impedance so that the oscilloscope will have as little effect on the measured circuit as possible. This problem is solved in the same manner that was used in the electronic voltmeter. The signal must also be amplified to attain a sufficient electric field between the Y-plates to deflect the electron beam over the complete screen area. Although amplification takes place, it is most important that the gain of the amplifier is constant otherwise the vertical deflection would not be proportional to the input signal.

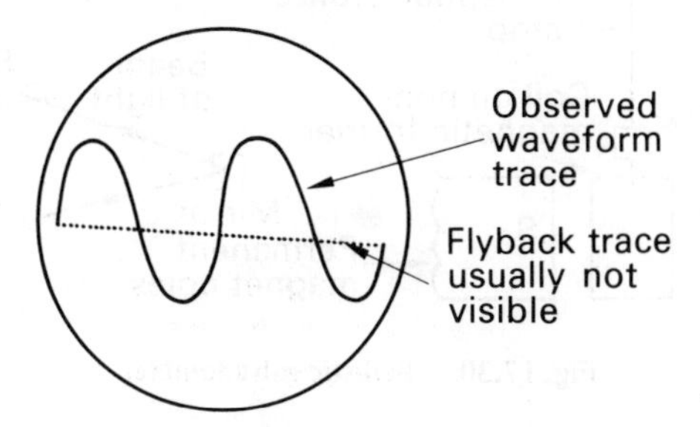

Fig. 17.28. Typical c.r.o. display

The X-plates are supplied from a time-base generator. This causes the beam to move across the screen from left to right at a constant speed. This beam is then returned to the beginning as quickly as possible and the process repeated. Provided this fly back is sufficiently fast, the beam will leave no trace on the screen. If the frequency of the time base waveform is adjusted so that it is equal to the frequency of the signal under observation then a steady trace will be observed on the screen. This will also occur if the signal frequency is an integral multiple of the time-base frequency in which case a number of waveforms can be observed. This is indicated in Fig. 17.28.

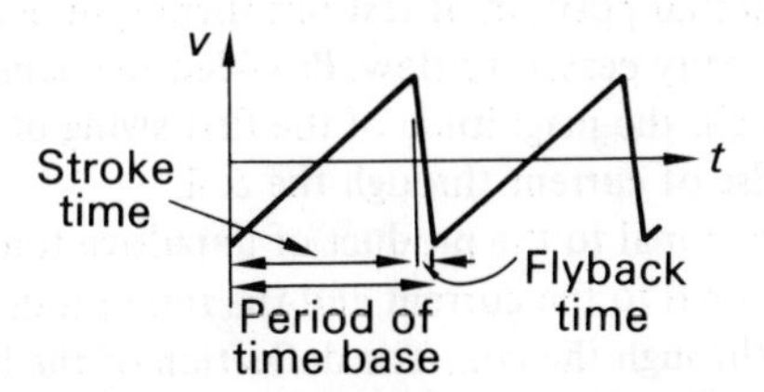

Fig. 17.29. Voltage waveform for a c.r.o. time base

A typical time-base waveform is shown in Fig. 17.29. It should be noted that this internally produced time base is not always used and that there are many other methods of applying the oscilloscope to electrical and other measurements.

17.16 Ballistic galvanometer

This instrument is similar in principle to the moving-coil instrument described in para. 17.4. However, by using the suggested fibre suspension system and by reflecting a beam of light to obtain the deflection display, the resulting movement has minimum damping and a relatively large inertia mass. It is thus able to indicate the charge passed through it. The basic construction is indicated in Fig. 17.30.

The coil is wound on a non-magnetic former to reduce any electromagnetic damping effect from the field of the permanent magnet. The coil current causes it to deflect, the extent of the deflection being observed from the displacement of the reflected beam of light on a scale. The further the scale is placed from the ballistic galvanometer, the more the indication is amplified.

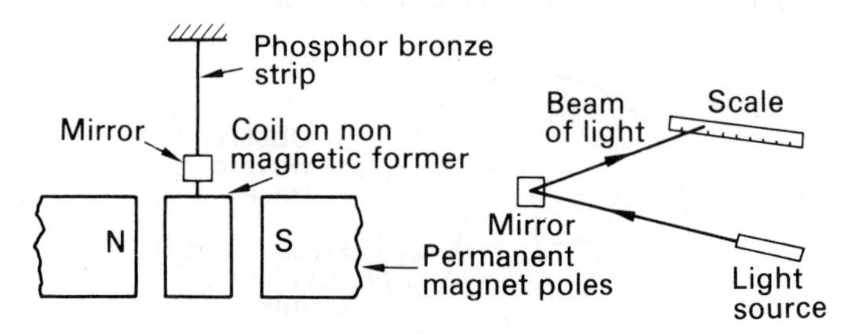

Fig. 17.30. **Ballistic galvanometer**

The current is led into and out from the coil by fine wires which exert an almost negligible control over the coil position. The torque of the suspension fibre or phosphor bronze strip results in a very small restoring torque. As a result, the coil oscillates when deflected from its normal position. A switch is usually connected across the coil. When the switch is closed, it short circuits the coil and damps the movement.

The deflecting torque at any instant is proportional to the coil current. In the moving-coil instrument, the current creates a continuous torque against the restraint of the springs. In the ballistic galvanometer, the current is passed through the coil for a short period of time thereby giving the movement an impulsive torque. This torque is applied whilst the movement is still effectively at rest. The impulse causes the coil to move from its initial position of rest but there is no longer a driving torque because the current has already ceased to flow. Provided the damping—due to the suspension restraint—is small, the magnitude of the first swing of the movement will be proportional to the pulse of current through the coil.

The deflection is proportional to the product of impulsive torque and time. However the torque is proportional to the current and the total product of current and time is the charge passed through the coil. The deflection of the ballistic galvanometer is therefore proportional to the charge of electricity passed.

It will be noted that this mode of operation only is valid provided the charge passes through the instrument in a very short period of time. After the first deflection, the indication will continue to oscillate for some considerable time due to the lack of damping and restraint. It is for this reason that the shunt switch is provided. Only the indication of the first swing may be used for measurement purposes.

Unlike other instruments, the ballistic galvanometer must be calibrated before every set of measurements is taken. This is necessary because the scale is not a fixed distance from the mirror: also because the resistance of the coil circuit must be taken into account. The calibration determines the quantity sensitivity or ballistic constant k of the ballistic galvanometer. This is the quantity of electricity in coulombs per unit of deflection.

Ignoring the resistance of the measured circuit, the ballistic galvanometer can be calibrated by either of the following methods. The most common method is by use of a standard solenoid. This consists of a long solenoid coil wound on a cylinder of insulating material. Usually the length is about 1 m long whilst the mean diameter is not more than 100 mm. In this way, the field is reasonably uniform at the centre of the solenoid. The secondary or search coil is placed axially within the solenoid and its dimensions must be accurately known. The calibration circuit is shown in Fig. 17.31.

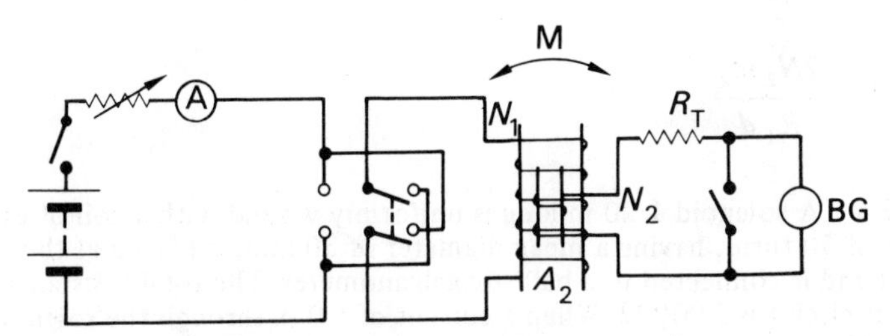

Fig. 17.31. Calibration circuit of ballistic galvanometer using a standard solenoid

In the following analysis , the subscript 1 applies to the primary solenoid coil whilst the subscript 2 applies to the circuit containing the search coil and the ballistic galvanometer. The area A_2 is the cross-sectional area of the search coil. Within the solenoid:

$$H = \frac{N_1 I_1}{l_1}$$

$$B = \frac{\mu_0 N_1 I_1}{l_1}$$

The flux passing through the search coil is therefore given by

$$\Phi_2 = \frac{\mu_0 N_1 I_1 A_2}{l_1}$$

Normally the flux linking the search coil per ampere of current in the primary coil is known in which case this flux can be determined before calibration. Assume that the current I_1 is reversed in a time Δt. The e.m.f. induced in the search coil is therefore

$$E_2 = N_2 \cdot \frac{\Delta \phi}{\Delta t}$$

$$= N_2 \cdot \frac{2\Phi_2}{\Delta t}$$

$$I_2 = \frac{N_2 . 2\Phi_2}{R_T . \Delta t}$$

where R_T is the total resistance of the secondary circuit. The total charge that flows is given by

$$Q_2 = \frac{N_2 . 2\Phi_2 . \Delta t}{R_T . \Delta t}$$

$$= \frac{2N_2 \Phi_2}{R_T}$$

$$= kd$$

where d is the number of divisions indicated by the deflection, caused by the current reversal.

$$k = \frac{2N_2 \Phi_2}{R_T d}$$

Example 17.6 A solenoid 1·20 m long is uniformly wound with a coil of 800 turns. A short coil of 50 turns, having a mean diameter of 30 mm, is placed at the centre of the solenoid and is connected to a ballistic galvanometer. The total resistance of the galvanometer circuit is 2 000 Ω. When a current of 5·0 A through the solenoid primary winding is reversed, the initial deflection of the ballistic galvanometer is 85 divisions. Determine the ballistic constant.

$$\Phi_2 = \frac{\mu_0 N_1 I_1 A_2}{l_1}$$

$$= \frac{4\pi \times 10^{-7} \times 800 \times 5 \times \pi \times 15^2 \times 10^{-6}}{1{\cdot}2}$$

$$= 2{\cdot}96 \times 10^{-6} \text{ Wb}$$

$$k = \frac{2N_2 \Phi_2}{R_T d}$$

$$= \frac{2 \times 50 \times 2{\cdot}96 \times 10^{-6}}{2000 \times 85}$$

$$= 1{\cdot}74 \times 10^{-9} \text{ C/div}$$

$$= \underline{1\,740 \text{ pC/div}}$$

The alternative method of calibration is similar to that already described. This time a standard mutual inductor is used, the effective construction being similar to that in Fig. 17.31. The mutual inductance is known, from which the secondary e.m.f. may be readily obtained as follows:

$$E_2 = M \cdot \frac{2I_1}{\Delta t}$$

$$I_2 = \frac{2MI_1}{R_T \Delta t}$$

$$Q_2 = \frac{2MI_1}{R_T}$$

$$= kd$$

$$k = \frac{2MI_1}{R_T d}$$

When the ballistic galvanometer is calibrated, the secondary circuit must include the search coil on the measured specimen. To illustrate the method of operation, consider the method of determining the B/H characteristic of a magnet material involving the use of the ballistic galvanometer. A suitable circuit is shown in Fig. 17.32.

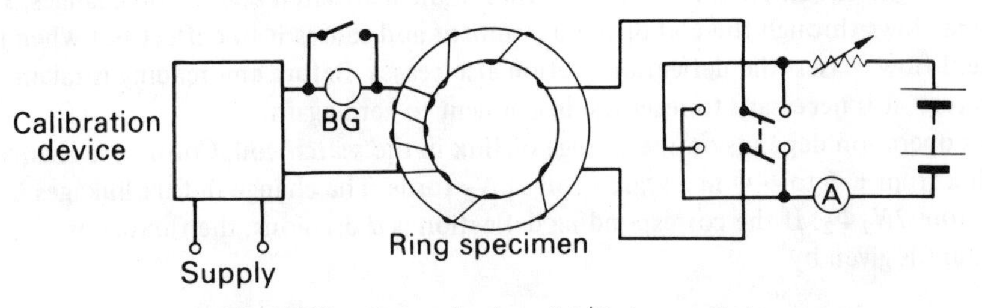

Fig. 17.32. Determination of B/H characteristic

Before any measurements are taken, the ring specimen is demagnetised by reversing the magnetising current repeatedly whilst it is gradually increased to its maximum value and back to zero. To protect the ballistic galvanometer it is short-circuited by the switch; it could otherwise receive repeated impulses which would damage the suspension. The gradual reduction of the magnetising current reduces the maximum flux density experienced with each reversal and the hysteresis loop gradually shrinks until the specimen is demagnetised.

With no magnetising current flowing, the ballistic galvanometer is calibrated by one of the above methods. The ballistic galvanometer circuit now includes the resistance of the ring specimen search coil.

The ring is again demagnetised and a small magnetising current is applied to the primary coil on the ring specimen. With the ballistic galvanometer short-circuited, the magnetising current is reversed several times. The short circuit is now removed from the ballistic galvanometer and the reversing switch operated once more. The deflection is noted and the galvanometer is short circuited. The primary current to the ring specimen is increased and the procedure is repeated. It should be noted that the final reversal of the switch must always take place in the same direction. This procedure is then continued until the magnetising current is such that saturation has taken place. This is indicated by very small deflections by the ballistic galvanometer.

Throughout the tests, the calibrating supply is open-circuited.

The results are evaluated from knowledge of the mean magnetic path length in the ring specimen, its cross-sectional area and the number of primary and secondary turns wound on it. For each value of magnetising current, the corresponding magnetic field strength can be calculated whilst the flux density is obtained via the expression for the ballistic constant. From the results, the B/H characteristic can be plotted.

This brief description gives some indication of the method of applying the ballistic galvanometer to electrical measurement. It is by no means the only method of using this instrument.

17.17 Fluxmeter

The fluxmeter is another instrument important in magnetic measurement. It is somewhat similar to the ballistic galvanometer except that the moving system can be pivotted or suspended by a silk thread. The moving coil is again supplied by fine wires that exert no control over its position.

The important difference is that the movement is rendered dead beat by electromagnetic damping, i.e. when the flux in the associated search coil changes, a current flows through the coil of the instrument and causes it to deflect but when the current flow ceases, the deflection motion also ceases. Before any reading is taken therefore, it is necessary to reset the instrument to zero again.

Its operation depends on the change of flux in the search coil. Consider a change of flux from $+\Phi$ to $-\Phi$ in a search coil of N_2 turns. The change in flux linkages is therefore $2N_2\Phi_2$. If the corresponding deflection is d divisions, the fluxmeter constant is given by

$$k = \frac{2N_2\Phi_2}{d}$$

In this case the constant is measured in weber-turns per division. The usual fluxmeter is of the Grassot type being self-contained and therefore not requiring calibration before use. The calibration initially is achieved by the methods used to calibrate the ballistic galvanometer.

Example 17.7 A coil of 120 turns is wound uniformly over a steel ring having a mean diameter of 1·0 m and a cross-sectional area of 500 mm². A search coil of 15 turns, wound on the ring, is connected to a fluxmeter having a constant of 300 μWb t/div. When a current of 6·0 A through the 120-turn coil is reversed, the fluxmeter deflection is 64 divisions. Calculate:
(*a*) The flux density in the ring.
(*b*) The corresponding value of relative permeability.

$$k = \frac{2N_2\Phi_2}{d}$$

$$300 \times 10^{-6} = \frac{2 \times 15 \times \Phi_2}{64}$$

$$\Phi_2 = 0{\cdot}64 \times 10^{-3}\ \text{Wb}$$

$$B = \frac{\Phi_2}{A_2} = \frac{0{\cdot}64 \times 10^{-3}}{500 \times 10^{-6}}$$

$$= \underline{1{\cdot}28\ \text{T}}$$

$$H = \frac{NI}{l} = \frac{120 \times 6}{1} = 720\ \text{At/m}$$

$$\mu_r = \frac{B}{\mu_0 H} = \frac{1{\cdot}28}{4\pi \times 10^{-7} \times 720}$$

$$= \underline{1400}$$

PROBLEMS ON MEASUREMENTS

1 A potentiometer test is applied to measure the resistance of a copper rod and the results are as follows:

Standardisation data: 0·209 V/mm.
Standard resistance 0·1 Ω (connected in series with the rod)
Length corresponding to the p.d. across the standard resistor: 250 mm.
Length corresponding to the p.d. across the copper rod: 400 mm.
Determine:

(*a*) The p.d. across the standard resistance.
(*b*) The p.d. across the copper rod.
(*c*) The resistance of the rod.

0·5225 V; 0·836 V; 0·16 Ω

2 A resistance is to be measured by the Wheatstone bridge method. At balance, AB = 100 Ω, BC = 10 Ω and AD = 85 Ω. A resistor of unknown value is connected across CD. A p.d. of 1·5 V is maintained across AC and the galvanometer is connected across BD. Sketch the circuit diagram and calculate:

(*a*) The value of the unknown resistance.
(*b*) The potential difference across AB and across AD.

8·5 Ω; 1·363 V; 1·363 V

3 A moving-coil galvanometer of resistance 5 Ω gives a full-scale reading when 15 mA passes through the instrument. Explain, with the help of circuit diagrams, how its range could be altered so as to read up to 5·0 A and also 150 V. Calculate the values of the resistances which would be required.

15·045 mΩ; 9 995 Ω

4 A type of moving-coil voltmeter in common use has a sensitivity of 500 Ω/N and 100 V gives full-scale deflection. What is the resistance of the meter and what current does it take for full-scale deflection?

A resistance of 25 kΩ is connected in series with a resistance of 50 kΩ across a 90-V supply and the above voltmeter is used to measure the voltage across the 50-kΩ resistor. What will be the reading on the voltmeter? What is the voltage across the 50-kΩ resistor before the voltmeter is connected? Why is there a difference in voltages?

50 kΩ; 2 mA; 45 V; 60 V

5 The 100-turn coil of a moving-coil instrument has an active length 30 mm and diameter 20 mm. The flux density in the air-gap is 200 mT. Full-scale deflection is obtained when the coil carries a current of 5·0 mA. For the conditions stated, calculate the total deflecting torque.

60 μN m

6 A moving-coil ammeter registers full-scale deflection at 20°C when the current in the coil is 10 mA. When the meter is fitted with a shunt, full-scale deflection at 20°C is obtained for a total current of 1·0 A. At a temperature of 40°C, the meter, with shunt, is found to read 7% low although the total current remains unaltered at 1·0 A. The shunt is constructed of German Silver with a temperature coefficient of resistance of $2{\cdot}5 \times 10^{-4}$/K at 20°C. Calculate the temperature coefficient of resistance of the moving coil at 20°C.

$40{\cdot}5 \times 10^{-4}$/K

7 Draw a circuit diagram to show a voltmeter, an ammeter and a wattmeter connected to measure the power input to a steel-cored coil. What information could be obtained from the readings?

If the readings on the instruments are 110 V, 2·5 A and 150 W respectively and the d.c. resistance of the copper winding is 15 Ω, calculate the inductance of the coil and the power loss in the steel core. The supply is sinusoidal and the frequency is 50 Hz.

0·118 H; 56·25 W

8 A moving-iron voltmeter of full-scale deflection 120 V has an inductance of 0·6 H and a total resistance of 2·4 kΩ. It is calibrated to read correctly on a 60-Hz circuit. What series resistance would be necessary to increase its range to 600 V?

2·660 kΩ

9 The capacitance of an electrostatic voltmeter reading 0 – 2 000 V increases uniformly from 45 to 55 pF as the pointer moves from zero to full-scale deflection. It is required to increase the range of the instrument to 20 000 V by means of an external air capacitor. Calculate the area of a pair of capacitor plates suitable for the purpose assuming a spacing of 10 mm.

8 500 mm^2

10 A steel ring of 250 mm mean diameter and 1 000 mm^2 cross-sectional area is split in two across a diameter and each pair of cut faces is separated by an air-gap 1·0 mm wide. There are two windings on one section of the ring: a primary winding of 500 t and 10 Ω resistance and a secondary winding of 10 t and 1 Ω resistance. The primary winding is connected to a 24-V d.c. supply and the secondary winding is connected to a ballistic galvanometer having a resistance 999 Ω. The two portions of the ring are now brought together. Neglecting flux leakage and assuming that the relative permeability of the steel remains constant at 500 over the range considered, calculate the total charge which flows through the galvanometer when the air-gaps are eliminated.

53·7 μC

11 A steel ring, 400 mm^2 cross-sectional area with a mean length 800 mm, is wound with a magnetising winding of 1 000 turns. A secondary coil with 200 turns of wire is connected to a ballistic galvanometer having a constant of 1 μC/div. The total resistance of the secondary circuit is 2 kΩ. On reversing a current of 1·0 A in the magnetising coil, the galvanometer gives a throw of 100 scale divisions. Calculate:

(*a*) The flux density in the specimen.
(*b*) The relative permeability at this flux density.

1·25 T; 796

12 A ballistic galvanometer having a circuit resistance of 5 kΩ and a constant of 0·1 μC per scale divisions is connected in turn with a coil of 2 turns wound over a field coil of a d.c. machine and one of 3 turns placed on the armature surface to embrace the total pole flux entering the armature. When the normal field current is reduced to zero, the galvanometer readings are 113 and 136 divisions respectively. Determine:

(*a*) The flux per pole.
(*b*) The air-gap flux per pole.
(*c*) The leakage coefficient.

28·25 mWb; 23·7 mWb; 1·25 (SANCAD)

Index